Advances in Plant Disease Management

Advances in Plant Disease Management: Volume II: Strategic and Applied Research is an invaluable compilation for researchers/students/stakeholders/policymakers in agriculture. This book aims to offer the latest understanding of how fundamental and basic research can be translated toward the engineering of biotic stress-resilient crops through applied and strategic management of plant diseases. Volume I clearly explained the updated knowledge on basic and applied phenomena of pathogen's interplay with the host, the host immune system, and crosstalks among downstream regulating molecules as unraveled through genomics, proteomics, metabolomics, bioinformatics, and molecular studies. This volume of the book equips readers with the knowledge and understanding to confidently employ this basic information in the formulation of management strategies for major crop plant diseases.

This book offers comprehensive coverage of the research advances in plant disease management, including:

- Newer insight into pest risk analysis (PRA) and its significance in international trade.
- Developments in eco-friendly green technologies that are safe for both humans and the environment to manage diseases.
- Use of AI tools for diagnosis, development of models for advanced prediction of the outbreak of epidemics, and need-based application of agrochemicals and their appropriate formulations for use through drones.
- The information regulation and use of biostimulants for biotic and abiotic resilience.
- Plant protection policies that support the agricultural production system from a global perspective.

Pranjib K. Chakrabarty, PhD, served as Assistant Director General (Plant Protection and Biosafety) and Member (Plant Sciences), ASRB, New Delhi. He did pioneer research on host–pathogen interaction, functional genomics, and diagnostics. He is an adjunct faculty of the University of Florida, Gainesville, and had postdoctoral research experience at the University of Florida, Gainesville, and the University of Missouri-Columbia, USA.

Kalyan K. Mondal, PhD, is Joint Director (Research) at ICAR-NIBSM, Raipur. He is a visiting fellow at Stanford University, California, USA, and he has significantly contributed to the understanding of the role of bacterial Type-3-Secretion-System (T3SS) effectors in the modulation of plant defense.

Mahender S. Saharan, PhD, is the Head of the Division of Plant Pathology, ICAR-IARI, New Delhi. He contributed significantly to the study of wheat pathology and actively participated in strategy implementation for managing wheat rusts and Karnal bunt. He worked in many international collaborative research projects, namely, ISCB, ACIAR, CIMMYT, and BBSRC.

Charudatta Digambarrao Mayee, PhD, is a legendary plant pathologist who served as Vice Chancellor-MAU Parbhani; Director-Central Institute of Cotton, Nagpur; and Agriculture Commissioner, Government of India, New Delhi, before superannuating as the Chairman, ASRB, Ministry of Agriculture & Farmers' Welfare, Government of India.

Jagdish Kumar, PhD, is an Alexander Humboldt Fellow of Germany. He served as Joint Director of the School of Crop Health Biology Research and Director of ICAR-NIBSM, Chhattisgarh. He has vast experience in developing plant disease forecasting models. He worked on the biochemical basis of host resistance in cereals.

Advances in Plant Disease Management

Volume II: Strategic and Applied Research

Edited by

Pranjib K. Chakrabarty, Kalyan K. Mondal, Mahender S. Saharan, Charudatta Digambarrao Mayee, and Jagdish Kumar

CRC Press
Taylor & Francis Group
Boca Raton London New York

CRC Press is an imprint of the
Taylor & Francis Group, an **informa** business

Designed cover image: Shutterstock

First edition published 2025
by CRC Press
2385 NW Executive Center Drive, Suite 320, Boca Raton FL 33431

and by CRC Press
4 Park Square, Milton Park, Abingdon, Oxon, OX14 4RN

CRC Press is an imprint of Taylor & Francis Group, LLC

ISBN: 978-1-032-87301-5 (hbk)
ISBN: 978-1-032-87306-0 (pbk)
ISBN: 978-1-003-53194-4 (ebk)

DOI: 10.1201/9781003531944

Typeset in Times
by codeMantra

Contents

Foreword

Crop protection is crucial to food and nutritional securities and plays a major role in achieving the targets of production. The task becomes more challenging in the current scenario of climate change, globalization of agricultural trade, and continual emergence of new pests and diseases. It is estimated that global losses of food crops due to biotic stresses range between 20% and 40%. The task of protecting crops from biotic stresses, particularly under the changing climate conditions, becomes highly challenging. The alarms are further exacerbated by the increasing human population to meet the demand, globalization of trade, introduction of invasive and emergence of new and aggressive pests and pathogens. Recent advances in science and technology have opened new avenues for novel and strategic deployment of technologies to efficiently address avoidable losses. Together with traditional technologies and enabling policies, the capabilities of crops to defend against intruding plant pathogens can be greatly enhanced. Coupling artificial intelligence for biotic stress monitoring and diagnosis can result in greater precision on stress surveillance. Future research may target novel and innovative methods contrived after deciphering molecular mechanisms of host–biotic stress interactions. The next generation of climate-resilient crop protection strategies would focus more on the use of smart formulations based on antimicrobial peptides, recombinant antibodies, dsRNA, gene silencing, CRISPR/CAS editing, etc. Host-induced gene silencing (HIGS) and spray-induced gene silencing (SIGS) through the application of miRNA and siRNA offer the best alternatives that could cut down the use of ecologically hostile chemo-therapeutants for the management of biotic stresses in agricultural and horticultural crops. However, many of the potential technologies developed using advanced tools have inherent regulatory limitations for deployment. A basic tenet for effective management lies in understanding various limitations so that rational ways and means can be developed for the deployment of such technologies in a field-ready condition. Today, plant pathologists have a greater responsibility to ensure plant health to sustain food and nutritional security. Books are important means to decipher updated views of the problems, technology advancements, and existing gaps to reorient future research, development, and practices. This is a continuous process as the information on new findings continuously pours in. The editors of this important book are renowned scientists, well-acclaimed in their respective areas of specialization. Incidentally, all are formally trained in plant pathology and contributed to the advances in technologies and strategies for effective disease and pest management. Besides, the first and the last editors were also engaged in framing policies governing plant protection on behalf of ICAR and the Government of India. I am confident that this book, *Advances in Plant Disease Management: Volume II: Applied and Strategic Approaches* published by the CRC

Press, will serve as a guide and inspiration for scientists, policymakers, and students for future research into the realms of plant protection.

Dr Himanshu Pathak
FNAAS, FINSA
Director General, International Crops Research Institute for the Semi-Arid Tropics (ICRISAT), ICRISAT - India (Headquarters), Hyderabad, Patancheru 502324, Telangana, India

Preface

Agriculture serves as a driving force in India's overall growth and development, providing food security for its 1.42 billion people and employing 45% of its workforce. Over a period of time, India has transformed itself from being a food deficit to a food sufficient and finally a food-surplus nation. To meet the demand for 1.6 billion healthy population at its peak, it is estimated that India needs to produce 50% more food grains by 2050. This monumental production needs a quantum leap through novel and scientific interventions in production and productivity, besides protecting the humongous losses due to biotic and abiotic stresses. However, despite the availability of modern technologies and tools, the threatening pests and diseases of crop plants are of serious concern for today's researchers working on plant protection. Newer insights into the basic understanding of host–pathogen interactions, unshielded the plethora of possibilities to increase food production through effective protection. In this context, this book entitled *Advances in Plant Disease Management: Volume II: Strategic and Applied Research* is an attempt to give first-hand references on how the basic concepts driving biological phenomena in plants, pathogens, and environmental microbes, described in Volume I, can be employed strategically as a mean to develop plants, resilient to biotic stresses and maneuver nature-friendly options to protect crop health. The primary aim of this book is to comprehend how, based upon the up-to-date information on basic biological phenomenon or modules, researchers could transform the knowledge to explore novel strategies of plant health management.

This book will serve as a reference book for stakeholders associated with agriculture, including students, scientists, researchers, and policymakers of crop production and protection technology. Volume II of this noteworthy book contains 15 chapters written by globally acclaimed scientists on different areas of plant protection sciences. Briefly, the current volume discusses the insights into Pest Risk Analysis (PRA) and its significance in international trade; recent developments in biopesticide use; fungicides resistance/residue management; off-label use of agrochemicals; drones in crop protection; diagnostic tools to identify and detect plant pathogens integrated management of emerging diseases including Phytoplasma; plant parasitic nematodes through bioinoculants; induced immunity in plants; and development, regulation, and use of biostimulants, to support eco-friendly and sustainable agriculture. The editors thank all the contributors for their valuable and timely contributions. They also wish to place on record the cooperation and painstaking efforts made by Mr. Selvakumaran, Deputy Manager and Ms. Yasmin, Executive of Dhanuka Agritech Ltd, in going through the MS critically and bringing it to the level it is presently at. The editors put their best effort into bringing out this book with minimum error and would look forward to learning from the esteemed readers about its practical utility.

Editors: Pranjib K. Chakrabarty, Kalyan K. Mondal, Mahender S. Saharan, Charudatta Digambarrao Mayee, and Jagdish Kumar

Editor Biography

Dr. Pranjib K. Chakrabarty has served ICAR in various capacities as Scientist (Plant Pathology), Principal Scientist (Biotechnology), and Head of the Division of Crop Improvement at the Central Institute for Cotton Research (CICR) Nagpur, before joining ICAR, Krishi Bhavan, New Delhi, as Assistant Director General (Plant Protection and Biosafety). He extensively studied and did pioneering research on general and molecular plant pathology with special reference to host–pathogen interaction, signal transduction for resistance signaling, functional genomics of pathogenicity genes, molecular diagnostics, and transgenic research. He did his post-doctoral research at the University of Florida, Gainesville, and at the University of Missouri-Columbia in Columbia, USA. He is also an adjunct faculty of the University of Florida, Gainesville. He visited several countries including Australia, UK, Canada, Brazil, Ecuador, Switzerland, Netherlands, China, Malaysia, Singapore, Nepal, Bhutan, Ethiopia, Bangladesh, etc., representing national and international causes on plant protection and biosafety. Besides his assignments as ADG plant protection in ICAR, he was the nodal officer on GM Biosafety and Biosecurity; member of the CIBRC, Ministry of Agriculture; Member of RCGM, DBT; Chairman of the technical committee of central sector scheme on monitoring pesticide residue at national level, DAC, MOA; Member, Task Force (Agricultural Biotechnology), DBT; Member Task Force DRDO, MEA; Chairman, Board of Institute of Pesticides Formulation Technology, Ministry of Chemicals & Fertilizers, GOI; and Member of Governing Body and Executive Body of NIPHM, Hyderabad. With his direct association with four National Plant Protection Institutes of ICAR and seven All India coordinated research projects on plant protection, as ADG (PP) besides policymakers in the Ministry of Agriculture, he has been associated with the development of policies and guidelines for enabling and ensuring crop biosecurity. Since April 2019, he is working as Member (Plant Sciences), ASRB, New Delhi, in the Ministry of Agriculture & Farmers' Welfare, GOI, New Delhi. He was President of the Indian Society of Mycology & Plant Pathology for two consecutive years from 2018–2019 to 2019–2020 and also of the Indian Phytopathological Society during 2020–2021. He is a Fellow of NAAS, New Delhi, besides several other Professional Scientific Societies.

Dr. Kalyan K. Mondal received both BSc (Ag) Hons and MSc (Ag) in plant pathology from the Bidhan Chandra Krishi Viswa Vidyalaya, West Bengal, in 1991 and 1993, respectively. He obtained his doctoral degree in plant pathology from ICAR-IARI, New Delhi, in 1999. After a stint (2001–2006) as Scientist at CSKHPKV, Palampur, he joined the Division of Plant Pathology, ICAR-IARI, New Delhi, as Senior Scientist (2006–2012) and continued as Principal Scientist (until August 2023); presently he is working as Joint Director (Research) at ICAR-National Institute of Biotic Stress Management, Raipur, CG. He was trained from Stanford University, California, USA, as a visiting scholar to have insight into the role of bacterial Type-3-Secretion-System (T3SS) effectors in the modulation of plant defense. He deciphered for the first time effector repertoire of two important bacterial pathogens, *Xanthomonas oryzae* pv. *oryzae* (Xoo) and *X. axonopodis* pv. *punicae* (Xap) infecting rice and pomegranate, respectively. He functionally demonstrated the role of crucial effectors of both the pathogens during disease development in respective crops. He generated full-genome sequence of Xoo race 4, a virulent population from India; and he was involved in the development and commercialization of seven blight resistance rice varieties, namely Pusa 1592, Pusa 1718 and Pusa 1728, Pusa 1847, Pusa 1885, Pusa 1886 and Samba 1853 having *R*- genes *xa13* and *Xa21*. Dr. Mondal became the Fellows of NAAS (2023), NABS (2020), ARRW (2021), SARP (2020), IPS (2013), and member of NASI (2013); He is the recipient of many prestigious awards including Nanaji Deshmukh ICAR Award (2021), Dr. B.P. Pal Memorial NABS Best Scientist Award (2022), Prof. JP Verma Memorial Award (2016), Prof MK Patel Memorial Young Scientist Award (2008); Prof MJ Narasimhan Academic Merit Award (2001). He served as a member of important academic and research committees including Task Force DBT (Agricultural Biotechnology), Project Reviewer DST-SERB, Member IMC, ICAR-NBAIM (2016–2019), Editor (2017–2019) for *Indian Phytopathology* and Jt. Secretary of Indian Phytopathological Society (2020–2022). He published >50 research papers in journals of international repute and authored a very popular text book *Plant Bacteriology* published by Kalyani publishers. He has delivered lead talks and guest lectures in international conferences across the world including the Philippines, Vietnam, China, Thailand, and USA.

Dr. Mahender S. Saharan did his PhD in plant pathology (Gold Medalist) from CCS HAU, Hisar, Haryana. Currently, he is the Head of the Division of Plant Pathology, ICAR-Indian Agricultural Research Institute, New Delhi. Earlier till 2016, he served in different capacities such as Scientist, Senior Scientist, Principal Scientist, and Principal Investigator, Crop Protection at ICAR-IIWBR, Karnal. He contributed significantly to the study of wheat pathology and actively participated in strategy implementation for managing wheat rusts and Karnal bunt in the main wheat belt of the country. He characterized *Fusarium* spp. causing head scab of wheat. He was also associated with draft genome sequencing of *Tilletia indica* and *Bipolaris sorokiniana*, two major pathogens of wheat. He has rich experience in the genetics of resistance and management aspects of plant diseases. He contributed significantly to the development and release of 16 wheat varieties in the country besides registering 23 wheat genetic stocks for rusts, Karnal bunt, head scab, and spot blotch resistance. He worked in many international collaborative research projects, namely, ISCB, ACIAR, CIMMYT, and BBSRC. He is a widely traveled scientist, and he visited Mexico, Switzerland, Russia, Japan, China, Bangladesh, Nepal, Kenya, Turkey, Malaysia, Australia, Italy, and the USA to participate and present invited talks in international conferences. He is a recipient of the prestigious BGRI Gene Stewardship Award and is also a Fellow of the National Academy of Agricultural Sciences, New Delhi, besides the Indian Phytopathological Society, New Delhi, and Society for Advancement of Wheat & Barley Research, Karnal.

Dr. Charudatta Digambarrao Mayee is a renowned Plant Protection Scientist. Currently, he is serving as the President of the South Asia Biotechnology Centre (SABC), New Delhi, and Chairman of the Agricultural Finance Corporation of

India, Mumbai. Born in a farming family of Maharashtra, Dr. Mayee obtained his agricultural degrees from Maharashtra and PhD from IARI in plant pathology. He commenced his career in plant pathology research at IARI and worked in various capacities at Central Rice Research Institute, Cuttack; Punjab Agricultural University, Ludhiana; the Ford Foundation, Delhi; and Maharashtra Agricultural University (MAU) Parbhani for nearly 30 years. The research, teaching, and extension experiences led him to serve as Vice Chancellor-MAU Parbhani; Director-Central Institute of Cotton, Nagpur; and Agriculture Commissioner, Government of India, New Delhi, before superannuating as the Chairman, Agricultural Scientists Recruitment Board, Ministry of Agriculture & Farmers' Welfare, Government of India.

Though specialized in plant pathology, Dr. Mayee committed himself for the overall growth of Indian Agriculture. He guided 20 PhD and more than 38 MSc students, wrote several books and monographs, published over 200 scientific publications in journals of repute, and served the cause of Agriculture Crop Protection through the development of the subject. He is a Fellow of several scientific committees and associations including Alexander Humboldt Fellow of Germany and the National Academy of Agricultural Sciences (NAAS). He is also a member of several international societies including ISAAA and ABNE, Africa. Biotechnology is his passion, and he was instrumental in the first commercial release of Bt cotton in India in 2002, a great feat that revolutionized cotton production in the country. He is a recipient of several honors and awards for agriculture development including the lifetime achievements awards and honorary Doctors of Philosophy from many universities. Currently, he is a Member on the Board of Management of many organizations and universities. His superadministrative capabilities and practical knowledge on plant protection issues and rational policy decisions on agriculture including those related to crops' biosecurity make him one of the much sought-after scientists in the academia and among policymakers. He has widely traveled and visited more than 50 countries in the world.

Dr. Jagdish Kumar did his PhD from Dr YSPUH&F, Solan, H.P., in mycology and plant pathology and is an Alexander Humboldt Fellow of Germany. He served ICAR in several capacities such as Principal Scientist, IIWBR, Karnal; Head of IARI Regional Research Stations at Wellington (T.N.) and at Shimla (H.P.); Joint Director, School of Crop Health Biology Research, ICAR-NIBSM, Raipur, Chhattisgarh (2015); and Director of ICAR – National Institute of Biotic Stress Management, Raipur, Chhattisgarh – a deemed to be university. He has vast experience in developing plant disease forecasting models. He worked on biochemical basis of host resistance in cereals besides earning specialization in breeding for disease resistance in wheat. He is associated with the development of 12 wheat varieties resistant to various diseases in India. He has rich experience conducting studies in plant pathology and molecular biology through international exposure during his research stay at CIMMYT, Mexico, and Justus Liebig University, Germany. He registered five genetic stocks with ICAR-NBPGR. He visited Germany, Mexico, Hungary, Belgium, China, and Nepal as part of his research assignments. Besides, having several awards, he is a Fellow of the Indian Phytopathological Society, New Delhi, and the Society for Advancement of Wheat & Barley Research, Karnal.

Contributors

Sundaram Aravindan
Scientist (Plant Pathology)
ICAR-National Institute of Seed Science & Technology
Mau, Uttar Pradesh, India

Davis Joseph Bagyaraj
INSA Hon. Scientist & Chairman
Centre for Natural Biological Resources and Community Development (CNBRCD)
Bangalore

Kaushik Banerjee
Director
ICAR- National Research Center for Grapes
Pune, India

Ashutosh Bhaik
Corteva Agriscience
Hyderabad

Bishwanath Chakraborty
Department of Biological Sciences
Aliah University
New Town, Kolkata, India

Usha Chakraborty
Department of Botany
University of North Bengal
Siliguri, Bengal, India

Sunil Chandra Dubey
Vice Chancellor
Birsa Agricultural University
Kanke, Ranchi, Jharkhand

Jean-Luc Genet
Corteva Agriscience

Raju Ghosh
Scientist, Legumes Pathology
International Crop Research Institute for the Semi-Arid Tropics (ICRISAT)
Hyderabad, Telangana

Ruchi Gupta
Pesticide Residue Analyst
Project Coordinating Cell, AINP-PR, ICAR-IARI
New Delhi.

Akansha Jain
Department of Mycology and Plant Pathology
Institute of Agricultural Sciences
Banaras Hindu University
Varanasi, India

Somasundaram Jayaraman
Head & Principal Scientist
Division of Soil Physics, ICAR-IISWC, Research Centre
Fernhill P.O., Udhagamandalam, Tamil Nadu, India

Aravindaram Kandan
Principal Scientist (Plant Pathology)
Division of Germplasm Conservation and Utilization, ICAR-National Bureau of Agricultural Insect Resources (NBAIR)
Bengaluru, India

Matiyar Rahaman Khan
Division of Nematology; ICAR-IARI
New Delhi

Anoop A. Krishnan
Assistant Director (Technical)
Export Inspection Agency
Kochi, Kerala, India

Dr Aundy Kumar
Principal Scientist, Division of Plant Pathology, ICAR-IARI
New Delhi.

Sanjay Kumar
Director
ICAR-National Institute of Seed Science & Technology (ICAR-Indian Institute of Seed Science)
Mau, Uttar Pradesh, India.

Andrew Leader
Corteva Agriscience

Mamadou Kane Mboup
Corteva Agriscience

Thangavelu Muthukumar
Professor
Department of Botany, Bharathiar University
Coimbatore, Tamil Nadu, India

D. K. Nagaraju
Joint Director (Ento)
Regional Plant Quarantine Station
Bengaluru

Anandkumar Naorem
Scientist - Soil Science
Division of Integrated Farming System, ICAR–Central Arid Zone Research Institute
Jodhpur, Rajasthan, India

Nidhi Rai
Department of Botany
Institute of Science
Banaras Hindu University
Varanasi, India

Sashi Pandey Rai
Department of Botany
Institute of Science
Banaras Hindu University
Varanasi, India

Raman Ramesh
Principal Scientist (Plant Pathology)
Crop Improvement and Protection (Crop Science)
ICAR-Central Coastal Agricultural Research Institute (CCARI)
Old Goa

Govind P. Rao
Division of Plant Pathology
ICAR-Indian Agricultural Research institute
New Delhi, India

Mediga Kasi Rao
Assistant Professor, Department of Plant Pathology
School of Agriculture, Mohan Babu University
Tirupati, Andhra Pradesh, India

Amitava Sanyal
Vice President- Product Development and Regulatory
Sumitomo Chemical India Ltd
New Delhi, India

Birinchi Kumar Sarma
Department of Mycology and Plant Pathology, Institute of Agricultural Sciences, Banaras Hindu University
Varanasi, India

S. Selvakumaran
Dhanuka Agritech Limited
Gurugram

Krishan Kumar Sharma
Former Network Coordinator
AINP-PR, ICAR-IARI
New Delhi

Mamta Sharma
Cluster Leader-Precision Phenotyping for Biotic, Abiotic Stresses and Nutrition and Principal Scientist II-Precision Phenotyping for Biotic Stresses, Accelerated Crop Improvement, International Crops Research Institute for Semi-Arid Tropics (ICRISAT), Patancheru-502324, Hyderabad, Telangana

O. P. Sharma
Former Principal Scientist (Plant Pathology)
ICAR- National Centre for Integrated Pest Management, LBS Building, IARI Campus
New Delhi

Harikesh Bahadur Singh
Department of Biotechnology
GLA University
Mathura, India

Jaiinder Pal Singh
Plant Protection Adviser
Government of India, Ministry of Agriculture and Farmers Welfare, DA&FW, Directorate of Plant Protection Quarantine and Storage, NH 4 Faridabad-121001.

Vishal Singh Somvanshi
Division of Nematology; ICAR-IARI
New Delhi

Katravath Srinivas
Scientist, REC Sub-Unit, Bidar
Central Sericultural Research and Training Institute (CSTRI)
Mysuru, Govt of India

Kesar Tandekar
Project Coordinating Cell, AINP-PR, ICAR-IARI
New Delhi

Avijit Tarfdar
Project Scientist, Legumes Pathology, International Crop Research Institute for the Semi-Arid Tropics (ICRISAT)
Patancheru-502324 Hyderabad, Telangana

Vandana Tripathy
Principal Scientist & Network Coordinator
All-India Network Project on Pesticide Residues (AINP-PR), ICAR-IARI
New Delhi

Rajeev K. Varshney
Director, Murdoch's Centre for Crop & Food Innovation, State Agricultural Biotechnology Centre, Food Futures Institute, Murdoch University, Murdoch 6150, Australia

Shekhar Varshney
South Asia Regulatory and Stewardship Leader, Corteva Agriscience, Gurugram, Haryana

Raman Kumar Walia
Former Project Coordinator
ICAR-All India Coordinated Research Project on Nematodes in Agriculture (AICRP on Nematodes), ICAR-IARI
New Delhi.

Suresh Walia
Division of Agricultural Chemicals, ICAR-Indian Agricultural Research Institute (IARI)
New Delhi

Yasmin
Dhanuka Agritech Limited
Gurugram

1 Pest Risk Analysis (PRA) and Its Significance in International Trade

Devalapura Kalasaiah Nagaraju and Jaiinder Pal Singh

1.1 INTRODUCTION

Over the past three decades, international trade in agricultural commodities, food products, and global travel has surged significantly. Consequently, the risk of pests and diseases spreading across borders has escalated. The responsibility for preventing the entry and establishment of such pests and diseases in new geographical areas lies with Plant Quarantine authorities. Agricultural scientists are actively debating how best to manage these risks associated with trade and travel while harmonizing phytosanitary measures practiced by different countries. The Food and Agriculture Organization (FAO) of the United Nations, as the custodian of the International Plant Protection Convention (IPPC), plays a key role in seeking harmonization. These organizations recognize that assessing risks related to trade in plant and plant products should be grounded in sound biological principles. The most effective approach to managing these risks is through "pest risk analysis" (PRA) (Devorshak, 2012).

PRA is a dynamic field, and countries worldwide have adopted it as a tool to inform regulatory decisions for plant protection. It guides decisions related to domestic activities, such as prioritizing pests for surveillance and making quarantine decisions regarding imported products. Analytical support for decision-makers has become indispensable. Furthermore, under international treaties, regulatory plant protection decisions affecting imported goods must be supported by a thorough risk assessment (Devorshak, 2012).

1.2 CONCEPTS AND DEFINITION OF PRA

In the context of statutory plant health related to international quarantine, the term "pest" is broadly defined as "Any species, strain, or biotype of plant, animal, or pathogenic agent injurious to plants or plant products." This comprehensive definition encompasses insects, nematodes, molluscs, fungal pathogens, viruses, viroids, phytoplasmas, and weeds that may adversely affect plants or plant products. In plant biosecurity, the term risk is applied to *the probability or likelihood of an event occurring and the consequences of that event.* PRA is defined as "The process of evaluating biological or other scientific and economic evidence to determine

DOI: 10.1201/9781003531944-1

whether an organism is a pest, whether it should be regulated, and the strength of any phytosanitary measures to be taken against it (ISPM 5, FAO, 2010)."

Further, a quarantine pest is "A pest of potential economic importance to the area endangered thereby and not yet present there, or present but not widely distributed and being officially controlled (IPPC, 1997, FAO, 1997)." In other words, it is a pest that has the potential to have an impact if it is introduced. Introduction refers to "The entry of a pest resulting in its establishment." Establishment is defined as "Perpetuation, for the foreseeable future, of a pest within an area after entry."

In essence, PRAs serve several critical functions. They assess the potential status of an organism, determining whether it meets the defining criteria for a quarantine pest. PRAs also provide estimates of associated risks, evaluate the acceptability of those risks, and recommend appropriate risk management strategies. The strength of available risk management options varies. Ultimately, the conclusion of a PRA informs policymakers, guiding decisions on applying or modifying phytosanitary measures to effectively manage the risks posed by the identified pest.

1.3 INTERNATIONAL STANDARDS FOR PHYTOSANITARY MEASURES RELATED TO PRA

IPPC standards referred to as International Standards for Phytosanitary Measures (ISPMs) have been developed to assist National Plant Protection Organizations (NPPOs). The primary ISPMs relevant to PRA are:

ISPM 2: Framework for pest risk analysis (FAO, 2007)
ISPM 11: Pest risk analysis for quarantine pests (FAO, 2004a)
ISPM 21: Pest risk analysis for regulated non-quarantine pests. (FAO, 2004b)

While ISPMs offer guidance on the factors to consider during PRA, they do not provide explicit instructions on how to conduct such analyses. However, several countries, including India, Australia, New Zealand, and the USA, have developed specific procedures for assessing the pest risks associated with importing plant commodities.

1.4 PURPOSE OF CONDUCTING PEST RISK ANALYSIS

PRA is initiated for various reasons. First, it often arises from an exporting country's request to export a commodity to an importing country. In such cases, the importing country assesses the risk of pests associated with the given commodity. Second, PRAs evaluate new pests that have the potential for introduction or have already been introduced into a country. Here, the goal is to determine whether specific actions are needed to prevent their introduction or to manage and eradicate them if already present. Lastly, changes in policy—such as revisions to requirements for wood packing material or nursery stock—may prompt the need for a PRA to identify appropriate measures for preventing pest entry and establishment.

PRA is most commonly applied for five different purposes, namely:

1. Analyzing risks for organisms
2. Analyzing risks for pathways
3. Analyzing risks for commodities
4. Supporting new policies or changes to existing policies
5. Prioritizing resources

At its simplest level, a PRA may focus on a single pest—an "organism-specific PRA." For imports and exports of agricultural products, PRAs are conducted for the specific commodities involved. In this context, commodity-based PRAs consider multiple organisms that may accompany a particular traded commodity. Alternatively, PRAs can take a broader approach by examining pathways. A pathway-based PRA explores various means by which pests can spread from one location to another. The complexity of PRAs varies based on the problem, the intended audience, and the availability of scientific data to support the analysis.

1.5 METHODOLOGY OF PEST RISK ANALYSIS

ISPM 11 (*Pest risk analysis for quarantine pests*) FAO, 2004a provides more information about each stage of PRA for quarantine pests. However, the process of PRA is briefly presented here. The PRA process consists of the following three stages (Figure 1.1):

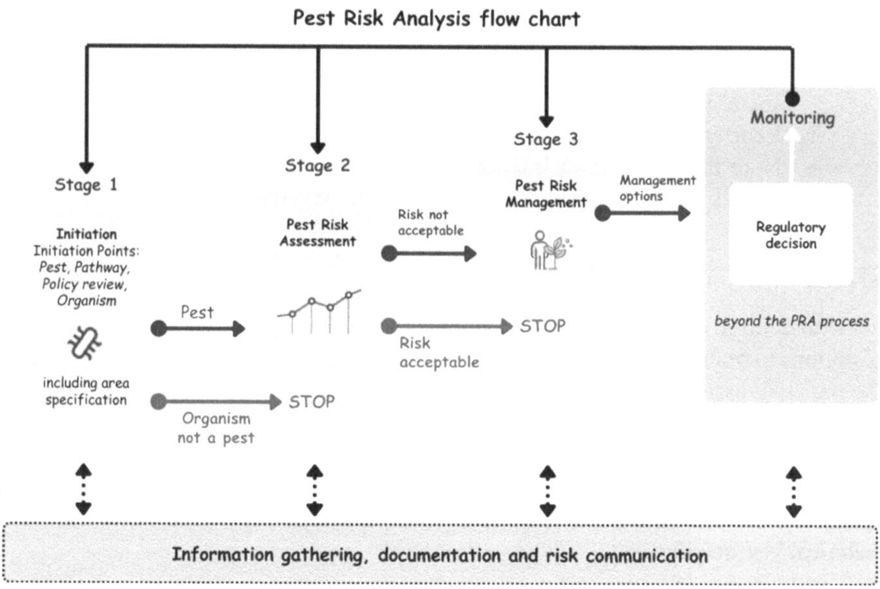

FIGURE 1.1 Pest Risk Analysis Flow Chart.

1. Initiation
2. Pest risk assessment
3. Pest risk management

STAGE 1: INITIATION STAGE: IDENTIFYING CONCERNS

a. The initiation stage aims to identify pests and pathways relevant to quarantine concerns within the specified PRA area.

STAGE 2: PEST RISK ASSESSMENT

There are three steps to the assessment of pest risk.

Step 1: Pest Categorization

i. Quickly assess whether a pest meets quarantine criteria.
ii. Eliminate pests early if they don't warrant further analysis.

Step 2: Assessing Pest Entry, Establishment, and Spread

iii. Evaluate pathways associated with the pest (e.g., imported commodities).
iv. Estimate the likelihood of pest establishment based on biological factors and environmental conditions.
v. Consider natural mobility, spread via vectors, and human activities.

Step 3: Assessing Potential Consequences

vi. Identify, describe, and quantify potential economic, environmental, and social impacts.
vii. Learn from impacts in invaded areas.
viii. Note that absence of impact in the pest's origin doesn't guarantee no impact in the PRA area due to unpredictable environmental effects.

STAGE 3: PEST RISK MANAGEMENT

For a quarantine pest, pest risk management is the process of evaluation and selection of options to reduce the risk of introduction and spread of the pest. Conclusions from the pest risk assessment (Stage 2) are used to support decisions regarding the level of risk presented by the pest. If a pest is judged to present an unacceptable risk, then phytosanitary measures should be identified that will reduce the risk to an acceptable level. Phytosanitary measures should accord with IPPC principles of necessity, managed risk, minimal impact, transparency, harmonization, non-discrimination, and technical justification.

1.6 LEVEL OF DETAIL REQUIRED

The level of detail in a PRA will be limited by the amount and quality of information available, the tools, and the time available before a decision is required. Quantitative and qualitative techniques are used in PRA, but PRA need only be as complex as is required by the circumstances to support a phytosanitary decision and provide the necessary technical justification to defend decisions regarding phytosanitary measures. Nevertheless, a PRA should be based on sound science, be transparent, and be consistent with other PRAs conducted by the NPPO.

1.7 UNCERTAINTY

Estimating the likelihood of pest introduction and the consequences that could result involves many uncertainties. Uncertainty is always part of PRA very often there is a lack of data necessary to reach secure conclusions. The subjective nature of PRA is also a source of uncertainty. ISPM 11 (FAO, 2004a) recognizes that PRA involves many uncertainties, largely since estimates and extrapolations are made from real situations where the pest occurs to a hypothetical situation in the PRA area. In most cases, analyses performed during PRA use historical data to forecast potential future events. It is important to document the areas of uncertainty and the degree of uncertainty in the assessment and to indicate where expert judgment has been used. This is necessary for transparency and may also be useful for identifying and prioritizing research needs.

1.8 CRITICISM OF PEST RISK ANALYSIS

PRA has faced criticism for being reactive, often implemented only after a pest problem has emerged. Many pests now under phytosanitary measures were recognized as risks only after they had spread beyond their original regions and caused damage elsewhere. Some newly escaped organisms were previously unknown to science, making current PRA standards inadequate for assessing unknown risks.

Additionally, PRA typically focuses on pests at the species level, assuming genetic stability. This can be an oversimplification, especially for pests with short generation times and rapid genetic changes. Impact assessments often consider only known hosts or habitats, potentially underestimating impacts if pests adapt to new environments. Furthermore, risk analyses for invasive species are criticized for being narrow, subjective, arbitrary, and susceptible to political influence.

1.9 DESIRABLE PROPERTIES OF A PEST RISK ANALYSIS

Analysts and stakeholders seek accuracy in PRA predictions and the effectiveness of proposed risk management options. However, measuring PRA effectiveness is challenging due to time lags between pest establishment and discovery, often making it difficult to trace incursions back to regulated pathways. Viewing PRAs as predictions of organism presence along the exposure pathway allows for validation through strategic audit and inspection data.

Here are some desirable properties for a PRA based on expert experience and ISPM 11 recommendations (Burgman et al., 2014):

- Clearly identify assessed pathways and specify how trade frequency and volume are accommodated.
- Define terms for likelihood and consequence as precisely as possible.

- Ensure rules for combining likelihoods are consistent with probability principles.
- Include transparent economic, social, and environmental impact assessments.
- Provide guidelines and examples for presenting uncertainties.
- Consider incorporating sensitivity analyses for both qualitative and quantitative elements.
- Avoid making conservative or risk-averse judgments without evidence, reserving interpretations for a transparent step at the end of the analysis.
- Express the importing country's attitude to uncertainty through its interpretation of likelihood and consequence bounds, relative to its Appropriate Level of Protection (ALP) and Acceptable Level of Risk (ALR).
- Explicitly state the ALP to help stakeholders understand and interpret the PRA.

1.10 RESOURCES AND PEST RISK ANALYSIS IN INDIA

Clause 3(7)(i) of Chapter II of Plant Quarantine (Regulation into India) Order 2003 (PQ Order 2003) mandates PRA for any plant or plant materials to be imported into India. The National Institute on Plant Health Management (NIPHM), Hyderabad, is the nodal institute for conducting training programs on PRA. The institute in collaboration with USDA-APHIS has conducted a number of international training programs. The PRA trainings are being imparted to Scientists, Teaching Faculty, Extension Officials, and Officials of the Directorate of Plant Protection Quarantine & Storage, State Agricultural Universities, and Indian Council of Agricultural Research. In addition, many officers have undergone the Risk Analysis Monitoring Program (RAMP) at the Plant Epidemiology and Risk Analysis Laboratory (PERAL), Raleigh, North Carolina, USA.

The country intending to export any agricultural commodity to India shall send a request through their NPPO in **PQ Form 23 (*Request for Pest Risk Analysis for import of Agricultural commodities into India*)** along with **PQ Form 24 (*Technical Information Required for Pest Risk Analysis*).** The request shall be sent to the Plant Protection Adviser (PPA), Government of India, Directorate of Plant Protection, Quarantine and Storage, N.H.-IV, Faridabad -121 001, Haryana, India (DPPQS, 2022b) (Figures 1.2 and 1.3).

The technical information provided by the exporting country will undergo critical examination, and a PRA will be conducted by qualified and trained analysts according to relevant ISPMs. The associated risks and mitigation measures will be communicated to the NPPO of the exporting country. Upon mutual agreement, the commodity will be granted market access with specified import conditions, which will be officially notified and listed in the PQ Order, 2003. The point of entry will be determined based on the import purpose, such as propagation, processing, or consumption (Figure 1.4) (Tables 1.1, 1.2 and 1.3).

PQ Form 23

Application for Pest Risk Analysis for Import of agricultural commodities into India

1. **Details of Applicant**
   1.1 Name/ Organisation..
   1.2 Address.. Postcode...................
   1.3 Phone ....Fax .... E-mail ....

2. **PRA General Parameters**
   2.1 Scientific& Common name of the product..
   2.2 Country/ countries of origin...
   2.3 Quantity/ Volume ....

3. **Product Type (circle one or more)**
   3.1 Processed/ Non-processed — 3.2 Living/ non- living
   3.3 Plant/ Animal — 3.4 Genetically modified/ non-genetically modified
   3.5 Seed/ plant/ soil — 3.6 Culture / non-culture
   3.7 Other...

4 **Product Processing (if applicable)**
   4.1 If seed: ground/ kibbled/ whole/ preserved
   4.2 If plant: fresh/ dried/ freeze dried/ preserved
   4.3 Processing refinement: cooked/ frozen/ pulped/ steamed
   4.4Specify treatment details ....

5 **Product Origins (please state if question not relevant)**
   5.1 Source location (by country, origin & locality) ....
   5.2 Production method, Certification scheme and / or accreditation type? ....
   ....

6 **End Use (circle one or more)**
   6.1 Human consumption / Processing/ Stock feed/ Pet food/ Fish food/ Seeds for sowing/ Nursery stock/ Multiplication/ Post-entry Quarantine/ Therapeutic/ Fertilisers/ *In-vivo / Invitro*
   6.2 Other ....

7 **End Destination (circle &/or specify)**
   7.1Rural/ urban — 7.2 Multiple locations/ single
   7.3Specify Country, State & / or region (PRA defined area) ....

8 **Entry (circle one or more)**
   Ship/ Air/ Ground transport/ Rail/Other...

9 **General Comments** (any further general comment or notes that need to be made, please make here)...

PRA request from may be submitted to:
Plant Protection Adviser, DPPQS, Faridabad-121001(Haryana) or
Joint Secretary (PP), DAC, Krishi Bhavan, New Delhi -110001

FIGURE 1.2 PQ Form 23: request for pest risk analysis for import of agricultural commodities into India.

PQ Form 24

Technical Information Requirement for Pest Risk Analysis (PRA)

1. Plant and Plant Product
1.1 Common name;
1.2 Scientific (genus & species/strain/variety/cultivar) name;
1.3 Resistant or non-resistant varieties;
1.4 Countries that have already imported;
1.5 Plant part to be imported (whole plant/seed/cutting/sapling/ budwood/bulb/fruit etc.);

2. Production Area
2.1 Place of production on map (country and province);
2.2 Production and Export (tons/year);

3. Cultivation practices
3.1 Harvest method and time;
3.2 Plant protection measures (to control and eradicate the pests);

4. Pest List (separately for all the pests)
4.1 Scientific & Common name;
4.2 Pest biology;
4.3 Plant parts affected;
4.4 Symptoms;
4.5 Distribution and pest free areas;
4.6 Pest status (prevalence);
4.7 Management practices;
4.7.1 Cultural practices;
4.7.2 Biological (use of biological control agents, resistant varieties, crop skipping…);
4.7.3 Chemical (type, method, time and number of pesticide use…)
4.8 Database and reference

5. Packaging
5.1 Method of packaging;
5.2 Inspection procedure;
5.3 Post harvest treatment;
5.4 Conditions and security of storage place.

6. Export program (policy/activity)
6.1 Trading partners;
6.2 Existing procedure for issuing phytosanitary certificates (including additional declaration).

7. Copies of relevant supporting documents.

FIGURE 1.3 PQ Form 24: technical information required for pest risk analysis.

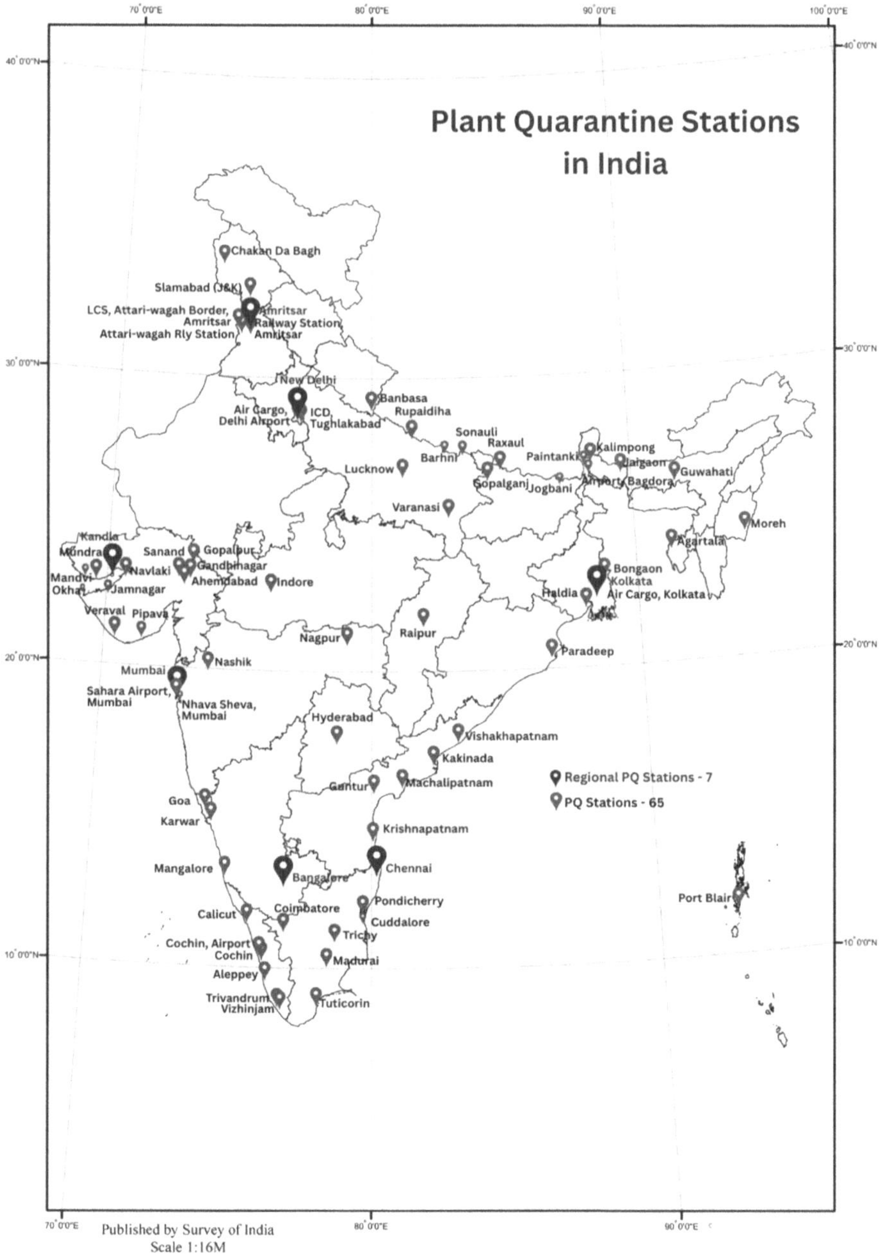

FIGURE 1.4 Directorate of Plant Protection Quarantine and Storage and its plant quarantine station.

TABLE 1.1
Point of Entries of Imported Agricultural Commodities

Schedule of PQ Order & Point of Entries	Number
Schedule-I	
Seaports	47
Airport	25
Land frontier station	24
Schedule-II	
Inland container depots and container freight stations	80
Schedule-III	
Foreign post offices	11

TABLE 1.2
Market Access Granted for Plants and Plant Products by NPPO, India (up to September 2022)

Schedule	# Commodities	Status
Schedule-IV	15	Prohibited commodities
Schedule-V	17	Permitted to import with recommendation of authorized institutions
Schedule-VI	701	Imported with certain conditions
Schedule-VII	519	Least risk commodities

TABLE 1.3
Regulated pests due to threat to Indian Biosecurity

Pest	# species	Remarks
Insects		
Fungus		
Viruses		
Nematodes		
Weeds	57	

1.11 DESIGNATED INSPECTION AUTHORITIES AND POST ENTRY QUARANTINE IN INDIA

The import of growing materials poses a high risk of pest invasion, with over 60% of exotic pests established in India linked to plant-propagating materials, fresh fruits, and vegetables. To mitigate this risk, imported plant-propagating materials are kept in post entry quarantine (PEQ) facilities for an extended period under the supervision of Designated Inspection Authorities (DIAs).

The heads of Plant Pathology Departments at State Agricultural Universities and Indian Council of Agricultural Research institutes are notified as DIAs. They are responsible for approving PEQ facilities and monitoring imported plant-propagating materials during the quarantine period. As of 2022, there are approximately 40 DIAs notified in India (DPPQS, 2024) (Figure 1.5).

India has a very robust interdisciplinary Plant Quarantine system in place with required regulatory backup. Indian Plant Biosecurity umbrella is given below (Table 1.4).

1.12 INSPECTION CLEARANCE OF IMPORTED CONSIGNMENTS

The Plant Quarantine (Regulation of Import into India) Order 2003 issued under Destructive Insects & Pests Act, 1914 (Act 2 of 1914) regulates the import of all agricultural commodities into India. Agricultural commodities imported through different point of entries are being inspected by concerned Plant Quarantine Stations as per the provisions of PQ Order 2003 and the procedures laid down in SOP for Phytosanitary Inspection and Plant Quarantine Clearance of Plants/Plant Products & other Regulated Articles.

Quarantine is the first line of defense against the invasion of non-indigenous pests, failure in the systems results in. Entry of non-indigenous pests through regulated international trade is the consequence of a failed quarantine system. Many countries including India have witnessed a number of invasions in the past, and it is quite difficult to pinpoint the pathways of entry since countries share porous borders with many neighboring countries. Upon entry, subsequent establishment depends on the availability of a suitable host and environment. The exotic pests that confront all these manifest in large numbers with conspicuous damage in an introduced environment. Subtle species having overlapping damage symptoms with already established pests are to be detected by a professional taxonomist in the pursuit for faunal studies. Invasions of non-indigenous weeds belong to the latter group, which come to the limelight after years of introduction and establishment.

Therefore, personnel involved in the phytosanitary inspection and clearance of imported agricultural commodities are to be qualified and well trained in the inspection and detection of pests in a commodity with adequate regulatory backup and state-of-the-art laboratory facilities. All the stakeholders including researchers, extension functionaries, farmers, general public, etc., should be sensitized toward implementing plant quarantine regulations. Interception and salvaging one consignment

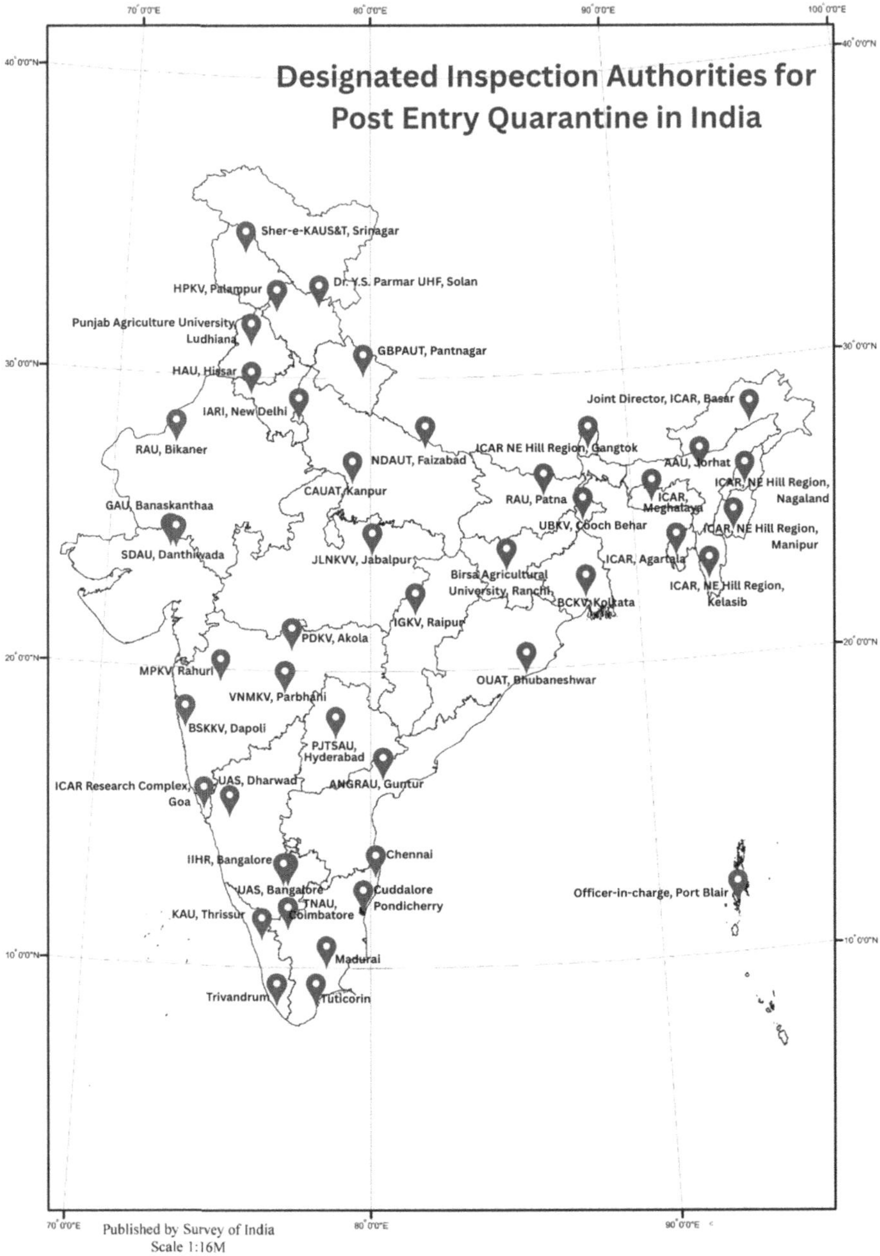

FIGURE 1.5 Designated inspection authorities in India.

TABLE 1.4
Indian Plant Biosecurity Umbrella

Sl. No.	Organization[a]	Activities
01.	Directorate of Plant Protection Quarantine and Storage (DPPQS)	Apex body for implementing Destructive Insect Pest Act 1914 (DIP Act 1914) and Plant Quarantine Order (Regulation of Import into India) Order 2003. Duties and responsibilities of National Plant Protection Organization (NPPO) as per Article IV of IPPC.
02.	Central Board of Indirect Taxes and Customs, Department of Revenue under the Ministry of Finance, Government of India.	Section 4 of the DIP Act 1914 empowers the Customs to operate the notification issued under DIP Act, as it has the same power issued under Section 11 of Customs Act 1962. Implementing ISPM 15 (International Standards for Phytosanitary Measures No. 15 - Regulation of wood packaging material in international trade).
03.	Indian Council of Agricultural Research (ICAR) and its institutions	Certification of import of Germplasm/Transgenics/ Genetically Modified Organisms (GMOs) Recommendation for import and monitoring of commodities listed in Schedule V of PQ Order 2003. Approval for post entry quarantine facilities for live plant materials and monitoring and release of plants in PEQ. Pest survey and surveillance. Introduction and evaluation of exotic insects for biological control of insect pests.
	ICAR-National Bureau of Plant Genetic Resources (NBPGR), New Delhi.	Issuance of Import Permit for Germplasm/Transgenics/ Genetically Modified Organisms (GMOs) for the purpose of agricultural research or experimentation purpose. The permit issued with the approval of Genetic Engineering Approval Committee (GEAC) or as the case may be, the Review Committee on Genetic Manipulation (RCGM) set-up by Department of Biotechnology under the provisions of sub-rule (2) of rule 4 of the Rules for the manufacture, use, import, export and storage of hazardous micro-organisms, Genetically engineered organisms or cells made under Sections 6, 8 and 25 of the Environment (Protection) Act, 1986 (29 of 1986) and subject to such restrictions and conditions prescribed thereof. Inspection and supervision and release of such consignments. Maintenance of accessions/varieties in gene bank as per Schedule XII of PQ Order 2003.
	ICAR- National Bureau of Agricultural Insect Resources (NBAIR)	Introduction and evaluation of exotic insects for biological control of insect pests.
	NBPGR, New Delhi/Director, Directorate of Rice Research, Hyderabad.	Seeds of Rice (*Oryza sativa*) for sowing

(Continued)

TABLE 1.4 (*Continued*)
Indian Plant Biosecurity Umbrella

Sl. No.	Organization[a]	Activities
	NBPGR, New Delhi/Directorate of Wheat Research, Karnal, Haryana.	Seeds of Wheat (*Triticum* spp.) for sowing
	ICAR- National Research Centre on Banana, Tiruchirappalli, Tamil Nadu	Rhizomes of Banana, Plantain and Abaca (*Musa* spp.) for propagation.
	ICAR-Central Tuber Crops Research Institute, Sreekaryam, Kerala.	Stem cuttings of Cassava or tapioca (*Manihot esculenta*), Stem (vine) cuttings rooted or un-rooted/tubers of Sweet potato (*Ipomoea* spp.) and tubers of Yam (*Dioscorea* spp.) for planting or propagation.
	ICAR-National Research Centre on Citrus, Nagpur, Maharashtra	Grafts/Bud wood/Plants of *Citrus* spp. (lemon, lime, orange, grapefruit, mandarins, etc.) and other Rutaceous hosts for propagation
	ICAR-Central Plantation Crops Research Institute, Kasaragod, Kerala	Seeds (beans)/pods/bud wood/rootstock of *Theobroma cacao* (Cocoa) and related species for propagation. Seed nuts/Seedlings/Pollen of Coconut (*Cocos nucifera*) and related species for propagation
	ICAR-Central Cotton Research Institute, Nagpur, Maharashtra	Seeds of Cotton (*Gossypium* spp.) for sowing
	ICAR-National Research Centre on Groundnut, Junagadh, Gujarat and International Crops Research Institute for Semi-Arid Tropics (ICRISAT), Patancheru, Andhra Pradesh.	Seeds/ Stem Cuttings/Plants of Groundnut (*Arachis* spp.) for propagation
	ICAR-Central Potato Research Institute, Shimla, Himachal Pradesh	Tubers and other planting material of Potato (*Solanum tuberosum*) and other tuber-bearing species of Solanaceae for propagation
	ICAR-Sugarcane Breeding Institute, Coimbatore, Tamil Nadu	Setts of Sugarcane (*Saccharum* spp.) for planting
	ICAR-Central Tobacco Research Institute, Rajahmundry, Andhra Pradesh	Seeds of Tobacco (*Nicotiana* spp.) for sowing
04.	Central Coffee Research Institute, Ministry of Commerce and Industry, Balehonnur, Chikkamagaluru, Karnataka	Seeds (beans) and berries (freshly harvested)/ Grafts / Bud wood/Seedlings/Rooted cuttings Coffee (*Coffea* spp.) and related species of Rubiaceae for propagation
05.	Indian Council of Forestry Research and Education and its institutions	Seeds/Plants/Fruits/Grafts and other planting material of Chestnut (*Castanea* spp.), Elm (*Ulmus* spp.), Oak (*Quercus* spp.), Pine (*Pinus* spp.) Poplar (*Populus* spp.), Walnut (*Juglans* spp.), and other coniferous species for propagation.

(*Continued*)

TABLE 1.4 (*Continued*)
Indian Plant Biosecurity Umbrella

Sl. No.	Organization[a]	Activities
06.	National Institute of Plant Health Management (NIPHM), Hyderabad	Nodal agency for capacity building in Agricultural Biosecurity
07.	State Agriculture Universities (SAUs)	Approval post entry quarantine facilities and monitoring and release of plants in PEQ. Export certification, Survey and Surveillance, etc.
08.	State Agricultural/Horticultural Departments	Export certification, Survey and Surveillance, etc.
09.	Private agencies/Accredited Treatment Providers	Disinfection of export/import agriculture commodities through phytosanitary treatments as per NSPM 9, 12, 15, 18, 20, 21, 22.

[a] Indicative of inter-ministerial/departmental cooperation. NSPM, National Standard for Phytosanitary Measures.

are a prevention of one potential invasion of a quarantine pest. Therefore, private accredited treatment providers in the country should also be sensitized about the importance of salvaging an infested consignment.

1.13 PLANTS FOR PROPAGATION AND POST ENTRY QUARANTINE

Guidelines for PEQ facility certification and PEQ Inspection are notified vide Gazette Notification No. SO.2286(E) dated 04.06.2018. As per the guidelines, the importer shall submit an application in duplicate for certification or renewal in PQ Form 18 of PQ Order, 2003 to the concerned Designated Inspection Authority (DIA/IA) and to the concerned Regional Plant Quarantine Station (RPQS). DIA along with an officer nominated by RPQS will conduct the joint inspection of the PEQ facility as per the guidelines specified in SOP for PEQ Inspection. DIA will issue the Certificate in PQ Form 19 of PQ Order, 2003. The importer may have to get prior permission from EXIM committee for the import of planting materials of coarse cereals, oil seeds, pulses, fodder, and fruit plants, and a joint inspection report is mandatory to get the EXIM committee approval.

Plants for propagation shall only be imported through notified Regional Plant Quarantine Stations. On arrival of the shipment, RPQS after inspection of consignment, if satisfied, shall accord "provisional clearance" to grow the imported materials under the PEQ facility, and the same shall be intimated to DIA in Annexure-5B of SOP for PEQ. Annexure-5B is also sent to the Director of Horticulture/Agriculture of the concerned State/UT for information and monitoring for exotic pests near the PEQ facility where the plants will be grown. Upon detection of any exotic pests by the State Department, the same has to be intimated to the Directorate of Plant Protection Quarantine and Storage (DPPQS) and DIA immediately.

The importer has the responsibility to intimate DIA in advance about the date of planting of imported material referred for PEQ; not to transfer or part with or dispose of consignment during the PEQ period; to provide complete access to DIA and PQ officers to PEQ facility and plants and extend necessary facilities at all times and abide by the instructions of DIA & PQ Officers; to maintain requisite conditions as stipulated in SOP for PEQ Inspection & Chapter-IV, Clause 11 of PQ Order, 2003; and to arrange for destruction of any part or whole of plant population when ordered for destruction in the event of quarantine/exotic pest detection.

The inspection of the consignment during the PEQ period shall be carried out jointly by DIA with a nominated officer of RPQS. The frequency of the inspections shall be decided considering the growing period, and a minimum of two inspections shall be carried out. Out of this, one inspection shall be invariably at the end of the PEQ period. At the end of the final inspection, the DIA shall forward a copy of the PEQ inspection report [Annexure-10A (Part1&2)] duly signed by him to the concerned RPQS. Decision regarding the final release of the consignment shall be granted by RPQS after taking into consideration the joint inspection report submitted by the DIA & PQ Officer. The concerned RPQS to endorse a final clearance to PPA and concerned Customs for intimating the final clearance from PQ of the imported material, since only a "provisional clearance" is granted at the time of import (DPPQS, 2022) (Figure 1.6).

1.14 IMPORT AND INSPECTION OF PERISHABLE CARGO

Apples and many such commodities are being imported into India from different countries with mandatory pre-shipment cold treatment as specified in Schedule VI of Plant Quarantine (Regulation of Import into India) Order, 2003. The fruits should be treated at below 0°C or 0.55°C or 1.1°C for 10 or 11 or 12 days, respectively, with in-transit refrigeration. The pre-shipment cold treatment is to mitigate the risk associated with exotic insects, namely, Mediterranean fruit fly, *Ceratitis capitate* (Wiedemann); Apple maggot, *Rhagoletis pomonella* Walsh; Queensland fruit fly, *Bactrocera tryoni* (Froggatt); Natal fruit fly, *Ceratitis rosa* Karsch (all Diptera: Tephritidae); Raspberry beetle, *Byturus tomentosus* (De Geer) (Coleoptera: Byturidae); Red plum maggot, *Grapholita funebrana* Treitschke; Oriental fruit moth, *Grapholita molesta* (Busck); and Manchurian fruit moth, *Grapholita inopinata* Heinrich (all Lepidoptera: Tortricidae). However, these cold treatments are not effective on many insects and plant pathogens as seen in the case of apples imported from China, which were intercepted with live mealy bugs. Now, apple shipments from Belgium, Greece, Italy, Japan, Turkey, New Zealand, and Chile are intercepted with *Venturia inaequlis*. Therefore, strict quarantine inspections are mandatory at the point of entry, and all the plant pathogens intercepted should be carefully examined in the laboratory. PRA should be re-looked into in such cases, as cold treatment is not meant for pathogens (Nagaraju et al., 2021c) (Figure 1.7).

Fruits such as Longan, Dragon Fruit, Rambutan, and Mangosteen are being imported from neighboring countries in carton boxes without any pre-shipment or in-transit cold treatment. Such shipments are being regularly intercepted with exotic pests (Figure 1.8).

a. Approved PEQ facilities of Indo-American Hybrid Seeds at Bangalore

b. An approved PEQ unit

c. Grafted Catci imported from South Korea in the PEQ facility

d. *Poinsettia* imported from Spain in the PEQ facility

e. Poinsettia showing disease like symptoms

f. Suspected plant segregated and kept within the facility for further observation

FIGURE 1.6 Imported plants for propagation in post entry quarantine (PEQ) facilities. (a) Approved PEQ facilities of Indo-American hybrid seeds at Bangalore. (b) An approved PEQ unit. (c) Grafted Catci imported from South Korea in the PEQ facility. (d) *Poinsettia* imported from Spain in the PEQ facility. (e) Poinsettia showing disease-like symptoms. (f) Suspected plant segregated and kept within the facility for further observation.

FIGURE 1.7 Apple scab, *Venturia inaequlis* interceptions in the imported shipments.

a. Mealy bug intercepted on Longan

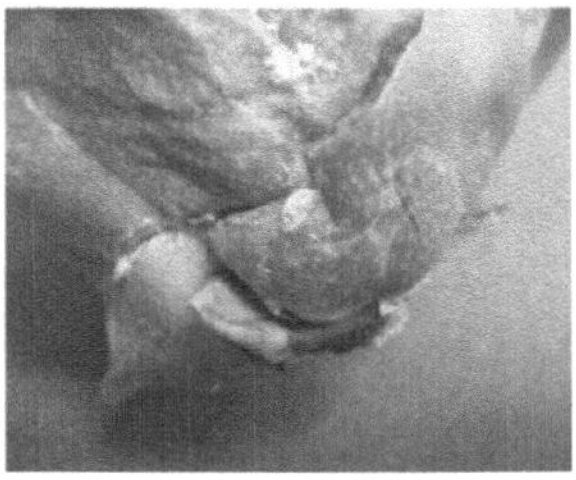

b. Mealy bug intercepted on Dragon fruit

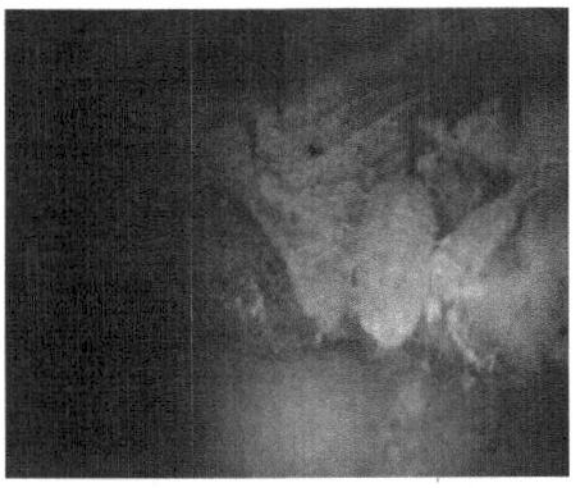

c. Mealy bug intercepted on Dragon fruit

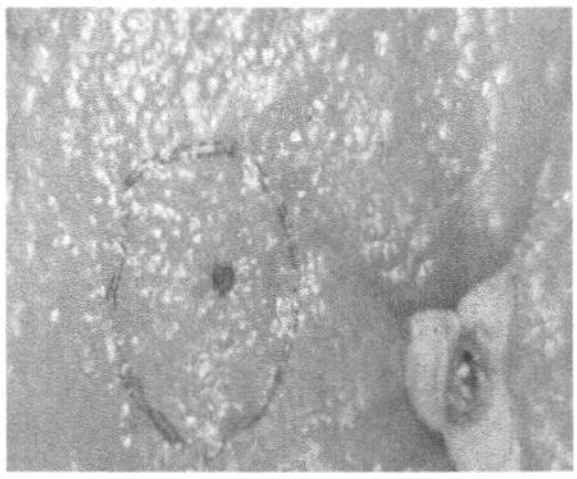

d. Scale insect intercepted on Citrus fruit

e. Mealy bug intercepted on Rambutan

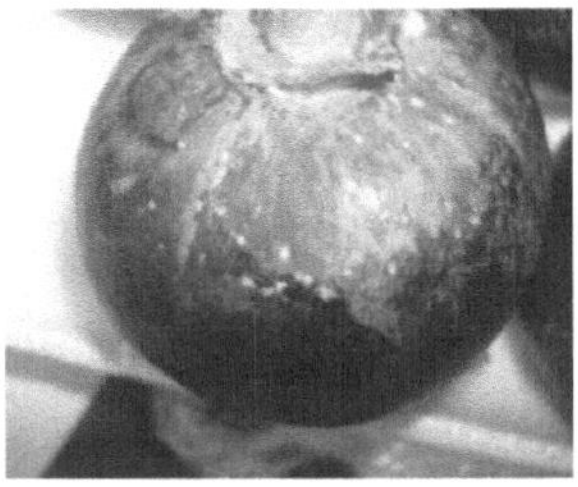

f. Scale insect intercepted on Mangosteen

FIGURE 1.8 Exotic pests intercepted on perishable cargo. (a) Mealy bug intercepted on Longan. (b) Mealy bug intercepted on Dragon fruit. (c) Mealy bug intercepted on Dragon fruit. (d) Scale insect intercepted on Citrus fruit. (e) Mealy bug intercepted on Rambutan. (f) Scale insect intercepted on Mangosteen.

FIGURE 1.9 Inspection and sampling imported shipments. (a) Bulk shipment of peas. (b) Bulk shipment of raw cashew nuts. (c) Bulk shipment of wheat. (d) Sampling.

1.15 IMPORT INSPECTION OF NON-PERISHABLE CARGO

Shipments of cereals, pulses, spices, and similar commodities should be sampled according to Standard Operating Procedures (SOP) based on international sampling guidelines. Representative samples are sieved onto a uniformly spread white sheet using sieves of various mesh sizes to capture all possible sizes of seeds and plant materials contaminating the commodity.

The collected seeds and plant materials, both on the white sheet and retained in the sieve, are then examined. Parameters such as size, color, texture, and the presence of any attachments are used to separate foreign materials from the main commodity. Extremely small seeds and plant materials are examined under a stereo binocular microscope to detect weed seeds and other extraneous materials (Figures 1.9 and 1.10).

1.16 INTERCEPTION OF EXOTIC WEED SEEDS

Four weed species non-indigenous to India were intercepted in imported lentil and lentil dust consignments. *Raphanus raphanustrum* (Brassicaceae) in lentils

a. Animal feed shipment in gunny bags in an vessel

b. Bulk shipment of an animal feed shipment in a vessel

FIGURE 1.10 Inspection of bulk shipments of animal feed. (a) Animal feed shipment in gunny bags in a vessel. (b) Bulk shipment of an animal feed shipment in a vessel.

from Australia, *Polygonum lapathifolium* (Polygonaceae) and *Thlaspi arvense* (Brassicaceae) in lentils from Canada and the USA, and *Echinochloa crus-pavonis* (Poaceae) in lentil dust imported from Sri Lanka were intercepted.

The interception of indigenous and non-indigenous weed species highlights lapses in weed management practices at the farm level and during the cleaning process at farms and export elevators in the exporting country. Lentil cultivation is mechanized in all three major lentil-exporting countries. Since lentils are low-growing plants harvested close to the ground using combines, both crops and weeds are harvested together, leading to weed contaminants in import shipments.

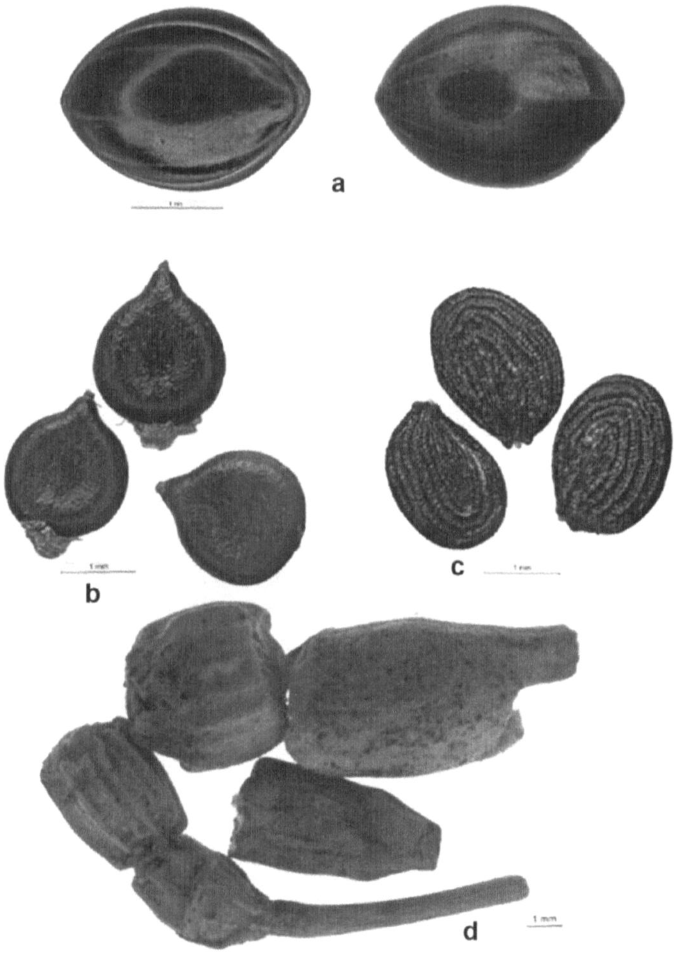

FIGURE 1.11 (a) *E. crus-pavonis*, (b) *P. lapathifolium*, (c) *T. arvense*, (d) *R. raphanistrum*: exotic weed seeds intercepted in lentil and lentil husk shipments from the USA, Canada, Australia, and Sri Lanka.

Mack (2000) noted that there are no agreed criteria for the minimum damage, spread, or population size needed for an alien species to be considered invasive. Even a small number of individuals with limited genetic variation from their native range can cause significant environmental damage in a new environment through reproduction and spread.

India's diverse climate and rich biodiversity make it particularly vulnerable to invasive alien species. The introduction, spread, and establishment of these species can have severe economic consequences (Nagaraju et al., 2021a) (Figure 1.11).

1.17 EXOTIC STORAGE PESTS INTERCEPTED IN RAW CASHEW NUTS ANIMAL FEED SHIPMENTS

The raw cashew nuts and animal feed shipments imported from different countries are being intercepted with exotic storage pests, namely, *Ahasversus advena* (Waltl.) (Coleoptera: Silvanidae), *Carpophilus* sp. (Coleoptera: Nitidulidae), *Cryptolestes ferrugineus* (Stephens) (Coleoptera: Laemophloeidae), and *Ephestia* sp. (Lepidoptera: Pyralidae) (Nagaraju et al., 2022a). Both the commodities are supposed to be risk-free and are listed in Schedule-VII of the PQ Order.

Animal feed consignments are kibbled–crushed seeds/pellets/dried cake form thereby denatured. Lentil husk shipment imported from Sri Lanka as animal feed was also intercepted with exotic weed seeds. Lentils are not grown in Sri Lanka, and the country's requirement is met only through imports. The intercepted weed might have contaminated lentils imported to Sri Lanka from any of the three exporting countries, such as Australia, Canada, and the USA. The imported lentils are processed in dhal mills, and husk, a by-product of processing industries, is being exported to India. Interception of *E. crus-pavonis* in lentil husk shipment from Sri Lanka establishes the ability of a weed species to escape through multiple quarantine inspections at least at three levels such as country of export, country of import, and country of re-export. It is also interesting to note that the whole system of processing could not eliminate the weed infestation, which is scary. New interceptions on any shipment lead to a review of the existing PRA and a new set of guidelines to be implemented for import-intercepted consignments (Figure 1.12 to 1.15).

1.18 INSPECTION WOODEN LOGS

1.19 INTERCEPTION OF TWO POTENTIAL WOOD FEEDING INVASIVE *COPTOTERMES* TERMITE SPECIES IN INDIA

Several termite species are significant economic pests, causing damage to wood in both man-made structures and also in forest ecosystem. The species belonging to genus *Coptotermes* (Rhinotermitidae: Isoptera: Blattodea) are widespread, and few of them are among the world's most dangerous wood pests. Two species of *Coptotermes* intercepted and described from yemene, mora, and tali logs are imported from Columbia, Suriname, and Equitorial Guinea. The intercepted termites were two non-native species of *Coptotermes*: *Coptotermes testaceus* (Linnaeus) and *Coptotermes sjöstedti* Holmgren (Raju et al., 2019). Identification was initially made with morphological characters and then with molecular DNA barcoding for a 348 bp fragment of 16S rRNA gene. Maximum likelihood phylogenetic analysis results showed that data recovered as monophyletic confirms the species identity. *Coptotermes testaceus* and *C. sjöstedti* are in single clade. This first interception suggests possible chances of entry and establishment in India and provides an important alert for authorities concerned to take the necessary steps in monitoring and preventing its possible introduction. There are no records of invasive termites being introduced into India so far, but robust quarantine measures are in place to prevent unexpected entry. Further, the interception of quarantine pests triggers the revision of PRA and the updation of import regulations (Nagaraju et al., 2020) (Figure 1.16 and 1.17).

Inspection wooden logs:

FIGURE 1.12 Inspection and interception of pests in imported wooden logs. (a) Containerized cargo of pine logs. (b) Powdery exudates from the infested logs. (c) Galleries on the logs and below the bark. (d) Inspection of a bulk shipment of wooden logs.

1.20 INTERCEPTION OF QUARANTINE PESTS AND REVISION OF PEST RISK ANALYSIS

Many a time, pest profile will be changing over a period of time due to a number of factors affecting the production stage. Due to this, there are more possibilities that a frequent number of quarantine pest interceptions can be expected. In accordance with that, NPPO, India, reported a number of live pest interceptions in the shipments of plants and plant products imported (Table 1.5 and 1.6).

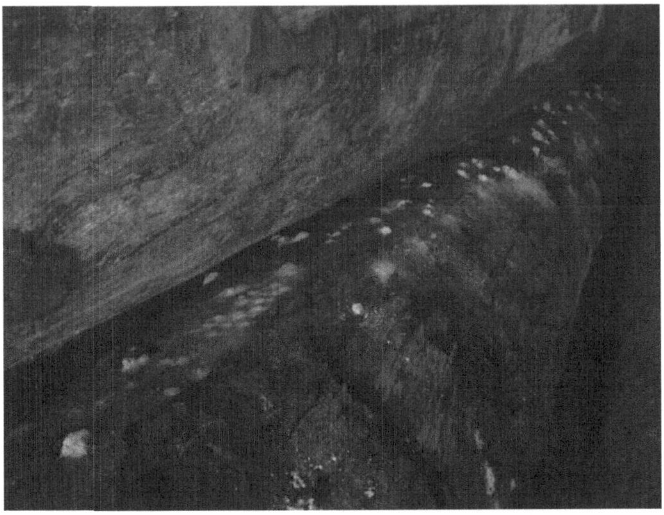

FIGURE 1.13 Interception of live Scolytidae and other Coleopteran insects in tali logs from Cameroon.

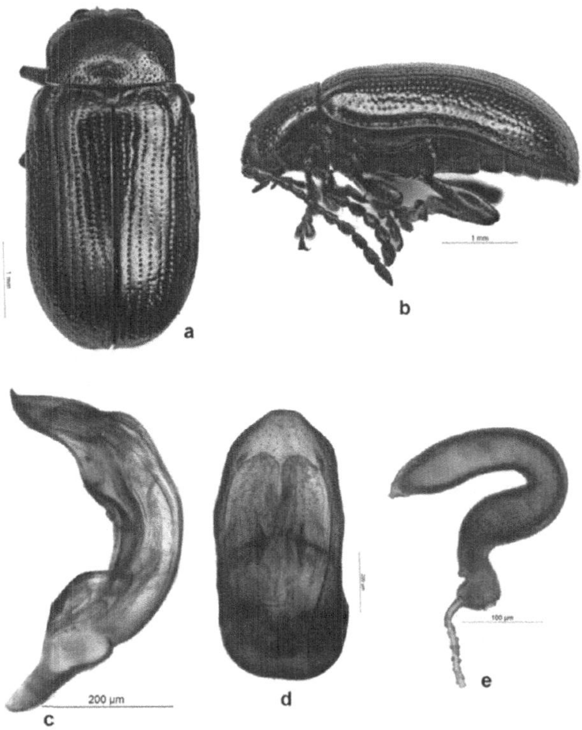

FIGURE 1.14 *Phratora laticollis* (Coleoptera, Chrysomelidae, Chrysomelinae) intercepted on poplar logs imported from Belgium and Germany; (a) Male dorsal view, (b) Lateral view, (c) Aedeagus in lateral view, (d) Apex of Aedeagus, (e) Spermatheca.

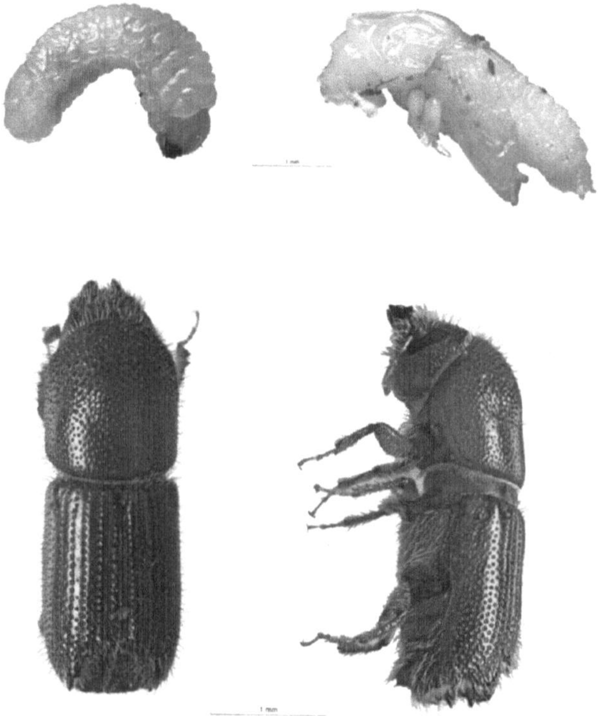

FIGURE 1.15 Larvae, pupa, and adults of *Ips* sp. intercepted on pine logs imported from the USA.

1.21 WAY FORWARD

PRA is a dynamic process. The interception of live insects in consignments indicates a failure in export inspection and fumigation. Consequently, all imported containers and bulk shipments with live insects, whether quarantine or non-quarantine, are fumigated with methyl bromide at 48 g/m^3 for 24 hours at normal atmospheric pressure (NAP) and re-inspected to ensure pest freedom.

Noncompliances are reported to trading partners for each interception, following ISPM-13 guidelines (FAO, 2001). However, consignments are released for use according to IPPC guidelines. There are nearly 688 and 648 accredited fumigation agencies per NSPM 12 and NSPM 22 to carry out methyl bromide and aluminum phosphide fumigation, respectively, in the country. Consignments intercepted with diseases are issued deportation orders, while those with exotic weed seeds are salvaged. Each interception triggers a review of the PRA and the notification of additional checkpoints and treatments. Reporting noncompliances allows the exporting country to investigate and take corrective actions to prevent future issues, thereby avoiding potential import prohibitions.

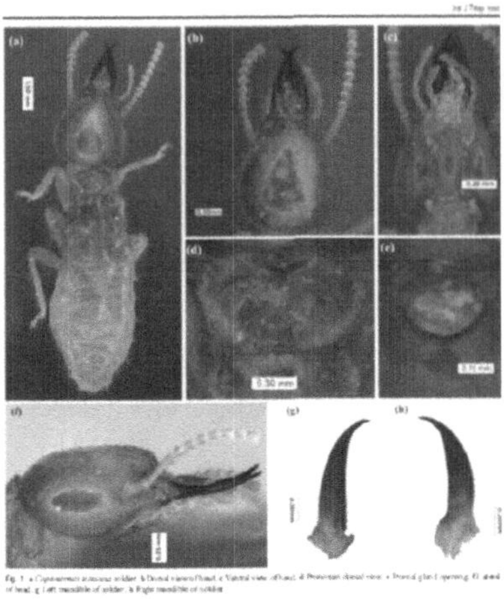

a. ***Coptotermes testaceus* (Linnaeus)**

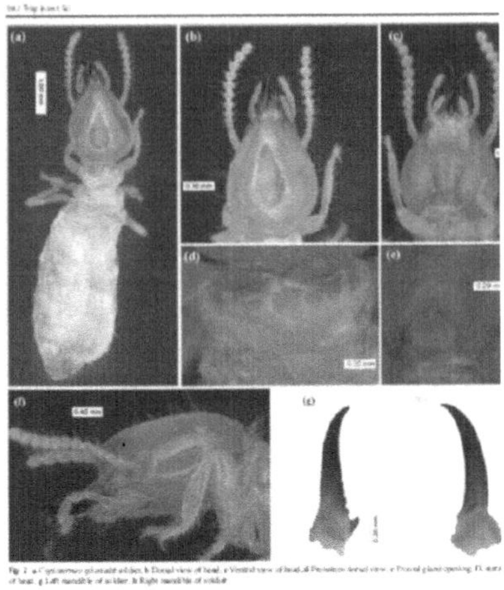

b. ***Coptotermes sjöstedti* Holmgren**

FIGURE 1.16 Termites intercepted on imported wooden logs. (a) *Coptotermes testaceus* (Linnaeus). (b) *Coptotermes sjöstedti* Holmgren.

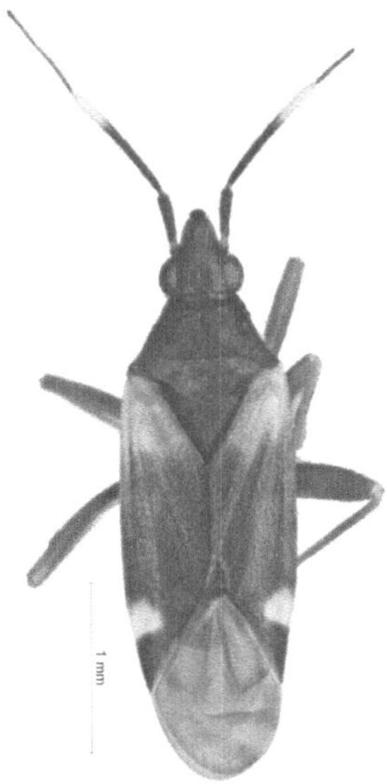

a. ***Fulvius anthocoroides* (Reuter, 1875) (Hemiptera: Miridae) wooden logs imported from Cameroon**

b. ***Hololepta plana* (Sulzer) (Coleoptera, Histeridae) intercepted on poplar logs imported from Belgium**

FIGURE 1.17 Non-quarantine insects and hitchhikers intercepted on wooden log imported from different countries (Nagaraju et al., 2021b, 2022b). (a) *Fulvius anthocoroides* (Reuter, 1875) (Hemiptera: Miridae) wooden logs imported from Cameroon. (b) *Hololepta plana* (Sulzer) (Coleoptera, Histeridae) intercepted on poplar logs imported from Belgium.

TABLE 1.5
Pests Intercepted in Imported Commodities During Last 5 Years

Year	# Interceptions
2018–2019	1,118
2019–2020	3,023
2020–2021	1,051
2021–2022	1,781
2022–2023	2,795
2023–2024	1,487

TABLE 1.6
Diversity of Pests Intercepted in Imported Commodities

Organism	Number of Species	Number of Commodities	Country of Origin
Bacteria	15	13	USA, France, the Netherlands, Spain, Italy, Israel, Kenya, Germany, Ireland, Singapore (10)
Fungi	158	154	USA, UK, Australia, New Zealand, Japan, South Korea, Canada, China, France, Germany, Denmark, Taiwan, Bhutan, the Netherlands, Singapore, Sri Lanka, Thailand, Sudan, Costa Rica, Cuba, Fiji, Malaysia, Papua New Guinea, Solomon Island, Tahiti, Russia, Israel, Kenya, Pakistan, Zimbabwe, Poland, S. Africa, Mexico, Peru, Hong Kong, Philippines, Turkey, Spain, Italy, Bhutan, Egypt, Venezuela, Burkina Faso, Kuwait, Indonesia, Hungary, Ivory Coast, French Polynesia, Scotland (49)
Viruses	7	12	USA, France, the Netherlands, Germany, Israel, Australia, Malaysia, Singapore, Thailand, Costa Rica (10)
Nematodes	32	45	USA, UK, Turkey, Singapore, Thailand, the Netherlands, France, Germany, Malaysia, Sri Lanka, Ireland, Scotland (12)
Insects	98	32	USA, UK, China, Japan, Vietnam, Kenya, Malaysia, Singapore, Australia, Austria, Colombia, Indonesia, Ivory Coast, Myanmar, Nigeria, Panama, Togo, the Netherlands, Fiji, Costa Rica, Lagos, Zimbabwe, Turkey, Russia (24)
Weeds	112	7	USA, Australia, Canada, France, Turkey, Myanmar (6)

Source: http://plantquarantineindia.nic.in/PQISPub/html/pestIntercept-con.htm; Nagaraju et al. (2020, 2021c, 2022b), Raju et al. (2019, 2022), Sushil (2016, 2018), and Sushil et al. (2021).

1.22 SUMMARY

International trade is expanding annually, and it is a significant pathway for the introduction of exotic pests. As trade grows, the risk posed by these pests increases, leading to substantial economic and environmental losses. To ensure the safe trade of plants and plant products, NPPOs should implement scientifically justified phytosanitary measures based on PRA.

Countries involved in trading plants and plant products must invest in robust phytosanitary regulatory systems. Effective strategies include targeted inspections and testing of specific pathways rather than inspecting all plants and products. This approach allows for better utilization of limited resources by focusing on high-risk plants and products while conducting fewer checks on low-risk items.

Risk levels vary with different pathways; for example, propagative materials pose a higher risk than produce intended for consumption. Propagative materials are not processed and are intended for propagation, increasing their potential to introduce or spread pests. Additionally, pathogens are often microscopic and cannot be easily detected through visual screening. Understanding these risks makes a targeted inspection approach more practical and manageable.

REFERENCES

Burgman, M., Roberts, B., Sansford, C., Griffin, R. and Mengersen, K. 2014. The role of pest risk analysis in plant biosecurity. In: G. Gordh and S. McKirdy (eds.), *The Handbook of Plant Biosecurity*. Springer, Dordrecht, pp. 235–267.

Devorshak, C. 2012. *Plant Pest Risk Analysis: Concepts and Application*. CAB International, British Library, London, UK.

DPPQS. 2022a. https://plantquarantineindia.nic.in/PQISPub/html/pestIntercept-con.htm. Accessed: 5 December 2022.

DPPQS. 2022b. https://pqms.cgg.gov.in/pqms-angular/homeGuidelines. Accessed: 5 December 2022.

DPPQS. 2024 https://pqms.cgg.gov.in/pqms-angular/homeGuidelines/Import PLANT QUARANTINE (REGULATION OF IMPORT INTO INDIA) ORDER, 2003. Accessed: 18th February 2025.

FAO. 1997. *International Plant Protection Convention (IPPC)*, Second Edition. U.S. Government Printing Office, Rome.

FAO. 2001. International standards for phytosanitary measures publication no. 13. Glossary of phytosanitary terms. In: *Secretariat of the International Plant Protection Convention, Food and Agriculture Organisation of the United Nations*, Rome.

FAO. 2004a. International standards for phytosanitary measures publication no. 11: pest risk analysis for quarantine pests, including analysis of environmental risks and living modified organisms. In: *Secretariat of the International Plant Protection Convention, Food and Agriculture Organisation of the United Nations*, Rome.

FAO. 2004b. International standards for phytosanitary measures publication no. 21: pest risk analysis for regulated non-quarantine pests. In: *Secretariat of the International Plant Protection Convention, Food and Agriculture Organisation of the United Nations*, Rome.

FAO. 2007. International standards for phytosanitary measures publication no. 2: guidelines for pest risk analysis. In: *Secretariat of the International Plant Protection Convention, Food and Agriculture Organisation, United Nations*, Rome.

FAO. 2010. International standards for phytosanitary measures publication no. 5. Glossary of phytosanitary terms. In: *Secretariat of the International Plant Protection Convention, Food and Agriculture Organisation of the United Nations*, Rome.

Govt. of India. 1914. The Destructive Insects and Pests Act (Act No. II of 1914) passed by Governor General of India in Council on 3rd Feb 1914 and subsequent modifications. 4p.

Govt. of India. 2003. Plant Quarantine Order (Regulation of Import into India). Gazette notification S.O. 1322 (E), dated 18th Nov., 2003 and subsequent modifications. 331p.

Mack, R. N. 2000. Assessing the extent, status and dynamism of plant invasions: current and emerging approaches. In: H. A. Mooney and R. J. Hobbs (eds.), *Invasive Species in Changing World*. Island Press, Washington DC, pp. 384.

Nagaraju, D. K., Iyyanar, D., Maharaj, S., Esakkirani, B., Venkatareddy, K., Keshavamurthy, G. M., Kapoor, K.S., Om Prakash, V., Ravi, P. and Singh, M. C. 2021a. Interception of non-indigenous weed seeds in lentil and lentil husk shipments imported *from* Australia, Canada, U.S.A., and Sri Lanka to India. *Indian Journal of Weed Science*, 53(4): 417–420.

Nagaraju, D. K., Iyyanar, D., Maharaj, S., Prema Ranjitham, T., Kasturi, N., Esakkirani, B., Om Prakash, V. and Ravi, P. 2022a. Exotic storage pests intercepted in raw cashew nuts shipments imported into India and cost of salvaging the intercepted material. *Insect Environment*, 25(2): 189–196. https://doi.org/10.55278/URCL8831.

Nagaraju, D. K., Jain, R. K., Iyyanar, D., Maharaj, S., Kasturi, N., Prema Ranjitham, T., Om Prakash, V. and Ravi, P. 2022b. Interception of live Phratora laticollis (Suffrian) (Coleoptera: Chrysomelidae) on poplar logs imported from Belgium and Germany. *Indian Journal of Entomology*, 85(1): 1–3. https://doi.org/10.55446/IJE.2022.441.

Nagaraju, D. K., Kalleshwaraswamy, C. M., Iyyanar, D., Maharaj, S., Jain, R. K., Kasturi, N., Ranjith, M., Mahadevaswamy, H. M. and Asokan, R. 2020. First interception of two wood feeding potential invasive *Coptotermes* termite species in India. *International Journal of Tropical Insect Science*, 41: 1043–1052. https://doi.org/10.1007/s42690-020-00287-5.

Nagaraju, D. K., Kasturi, N., Maharaj, S., Iyyanar, D., Om Prakash, V. and Ravi, P. 2022c. Interception of Hololepta plana (Sulzer) (Coleoptera, Histeridae) a rare beetle on poplar logs imported from Belgium. *Insect Environment*, 25(1): 18–20 https://doi.org/10.55278/CBMZ7758.

Nagaraju, D. K., Prema Ranjitham, T., Esakkirani, B., Kapoor, K. S., Om Prakash, V. and Ravi, P. 2021b. Interception of Saproxylic Hemiptera, Fulvius anthocoroides (Reuter, 1875) on wooden logs imported from Cameroon to India, a new report for Cameroon. *Insect Environment*, 24(3): 423–426.

Nagaraju, D. K., Ravichandaran, S., Prakasam, V., Ramesh, H. P., Mohanta, P., Bohria, A. K., Om Prakash, V. and Ravi, P. 2021c. Interception of *Venturia inaequalis*, a pathogen of quarantine significance in apple shipments imported into India. *Pest Management in Horticultural Ecosystems*, 27(2): 157–161.

Raju, J., Gokulraam, M., Mohan, S. M., Keshavamurthy, G. M., Nagaraju, D. K. and Geetha, S. 2019. Interception of live exotic species *Cordylomera spinicornis* (Fabricius) (Coleoptera: Cerambycidae) in Tali wood imported from African countries, *Journal of Entomology and Zoology Studies*, 7: 432–435.

Raju, J., Nagaraju, D. K., Priti, S., Kalleshwaraswamy, C. M. and Sundararaj, R. 2022. Invasion of wood degraders through wood import and need to strengthen the plant quarantine measures in India, In: R. Sundararaj (eds.), *Science of Wood Degradation and its Protection*. Springer, Singapore, pp. 709–744.

Sushil, S. N. 2016. National agricultural bio-security system in India – Strengths and challenges. In: *International Conference on "Plant, Pathogen and People: Challenges in Plant Pathology to Benefit Humankind"*, February 23–27, 2016, NASC, New Delhi.

Sushil, S. N. 2018. Plant protection regulations in India - Strengths & challenges. In: *XVI AZRA International conference on Applied Zoological Research for Sustainable Agriculture and Food Security*, February 9–11, 2018, Banaras Hindu University, Varanasi.

Sushil, S.N., Nagaraju, D. K. and Srivastava, R. P. 2021. Safeguarding Indian agriculture through plant quarantine regulations: emerging issues and way forward. *Journal of Eco-Friendly Agriculture*, 16(2): 97–105.

2 Fungicide Resistance Management

Ashutosh Bhaik, Mamadou Kane Mboup, Jean-Luc Genet, and Andrew Leader

2.1 INTRODUCTION

Diseases are a major source of biotic damage that can be caused to the plant by plant pathogenic (disease-causing) organisms. Fungi are among the most important organisms that cause crop loss worldwide. Other sources of crop losses can be abiotic (nutrition and air) and biotic (bacterium, viruses, etc.). Diseases are a common occurrence in plants, often having a significant economic impact on yield and quality; thus, managing diseases is an essential component of production for most crops. Fungicides are the crop protection products used to control fungal disease by inhibiting or killing the fungus-causing plant disease. Broadly, there are three main reasons why we use fungicides:

- To control a disease during the establishment and development of a crop.
- To increase the productivity of a crop and to reduce blemishes. Diseased food crops may produce less because their leaves, which are needed for photosynthesis, are affected by the disease. Blemishes can affect the edible part of the crop and, in the case of ornamentals, their attractiveness, which both can affect the market value of the crop.
- To improve the storage life and quality of harvested plants and produce. Some of the greatest disease losses occur post-harvest. Fungi often spoil stored fruits, vegetables, tubers, and seeds. A few that infect grains produce toxins (mycotoxins) capable of causing severe illness or even death in humans and animals when consumed.

Through plant diseases are managed by integration of several control practices including crop rotation, selection of disease-tolerant or disease-resistant crop cultivars (cultivars genetically less susceptible), modifying time of planting, scheduling fertilization, micro-climate modification, sanitation, and application of fungicides. Fungicides are more often a vital part of disease management as they have the following advances over other methods:

a. They control most of the diseases satisfactorily.
b. Cultural practices often do not provide adequate disease control.

DOI: 10.1201/9781003531944-2

c. Resistant cultivars are not available or not accepted in the marketplace for many diseases.
d. Certain high-value crops have an extremely low tolerance for disease symptoms.

Fungicides are used as a formulated product consisting of an active ingredient plus inert ingredients that improve the performance of the product. Fungicides are typically mixed with water and then applied by spraying. Commercially, fungicides are available as dust, granules, gas, and, most commonly, liquid. They are applied to:

a. Seed, bulbs, roots of transplants, and other propagative organs. These treatments are usually done by the seed company. Some treatments need to be done by the grower on-site at the time of planting. The goal is to kill pathogens that are on the planting material or to protect the young plant from pathogens in the soil.
b. Soil in-furrow at planting, after planting as a soil drench (including through drip irrigation), or as a directed spray around the base of the plant.
c. Foliage and other aboveground parts of plants by means of a sprayer.
d. Inside of trees via trunk injection.
e. Harvested produce, as a dip or spray in the packinghouse.

The rapid advances in chemical disease control in agriculture have led to few concerns. Chemical products are being phased out or restricted due to safety and environmental concerns. Second, resistance development in the target population is building due to the repeated use of same mode of action (MoA) chemicals. The loss of a fungicide to agriculture through resistance is a problem that affects us all. It may lead to unexpected and costly crop losses to farmers, causing local shortages and increased food prices. Manufacturers lose revenue vital to funding the enormous development costs of new products. Without reinvestment, there would be no new compounds. This would cause serious disease management problems and endanger the world's food supply.

FUNGICIDES: TERMINOLOGY

Fungicide: a chemical or physical agent that kills or inhibits the growth of fungi. Fungicides have at least three names, all of which can be found on the label:

Chemical name: the name of the active ingredient (a.i.) in a fungicide (e.g., 5-methyl[1,2,4]triazolo[3,4-b][1,3]benzothiazol-1-ium).

Common name: a less technical term for the active ingredient (e.g., Tricyclazole).

Trade name: the patented name under which a product is commercially available (e.g., Beam).

Active ingredient (a.i.): the active component of a fungicide. A single active ingredient may be marketed under several different trade names.

Classification of fungicides: Fungicides can be classified in several different ways, including (1) mobility in the plant, (2) role in the protection of plants, (3) breadth of activity, (4) MoA, and (5) chemical group.

(1) MOBILITY IN THE PLANT

Contact fungicide: a fungicide that remains on the surface where it is applied but does not go deeper; these fungicides have no after-infection activity. Repeated applications are needed to protect the new growth of the plant and to replace material that has been washed off by rain or irrigation or degraded by environmental factors such as sunlight.

Systemic fungicide: a fungicide that is absorbed into plant tissue and may offer some after-infection activity. Very few fungicides are truly systemic (i.e., move freely throughout the plant); however, some are upwardly systemic (i.e., move only upward in the plant through xylem tissue), and some are locally systemic (i.e., move into treated leaves and redistribute to some degree within the treated portion of the plant.

(2) ROLE IN PROTECTION (SOME FUNGICIDES CAN FALL INTO MORE THAN ONE OF THESE CATEGORIES)

Preventative activity: occurs when a fungicide is present on the plant as a protective barrier before the pathogen arrives or begins to develop so that it prevents infection from occurring (also referred to as a protective activity).

Early-infection activity: occurs when the active ingredient of a fungicide can penetrate the plant and stop the pathogen in the plant tissues, usually most effective 24–72 hours after infection occurs, depending on the fungicide. Most fungicides that have early infection activity also have preventative activity and are most effective when applied before infection occurs. (also referred to as a curative activity, although we are going to avoid this term because many of these fungicides do not "cure" the plant).

Eradication: the ability to stop disease development after symptoms have developed. Very few fungicides have this capability, and growers must not rely on this as a means of disease control.

Anti-sporulant activity: the ability to prevent spores from being produced. In this case, the disease continues to develop (e.g., lesions continue to expand), but spores are not produced or released, so the amount of inoculum available to infect surrounding plants is reduced.

(3) BREADTH OF ACTIVITY

Single-site fungicide: fungicide active against only one point in one metabolic pathway in a fungus or against a single critical enzyme or protein needed by the fungus. These fungicides are less toxic to plants and tend to have systemic properties.

Multi-site fungicide: fungicide that affects a number of different metabolic sites within the fungus.

(4) MODE OF ACTION

Mode of action: how a fungicide acts on a target fungus, which is the specific process in the metabolism of the target fungus that is affected by a fungicide. Examples are damaging cell membranes, inactivating critical enzymes or proteins, or interfering with key processes such as energy production or respiration.

(5) CHEMICAL GROUP OR CLASS

Chemical group or class: the name given to a group of chemicals that share a common biochemical MoA and may or may not have similar chemical structure. Fungicides approved for use on field crops in Iowa fall into a few different groups: Quinone Outside Inhibitor (QoI) fungicides (which include strobilurins), triazoles, and a few miscellaneous others.

2.2 HISTORY OF FUNGICIDE RESISTANCE

Humans have battled plant diseases since ages by using different chemicals, the oldest report being the use of lime sulfur sprays (Forsyth, 1802). The use of Bordeaux mixture by Millardet (1885) marked the beginning of the use of copper compounds as fungicide for control of more dreaded diseases such as grape downy mildew (*Plasmopara viticola*) and potato late blight (*Phytophthora infestans*). Later research led to the development of organic compounds, especially dithiocarbamates (mancozeb), phthalimides (captafol), and chlorothalonil, which are now very widely used as non-systemic protectant fungicides. Owing to their multi-site modes of action, these chemicals are still prevalent. Despite extensive use over many years, resistance has not been an issue with these fungicides.

The last few decades of the 20th century saw the rapid development of new-generation fungicides, which are site-specific, selective, more systemic, and have more potency against targeted pathogens. A new set of criteria for fungicide development has been followed: (i) development of pesticides that are effective at an extremely low dosage, (ii) development of pesticides that are readily degradable and less residual in the environment, and (iii) development of selectively toxic agrochemicals. Even with these development strategies, the risk of resistance to these chemistries has not come down rather the cost of product development has increased many times, thereby limiting the number of new chemistries in the market. To date, 50 different modes of action (FRAC, 2022) are identified for fungicides.

During the first half of the 20th century, resistance against fungicides was unknown. It was only during 1960s when reports against performance failure started being reported from across the world. Usually, the first indications of a resistance problem came from reports from growers of failure of disease control following fungicide treatment.

During the late 1980s and early 1990s, the concept of fungicide resistance was well documented and perceived by the farmers as well as the academia. It was during 1990s when Fungicide Resistance Action Committee (FRAC), an industry committee, was formed to manage the situation after the onset of phenylamide resistance. Phenylamides Working Group was established, which soon after issued a set of guidelines for resistance management. The recommendations included using mixtures for foliar application, avoiding curative use, and limiting the number of sprays per season. These guidelines were implemented by all the companies involved, and this fungicide class successfully continued in use against all target diseases (Staub, 1994).

Nowadays, manufacturing companies as well as regulators are concerned about the resistance risk of new actives even before the registration of the compound against a particular disease. For companies, it is about extending the product life cycle as the cost of development has increased many folds when compared to what it was during the last century. European regulations make it mandatory to conduct resistance risk assessment and, if appropriate, systems for risk management can be proposed, in the context of official registration of plant protection products. FRAC has also established fungicide resistance management practices, which serve as a guideline for the use and labeling of the new crop protection products. Even novel MoA chemistries such as group 49 has its own Oxysterol Binding Protein Inhibitor (OSBPI) working group, which accesses the resistance situation annually and provides use recommendations.

2.3 RESISTANCE DEFINED

The loss of performance of a plant protection product because of the development of resistance in the target pest can be costly to the grower, the crop protection company, and the environment. First reports from the growers' field during the early 1960s–1970s were of failure of disease control following fungicide applications. Many other reasons than resistance could be attributed such as incorrect application (doze, time of application), use of expired or deteriorated or wrong products, wrongly identified pathogen, or application at the time of exceptionally heavy disease pressure, and at times, growers and advisers attribute difficulties of disease control to fungicide resistance in the absence of evidence that resistance was the main cause. However, in many cases, there was no other obvious explanation, and loss of control was soon shown to be associated with greatly decreased sensitivity of the pathogen, as revealed by laboratory or glasshouse tests on samples taken from the problem sites (FRAC, 2022).

The term fungicide resistance, as used by the FRAC, refers to an acquired, heritable reduction in sensitivity of a fungus to a specific anti-fungal agent (or fungicide). To manage resistance effectively, scientists study fungicide resistance on many different levels including the cellular, organismal, or population/field level. Reports of "resistance" from the field (i.e., where growers observed reduced efficacy of a product that has been effective against a particular pathogen) must be confirmed by studies at the organismal level showing a reduction in sensitivity of the fungal isolate(s) to the specific fungicide. Some scientists use the terms reduced sensitivity or tolerance when referring to smaller reductions in sensitivity, which may have little or no impact on fungicide usage in the field and save the term "resistance" for large reductions in sensitivity of individual isolates, which are likely to affect the efficacy of a specific fungicide under field conditions if the resistant isolates become widespread in the pathogen population. The term field resistance may also be used to indicate this loss of control under field conditions.

FRAC states that the term "resistance" should only be used in situations where:

- Development of resistance leads to failure of disease control under practical field conditions following proper labeled use of a fungicide AND
- Demonstration that loss of control is due to the presence of pathogenic strain(s) with reduced fungicide sensitivity

The development of fungicide resistance is a population evolutionary process. Fungi, like other organisms, are constantly changing. Occasionally, under certain conditions, these changes provide an advantage or disadvantage in terms of the progeny's ability to survive and reproduce. Advantageous changes allow the individual containing the change to survive and reproduce resulting in their progeny constituting a greater percentage of the population over subsequent generations. This can happen relatively rapidly in fungi as their **reproductive frequency** (i.e., the number of progenies produced from a single individual and the speed with which they complete their life cycle) is high. For example, a single *Phytophthora infestans* lesion can produce thousands of spores, and a spore can produce a new sporulating lesion in 3–5 days. The change may be evolutionarily neutral, or even slightly disadvantageous, under most conditions, and only be advantageous when certain factors are present. This is the case with fungicide resistance. In most cases of fungicide resistance, the change leading to reduced sensitivity is evolutionarily neutral except when the specific fungicide is applied. The fungicide is exerting **selection pressure** on the pathogen population since it is killing the initial (or **wild type**) population but does not kill the changed (or **mutant**) population (Figure 2.1). When changes are slightly disadvantageous under normal conditions (i.e., in the absence of the fungicide), the frequency of the changed population may decrease when the selection pressure is removed. This disadvantage is termed a **fitness penalty**.

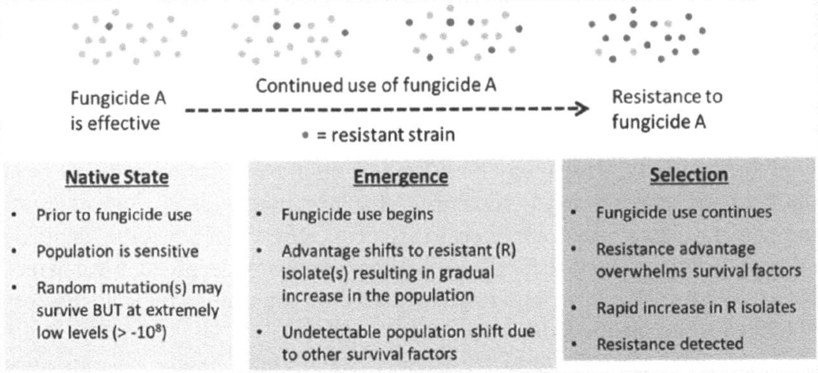

FIGURE 2.1 Different stages of Fungicide resistance when fungicide A is used repeatedly over several generations.

SOME DEFINITIONS

Qualitative resistance describes when fungicide resistance results from modification of a single major gene, and pathogen subpopulations are either sensitive or fully resistant to the pesticide. Resistance in this case is seen as a complete loss of disease control that cannot be regained by using higher rates or more frequent fungicide applications, e.g., resistance developed by several pathogens to the strobilurin (FRAC code 11) fungicides.

Quantitative resistance describes when fungicide resistance results from the modification of several interacting genes, and pathogen isolates exhibit a range in sensitivity to the fungicide depending on the number of gene changes. Variation in sensitivity within the population is continuous. Resistance in this case is seen as an erosion of disease control that can be regained by using higher rates, more frequent applications, or a fungicide in the chemical class that has inherently higher activity. Long-term selection for resistance in the pathogen by repeated applications may eventually result in the highest labeled rates and/or shortest application intervals not being able to adequately control the disease. An example is resistance in the cucurbit powdery mildew pathogen to the DMI (FRAC code 3) fungicides.

Cross resistance describes fungal isolates that are resistant to one fungicide and also resistant to other closely related fungicides, even when they have not been exposed to these other fungicides, because these fungicides all have similar MoA. Sometimes, fungicides in the same chemical class act at a slightly different point in the biosynthetic pathway in the pathogen, which sufficiently limits cross resistance.

Fungicides with **single-site MoA** (targeted) are generally at medium to high risk for resistance development. Single-site means the fungicide acts at a specific point in a biosynthetic pathway in the pathogen. These fungicides are at risk for resistance development because a change in the pathogen at this point can render the fungicide less effective or ineffective. A simple change of just one base pair in the DNA molecule can be sufficient to lead to full resistance (fungicide completely ineffective), as occurs with the strobilurin (FRAC code 11) fungicides.

Most fungicides being developed today have a single-site MoA because this is associated with lower potential for negative impact on the environment, including non-target organisms. Their targeted activity means that they can safely move into the plant (fungicide is not toxic to the plant), resulting in better rain fastness than contact fungicides and better activity as they can move across lower surface where pathogens often develop best. A few modern fungicides have systemic activity and can move more widely in plants, especially when applied to roots. Older fungicides, such as copper and chlorothalonil, have low potential for resistance to develop because they have **multi-site MoA**.

2.4 MECHANISM OF RESISTANCE

Fungi have been described as "a mutable and treacherous tribe", but that even this is something of an understatement is abundantly evident…

E. W. Buxton, Heterokaryosis, Saltation and Adaptation (1960)

There are four main mechanisms by which fungi can become resistant to fungicides:

2.4.1 Alteration of the Target Site so that Sensitivity to the Fungicide Is Reduced

By far the most common way that fungi can become resistant to a specific fungicide is via a change at the target site. As fungi grow, their DNA is replicated when new cells are created. This process of replication is imperfect, and errors can occur. These errors are known as mutations. Since DNA is the code used to produce enzymes in the cell, some mutations result in changes to the amino acid sequence of the target site, which in turn alters the shape of the lock/target site. The fungicide/key may not fit as well anymore or may not fit at all in the target site/lock. This results in a reduction in sensitivity that may range from small to very large.

2.4.2 Detoxification or Metabolism of the Fungicide

The fungal cell contains a vast array of metabolic machinery for normal cellular processes. This metabolic machinery may be able to modify the fungicide to a non-toxic form that is no longer harmful to the cell. Some fungicides are applied as inactive pro-fungicides, which require further metabolism by the fungal cell to become the active form. If fungal metabolism is altered such that the activation step does not occur, the active form of the fungicide is not produced.

2.4.3 Overexpression of the Target

As discussed earlier, the fungicide is "competing" with the natural substrate for the target site. As more and more fungicides enter the cell, it out-competes the natural substrate for the target and, as a result, shuts down critical cellular processes. The production of additional target site enzyme (i.e., overexpression of the target) may increase the likelihood that enough of the fungal substrate will be able to bind with the target site enzyme such that cellular processes such as respiration can occur to some degree. Higher doses of the fungicide in *in vitro* experiments may restore the balance in favor of the fungicide, but higher doses may not always be practical under field conditions.

2.4.4 Exclusion or Expulsion from the Target Site

Efflux pumps exist naturally within the cell to exclude or expel foreign substances or to export endogenous substances. In fungi, the most common efflux pumps are ATP-Binding Cassette (ABC) and Major Facilitator superfamily (MFS) transporters. Despite these efflux pumps, most fungicides can reach effective concentrations inside the cell and inhibit cellular processes. Occasionally, these transporters are successful in expelling enough of the fungicide such that an isolate has reduced sensitivity. The fungicides expelled from the cell by a specific transporter may or may not be active at the same target site; i.e., there is not a direct relationship between the transporter that expels a specific fungicide and the target site of the fungicide. Multidrug resistance (MDR) develops when a specific transporter is able to exclude multiple fungicides from different target site groups. Application of the fungicides in question may exert enough selection pressure that isolates containing these fungicide-exporting transporters become more prevalent in the population as is the case in *Botrytis cinerea* (Kretschmer et al. 2009).

2.5 EVALUATING RESISTANCE RISK

Risk of resistance buildup and loss of efficacy to fungicide is of greater concern to all the new actives that are commercialized by crop protection companies. Resistance risk assessment therefore is a regulatory requirement in European Union. Risk is assessed by combining risk values for the fungicide, pathogen, and agronomic system under consideration, using a risk matrix (triangle) defined by Kuck and Russel (2006). The risk of resistance developing to a fungicide is driven by three components (Figure 2.2). The first is the inherent pathogen risk. Pathogens are classified as high risk if they have a history of developing fungicide resistance, or if they would be expected to develop resistance rapidly, based on their biological characteristics. These are generally pathogens with short life cycles, that have multiple generations in a single crop season, and that produce spores in huge numbers.

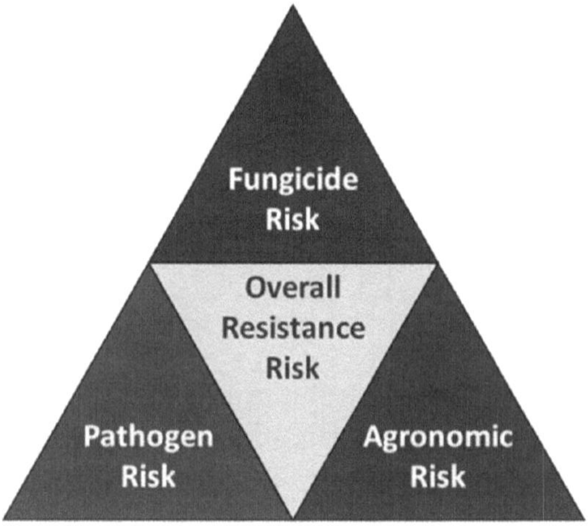

Fungicide risk

- Efficacy
- Site of action (single vs multisite)
- Novel mode of action
- Resistance mechanisms

Pathogen risk

- Epidemiology (cycles per season)
- Opportunity for recombination
- History with other fungicides

Agronomic factors

- Environment influence resistance development
- Cultural effect specific to crop and geography

FIGURE 2.2 Fungicide resistance risk triangle.

The second consideration is the risk associated with the MoA of the active ingredient. For example, multi-site fungicides, which interfere with multiple biochemical pathways, are considered as low risk, and resistance to these active ingredients is less likely to evolve. The final consideration is the agronomic risk, which encompasses the environment in which the fungicide is used, including farmer practices for managing disease.

The scheme has been refined lately by including numerical values for "fungicide risk" and "disease risk." These risk values are multiplied together to get the combined risk value. The resistance risk scheme was further refined by adding Agronomic risk as the third factor in the matrix (Figure 2.3). These ratings are based on the experience over the year of resistance development against the actives and published and/shared data during the meetings. The agronomic risk is disproportionately high in many Asian countries. In tropical and subtropical climates, the warm and humid environment is conducive to pathogen development, and farmers may exacerbate this favorable environment through practices that allow the pathogen to survive continuously, such as planting the same crop in consecutive seasons, with limited crop rotation. Up-to-date assessments of fungicide-associated risk are presented on the FRAC website for all fungicides in current use.

2.6 DETECTING AND MONITORING RESISTANCE

The first and most obvious detection of resistance is from the farmer field where he observes a decline in fungicide performance, a response for which is increasing dose rate and/or frequency applications of fungicides. Poor performance though can be attributed to several causes such as poor application and timing, wrong dose rate, or very exceptional disease pressure. If resistance is a problem, it needs confirming

Fungicide Risk	Risk	Combined Risk			Agronomic Risk
Benzimidazoles	High = 3	3	6	9	High = 1
Dicarboximides		1.5	3	4.5	Medium = 0.5
Phenylamides		0.75	1.5	2.25	Low = 0.25
QoI fungicides					
Carboxanilides	Medium = 2	2	4	6	High = 1
SBI fungicides		1	2	3	Medium = 0.5
Anilinopyrimidines		0.5	1	2.25	Low = 0.25
Phenylpyrroles					
Multi-site fungicide	Low = 1	0.5	1	1.5	High = 1
MBI-R inhibitors		0.25	0.5	0.75	Medium = 0.5
SAR inducers		0.125	0.25	0.3	Low = 0.25
		Low = 1	Medium = 2	High = 3	Pathogen Risk
		Seed borne and soil borne pathogens, Rust fungi	Eyespot, Rhynchos-porium, Septoria	Blumeria, Botrytis, Plasmopara, Magoporthe, Venturia	

FIGURE 2.3 A "risk matrix" fungicide resistance risk assessment scheme. (From Kuck and Russell (2006).)

trials of more than just a single season's work. So, anecdotal evidence from growers must be backed up by a program of field work supported by glasshouse and laboratory assays.

2.6.1 Bioassays

Bioassays are designed to reveal the difference in phenotypical response to pesticide molecule. Live pest colonies, lines, populations, or isolates of interest are exposed to pesticides and then compared to those of sensitive references. A most common approach to confirm resistance is by comparing the sensitivity of isolates obtained from sites where performance has eroded with the sensitivity of isolates never exposed to the at-risk fungicide. Ideally, the existence of a sensitivity distribution of the target fungal population established prior to widespread use of a new fungicide will allow a meaningful confirmation of resistance. The benefit of a "baseline" sensitivity distribution in various aspects of resistance management was described in detail by Russell (2005), and its importance is recognized in many countries where a baseline sensitivity distribution is a requirement for the registration of a new fungicide. The ability to confirm resistance through comparison with a baseline sensitivity will depend on the sample size from the suspected resistant population and inclusion of at least one reference isolate to check for variation between assay tests. In practice where baseline sensitivity data do not exist, comparisons can be made between isolates obtained from at-risk sites with those collected from untreated areas. Often researchers obtain baseline data using "historic" isolates, which have been maintained in culture collections, sometimes for many years, and which were isolated before the at-risk fungicide was used.

2.6.2 Molecular Assays

Molecular, nucleic acid-based assays detecting genes or mutations involved in resistance. The starting material can be living or dead population. Sufficient DNA or RNA of suitable quality should be extracted for further analysis. Literature is full of different molecular techniques used to monitor resistance, and certainly, the most well documented is perhaps the detection of the mutation generating the G143A amino acid change in the target b-type cytochrome of complex III of respiration, causing resistance to QoI fungicides.

2.6.3 Biochemical Assays

Most of the pesticides bind to and inactivate a vital protein (enzyme) responsible for some essential biochemical reaction in the pathogen, the efficiency is dependent on the pesticide reaching the binding site. Biochemical assays are usually used in a situation where the mechanism of resistance has been elucidated.

2.7 FUNGICIDE RESISTANCE MANAGEMENT APPROACHES

Fungicide resistance management strategies aim to delay the evolution and spread of resistance in a sensitive pathogen population, while ensuring effective disease control. If fungicide resistance is confirmed or highly suspected, diverse approaches to managing resistance need to be incorporated into disease management strategies.

2.7.1 Managing the Application Dose

- Most of the evidence suggests that an increased dose of fungicide increases selection for fungicide resistance (but note here that the primary aim of effective disease control may make it impossible to reduce fungicide dosages).
- Several possible mechanisms by which an increased dose may reduce selection have, however, not been studied. Partial resistance and multi-gene/multi-mutation cases are the key examples of this.

2.7.2 Managing the Number of Sprays

- All current evidence suggests that increasing the number of fungicide applications increases the selection for fungicide resistance.
- Most evidence suggests that splitting fungicide dosage between two or more applications increases selection.

2.7.3 The Use of Fungicide Mixtures

- Much of the evidence shows that adding a mixing partner (having different/multisite mode of action) to a high resistance risk fungicide reduces selection for fungicide resistance, even when the dose of the high-risk fungicide stays the same in the mixture.

- Mixture partners with multisite activity (e.g., low risk) are preferred.
- Adding a mixing component to a high-risk fungicide and reducing the dose of the high-risk fungicide further reduce selection for fungicide resistance.
- There is too little evidence on the use of mixtures of two at-risk fungicides, and work in this area is needed. The evidence that does exist suggests that mixing two at-risk fungicides is a valid anti-resistance strategy if no resistance has evolved to the mixing components.

2.7.4 The Use of Fungicide Alternations

- Limited evidence suggests that alternating with a fungicide that has a different mode of action does not alter selection for the high-risk fungicide, if the number of applications of the high-risk fungicide remains constant with and without alternation.
- The evidence suggests that replacing part of the fungicide program with a fungicide with a different MoA reduces selection.

2.7.5 Alternations versus Mixtures

- It depends on the balance between increased selection due to dose splitting and decreased selection due to mixing whether mixing reduces selection to a greater or lesser extent than alternation. The experimental and modeling evidence shows that in many cases, mixing is the better strategy, but for any single case, this needs to be established before conclusions can be reached. In practice, it is best to combine both (mixtures and alternation).

2.7.6 Protective versus Curative Use

Disease management strategies that rely heavily upon the curative application of fungicides often lead to more resistance problems due to (i) the large size of the pathogen population when the application is made from which resistant individuals are being selected and (ii) the difficulty in eradicating a pathogen entirely from inside the plant.

- Protective use consistently results in a lower rate of selection for fungicide resistance (but note that protective fungicide applications may be needed for effective disease control) because eradicant sprays will increase the opportunity for selection (many spores to select from).
- The existing evidence suggests that the specific circumstances will determine whether a shift in spray timing will increase or decrease selection for fungicide resistance.

Resistance management should be based on evidence interpreted within a sound experimental, theoretical, and practical framework. In discussions on resistance management, it is often not explicit what the evidence is. With this chapter, we hope to contribute to evidence-based resistance management (Adopted from van den Bosch et al., 2015).

2.7.7 Action Threshold or Forecasting Methods

What is recommended to the growers is the decision of fungicide application based on forecasting systems or action thresholds, when these are available, to ensure fungicides are applied when needed and to avoid the expense and possible environmental impact of unnecessary applications. Forecasting systems have been developed for several diseases based on an understanding of the environmental factors favorable for their development. Typically, these are based on temperature and relative humidity or leaf wetness in the area where the crop is grown. Threshold-based fungicide programs involve routinely scouting the crop for symptoms, then applying fungicides when the amount of symptoms reaches a critical level, beyond which the disease cannot be controlled adequately. An example of a critical level is one disease spot per five leaves examined. Knowledge of the disease cycle of the pathogen is important when developing and using forecasting systems and thresholds. Important aspects of the disease cycle include whether the disease is monocyclic (one generation per year) or polycyclic (multiple generations) and latent period (time between infection and production of new inoculum).

2.8 FUNGICIDE RESENTENCE ACTION COMMITTEE

Industry recognizes its responsibility in safeguarding new chemistries that are brought to market. Through FRAC and its Working Groups, companies are striving to establish more effective communications to alert all those involved in the research, production, marketing, registration, and use of fungicides to the problems of resistance.

With enlightened stakeholders, effective strategies can be conceived and adopted. Cooperative action is essential if we are to preserve the option of chemical disease control for our crops.

FRAC is a Specialist Technical Group of CropLife International (CLI; Formerly Global Crop Protection Federation, GCPF). As such, we work within the legal framework defined by CLI and take care to ensure that strict anti-trust guidelines are observed.

2.8.1 Purpose

The purpose of FRAC (www.FRAC.info) is to provide fungicide resistance management guidelines to prolong the effectiveness of "at-risk" fungicides and to limit crop losses should resistance occur.

The main aims of FRAC are to:

1. Identify existing and potential fungicide resistance problems.
2. Collate information and distribute it to those involved with fungicide research, distribution, registration, and use.
3. Provide guidelines and advice on the use of fungicides to reduce the risk of resistance developing and to manage it should it occur.

4. Recommend procedures for use in fungicide resistance studies.
5. Stimulate open links and collaborations with universities, government agencies, advisors, extension workers, distributors, and farmers.

2.8.2 FRAC Guidelines

FRAC Guidelines for resistance management are produced by the individual FRAC Working Groups and Expert Fora. These Guidelines provide information on how to use specific areas of fungicide chemistry for the control of plant diseases on various crops while maintaining a good anti-resistance strategy.

The Guidelines should be regarded as the minimum resistance management strategy required, and it is possible that a more stringent strategy should be used in individual cases. FRAC recommends that you seek advice from your local resistance management organization (e.g., local country FRAC or FRAG), your local crop advisor or extension agent, or the manufacturer or distributor of the product to see if a more restrictive strategy is recommended.

The FRAC Guidelines deal only with areas of fungicide mode of action. FRAC is not allowed to make recommendations for the use of individual products. If you require advice on which active ingredients to use in your disease control program, please consult your local crop advisor or extension agent, or the distributor or manufacturer of potential products.

REFERENCES

Buxton, E.W. Heterokaryosis, saltation and adaptation. In: Horsfall, J.G.; Dimond, A.E. (eds), *Plant Pathology*. Academic Press, New York, 1960, Vol. 2, pp. 359–405.

Forsyth, W. *A Treatise on the Culture and Management of Fruit Trees*. Nichols, London, 1802.

FRAC (Fungicide Resistance Action Committee). FRAC List of Plant Pathogenic Organisms Resistant to Disease Control Agents. HYPERLINK "http://www.frac.info"www.frac.info. Access 15 January 2022.

Kretschmer, M.; Leroch, M.; Mosbach, A.; Walker, A.-S.; Fillinger, S.; Mernke, D., et al. Fungicide-driven evolution and molecular basis of multidrug resistance in field populations of the grey mould fungus *Botrytis cinerea*. *PLoS Pathog*. 2009, 5(12): e1000696. https://doi.org/10.1371/journal.ppat.1000696.

Kuck, K.H.; Russell, E.P. FRAC: combined resistance risk assessment. *Asp Appl Biol*. 2006, 78: 3–10.

Millardet, P.M.A. Traitement du mildou par le melange de sulphate de cuivre et de chaux. *J d'Agric Practique*. 1885, 49: 513–516 (In French).

Russell, P.E. *Sensitivity Baselines in Fungicide Resistance Research and Management. FRAC Monograph No. 3*. Crop Life International, Aimprint, UK, 2005.

Staub, T. Early experiences with phenylamide resistance and lessons for continued successful use. In: Heaney, S.; Slawson, D.; Hollomon, D.W.; Smith, M.; Russell, P.E.; Parry, D.W. (eds), *Fungicide Resistance*. British Crop Protection Council, Farnham, Surrey, 1994, pp. 131–138.

van den Bosch, F.; Paveley, N.; Fraaije, B.; van den Berg, F.; Oliver, R. Evidence-based resistance management: a review of existing evidence. In: Ishii, H.; Hollomon, D.W. (eds), *Fungicide Resistance in Plant Pathogens: Principles and a Guide to Practical Management*. Springer, Tokyo, 2015, pp. 63–76.

3 Management of Pesticide Residues in Agriculture for Safer Food in the 21st Century

Kaushik Banerjee and Anoop A. Krishnan

3.1 INTRODUCTION

India is a significant exporter of fresh fruits and vegetables. It follows China in the production of fruits and vegetables worldwide. The nation is the world's top producer of banana, papaya, mango, lemon, and lime. India is also the second-largest producer of okra and ginger, and it comes in first place among nations that produce potatoes, onions, cauliflower, brinjal, cabbage, to name some. The enormous production base provides India with a high export potential. India exported fruits and vegetables worth Rs. 9,940.95 crores/1,342.14 USD million in 2020–21, of which Rs. 4,971.22 crores/674.53 USD million was for fruits and Rs. 4,969.73 crores/667.61 USD million was for vegetables (APEDA, 2022). Among the horticultural commodities, the major exported fruits include table grape, mango, and pomegranate, and the vegetables include okra, green chilli, and bitter gourd. The export of table grapes alone is about 313 million USD. The volume of mangoes exported from India increased from about 21,000 metric tons in the fiscal year 2021 to 28,000 metric tons in 2022 (Statistica, 2022) (https://www.statista.com/statistics/1042173/india-volume-of-mango-exports/). Despite these facts, India only accounts for ~2% of the world food market, primarily due to difficulties in meeting the demanding quality control criteria in importing nations such as the European Union, the United States, and the Gulf Cooperation Council. As a thumb rule, for export, the international food standards must be followed if a food business operator wants to participate more in export. Although the Codex standard is supposed to be followed for regulation of the export consignments, many countries have implemented their own quality standards, which vary from the Codex standards. This mostly happens due to the difference in food consumption patterns, average body weight, and associated risks to consumers.

Due to recent government initiatives through the Food Safety Standards Authority of India (FSSAI) on building up quality and safety standards for various food products at the national level, Indian consumers are becoming more aware of the importance of eating food that is safe. Since its inception, FSSAI has been closely working with the ICAR institutions and state agricultural universities to develop the food

DOI: 10.1201/9781003531944-3

quality and safety standards of India and harmonize the maximum residue limits (MRLs) at the global level.

A vast variety of chemical pesticides are used in agriculture all over the world and have been for many years. Given that India is a tropical nation with frequent insect pest and disease infestations, continuous usage of plant protection chemicals is essential to protect the crops. However, indiscriminate and uninformed use of pesticides is likely to harm people's health by exposing them to residues in food and water, in addition to occupational, bystander, and residential exposure. It also causes contamination of the environment through wind drift, leaching, and runoff, to name a few. Once ingested through consumption of contaminated agricultural/food crops, pesticides get metabolized by human bodies through various pathways to relatively less toxic metabolites, which eventually are removed via the excretion process. Processing could be a means to reduce the pesticide residue loads that are initially present in fresh or raw agricultural commodities, which is a challenging undertaking to manage. A broad review of pesticide residues, their monitoring, and management in fruits and vegetables of Indian origin is provided in this chapter.

3.2 PESTICIDES AND THEIR RESIDUES

3.2.1 Pesticides

The term pesticide has been defined by the World Health Organization (WHO) as any substance or mixture of substances intended for preventing or controlling any unwanted species of plants and animals and includes any substances or mixture of substances intended for use as a plant-growth regulator, defoliant, or desiccant. It also includes any substance used for the control of pests during production, storage, transport, marketing, or processing of food for man or animals or which may be administered to animals for the control of insects or arachnids in or on their bodies. It, however, does not apply to antibiotics or other chemicals administered to animals for other purposes, such as to stimulate their growth or to modify their reproductive behavior; nor does it apply to fertilizers. IUPAC also adopted the same definition for pesticide.

3.2.2 Pesticide Residue: A Potential Trade Barrier

Pesticide residue is any substance or mixture of substances in food for man or animals resulting from the use of a pesticide and includes any specified derivatives, such as degradation and conversion products, metabolites, reaction products, and impurities, which are considered to be of toxicological significance. The term also includes residues from unknown sources (i.e., background residues) as well as those from known uses of any chemical in question.

As long as they do not discriminate or use this as a cover for protectionism, governments are permitted to regulate trade to safeguard human, animal, or plant life or health under Article 20 of the General Agreement on Tariffs and Trade (GATT, 1947). Sanitary and Phytosanitary Measures Agreement (SPS) permits nations to establish their own standards. The "precautionary principle" may be used, but rules must be based on science. The Technical Barriers to Trade Agreement (TBT) aims to prevent unjustified trade barriers from being created through rules, standards, testing, and certification processes. It encourages nations to respect one another's methods for determining if a product complies with expectations. TBT promotes the removal of unnecessary trade barriers by avoiding overly strict regulations, but it also acknowledges that individual nations have the right to adopt their own standards that they may deem appropriate with regard to human life and health, variations in dietary habits (dietary exposure), the nature of the animal population, plants, and environment.

MRLs are trade standards that are established so that public health is unaffected, particularly with reference to vulnerable subpopulation groups (as children and the unborn). Good Agricultural Practices (GAPs) take into account the application of minimum quantities of pesticides necessary to achieve adequate pest control in such a manner that the amount of residues remaining in food is the smallest possible. At the international level, the Codex Alimentarius Commission of the FAO/WHO decides the MRL (Codex, 2022). The Codex Committee on Pesticide Residues (CCPR) was formed by the United Nations with the primary mandate to establish MRL for pesticides in food. In India, the MRL of different pesticide–commodity combinations are prescribed under the Food Safety and Standards Act, 2006 (FSSAI, 2006).

Although the Codex MRLs are applicable to all signatory nations, individual countries are free to have their own MRL regulations. For agro-export, it is thus essential that the commodities should comply with the latest MRL regulations of the importing country. For example, the MRLs set by the European Union (EU) have to be complied with for the export of any food commodity (e.g. fruits and vegetables) from India to any of the EU member countries (EU, 2022). Earlier, before September 2008, individual EU countries used to have their own MRLs. Since September 1st, 2008, the MRLs have been harmonized across the member countries of the European Union, which has expanded the market access across Europe and minimized trade barriers out of conflicting MRLs within the same territory. In addition to such Official MRLs, the Indian exporters often have to comply with specific private standards imposed by supermarkets, where MRLs are typically at 50%–70% of the official value rendering product compliance further challenging. Such private standards often raise apprehensions of combined toxicity in case of multiple pesticide detections in the same sample and even when the individual residue levels are well below the MRLs set.

3.3 PESTICIDE RESIDUES: ISSUES TO BE ADDRESSED IN AGRICULTURAL CROPS

Some of the key issues identified below need to be addressed on priority for the effective and scientific management of pesticide residues in horticulture crops.

3.3.1 Stringent and Ever-Changing Country-Specific MRLs

The TBT encourages the removal of unnecessary obstacles to trade flow by avoiding over-stringent regulations, it simultaneously recognizes individual countries' rights to adopt their own standards, which they might consider appropriate with respect to the life and health of human being, animal population, plants, and environment.

Although the Codex MRLs are applicable to all nations, individual countries might have their own MRL regulations. The MRL established for a pesticide varies with crop and the country. For agro-export, it is thus essential that the commodities should comply with the latest MRL regulations of the importing country. As the pre-harvest interval (PHI) or waiting period established for various pesticides varies with MRLs, it is almost impossible to adopt a universal PHI for a pesticide for market-specific residue-complied production.

3.3.2 Lack of Label Claim Chemicals for Use

As per the regulation of CIB&RC, a pesticide can be officially recommended or allowed to use against a particular pest or disease in a specific crop only if the pesticide has a label claim with CIB&RC for the use against that particular pest in the specific crop. The pesticides that have a label claim for use in specific horticulture are very few, and they are not sufficient to manage all pests and diseases in that particular crop. In such cases, the growers are left with no option other than to rely on the available chemicals in the market to protect their crop, and their nonscientific use leads to more residues in the harvested products.

3.3.3 Spurious Pesticides in Market

As per Pesticide Management Bill, 2020 (http://164.100.47.4/BillsTexts/RSBillTexts/asintroduced/Pesticide-as%20intr-23%203%2020.pdf) and 2008 (http://164.100.47.4/BillsTexts/RSBillTexts/asintroduced/Pesticide-as%20intr-23%203%2020.pdf), a pesticide shall be deemed spurious if it is not registered or licensed in the manner required; if it shows active ingredient higher or lower the limits prescribed; if it is an imitation of another pesticide; if the container bears the name of the manufacturer either fictitious or does not exist; if the chemical composition as approved by the Registration Committee is not adhered; if it has outlived its shelf-life; or if its import, manufacture, use, or sale is prohibited. The presence of spurious pesticides in the market and the lack of awareness among the farmers about them became a major reason for pesticide residues. For example, every year, a large number of grape samples get detected for the residues of the insecticide profenophos or its primary metabolite, 4-bromo-2-chlorophenol, mainly from spurious formulations (source: unpublished GrapeNet residue monitoring data).

3.3.4 Uncontrolled Use of "Bio-Pesticides"

The use of bio-pesticides in conventional as well as organic horticulture has an increasing trend. Many a time, the bio-pesticide formulations that are available in the market have deliberate contaminations with chemical or spurious pesticides. They are hardly tested or monitored for chemical pesticides before field applications. A survey reported by FICCI (2015) (https://www.google.com/url?sa=t&rct=j&q=&esrc=s&source=web&cd=&cad=rja&uact=8&ved=2ahUKEwim1eLM--b7AhU8k1YBHa6zBH0QFnoECA4QAQ&url=https%3A%2F%2Fcroplife.org%2Fwp-content%2Fuploads%2F2015%2F10%2FStudy-on-sub-standard-spurious-counterfeit-pesticides-in-India.pdf&usg=AOvVaw3fJB2FVGgu1Fr7UKAkt9y2) has reported contamination of several biopesticide formulations to be laced with chemical pesticides, which included avermectin, acetamiprid, buprofazin, carbendazim, dichlorvos, dimethomorph, imidacloprid, metalaxyl, thiamethoxam, etc. The use of such bio-pesticides at later stages of crop or near to harvest results in the detection of these pesticide residues in harvested produce.

3.4 PESTICIDE RESIDUE MANAGEMENT

Considering the national and international food safety issues, it is very important to adopt residue management strategies to tackle the pesticide residue problems in fruits and vegetables and thereby increase the export of these commodities and also to ensure safety to domestic consumers. Some strategies for pesticide residue management in fruits and vegetables are described below.

3.4.1 Pre-Harvest Interval a Tool for Residue Management

When a pesticide is applied in the field, the initial residue (on the day of application present on crop) degrades due to interventions of various factors such as chemical degradation (hydrolysis, oxidation, reduction), microbial degradation, and by physical means (photolysis, etc.), and ultimately, the residue degrades/dissipates to below the quantifiable/detectable limits. PHI is most important in field-level management of pesticide residues. On the basis of the MRL, the PHIs of pesticides are estimated as the duration (in days) that must be provided between the last application of a pesticide and harvesting of the produce so that its residue level at harvest reaches below the MRL. Estimating the PHI of a pesticide ideally involves multi-location field trails wherein the pesticides are applied following the guidelines of GAP.

3.4.2 Adoption of Techniques of Calibrated/Precision Spraying

Adoption of appropriate spraying technologies improves field coverage of pesticides, and for this, it is important to choose the correct spraying equipment and plan the spraying operation in such a way, so that the spray coverage in a crop field is uniform. Calibration of the sprayers in this context plays a significant role, which prevents localized accumulations of spray droplets and allows a uniform distribution of the sprayed formulations.

3.4.3 Integrated Pest Management with Minimal Pesticide

To some extent, IPM can reduce the usage of chemical pesticides in the field. As much as possible, use environmental-friendly strategies and involve the less persistent synthetic pesticide in the IPM package.

3.4.4 Field-Level Management

The major strategies to take care at field-level management are as follows:

- Use only the recommended pesticides that have label claims
- Do not spray banned or non-recommended chemicals

- Strictly adhere to the recommended dose of application including the recommended volume of water to be sprayed per hectare or per acre.
- Strictly maintain recommended PHI to minimize consumer risk regarding the detection of residues
- Integrated pest management involving minimal usage of highly toxic chemicals
- Sequencing the pesticides in relation to their persistence, particularly in the last 2-month period before harvest
- Adoption of bio-control measures during the last month before harvest
- "Combination Products," which contain two or more active ingredients as given in the above list, can be used. However, the dose of the individual active ingredient should not exceed the recommended dose

3.4.5 Regulatory Control

Knowingly or unknowingly, farmers keep applying banned or restricted use chemicals for pest and disease management. It is extremely important to replace the usage of highly toxic chemicals with those having safer mammalian and environmental toxicity profiles. In many cases, the detection of pesticide residues in fruits and vegetables includes restricted use chemicals or those that do not have a label claim. Strict regulatory control needs to be established to minimize the use of such restricted or unauthorized (without official label claim) chemicals. The regulatory authority should ensure the availability of label claim for optimal control of various pests and diseases in targeted crops, which will avoid violation of GAP recommendations. This can be achieved through extension programs involving all the stakeholders, e.g., farmers, state government agricultural/horticultural department officials, and exporters under the overall supervision of the Food Safety and Standard Authority of India and Central Insecticide Board.

3.5 TRACEABILITY OF PRODUCES

Traceability is the ability to trace and follow a food or substances intended to be or expected to be incorporated into a food through all stages of production, processing, or distribution. The first comprehensive product traceability was established through the European Commission regulation EC 178/2002. The food safety/traceability systems under Grapenet and Hortinet were largely developed in the same line as that of this EU regulation as explained below.

3.6 GRAPENET

This is a pre-harvest pesticide residue monitoring program, which is unique in the world. Here, pesticide residues are monitored when the crop is standing in the field. Grape cultivation in India is spread predominantly in the peninsular region of Maharashtra, Karnataka, Andhra Pradesh, and Tamil Nadu states. India started exporting table grapes to the European market since 1990, initially in small quantities, which increased steadily from the year 2000. In those days, the residue testing in grapes was voluntary. The samples used to be tested only for the residues of chlorinated hydrocarbons. In the year 2003, the export of Indian grapes faced a serious trouble in the European market due to the detection of monocrotophos (an organophosphorus insecticide) and methomyl (a carbamate insecticide) above the EU- MRL. This significantly affected the trade of grapes. To manage this export crisis, a food safety-traceability system called Grapenet was introduced by APEDA, which ensures that the table grape from India is exported through a unique internet-based pre-harvest residue monitoring procedure that puts in place an end-to-end system for monitoring the agrochemical residues, achieves product quality standardization, and facilitates tracing back from retail shelves to the farm of the Indian growers through various stages of sampling, testing, certification, and packing. GrapeNet integrates all stakeholders in the supply chain of grape export, namely, Farmers, State Government Horticulture departments, National Referral Laboratory (NRL), Residue Testing Laboratories, Agmark Quality Certification Department, Phytosanitary Department, Pack houses, Exporters, and APEDA, with each having specified duties to perform for promoting export of table grapes to the EU countries. The NRL coordinates all the technical activities related to residue testing and decision-making on MRL compliance. Every consignment of fresh grapes from India to the EU is monitored through this system, which starts at the very root of the process; registration of farmers up to plot (up to 1.2 ha) level with the location coordinate data recording at district headquarters by the State Horticulture departments, and subsequently issuing a registration certificate to the farmers. Each farmer, who intends to export directly or supply fresh grapes to an exporter, should get the registration of his/her farm and plot(s) with a unique registration number. One plot constitutes a maximum area of 1 Ha or part thereof and is extendable up to 1.2 ha.

To manage the above activities, ICAR-National Research Centre for Grapes, Pune, was declared by APEDA as the NRL in the year 2003, and the NRL started being operational in early 2004 (Grapenet, 2022). The contributions of ICAR-NRCG as the NRL for the export facilitation of grapes are summarized below:

- Every year, the package of practices is decided by the institute in consultation with the stakeholders of the Indian grape industry, which provides a list of recommended chemicals having label claim for usage in grapes (as per approval of Central Insecticides Board and Registration Committee, Ministry of Agriculture, Govt. of India), target pests and diseases, application rate, and PHI to minimize residue accumulations at harvest. This is a dynamic system that started with a recommendation of only 28 agrochemicals in 2003–2004 and has reached the number of 69 recommended agrochemicals in 2021–2022. The initiatives of the institute have resulted in the highest number of CIB&RC label claims among the fruit crops for grapes.
- NRL has established a statistically validated pre-harvest sampling technique for grapes, where 5 kg samples comprising small bunchlets are drawn randomly from all over a vineyard area of 1 ha. This technique reduced the sampling-induced variations in residue test results from ~60% to 70% (bunch-to-bunch or vine-to-vine RSD) in 2004 to the current level of less than 12%, which is satisfactory considering natural variations.

- The institute has developed fit-for-purpose analytical methods for pesticides, which allow multiresidue analysis of hundreds of pesticides by a single chromatographic run. The institute has also developed sensitive and selective methods for the residue estimations of plant growth regulators, and ionic herbicides, making the scope of residue monitoring quite comprehensive.
- The institute has significantly contributed to the capacity building of the food testing laboratories of the country by sharing validated methods, assessing their technical capabilities, and improving their performance by organizing proficiency tests (PT) at a frequent interval. The institute has guided these laboratories to get equipped with high precision and calibrated equipments, including the latest-generation gas chromatography-mass spectrometry (GC-MS/MS) and liquid chromatography-mass spectrometry (LC-MS/MS) systems. The institute has also implemented stringent analytical quality control procedures in the country that ensure the laboratories perform the residue tests with equivalent proficiency to the rest of the world. With these endeavors, the test results of the APEDA-nominated testing laboratories of the country are receiving wider acceptance from the stakeholders.
- The NRL of the institute updates the list of chemicals for monitoring on a regular basis. This list has been expanded from 34 pesticides in 2004 to the current list of 271 agrochemicals, including the pesticide metabolites of toxicological concern. This list covers all the insecticides, fungicides, herbicides, and plant growth regulators that are currently available in the Indian market.
- The institute has ensured that the traceability system of GrapeNet automatically finds out from the test measurements entered whether a sample qualifies for export to the EU and generates their test reports. In case a sample fails for the EU-MRL compliance, the NRL examines the case and issues an internal alert restricting it from export. The exporters can only source the produce from those plots of the farmers whose samples have complied with all MRLs.
- This institute has also contributed to upgrading the AGMARK certification for quality assessment and gradation of grapes in terms of "Extra Class," "Class I," "Class II," etc.

Over the years, the endeavors of this institute have significantly contributed to ensuring food safety in table grapes with the following realistic outcomes.

- Detections of banned/restricted chemicals almost nullified.
- Significant reduction in the detections of unauthorized (and non-label claim) chemicals.
- An overall quality consciousness has emerged. Grapes in the domestic markets are appearing quite safe.
- The internal sample failure for the EU-MRL noncompliance has reduced from ~24% in 2004 to 2%–5% in the past 10 years. This improvement in quality has in turn enhanced the export of grapes from around 300 containers a year in 2004 to the current volume of ~8,000–9,000 containers per year.

3.7 HORTINET – THE FOOD SAFETY/TRACEABILITY SYSTEM FOR HORTICULTURAL CROPS

With the success of Grapenet in managing pesticide residues in grapes, the scope of the traceability system was expanded by including other crops of commercial significance. These include mango, pomegranate, citrus, betel leaves, and various vegetables, e.g. okra, onion, green chilli, and curry leaf. A case study for the process of vegetable export (https://traceability.apeda.gov.in/hortinet/PSC/PSCLogin.aspx#) is explained below:

For vegetables, exporters are permitted to export if they procure the fruits or vegetables from registered farmers. The State Horticulture Department or Pack House can register the farmers. The exporter then constructs a "lot" in a pack house that has received approval from APEDA and enters the information about the products that were purchased from the farmer or suppliers. After the lot is formed, the exporter enters each lot's packaging information into the online system, and then it is forwarded for residue testing by an APEDA-accredited food testing laboratory. The laboratory collects samples from the batches that the exporter has sent into the system, analyzes the residues as per the list of chemicals defined by NRL-NRCG, and sends the exporter the test report. The exporter creates a consignment from multiple lots and forwards an application for Agmark Certification to the Laboratory, if applicable. Once the Agmark certificate is generated, the consignment can then be forwarded for a Health Certificate. The Health Certificate is then issued by the Laboratory on the forwarded application. After the Health Certificate is issued, exporter will apply for Phytosanitary Certification. The PSC Authority is responsible for issuing PSC. After receiving the PSC, the consignment can be exported.

3.8 EIC'S ROLE IN EXPORT INSPECTION AND FACILITATION

The Export Inspection Council (EIC) was set up by the Government of India under Section 3 of The Export (Quality Control and Inspection) Act, 1963 (22 of 1963), in order to ensure sound development of the export trade of India through Quality Control and Inspection and for matters connected thereof. EIC works as an advisory body to the Central Government, which is empowered under the Act to:

i. Notify commodities that will be subjected to quality control and /or inspection prior to export,
ii. Establish standards of quality for such notified commodities, and
iii. Specify the type of quality control and/or inspection to be applied to such commodities.

EIC assures the quality and safety of the products specified under the Export (Quality Control and Inspection) Act 1963 through either quality assurance or food safety management-based certification or a consignment-wise inspection. Under EIC, Export Inspection Agencies (EIAs) are located in Delhi, Mumbai, Kolkata, Chennai, and Kochi, with a network of 24 sub-offices that are backed by eight NABL-accredited laboratories at various places. EIC provides mandatory certification for various food items, namely fish and fishery products, dairy product, honey, egg products, meat and meat products, poultry meat products, animal casing, gelatin, ossein and crushed bones, and feed additive and pre-mixtures while other food and non-food products are certified on voluntary basis.

EIC has a broad vision to facilitate worldwide access for Indian exports through a credible and efficient inspection and certification system and earn global recognition as India's premier organization for certifying quality and safety to meet international norms. It has an experience of more than five decades in inspecting, testing, and certifying export items based on the importing country's requirements. Notably, EIC is the only organization in India with a global acceptance in the field of certification.

Besides its advisory role, EIC exercises technical and administrative control over the five-field organization, i.e., EIAs, under Section 7 of the Act for the purpose of implementing various measures and policies formulated by EIC. Through this process, the quality control and inspection in respect of products notified under the Export (Quality Control and Inspection) Act, 1963 are managed to ensure that the products meant for export are in compliance with the importing country's requirements. The type of quality control and inspection could be consignment-wise or Food Safety Management System based. The EIC has three-tier surveillance mechanism to exercise control of the approved establishments. Product testing is integral to the export inspection and certification services being rendered by EIC/EIAs, and therefore, EIAs have established a network of laboratories to cater to the inflow of samples generated during the operation of its export inspection and certification systems. However, the need for independent testing facilities arises from time to time to supplement its own laboratory network either due to a shortage of capacity or the absence of specific testing capability as per standards for export, which are normally the international standards or regulatory standards prescribed by the importing countries. Such laboratories are approved by EIC through the integrated assessment done by the National Accreditation Board for Testing and Calibration (NABL) in compliance with ISO/IEC 17025:2017 and the requirements of various importing countries.

3.9 WHAT HAPPENS WHEN AN EXPORT CONSIGNMENT FAILS FOR MRL COMPLIANCE?

In the past, Indian consignments to markets such as the United States and the EU have faced information notifications, rejections, and alerts due to the presence of higher-than-approved levels of residues and contaminants in various food commodities. Many developed countries have a traceability mechanism for monitoring the alerts, namely, Rapid Alert System for Food and Feed (RASFF) and FDA Import Refusal. In such cases, the exporters find options of diverting their products to other markets with less restrictive MRLs, or those are returned to India. If market diversion is not possible, the whole consignment is destroyed in the importing country. This mostly happens if the product has a low shelf life.

EIC has an effective complaint-handling mechanism (https://www.eicindia.gov.in/WebAppl/) in place. When a complaint is received from the importing country on food safety grounds such as product contamination with microorganisms, enzymes, and toxins, or with residues (antibiotics, pesticides, heavy metals such as cadmium and mercury), the competent authority (EIC/EIA) follows a laid down procedure as given below. EIC scrutinizes the complaint received from foreign countries or any other source and informs the EIA concerned immediately to take action as per the complaint handling procedure. EIC may seek clarification from the importing country/Health Authorities, if required.

On receipt of information regarding the rejection of a cargo by overseas health authorities from any importing country, the exporter concerned is expected to immediately contact EIC with the requisite details. In case of a merchant exporter, a copy of the communication is also sent to the manufacturer/processor. The EIC refers the complaint immediately to the concerned EIA. As a case study, in case of fruits, vegetables, fish, and fishery products, on receipt of a complaint from an importing country, EIA places the concerned processing unit under "Internal Alert" immediately after the receipt of the information about the rejection. Based on the nature of the rejection/complaint, a detailed assessment of the establishment is carried out to determine the cause of contamination/rejection. During the assessment, remedial measures are also suggested to prevent further rejection. Assessment of the processing facility is carried out by an Assessment Panel of Experts (APE) comprising officials from EIC/EIA/Commodity-specific institutes/Empaneled members by EIA. The quorum of the panel is a minimum of two. The main objective of the assessment is to determine the root cause of the complaint/rejection. A detailed root-cause analysis is carried out by the panel (including audit of the primary production facilities) to ascertain the actual cause of rejection. The audit is carried out to address the source of a specific contaminant or cause of rejection. The APE audits the unit depending on the type of rejection. In case of rejection due to microbiological parameters, a lab expert is included for auditing the functioning of the in-house laboratory.

3.10 CONCLUSION

With the advent of the information explosion, consumer awareness is growing with respect to the right to have safe food. No one in the society is ready to tolerate the menace of the residues of agrochemicals in the food intended for his/her nurture. On the other hand, being in the tropical belt we do not have the luxury of avoiding the use of agrochemicals in pest management. Agrochemicals are used purposefully in the environment to ensure a good harvest. It is the dose and time of application, which differentiate the safe and unsafe use of a pesticide. ICAR-NRCG believes that if the recommendations of good agricultural practices are followed sincerely, it is definitely possible to minimize the residue load of agrochemicals in grapes, which in turn will ensure the safety and the health of the consumers and environment as a whole. The implementation of the residue monitoring system through GrapeNet has indeed emerged as a model system in the whole country and is being expanded to cover all the horticultural commodities on a timescale.

REFERENCES

APEDA (2022). agriXchange. https://agriexchange.apeda.gov.in/indexp/reportlist.aspx

Codex (2022). https://www.fao.org/fao-who-codexalimentarius/codex-texts/dbs/pestres/en/

EC (2002). https://www.google.com/url?sa=t&rct=j&q=&esrc=s&source=web&cd=&cad=rja&uact=8&ved=2ahUKEwjwpau_jOj7AhWsyXMBHdWgDggQFnoECBQQAQ&url=https%3A%2F%2Feur-lex.europa.eu%2Flegal-content%2FEN%2FTXT%2FPDF%2F%3Furi%3DCELEX%3A02002R0178-20140630%26rid%3D1&usg=AOvVaw3A6_-YZz3__ahO2-zg2Jlc

EU (2022). EU Pesticides Database. https://food.ec.europa.eu/plants/pesticides/eu-pesticides-database_en

FICCI (2015). Study on Sub-Standard, Spurious/Counterfeit Pesticides in India 2015-Report. https://www.google.com/url?sa=t&rct=j&q=&esrc=s&source=web&cd=&cad=rja&uact=8&ved=2ahUKEwim1eLM--b7AhU8k1YBHa6zBH0QFnoECA4QAQ&url=https%3A%2F%2Fcroplife.org%2Fwp-content%2Fuploads%2F2015%2F10%2FStudy-on-sub-standard-spurious-counterfeit-pesticides-in-India.pdf&usg=AOvVaw3fJB2FVGgu1Fr7UKAkt9y2

FSSAI (2006). https://fssai.gov.in/cms/food-safety-and-standards-act-2006.php

GATT (1947). https://www.google.com/url?sa=t&rct=j&q=&esrc=s&source=web&cd=&ved=2ahUKEwi7rIq-kOj7AhXH63MBHWmBCsgQFnoECAsQAw&url=https%3A%2F%2Fwww.wto.org%2Fenglish%2Fdocs_e%2Flegal_e%2Fgatt47_02_e.htm%23%3A~%3Atext%3D20.%2C21.&usg=AOvVaw3E4s-Ng2uyFYDClWAjtm1W

Grapenet (2022). https://apeda.gov.in/apedawebsite/Grapenet/Hortinet.htm

Statistica(2022). https://www.statista.com/statistics/1042173/india-volume-of-mango-exports/

The Pesticide Management Bill (2008). https://164.100.47.4/BillsTexts/RSBillTexts/asintroduced/Pesticide-as%20intr-23%203%2020.pdf

The Pesticide Management Bill (2020). https://164.100.47.4/BillsTexts/RSBillTexts/asintroduced/Pesticide-as%20intr-23%203%2020.pdf

4 Off Label Use of Agrochemicals
Risks and Trade Implications

Vandana Tripathy, Krishan Kumar Sharma, Ruchi Gupta, Kesar Tandekar, and Suresh Walia

4.1 INTRODUCTION

Agriculture plays a vital role in India's economy. Nearly 70% of the rural population in India is dependent on agriculture (FAO India, 2022). According to the Economic Survey 2021–2022, the contribution of agriculture to the Indian GDP stood at 19.9%. The country today is among the top ten producers of food in the world and contributes a good share to the world's food basket. India in 2021–2022 produced 314.51 million tons of food grains, 200.45 million tons of vegetables, and 102.48 million tons of fruits (APEDA, 2022a,b). The country generated revenue of US$ 14,400.24 million by exporting 32.3 million tons of cereals and 3.37 million tons of fresh fruits and vegetables (APEDA, 2022c,d). It has been estimated by the FAO that each year 20%–40% of global crop production is lost due to pests and diseases. To prevent such losses, chemical pesticides are employed to protect crops and food commodities from pests, weeds, and diseases and in veterinary and public health to control insect pests and vector-borne diseases. While pesticides have played a key role in improving agricultural productivity by preventing large crop losses, raising agricultural output, and ultimately increasing farm income, their misuse has led to pesticide residues in food, feed, and the environment; pest resistance; pest resurgence; outbreaks of secondary pests; and killing of non-target organisms including natural enemies, pollinators, etc. When a pesticide is approved by the regulator for use on a crop, it gets a "label claim" for that crop implying the legal sanction for its approved use in the country. In spite of the approved usage of a pesticide on a specific crop/commodity, there are cases when a pesticide is detected on the crop/commodity for which it has not been registered/approved for use. Such usage of pesticides is considered an off-label use and is not legally permitted. This chapter gives a brief account of the pesticide regulations in India, the use of off-label pesticides, risk assessment, trade implications, and using internationally harmonized programs and crop grouping approach to overcome the challenges.

DOI: 10.1201/9781003531944-4

4.2 AGROCHEMICAL USE AND REGULATION IN INDIA

The worldwide consumption of pesticides is about 2 million tons per year, out of which India's share is less than 4%. The area under pesticide use in India is only 9% of the total cultivated area. India has emerged as the 12th largest user of pesticides in the world and 3rd largest in Asia (Nayak and Solanki, 2021). Some pesticides that are not manufactured in India are imported while others that are produced in excess are exported to various countries. In 2020, the country imported 1,40,388.37 metric tons (MT) and exported 5,05,215.09 MT of pesticides (FAOSTAT, 2020a). During the last seven decades, the pesticide consumption has increased from 2,353 MT in 1955–1956 to 58,720 MT in 2021–2022, but its share is meager 1% of the total global consumption (DA & FW, 2022). According to the FAO, the present per hectare consumption of pesticides in India is among the lowest in the world and stands at 0.37 kg/ha against 1.95 kg/ha in China, 2.48 kg/ha in the United Kingdom, 2.54 kg/ha in the USA, and 11.89 kg/ha in Japan (FAOSTAT, 2020b). It has been estimated that only 0.1% of the applied pesticide reaches the target organism, and the rest remains unutilized, which ultimately contaminates the soil, air, and aquatic environment (Carriger et al., 2006). Worldwide, the major categories of the consumed pesticides are herbicides (~40%), insecticides (~33%), and fungicides (~10%), but in India, the crop protection market is skewed more toward insecticides (~65%) followed by fungicides (~18%), herbicides (~16%), biopesticides (~3%), and other group chemicals (~3%). As per the Bureau of Industrial Costs and Prices (BICP) analysis, cotton and paddy are the major consumers of pesticides in India. Maharashtra and Uttar Pradesh are the major consumers of chemical pesticides. Out of total insecticide use in India, 50% is estimated to be diverted to cotton pest management (Mooventhan et al., 2020; Nayak and Solanki, 2021).

Before a pesticide is introduced in the country, it is mandatory to register it with the Central Insecticides Board and Registration Committee (CIB&RC) under the Insecticides Act, 1968 and Rules 1971. The Act regulates the import, manufacture, sale, transport, distribution, and use of insecticides to prevent risk to humans, animals, and the environment and authorizes the regulatory agency to implement control over product quality, packaging, labeling, and safety of the users. The CIB&RC is also responsible for recommending the minimum gap (pre-harvest interval (PHI) or waiting period) that should be present between the last pesticide application and harvesting of the crops. The registration process involves scientific evaluation of the bioefficacy, chemistry, and toxicology data generated under different agro-climatic locations to ensure that the pesticide does not adversely affect the human and environmental health. The Registration Committee (RC) confirms the claims of the manufacturer or importer regarding the effectiveness and safety of the pesticides. For securing regulatory approval for a pesticide–crop combination, multi-location supervised field trials are conducted for generating data on the efficacy of the pesticides against the target pests and analysis of the residues in/on the desired crop. In India, multi-locational supervised field trials are conducted by the All India Network Project on Pesticide Residues (AINP-PR) established by the Indian Council of Agricultural Research (ICAR) to develop uniform protocols for the safe use of pesticides by recommending "good agricultural practices" based on multi-location "supervised field trials" to evaluate the pesticide residues at harvest. Under the project, multi-location supervised field trials are conducted at its various coordinated centers in the country. The data generated are utilized for fixing MRLs and the PHIs/safe waiting periods.

The CIB&RC has made it mandatory to fix the maximum residue limit (MRL) of every pesticide on specific crops (label claim), before it could be allowed for registration and use in agriculture. An MRL is the highest level of a pesticide residue that is legally tolerated in or on food or feed when pesticides are applied correctly in accordance with Good Agricultural Practice. The issues concerning food safety due to pesticide residues are handled by the Food Safety and Standards Authority of India (FSSAI, 2018, 2022), a statutory body established under the Food Safety and Standards Act, 2006 and the Food Safety and Standards (Contaminants, Toxins and Residues) Regulation, 2011. The FSSAI lays down regulatory standards for food articles based on scientific assessment and regulates the manufacture, processing, distribution, sale, and import of food to ensure safe food for human consumption.

4.3 PESTICIDE LABELS AND HAZARD COMMUNICATION

After careful evaluation of the scientific data submitted by the pesticide company and weighing its benefits and risks, the CIB&RC approves label claims about the safe and effective use to control target pests on a particular crop. The label is affixed at the designated place on the pesticide package/container. It helps the users to quickly identify the product and its correct use/application. Pesticide labels are legally enforceable and contain correct information about the chemical name of the active ingredient, brand name, pesticide formulation type [e.g., emulsifiable concentrate (EC), flowable or liquid (F or L), solution (S), wettable powder (WP), soluble powder (SP), water dispersible granules (WDG), dry flowables (DF), water-soluble packets (WSP) etc.], net content and quantity, chemical composition, inert ingredients, targeted pests and crop types, bioefficacy, etc. It may also contain essential instructions for the storage, sale, handling, disposal, and important facts about symptoms of poisoning and first-aid measures such as antidotes required to avoid harm to human health and the environment. It is obligatory on the part of the pesticide manufacturer to also include information about the manufacturer's registration and license to sell, manufacturing unit location, batch number, manufacturing and expiry date, and precautionary statements about the hazards and risks of the product. Other hazard communication-related aspects already available in the Material Safety Data Sheet (MSDS) are, however, not required to be mentioned on the label when a pesticide is offered for sale in the market.

It is also the legal requirement to indicate the toxicity level of the pesticide on the label. As per the WHO classification, pesticides are categorized into four classes on the basis of their toxicity levels. The class Ia pesticides are indicated by the word "POISON" with the symbol of the skull and cross bones, while the class Ib, II, and III pesticides are indicated by "POISON," "DANGER," and "CAUTION" signals denoted in bright red, bright yellow, bright blue, and bright green color, respectively. From adult and children's safety point of view, the warning/cautionary statement and their inflammable nature, if any, should also be indicated on the label (Li, 2022). In India, the toxicity level of the pesticide is indicated as category 1 – Red Triangle (toxicity <50 mg/kg); category 2 – Yellow Triangle (toxicity 51–500 mg/kg); Category 3 – Blue triangle (toxicity 501–5,000 mg/kg); and category 4 (Green Triangle (toxicity >5,000 mg/kg) (Table 4.1).

TABLE 4.1
Acute Toxicity of Pesticides According to WHO Classification

WHO Class	Classification	LD_{50} for the Rat (mg/kg bw)	
		Oral	Dermal
Ia	Extremely hazardous	<5	<50
Ib	Highly hazardous	5–50	50–200
II	Moderately hazardous	50–2,000	200–2,000
III	Slightly hazardous	>2,000	2,000–5,000
U	Unlike to present acute hazard	>5,000	>5,000

Source: WHO (2010).

The pesticide manufacturers and users are required by law to follow all the instructions and directions contained in the label. Farmers/users must be made aware of the fact that the pesticide labels/leaflets/pamphlets/brochures are not mere papers but legal documents that contain important information about their safe use. The instructions given on the label should therefore be strictly followed during the application of the pesticide formulation on a crop. Failure to display a mandatory warning on the label of a pesticide product is considered as "misbranding" and is liable for prosecution under Sec 29 (1) (a) of the Insecticides Act, 1968. It is unlawful for any individual to sell a misbranded pesticide that contains a false or misleading statement about the product's efficacy and toxicity on the label.

4.4 OFF-LABEL USE OF PESTICIDES

Off-label use is encountered when a registered chemical is used in a manner that is not specified on the product label (for example, to control a different pest, protect a different crop, use in a different manner, and/or using a different dose). The process of obtaining label claim approval for a particular crop–pesticide combination is time-consuming and quite demanding in terms of effort and monetary input. The label claims on many crops, especially minor crops, are usually not pursued in cases where extending label uses may be uneconomical for chemical companies (for example, minor crops).

Worldwide, countries have different guidelines to regulate the use of off-label pesticides. In Australia, a permit is required from the Australian Pesticides and Veterinary Medicines Authority (APVMA), to use an unregistered chemical or a "restricted use" chemical in an off-label manner. However, APVMA permit is not required when using registered agricultural chemical products in low-risk circumstances such as to control a pest that is different from the pest stated on the approved label for that use, the registered agricultural chemical product is mixed with another registered agricultural chemical product, at a rate that is lower than the rate for that use as stated on the approved label, at a concentration that is lower than the concentration for that use as stated on the approved label, at an interval that is less frequent than the interval stated on the approved label for that use, or in accordance with a use provided on the approved label for two or more other states or territories (Agriculture Victoria, 2022).

Presently, there are no guidelines in India for off-label pesticide use. As of October 1, 2022, a total of 318 pesticides are registered in the country and about 802 pesticide formulations are approved for use in agriculture. More than 500 crops are grown, of which the labeled use is restricted to less than 100 crops (~18%) (DPPQS, 2018, 2022a). Some major crops have approved label claim for a large number of pesticides, while others have approved label claim for very few number of pesticides. This limits the number of approved crop protection products to many crops including the minor crops (DPPQS, 2022b,c). If a pesticide is highly effective for the control of certain pests, it is commonly used by farmers on other crops affected by that pest, irrespective of their label claims.

In India, under the Department of Agriculture & Farmers Welfare, central sector scheme, Monitoring of Pesticide Residues at National Level (MPRNL), residues of off-label pesticides have been detected in 15%–20% samples.

4.5 CHALLENGES AND PROSPECTS

4.5.1 Risk Assessment

Pesticide label directions are based on substantial scientific testing and rigorous evaluation to ensure that products can be used without harm to workers, consumers, and the environment. When a pesticide is used on a crop for which it is not approved or in a manner that is inconsistent with the label, it may pose real health risks to consumers. Since there are no recommendations for the off-label use of pesticides on a particular crop, the farmers lack guidance on the dosage of application resulting in higher residues on the crop, adverse effect on the environment and the non-target organisms, and/or phytotoxicity to the crops. Pesticides are inherently toxic, and studies have shown that pesticide exposure may be linked to dermatological, gastrointestinal, and neurological dysfunctions, cancers, and various diseases in humans (Mostafalou and Abdollahi, 2012; Kim et al., 2017; Rani et al., 2021). During the registration process of pesticides on food commodities, dietary risk assessments are conducted by national or regional authorities for the establishment of legal MRLs. The residue levels that support the chronic and acute assessment and MRL setting are obtained from supervised residue trials conducted with the pesticide applied according to GAPs.

An implication that alleviates the issue of off-label use of pesticides is the non-existence of MRLs of pesticides. The non-judicious use of pesticides and their persistence in the environment can also be detrimental to soil health (Hussain et al., 2009; Ataikiru et al., 2019) and various beneficial and non-target species of animals, plants, and insects (Ndakidemi et al., 2016; Mahmood et al., 2016). From an international perspective, food safety and consumer protection are based on the guiding principle of the risk analysis of the Codex Alimentarius Commission (CAC) and the Joint FAO/WHO Meeting on Pesticide Residues (JMPR), an expert ad hoc body administered jointly by FAO and WHO. Consumer protection is ensured through toxicologically based safety limits, namely the ADI (Acceptable Daily Intake), which refers to the maximum quantity of a substance, which can be consumed every day for a lifetime, and the ARfD (Acute Reference Dose), which refers to the maximum quantity of a substance that can be ingested in a single day without any appreciable risk to the consumer. In case of off-label pesticide detection in food commodities, risk assessment studies may be conducted, and the pesticide residue data from the national monitoring programs may be used to assess the risks to consumers. The WHO/FAO Joint Meeting on Pesticide Residues (JMPR) considers the monitoring data on spices for evaluation of Codex MRLs. The monitoring data generated under the MPRNL scheme were submitted to FAO/WHO/JMPR for risk assessment and fixation of MRL. Based on these data, the CCPR established and adopted Codex MRLs (CXLs) for 37 different pesticide-crop combinations (JMPR Report, Pesticide Residues in Food, 2019).

Globally various studies are reported where pesticide residue monitoring data have been used for dietary risk assessment and consumer safety. In China, Li et al. (2021) investigated the levels of 18 pesticides in 6,554 vegetable samples collected from 31 regions during 2014–2017, and the risk assessment showed that the detected pesticides did not cause any significant chronic risk to consumers. In another study performed in the mid-western region of China, 123 vegetable samples were collected and analyzed, and the detected residues indicated an acceptable safety risk of the pesticide residues (Qin et al., 2021). Yu et al. (2016) investigated organophosphorus pesticides' (OPs) concentrations in 217 fresh vegetable samples and estimated the potential health risks to inhabitants of northeast China. The study concluded that the exposure to OPs did not pose any health risk. In a Korean study, Yi et al. (2020) assessed the exposure to pesticide residues from 34,520 samples of 96 vegetable types collected from the Seoul market during 2010–2014. The study showed that despite the high-level detection of some pesticide residues, they did not present any serious public health problem. In another study, Galani et al. (2020) monitored residues of 81 pesticides in 160 samples of agricultural products collected from three cities of Cameroon. The cumulative risk assessment showed that the agricultural products were safe for consumption. Bhandari et al. (2019) analyzed residues of 23 pesticides in 86 vegetable samples collected from local farmer's field in southern Nepal. The risk assessment showed that the high dietary risk was mainly associated with the residues of OP pesticides and triazophos (detected in tomatoes).

In a recent study from Egypt, Ibrahim et al. (2022) analyzed 175 samples of locally produced fruits and vegetables for pesticide residues. In another Egyptian study reported by Gad Alla et al. (2015), 116 vegetable samples were collected from different regions of Egypt and analyzed. In another study from Egypt, Ibrahim et al. (2018) analyzed 103 pepper and 104 cucumber samples for 450 multi-class pesticides. For all three studies, the risk assessment of pesticide residues showed that the long-term exposure of Egyptian consumers to pesticide residues through the consumption of raw fruits and vegetables was not associated with any health risk.

Indian studies have also been reported on the monitoring of pesticide residues in vegetables for dietary risk assessment. Sharma et al. (2022) carried out a systematic long-term study of 155 multiclass pesticide residues in 966 vegetable samples grown in the North and North-Western parts of India from 2013 to 2020. The chronic risk assessment due to the dietary exposure of the pesticides showed that the detected residue levels in vegetables were within safe limits, and their consumption did not pose any dietary risk to the consumers. In another study, Tripathy et al. (2022) monitored pesticide residues in 2,319 samples of fruiting vegetables grown in the Northern and Western regions of India. Dietary risk assessment showed that the pesticide residue levels in the analyzed vegetables were within safe limits, and their consumption will not pose any dietary risk to the consumers.

4.5.2 Trade Implications

Presence of the residues above the permissible level has become a major bottleneck in the trade of food commodities by importing countries under the World Trade Organization (WTO) agreement on the Application of Sanitary and Phytosanitary Measures (SPS Agreement). The legal standard for pesticide residues worldwide is the maximum residue limit (MRL), which is established by the regulatory bodies of the respective countries based on their GAP. Variations in the MRLs of the same pesticide–crop combinations in the country of import and export may lead to non-tariff trade barriers. The issue gets further aggravated if off-label pesticides are detected in the food commodities. Since MRLs for off-label pesticides on crops are not fixed, the importing countries may set either default MRLs or fix stringent MRLs based on their capability of detecting the pesticide residues at the lowest level. Setting MRLs as low as the Limit of Detection (LOD) may lead to the rejection of the produce.

India is the leading producer and exporter of agricultural and horticultural crops such as rice, tea, coffee, cashew, grapes, fresh vegetables, chilli powder, and other spices to various countries. About 40% of the crops are prevented from being included in the trade chain due to the non-tariff trade barriers, which arise because of limited regulatory approval of safe and effective plant protection chemicals. The low number of registered pesticides for a commodity, absence of domestic MRLs, and detection of off-label pesticides in exported commodities have led to the rejection of the export consignments from the port of entry causing huge economic losses and thus hampering smooth trade between India and the importing countries. To overcome barriers related to off-label pesticides detection, the importing countries may consider the MRLs of the exporting country or Codex MRLs or perform a risk assessment of the residue data from the exporting country.

4.5.3 Minor Use Pesticide Programs

The term "minor use" of a pesticide refers to the crop–protection treatments used on low acreage high-value crops referred to as "minor crops" or in major crops where a particular pest occurs only in limited situations or where pest control is only needed on a small portion of the overall crop acreage. The minor use crops may be appropriately grouped into subgroups, and certain cost-effective off-label pesticides after due diligence are identified to protect such crops from pest damage. Each country may have its own list of minor use crops, but some prominent ones include coconut, areca nut, tobacco, sweet potato, basil, and various spices crops such as cardamom, coriander, fennel, ginger, and turmeric. Minor use crops are identified based on the area under cultivation, pest types, food production (tonnage/ acreage), dietary intake of such foods at the local level, and economic returns to the crop growers and manufacturers of such products. Most of the pesticides are registered against the major crops to control specific pests but minor crops, which are cultivated in smaller geographies and require small quantities of pesticides, do not attract the commercial interest of the manufacturers to seek registration of the pesticides for minor crops. Further, owing to the higher costs involved in the generation of bio-efficacy, toxicology, and residue data to support a registration for minor uses, the industry is unwilling to pursue this exercise as they find it economically unviable.

Challenges associated with minor uses are common to both developed and emerging economies. Non-availability of products for minor uses negatively impacts international/national/regional economies and leads to a surge in the cases of off-label pesticide use. Though, concerted efforts are being made globally for the implementation of "Minor Use Pesticide Programmes" in operation since 2002 by the Agriculture and Agri-Food *Canada* (AAFC) and Health Canada's Pest Management Regulatory Agency (PMRA) to increase grower competitiveness by improving access to new and effective crop protection tools and technologies. The Global Minor Use Summits (GMUS) sponsored by the FAO, USDA, US-EPA, Canadian-PMC, and IR-4 are organized from time to time to improve communications and information exchange, harmonization of data requirements, capacity building for the conduct of residue and efficacy trials, and engaging with CODEX for better support for minor uses, crop grouping, and related projects. Keeping in view the various developments at the national and international levels, the stakeholders group comprising growers, regulators, industry, and exporters should work together to find a viable solution to overcome the challenges related to minor use pesticides.

4.5.4 Crop Grouping to Minimize Off-Label Pesticide Use

Crop grouping aims at uniting similar types of crops into a group or subgroup to facilitate the use of pesticides in as many crops as scientifically possible. Crops are grouped based on their similarities in botanical classification, morphology, cultural practices, growing seasons, locations or growth habit, edible portion of the commodity, as well as potential for pesticide residues. Crop grouping enables extrapolation of data generated for a major/target crop to other related crops of the same crop group eliminating the need for fresh data for each individual crop in the group.

The process of obtaining label claim approval for a particular crop–pesticide combination is time-consuming and quite demanding in terms of efforts and monetary input due to which label claims on many crops especially minor crops are not pursued, by the industry. Concerted efforts have been made globally for the implementation of the crop grouping concept so that the growers have access to new and effective crop protection tools and technologies. Each crop group is indicated by a representative crop, which is generally the most economically important commodity in production/consumption in the group/subgroup. The commodity is chosen such that it indicates the upper range of residues that can be encountered for the group/subgroup based on the same or comparable GAPs and other available information. An MRL for the group may be estimated from the highest residue level for any of the individual representative commodities or from the larger combined dataset (Codex, 2017).

According to the Codex crop grouping, the residue levels on a representative commodity can be used to estimate the residue levels on related commodities present in the same group/subgroup for which trials have not been conducted through the method of residue extrapolation.

The limited number of label claims on many crops in India, namely curry leaves, herbs, capsicum, leafy vegetables, and spices, have left farmers with insufficient pest management options. These commodities hence run the risk of being rejected in international trade on account of off-label use of the pesticide and the absence of country MRLs. Shifting to a crop grouping approach would therefore lessen the burden of obtaining label claim for each and every crop–pesticide combination as it allows researchers to apply residue data collected from one crop to other similar crop without having to conduct additional research. Practicing crop grouping can thus be highly advantageous as it gives farmers a range of pesticides that can be used for a crop without stressing about its label claim. In addition, it is economical for the industries as the need to generate data across multiple crops in the same crop group is removed.

The issue of off-label use of pesticides in India has been identified as a major Sanitary and Phytosanitary (SPS) and Technical Barriers to Trade (TBT) concern. To address this concern, the DAC constituted a committee in 2013 to study the aspects of crop grouping within the draft principles of Codex Committee for Pesticide Residues (CCPR) and develop a national document on crop grouping based on the Codex guidelines of crop grouping focusing on the principles and guidance on the selection of representative commodities and classification of crop commodities with representative commodities. The report submitted in 2015 identified five representative crop types (fruits, vegetables, grasses, nuts and seeds, and herbs and spices) under varying situations of trade patterns and perception of risks on pesticide residues in the country and recommended extrapolation of existing MRLs of the given pesticide substance to those crops within the same crop group/subgroup of crops and termed as Group MRL (Table 4.2).

The representative crops identified under the crop commodity group are, however, subject to review under varying situations of trade patterns, perception of risks on pesticide residues, and the introduction of new crops in the country. To recommend the procedure to implement crop grouping principles within the ambit of the Insecticides Act, a subcommittee was constituted in 2017 by the RC to deliberate on the modalities and guidelines for implementation of crop grouping principles with regard to data generation and setting of group MRLs. In consonance with the latest Codex crop grouping classification, the committee improvised the existing crop grouping and proposed the revised requirements for bio-efficacy and residue data generation for representative and member crops, which were accepted by the RC.

Thus, the implementation of crop grouping concept in India has facilitated the label claim expansion and fixation of MRL of pesticides on all crops especially minor crops and those that are commercially less important and ultimately overcome the trade barriers. At the international level, Crop Grouping Consulting Committee (ICGCC) is also working to harmonize crop groupings to update the CODEX classification of foods and feeds.

TABLE 4.2
Representative Crop Groups of Various Commodities

Crop Code (Group/Subgroup)	Crop Group	Representative Crop	Commodities
		Crop Type 01 – Fruits	
Group 001 (FC)	**Citrus Fruits**		
Subgroup 001A	Lemons and Limes	**Lemon**	Citron; Kaffir Lime; Lemon; Lime; Lime, Sweet; Limequats; Mexican lime, Rough Lemon
Subgroup 001B	Mandarins	Mandarins	Mandarin, Mediterranean Mandarin, Willow leaf mandarin, Kinnow[a]
Subgroup 001C	Oranges, Sweet, Sour	Orange	Orange Sour; Orange, Sweet; Indian wild orange; Golden Lime
Subgroup 001D	Pummelos	Pummelos or Grapefruit *(Chakotara)*	Grapefruit; Pummelo; Tangelo
Group 002 (FP)	**Pome fruits**		
Subgroup002A	Pomes	Apple or pear	Apple; Crab-apple; Loquat; Medlar; Pear; Quince
Group 003 (FS)	**Stone fruits**		
Subgroup 003A	Cherry	Cherry (sweet or sour)	Cherry black; Cherry Nanking; Cherry Sour; Cherry Sweet; **Spanish cherry/ Maulsari**
Subgroup 003B	Plums	Plum	Bullace; Cherry plum; Jujube Chinese; Plum, Plum beach; Prune Plum; **Ramontchi, Governor's plum; Indian plum**
Subgroup 003C	Peaches	Peach or Apricot	Apricot; Japanese apricot; Nectarine; Peach
Group 004 (FB)	**Berries and other small fruits**		
Subgroup 004A	Cane berries	Blackberry or Raspberry	Cane berries; Blackberries; Raspberries (Red, Black, yellow), Rose leaf bramble; *Rubus cooperi*; *Rubus polyodontus* Hand; ***Rubus quinquefoliolatus*; Yellow Himalayan raspberry; West Indian raspberry; Stone bramble**
Subgroup 004B	Bush berries	Gooseberry	Gooseberry; European barberry

(Continued)

TABLE 4.2 (*Continued*)
Representative Crop Groups of Various Commodities

Crop Code (Group/Subgroup)	Crop Group	Representative Crop	Commodities
Subgroup 004C	Large shrub/berries	Mulberries or Phalsa	Mulberries; Phalsa
Subgroup 004D	Small fruit vine climbing	Grapes	Grapes, Table grapes; Wine grapes
Subgroup 004E	Low growing berries	Strawberry	Strawberry; Strawberries, Wild; **Mock strawberry/Indian strawberry**
Group 005 (FT)	**Assorted tropical and sub-tropical fruits - edible peel**		
Subgroup 005A	Assorted tropical and sub-tropical fruits – edible peel – small	Olives	Almondette, Arbados cherry; Carandas plum; Ceylon iron wood; Ceylon olive; Chinese olive, Chirauli nut; False sandalwood; Gooseberry; Abyssinian; Ceylon; Fragrant Manjack; Hog plum; Illawarra plum; Jamaica cherry; Jambolan; Java apple; Karanda; Kapundung; Otaheite gooseberry; Rumberry; Sea grape; Surinam cherry; Table olives; Black Dammar; Mootikaya
Subgroup 005B	Assorted tropical and sub-tropical fruits – edible peel – medium to large	Guava or Indian (*Ber*)	Ambarella; Bilimbi; Carambola; Carob; Cashew apple, Fig, Gooseberry Indian, Guava, Guava Brazilian; Jujube Indian (*Ber*); Mombin Malayan, Monkeyfruit; *Noni;* Pomerac; Rose apple; Sentul
Subgroup 005C	Assorted tropical and sub-tropical fruits – edible peel – palms	Date	Date; Doum or Dum palm, Jelly palm; Desert date
Group 006 (FI)	**Assorted tropical and sub-tropical fruits – inedible peel**		

(*Continued*)

TABLE 4.2 (*Continued*)
Representative Crop Groups of Various Commodities

Crop Code (Group/Subgroup)	Crop Group	Representative Crop	Commodities
Subgroup 006A	Assorted tropical and sub-tropical fruits – inedible peel – small	Litchi	Bael fruit; Litchi; Longan; Madras- thorn; Mesquite; Tamarind (sweet varieties); Wampi; Burmese grape[a]
Subgroup 006B	Assorted tropical and sub-tropical fruits – inedible smooth peel - large	Mango or banana	Avocado; Banana; Cacao; *Kokam*, Langsat; Mango, Mangosteen; Papaya; Pomegranate; Sataw Tamarillo
Subgroup 006C	Assorted tropical and sub-tropical fruits – inedible rough or hairy peel - large	Pineapple or Sapota	Breadfruit; Champedak; Custard Apple; Durian; Elephant apple; Jackfruit; Monkey-bread tree; Pineapple; Rambutan; Sugar apple; Sapota (Sapodilla)
Subgroup 006D	Assorted tropical and sub-tropical fruits – inedible peel – cactus	Prickly pear	Prickly pear
Subgroup 006E	Assorted tropical and sub-tropical fruits – inedible peel – vines	Kiwifruit or Passion fruit	Granadilla Giant; Kiwifruit; Passion fruit
Subgroup 006F	Assorted tropical and sub-tropical fruits – inedible peel – Tropical palm fruits	Coconut (Tender and mature)	Coconut; Muriti; Palmyra Palm
		Crop Type 02 – Vegetables	
Group 009 (VA)	**Bulb vegetables**		
Subgroup 009A	Bulb Onions	Onion	Daylilly; Garlic; Great-headed Garlic; Lily; Onion Bulb; Shallot; Silverskin onion
Subgroup 009B	Green Onions	Green (spring) onion	Leek; Pearl onion; Spring onion
Group 010 (VB)	**Brassica vegetables**		
Subgroup 10A	Flower head Brassicas	Cauliflower	Broccoli; Cauliflower
Subgroup10B	Head Brassicas	Cabbage	Cabbages head; Brussels sprouts
Subgroup 10C	Stem Brassicas	Kohlrabi (Knolkhol)	Kohlrabi (Knolkhol); Stem mustard

(*Continued*)

TABLE 4.2 (*Continued*)
Representative Crop Groups of Various Commodities

Crop Code (Group/Subgroup)	Crop Group	Representative Crop	Commodities
Group011 (VC)	**Fruiting vegetables, Cucurbits**		
Subgroup 11A	Fruiting vegetables, Cucurbits – Cucumbers and Summer squashes	Cucumber and gourd (Bitter gourds or bottle gourd)	Balsam apple; Bitter gourd (Bitter melon); Ridge gourd; Bottle gourd; Cucumber; Casaba or Casaba melon; Chayote; Chieh qua, Citron melon; Courgette (Zucchini) ; Cushaws (*Cucubita mixta*); Cucumber; Gac; Gherkin; West Indian Gherkin; Gourd; Bitter snake Gourd; buffalo Gourd; Malabar Gourd; Pointed Gourd; Round gourd; Ivy gourd; Loofah; Angled; Loofah Smooth; Snake gourd; Summer Squash; Spine gourd **/Kankoda /Katole**
Subgroup 11B	Fruiting vegetables, Cucurbits – Melons, Pumpkins and Winter squash	Melon (musk melon)	Cantaloupe; Casaba melon; Melons; Muskmelon; Serpent Melon (Kakri); Watermelon; Butternut squash; Cheese pumpkin; Cushaws (*Cucurbita argyrosperma*); Giant pumpkin; Hubbard squash; Indian round gourd; Pumpkins; Wax gourd; Winter squash
Group 012 (VO)	**Fruiting vegetables, other than Cucurbits**		
Subgroup 12A	Tomatoes	Tomato	Tomato; Bush tomato; Cherry tomato; Cocona; Currant tomato; Garden huckleberry; Goji berry; Ground cherries; Strawberry tomato; Sunberry; Tomatillo
Subgroup 12B	Pepper and pepper-like commodities	Okra and Chilli Pepper	Bird chili peppers; Cherry Martynia; Okra;Peppers ; Chili Peppers; Sweet peppers (Bell pepper); Long peppers; Roselle

(*Continued*)

TABLE 4.2 (*Continued*)
Representative Crop Groups of Various Commodities

Crop Code (Group/Subgroup)	Crop Group	Representative Crop	Commodities
Subgroup 12C	Eggplant and eggplant-like commodities	Eggplant (Brinjal)	Eggplant; Pea Eggplant; Pepino; Tree melon; Aubergine
Group 013 (VL)	**Leafy vegetables (including Brassica leafy vegetables)**		
Subgroup 013A	Leafy greens	Spinach and Lettuce	African Eggplant leaves; Amaranth leaves; Aster Indian; Ayoyo; Barley shoot; Bitter leaf; Blackjack; Cat's Whiskers; Chamsuk; Chicory leaves; Chili pepper leaves; Chrysanthemum edible leaves; Common bean leaves; Cowpea leaves; Dandelion; Daylily leaves; Ebolo; Fame flower; Feather cockscomb; Glasswort common; Gomchwi; Goosefoot leaves; Jute; Lettuce bitter; Lettuce Head; Lettuce Leaf; Mallow leaves; Peanut leaves; Perilla leaves; Plantain leaves; Purslane; Purslane Winter; Sanmaneul leaves; Sowthistle; Soya bean leaves; Spinach; Spinach Indian; Violet Chinese
Subgroup 013B	Brassica Leafy vegetables	Mustard greens	Chinese cabbage (type Pak-choi); Cress Garden; Cress upland; Flowering white cabbage; Hanover salad; Kale; Kohlrabi leaves; Leaf mustard; Mustard greens; Mustard (tuberous rooted, Chinese); Purple-stem mustard; Rape greens; Radish leaves; Rucola; Turnip greens
Subgroup 013C	Leaves of root and tuber vegetables	Sweet Potato leaves or Taro (Arbi) leaves	Arrowroot leaves; Cassava leaves; Sweet potato; leaves; Taro leaves; Beet leaves (*chard*)

(*Continued*)

TABLE 4.2 (*Continued*)
Representative Crop Groups of Various Commodities

Crop Code (Group/Subgroup)	Crop Group	Representative Crop	Commodities
Subgroup 013D	Leaves of trees, shrubs, and vines	Roselle leaf or Grape leaves	*Acacia* shoots; Ben moringa leaves; Grape leaves; Monkey-bread tree leaves; Papaya leaves; Roselle leaves; White lead tree
Subgroup 013E	Leafy aquatic vegetables	Kangkung (water spinach)	Kangkung; Watercress; Water mimosa
Subgroup 013 F	Witloof	Witloof chicory (sprouts)	Witloof chicory (sprouts)
Subgroup 013G	Leaves of Cucurbit crops	Pumpkin leaves	Balsam pear leaves; Chayote leaves; Ivy gourd; *Kahurura*; Pumpkin leaves
Subgroup 013H	Young leaves	Leaf lettuce or any crop intended to use as baby leaves, (harvested up to 8 true leaf stage)	Leaves of crops under use as commodity (these are not often traded and so need not be under consideration at present; however, as and when arising, such relevant commodity can be viewed by competent agency for the purpose of fixing MRL value
Subgroup 013 I	Sprouts	Mungbean sprouts or chickpea sprouts or soybean sprouts	Alfalfa sprouts, Mungbean sprouts, Radish sprouts, Soybean sprouts, chickpea sprouts
Group 014 (VP)	**Legume vegetables**		
Subgroup 014A	Beans with pods	French bean (Common bean)	Broad bean; Catjang; Common bean / French bean/ *Phaseolus vulgaris*; Cowpea; Goa bean; Cluster bean (*Guar*); Jack bean; Lablab bean; Moth Mung bean; Rice bean; Scarlet runner bean; Soybean; Stink bean; Sword bean; Urd bean; Winged pea; Yard-long bean

(*Continued*)

TABLE 4.2 (*Continued*)
Representative Crop Groups of Various Commodities

Crop Code (Group/Subgroup)	Crop Group	Representative Crop	Commodities
Subgroup 014B	Peas with pods	Garden Peas or Pigeon pea (*for pods and succulent* immature seeds)	Garden pea; Grass pea; Lentil; Pigeon pea; Podded pea;
Subgroup 014C	Succulent beans without pods	Common bean or Cowpea or Soybean (succulent beans)	Beans without pods; Black-eyed peas; broad bean (shelled); Catjang: Common bean: Cowpea: Goa bean: Jack bean: Lablab bean: Limabean: Lupin: Moth bean: Scarlet runner bean: Soybean: Stink bean: Velvet bean
Subgroup 013 I	Sprouts	Mungbean sprouts or chickpea sprouts or soybean sprouts	Alfalfa sprouts, Mungbean sprouts, Radish sprouts, Soybean sprouts, chickpea sprouts
Group 014 (VP)	**Legume vegetables**		
Subgroup 014A	Beans with pods	French bean (Common bean)	Broad bean; Catjang; Common bean / French bean/ *Phaseolus vulgaris*; Cowpea; Goa bean; Cluster bean (*Guar*); Jack bean; Lablab bean; Moth Mung bean; Rice bean; Scarlet runner bean; Soybean; Stink bean; Sword bean; Urd bean; Winged pea; Yard-long bean
Subgroup 014B	Peas with pods	Garden Peas or Pigeon pea (for pods and succulent immature seeds)	Garden pea; Grass pea; Lentil; Pigeon pea; Podded pea;
Subgroup 014C	Succulent beans without pods	Common bean or Cowpea or Soybean (succulent beans)	Beans without pods; Black-eyed peas; broad bean (shelled); Catjang: Common bean: Cowpea: Goa bean: Jack bean: Lablab bean: Limabean: Lupin: Moth bean: Scarlet runner bean: Soybean: Stink bean: Velvet bean
Subgroup 014D	Succulent peas without pods	Garden Peas or Chickpea	Garden pea; others such as Chickpea, Lentil, Pigeon pea

(*Continued*)

TABLE 4.2 (*Continued*)
Representative Crop Groups of Various Commodities

Crop Code (Group/Subgroup)	Crop Group	Representative Crop	Commodities
Subgroup 014E,	Underground immature beans and peas	Peanut (immature)	Peanut (immature)
Group 015 (VD)	**Pulses**		
Subgroup 015A	Dry beans	Common bean and soybean (dry)	Adzuki bean (dry); Black-eyed pea; Broad bean (dry); Butter bean; Common bean/ kidney bean/ Rajma/ *Phaseolus vulgaris* (dry); Common vetch; Cowpea (dry); Goa bean; Guar beans (dry); Horse gram; Lablab bean (dry); Lupin (dry); Moth bean(dry); Mung bean (dry); Narbon bean; Rice bean (dry); Scarlet runner bean(dry); Soya bean (dry); Swordbean (dry); Tepary bean (dry); Tick bean; Urd bean (dry); Velvet bean (dry); Winged pea (dry);
Subgroup 015B	Dry peas	Chickpea or Pigeon pea or Field pea (dry)	Peas (dry), Chick-pea (dry), Field pea(dry), Grass-pea(dry), Lentil (dry), Pigeon pea (dry)
Group 016 (VR)	**Root and tuber vegetables**		
Subgroup 016A	Root vegetables	Carrot and Radish	Beetroot; Edible Burdock greater; Caraway black root; Carrot; Celeriac; Chicory roots; Dandelion root; Ginseng; Horseradish; Kudzu; Madeira vine; Parsley; Turnip-rooted Parsley; Radish; Radish (black); Radish (Japanese); Sugar beet; Ti palm; Turnip
Subgroup 016B	Tuberous and corm vegetables	Potato	Alocasia; American potato bean; Arrowroot (Guinea); Arrowroot (Polynesian); Canna (edible); Cassava (Tapioca); Chayote root; Chinese potato; Elephant foot yam; Gastrodia tuber; Goa bean root; Potato; Sweet potato; Taro; Tiger nut; Yams; Yam bean
Subgroup 016C	Aquatic root and tuber vegetables	Water chestnut (Singhara)	Lotus tuber (*kamal-kakri*); Water chestnut, Water bamboo, Foxnut

(*Continued*)

TABLE 4.2 (*Continued*)
Representative Crop Groups of Various Commodities

Crop Code (Group/Subgroup)	Crop Group	Representative Crop	Commodities
Group 017 (VS)	**Stalk and stem vegetables**		
Subgroup 017A	Stalk and stem vegetables - Stems and Petioles	Flowering stalk of Garlic or Celery	Burdock (edible tops); Celery; Lettuce; Fennel bulb; Flowering stalk of Garlic; Sweet potato (stems); Taro stems; Zuiki; **Garlic Scapes**
Subgroup 017B	Stalk and stem vegetables - Young shoots	Asparagus	Asparagus; Bamboo shoots; Ferns (edible)
Subgroup 017C	Stalk and stem vegetables – Others	Water celery	Palm hearts; Prickly pear pads; Water-celery
Group 018 (VF)	**Edible Fungi**	Button mushroom	Fungi Edible except Mushrooms; Button mushroom; Cep; Hirmeola; More; Net bearing Dictyophora; Oyster mushroom; Straw mushroom; Wood ears mushroom
		Crop Type 03- Grasses	
Subgroup 020A	Wheat, similar grains, and pseudocereals without husks	Wheat	Wheat, similar grains, and pseudocereals without husks (GC 2086): Amaranth, grain; Cañihua; Chia; Cram-cam; Huauzontle; Quinoa; Rye; Triticale; Wheat
Subgroup 020B	Barley, similar grains, and pseudocereals with husks	Barley	Barley, similar grains, and pseudocereals with husks (GC 2087): Barley; Buckwheat; Buckwheat, Tartary; Oats
Subgroup 020C	Rice cereals	Rice	Rice Cereals (GC 2088): Rice; Rice, African; Wild rice
Subgroup 020D	Sorghum Grain and Millet	Sorghum Grain	Sorghum Grain and Millet (GC 2089): Hungry rice; Job's tears; Millet; Sorghum Grain; Teff or Tef;
Subgroup 020E	Maize Cereals	Maize	Maize; Popcorn; Teosinte
Subgroup 020F	Sweet corns	Sweet corn (Corn-on-the-cob) (kernels plus cob with husk removed)	Sweet corns (GC 2090): Baby corn; Sweet corn (Corn-on-the-cob) (kernels plus cob with husk removed); Sweet corn (whole kernel without cob or husk)

(*Continued*)

TABLE 4.2 (*Continued*)
Representative Crop Groups of Various Commodities

Crop Code (Group/Subgroup)	Crop Group	Representative Crop	Commodities
Group 021 (GS)	**Grasses, for sugar or syrup production**	Sugarcane	Sugar cane, Sorgo or Sorghum (sweet)
		Crop Type 04- Nuts and Seeds	
Group 022 (TN)	**Tree nuts**	Almond or walnut; and Cashew nut	Almonds; Betel nut /Arecanut; Candle nut; Cashew nut; Coconut; Hazelnuts; Japanese horse-chestnut; Java almonds; Dhuna rata; Pine nuts; Pistachio nut; Tropical almond; Walnuts, Indian or Himalayan Horse Chestnut, Dhuna Rata, Black dammar
Group 023 (SO)	**Oilseed and oil fruits**		
Subgroup 023A	Small seed oilseeds	Rapeseed or Mustard,	Rape seeds; Mustard seeds; Linseed; Mustard seed (Field/ toria); Mustard seed (Indian); Perilla seed; Poppy seed; Sesame seed
Subgroup 023B	Sunflower seeds	Sunflower seeds	Jojoba seed; Niger seed; Safflower seed; Sunflower seed
Subgroup 023C	Cotton seed	Cotton seed	Cotton seed
Subgroup 023D	Other oil seeds	Peanut or Soybean seed	American oil palm seed; Argan Nut; Babassu seed; Ben Moringa seed; Castor bean; Hempseed; Kapok; Melon seed; Palm nut; Peanut; Pumpkin seed; Sea buckthorn seed; Shea nut; Soybean seed
Subgroup 023E	Oil fruits (fruits of palm trees)	Olives for oil production	American oil palm fruit; Maripa palm fruit; Olives for oil production; Palm fruit (African oil palm); Tucum fruit
Group 024 (SB)	**Seed for beverages and sweets**	Coffee bean	Coffee beans, Cocoa beans; Seeds of the sweet basil-sabja, tukmaria
Group 025	**Tree saps**	Any commodity in this subgroup	Birch sap; Hophornbeam sap; Manna sap; Maple sap; Nut sap; Palm sap; Sycamore sap;

(*Continued*)

TABLE 4.2 (*Continued*)
Representative Crop Groups of Various Commodities

Crop Code (Group/Subgroup)	Crop Group	Representative Crop	Commodities
		Crop Type 05 – Herbs and Spices	
Group 027 (HH)	**Herbs**		
Subgroup 027A	Herbs (herbaceous plants)	Coriander leaves or Mint leaves	Anise; Basil leaves ; Indian Borage; Calendula Flowers / leaves; Cumin (Carraway) leaves; Celery leaves; Coriander leaves; Costmary; Coverfern; Cilantro leaves; Dill leaves; Fennel leaves; Fenugreek leaves; Geranium leaves; Horehound; Lavender; Lemongrass; Marigold leaves; Marjoram; Mint; Nettle; Pandan, leaves; Parsley, leaves; Pennywort; Perilla, leaves; Rosemary; Sage and related Salvia species; Savor, Summer and Winter; Common Sorrel; Southernwood; Stevia; Thyme; Wasabi Stem; Wild beetle leaf bush; Psyllium seed husk / Isabgol
Subgroup 027B	Leaves of woody plants (leaves of shrubs and trees)	Curry leaf or Cinnamon (Tejpata leaves)	Aniseed myrtle; Boldo; Curry leaf tree (*Murraya koenigii*); Eucalyptus leaves; Laurel leaves; Pepper (leaves); Pepperbush leaves; Indian bay leaf (Tejpata leaves) *Cinnamomum tamala*
Subgroup 027C	Edible flowers	Any commodity in this subgroup	Calendula, flowers; Courgette, flowers; Daylily, flowers; Daisy, common, flowers; Geranium, flowers; Marigold, flowers; Nasturtium, flowers; Violet, flowers; rose flowers and other edible flowers
Group 028 (HS)	**Spices**		
Subgroup 028A	Seeds	Cumin seed or Coriander seed	Carom (Ajowan) - *Ajwain*; Angelica seed; Anise seed; Basil seed; Black caraway; Caraway seed; Celery seed; Coriander seed; Cumin seed; Dill seed; Fennel seed; Fenugreek seed; Nutmeg; Parsley seed

(*Continued*)

TABLE 4.2 (*Continued*)
Representative Crop Groups of Various Commodities

Crop Code (Group/Subgroup)	Crop Group	Representative Crop	Commodities
Subgroup 028B	Fruit or Berry	Cardamom or Black Pepper	Caper berries; Cardamom (pods and seeds); Cumin black; Cardamom black; Eucalyptus fruit; Gardenia fruit; Juniper berry; Nasturtium pods; Black Pepper; Pepper (Cubeb); Pepper (Long); Pepper bush berry; Pimento (fruit); Tamarind; Vanilla beans; pink pepper; green pepper
Subgroup 028C	Bark	Cinnamon	Cassia bark; Cinnamon bark; Eucalyptus bark; Mastic; Red cinchona
Subgroup 028D	Root or rhizome	Turmeric or Ginger	Coriander (root); Ginger (rhizomes); Liquorice (roots); Turmeric (root); Asafoetida; Zedoary; ashwagandha roots
Subgroup 028E	Buds	Cloves	Caper buds; Cassia buds; Cloves (buds)
Subgroup 028F	Flower or stigma	Saffron	Saffron
Subgroup 028G	Aril	Mace (*Javitri*)	Mace
Subgroup 028H	Citrus Peel	*Kafir* lime peel	Kaffir lime peel; Lemon, peel; Orange, peel; Satsuma mandarin, peel
Subgroup 028I	Dried chili peppers	Peppers, Chili, dried	Peppers, Chili, dried
Group 029- Derived product of plant origin			
Group 029	Tea	Tea	Black, green, fermented tea

[a] To be included in the Codex crop grouping.

4.6 CONCLUSION

Frequent detection of residues of off-label pesticides in food crops and commodities is a cause of growing concern. Apart from potential health implications, their detection in the exportable commodities adversely affects the economy due to the rejection of export consignments by the country of import. This happens because unlike in other countries, the number of registered pesticides in India is far less. Therefore, there is an urgent need to enhance the number of registered pesticides in India by adopting codex MRLs, fixing national MRLs on more number of major and minor crops as well as allowing the use of off-label pesticides by setting their MRLs on the desired crops. Further, the crop grouping approach, which provides an economical solution of expanding the label claims of the existing pesticides for various crop–pesticide combinations, also needs to be vigorously pursued. Therefore, bringing more pesticides for use on more crops (major and minor) under the ambit of labeled pesticides through label claim expansion and crop grouping would facilitate fair trade practices, help India move toward economic prosperity, and promote its role as one of the major producers and exporters of the food products to the world.

REFERENCES

Agriculture Victoria. (2022). Off-Label Use of Agricultural Chemicals. https://agriculture.vic.gov.au/farm-management/chemicals/offlabel-chemical-use/off-label-use-of-agricultural-chemicals (accessed on September 5, 2022).

APEDA (Agricultural and Processed Food Products Export Development Authority). (2022a). Fresh Fruits and Vegetables. https://apeda.gov.in/apedawebsite/six_head_product/FFV.htm (accessed on September 10, 2022).

APEDA (Agricultural and Processed Food Products Export Development Authority). (2022b). Cereals. https://apeda.gov.in/apedawebsite/six_head_product/cereal.htm (accessed on September 10, 2022).

APEDA (Agricultural and Processed Food Products Export Development Authority). (2022c). India Export of Agro Food Products, Cereals. https://agriexchange.apeda.gov.in/indexp/MainheadPrdGrp_prdwise.aspx?gcode=%270601%27%2c%270602%27%2c%270603%27%2c%270604%27%2c%270605%27 (accessed on September 10, 2022).

APEDA (Agricultural and Processed Food Products Export Development Authority). (2022d). India Export of Agro Food Products, Fresh Fruits and Vegetables. https://agriexchange.apeda.gov.in/indexp/MainheadPrdGrp_prdwise.aspx?gcode=%270201%27%2c%270202%27%2c%270203%27%2c%270204%27%2c%270205%27%2c%270206%27%2c%270207%27 (accessed on September 10, 2022).

Ataikiru, T. L., Okpokwasili, G. S., & Okerentugba, P. O. (2019). Impact of pesticides on microbial diversity and enzymes in soil. *South Asian Journal of Research in Microbiology*, 4(2), 1–16. https://doi.org/10.9734/sajrm/2019/v4i230104.

Bhandari, G., Zomer, P., Atreya, K., Mol, H. G. J., Yang, X., & Geissen, V. (2019). Pesticide residues in Nepalese vegetables and potential health risks. *Environmental Research*, 172, 511–521. https://doi.org/10.1016/j.envres.2019.03.002.

Carriger, J. F., Rand, G. M., Gardinali, P. R., Perry, W. B., Tompkins, M. S., & Fernandez, A. M. (2006). Pesticides of potential ecological concern in sediment from Florida canals: An ecological risk prioritization for aquatic arthropods. *Soil and Sediments Contamination*, 15(1), 21–45. https://doi.org/10.1080/15320380500363095.

Codex (Codex Alimentarius Commission). (2017). Principles and Guidance on the Selection of Representative Commodities for the Extrapolation of Maximum Residue Limits for Pesticides to Commodity Groups (Document No. CXG 84–2012).

DA & FW (Department of Agriculture & Farmers Welfare). (2022). Statistical Database: Consumption of Chemical Pesticides in Various States/UTs During 2017–18 to 2021–22. https://ppqs.gov.in/statistical-database (accessed on September 20, 2022)

DPPQS. (2018). Directorate of Plant Protection Quarantine and Storage, Minutes of 396th Meeting of Registration Committee held on 29.11.2018, New Delhi. https://ppqs.gov.in/sites/default/files/396_rc_minutes.pdf (accessed on March 31, 2022).

DPPQS. (2022a). Directorate of Plant Protection Quarantine and Storage, Pesticides and Formulations Registered for Use in the Country Under the Insecticides Act, 1968 (updated 01.04.2022). https://ppqs.gov.in/sites/default/files/registered_formulations.pdf (accessed on June 12, 2022).

DPPQS. (2022b). Directorate of Plant Protection Quarantine and Storage, Major Use of Pesticides (Fungicides). https://ppqs.gov.in/sites/default/files/major_use_of_pesticides_fungicides_as_on_31.05.2022.pdf (accessed on June 12, 2022).

DPPQS. (2022c). Directorate of Plant Protection Quarantine and Storage, Major Use of Pesticides (Insecticides). https://ppqs.gov.in/sites/default/files/major_use_of_pesticides_insecticides_as_on_31.05.2022.pdf (accessed on June 12, 2022).

FAO (Food and Agriculture Organization) in India. (2022). India at a Glance. https://www.fao.org/india/fao-in-india/india-at-a-glance/en/ (accessed on September 29, 2022).

FAOSTAT (Food and Agriculture Organization Corporate Statistical Database). (2020a). Pesticides Trade. https://www.fao.org/faostat/en/#data/RT (accessed on September 15, 2022).

FAOSTAT (Food and Agriculture Organization Corporate Statistical Database). (2020b). Pesticides Indicators. https://www.fao.org/faostat/en/#data/EP (accessed on September 29, 2022).

FSSAI (Food Safety and Standards Authority of India). (2018). Gazette Notification on Food Safety and Standards (Contaminants, Toxins and Residues) Amendment Regulation related to MRL of Pesticide. The Gazatte of India: Extraordinary, [Part III-Sec. 4], 24 December, 2018, pp. 1–51.

FSSAI (Food Safety and Standards Authority of India). (2022). https://www.fssai.gov.in/ (accessed on September 20, 2022).

Gad Alla, S. A., Loutfy, N. M., Shendy, A. H., & Ahmed, M. T. (2015). Hazard index, a tool for a long–term risk assessment of pesticide residues in some commodities, a pilot study. *Regulatory Toxicology and Pharmacology*, 73(3), 985–991. https://doi.org/ 10.1016/j.yrtph.2015.09.016.

Galani, Y. J. H., Houbraken, M., Wumbei, A., Djeugap, J. F., Fotio, D., Gong, Y. Y., & Spanoghe, P. (2020). Monitoring and dietary risk assessment of 81 pesticide residues in 11 local agricultural products from the 3 largest cities of Cameroon. *Food Control*, 118, 107416. https://doi.org/10.1016/j.foodcont.2020.107416.

Hussain, S., Siddique, T., Saleem, M., Arshad, M., & Khalid, A. (2009). Impact of pesticides on soil microbial diversity, enzymes, and biochemical reactions. *Advances in Agronomy*, 102, 159–200. https://doi.org/10.1016/S0065-2113(09)01005-0.

Ibrahim, M. A., Belal, M. H., Abdallah, I. S., & El-Sawi, S. A. M. (2022). Monitoring and risk assessment of pesticide residues in some locally produced vegetables and fruits. *Egyptian Journal of Chemistry*, 65(7), 429–439. https://doi.org/10.21608/EJCHEM.2021.106776.4920.

Ibrahim, N. M., Eweis, E. A., El-sawi, S. A. M., & Nassar, K. R. A. (2018). Monitoring and risk assessment of pesticide residues in some vegetables in Egypt. *Middle East Journal of Applied Sciences*, 8(2), 669–679.

JMPR (Joint FAO/WHO Meeting on Pesticide Residues). (2019). *Report of the Extra Joint Meeting of the FAO Panel of Experts on Pesticide Residues in Food and the Environment and the WHO Core Assessment Group on Pesticide Residues* Geneva, Switzerland, 17–26 September 2019, pp. 1–662.

Kim, K.-H., Kabir, E., & Jahan, S. A. (2017). Exposure to pesticides and the associated human health effects. *Science of the Total Environment*, 575, 525–535. https://doi.org/10.1016/j.scitotenv.2016.09.009.

Li, H., Chang, Q., Bai, R., Lv, X., Cao, T., Shen, S., Liang, S., & Pang, G. (2021). Simultaneous determination and risk assessment of highly toxic pesticides in the market–sold vegetables and fruits in China: A 4–year investigational study. *Ecotoxicology and Environmental Safety*, 221, 112428. https://doi.org/10.1016/j.ecoenv.2021.112428.

Li, Z. (2022). Improved pesticide product labeling information for household lawn management: Recommended safe durations in support of minimizing children's exposure to pesticides. *ACS Chemical Health and Safety*, 29(2), 230–239. https://doi.org/10.1021/acs.chas.1c00092.

Mahmood, I., Imadi, S. R., Shazadi, K., Gul, A. & Hakeem, K. R. (2016). *Effects of pesticides on environment.* In: K. R. Hakeem et al. (eds.), *Plant, Soil and Microbes Volume 1: Implications in Crop Science*, Springer International Publishing, Switzerland, pp. 253–269. https:// doi.org/10.1007/978-3-319-27455-3_13.

Mooventhan, P., Baskaran, R. K. M., Kumar, J., & Kaushal, P. (2020). *Technical Bulletin on Current Status and Guidelines for Safe Use of Pesticides in Agriculture*. National Institute of Biotic Stress Management, Chhattisgarh, pp. 1–32.

Mostafalou, S. & Abdollahi, M. (2012). Concerns of environmental persistence of pesticides and human chronic diseases. *Journal Clinical and Experimental Pharmacology*, S5, e002. https://doi.org/10.4172/2161-1459.S5-e002.

Nayak, P. & Solanki, H. (2021). Pesticides and Indian agriculture – A review. *International Journal of Research – GRANTHAALAYAH*, 9(5), 250–263. https://doi.org/10.29121/granthaalayah.v9.i5.2021.3930.

Ndakidemi, B., Mtei, K., & Ndakidemi, P. A. (2016). Impacts of synthetic and botanical pesticides on beneficial insects. *Agricultural Sciences*, 7(6), 364–372. https://doi.org/10.4236/as.2016.76038.

Qin, G., Chen, Y., He, F., Yang, B., Zou, K., Shen, N., Zuo, B., Liu, R., Zhang, W., & Li, Y. (2021). Risk assessment of fungicide pesticide residues in vegetables and fruits in the mid–western region of China. *Journal of Food Composition and Analysis*, 95, 103663. https://doi.org/ 10.1016/j.jfca.2020.103663.

Rani, L., Thapa, K., Kanojia, N., Sharma, N., Singh, S., Grewal, A. S., Srivastav, A. L., & Kaushal, J. (2021). An extensive review on the consequences of chemical pesticides on human health and environment. *Journal of Cleaner Production*, 283, 124657. https://doi.org/10.1016/j.jclepro.2020.124657.

Sharma, K. K., Tripathy, V., Sharma, K., Gupta, R., Yadav, R., Devi, S., & Walia, S. (2022). Long-term monitoring of 155 multi-class pesticide residues in Indian vegetables and their risk assessment for consumer safety. *Food Chemistry*, 373(Pt B), 131518. https://10.1016/j.foodchem.2021.131518.

Tripathy, V., Sharma, K. K., Sharma, K., Gupta, R., Yadav, R., Singh, G, Aggarwal, A., & Walia, S. (2022). Monitoring and dietary risk assessment of pesticide residues in brinjal, capsicum, tomato, and cucurbits grown in Northern and Western regions of India. *Journal of Food Composition and Analysis*, 110, 104543. https://doi.org/10.1016/j.jfca.2022.104543.

WHO (World Health Organization) & International Programme on Chemical Safety. (2010). The WHO Recommended Classification of Pesticides by Hazard and Guidelines to Classification 2009. https://apps.who.int/iris/handle/10665/44271.

Yi, Y.-J., Joung, H.-J., Kum, J.-Y., Hwang, S., & Kim, M. S. (2020). Pesticide residues in vegetables and risk assessment for consumers in Korea during 2010–2014. *Food Additives and Contaminants: Part A*, 37(8), 1300–1313. https://doi.org/10.1080/19440049.2020.1769198.

Yu, R., Liu, Q., Liu, J. S., Wang, Q. C., & Wang, Y. (2016). Concentrations of organophosphorus pesticides in fresh vegetables and related human health risk assessment in Changchun, Northeast China. *Food Control*, 60, 353–360. https://doi.org/10.1016/j.foodcont.2015.08.013.

5 Drones in Crop Protection

A Paradigm Shift in Application Technology

Shekhar Varshney, Amitava Sanyal, and Pranjib K. Chakrabarty

"Farming is a profession of hope" – the said maxim turns equally true for conventional as well as modern crop protection practices. With automation, use of computer systems and sensors, data management and many other progressive farming tools, the traditional ways of pest management emanate from existing methods to digital technology. As a matter of fact, farming in India is swerving through a significant metamorphosis, and drones turned out to be a part of this success story. Drone technology is one of the magnificent innovations of present times, which not only takes care of labour-intensive, recurring and menial tasks of spray applications to control insect pests, diseases and weeds in agri-ecosystem but also contributes significantly to the country's economy with better farm outputs.

With the advent of drone technology providing enormous benefits and opportunities to Indian farmers, precision agriculture and smart farming in India would soon be talks of the towns. Drones can provide real-time imagery and sensor data from farm fields, which cannot be quickly accessed manually (Malveaux et al., 2014). Global Positioning System (GPS) and customisable apps for smartphones and tablets have provided improved flight durations, reliability, ease of use and the ability to better utilise cameras and other sensors needed for applying drones in agriculture and natural resources (Hogan et al., 2017). The use of drones is becoming so intimately interwoven within the fabrics of most sectors of developed and developing economies, and losing control of it would have potentially far-reaching if not calamitous implications (Matthew et al., 2021).

A drone or an unmanned aerial vehicle (UAV) according to the definition by the International Civil Aviation Organization (ICAO, 2011) is an aircraft operated without a human pilot on board. ICAO (2015) defined the following terms: (i) Remotely Piloted Aircraft System (RPAS) – which is a remotely piloted aircraft, its associated remote pilot station(s), the required command and control links and any other components as specified in the type design; and (ii) Remotely Piloted Aircraft (RPA) – an unmanned aircraft piloted from a remote pilot station. All these terms referto a drone that is remotely controlled by an operator on the ground or pre-programmed to fly specific routes (Nelson and Gorichanaz, 2019).

 DOI: 10.1201/9781003531944-5

5.1 GOVERNMENT'S INITIATIVE OF INTRODUCING DRONES IN CROP PROTECTION

The Ministry of Civil Aviation, in August 2021, published the notification on Drone Rules, 2021, after thoroughly contemplating and giving due consideration to the industry viewpoint on its previously published drafts. This can be considered as an exemplary approach of the Indian Government for bringing the emerging technology of drones in our country. The "Drone Rules, 2021" are intended to be farmers-friendly, industry-friendly, liberal and easy to be adopted by the end users. This is one of the most progressive approaches of the Central Government, wherein "ease of doing business" has been given due consideration by exhibiting less governance with a single-portal registration process and simplification of the requirements for registration of drones. The number of channels (steps) that the operator has to deal with has been reduced to a minimum level to avoid delays in granting registrations of drones. Smaller drones, called nano-drones up to 25 kg weight, have been exempted from almost all compliance requirements under the regulations as these are considered to be safer.

5.2 SALIENT FEATURES OF NEW DRONE POLICY

5.2.1 Green, Yellow and Red Zones

The Drone Policy, brought by the Central Government, has been designed to steer an era of multi-fold agricultural growth with safety and security as one of the prime priorities. The Directorate General of Civil Aviation (DGCA) has provided a "Digital Sky Platform", which is an online platform hosted by DGCA for various activities related to the management of unmanned aircraft system activities in India. "Drone" as per the Government's regulation means an "unmanned aircraft system". There are various zones primarily categorised as follows:

"Green Zone" means the airspace of defined dimensions above the land areas or territorial waters of India, upto a vertical distance of 400 feet or 120 m that has not been designated as a red zone or yellow zone in the airspace map for unmanned aircraft system operations and the airspace upto a vertical distance of 200 feet or 60 m above the area located between a lateral distance of 8 and 12 km from the perimeter of an operational airport.

"Yellow Zone" means the airspace of defined dimensions above the land areas or territorial waters of India within which unmanned aircraft system operations are restricted and shall require permission from the concerned air traffic control authority. The airspace above 400 feet or 120 m in the designated green zone and the airspace above 200 feet or 60 m in the area located between the lateral distance of 8 and 12 km from the perimeter of an operational airport shall be designated as the yellow zone.

"Red Zone" means the airspace of defined dimensions, above the land areas or territorial waters of India or any installation or notified port limits specified by the Central Government beyond the territorial waters of India, within which unmanned aircraft system operations shall be permitted only by the Central Government.

5.2.2 Categorisation and Classification

The Ministry of Civil Aviation has categorised Unmanned Aircraft System (Drones) into the following three categories: (i) aeroplane, (ii) rotorcraft and (iii) hybrid unmanned aircraft system. These three categories shall be further sub-categorised as (i) remotely piloted aircraft system, (ii) model remotely piloted aircraft system and (iii) autonomous unmanned aircraft system.

The Unmanned Aircraft System (Drones) shall, based on the maximum all-up weight including payload, be classified as:

a. Nano unmanned aircraft system: weighing ≤ 250 g
b. Micro unmanned aircraft system: weighing > 250 g, but ≤ 2 kg
c. Small unmanned aircraft system: weighing > 2 kg, but ≤ 25 kg
d. Medium unmanned aircraft system: weighing > 25 kg, but ≤ 150 kg
e. Large unmanned aircraft system: weighing > 150 kg.

5.2.3 Operational Process

The drones shall not be operated without first registering it on the digital sky platform and obtaining a unique identification number, unless exempted from the requirement of a unique identification number under these rules. The Ministry of Civil Aviation has further specified that a registration record shall be maintained by the DGCA of all such unmanned aircraft systems to which a unique identification number has been issued under these rules. It shall be the responsibility of the person operating an unmanned aircraft system to ensure that such unmanned aircraft system conforms to a valid-type certificate.

The Remote Pilot Certificate will be issued by an authorised remote pilot training organisation to any individual under rule 34. DGCA or any other entity authorised by DGCA will issue a "type certificate", which will certify that the unmanned aircraft system of a specific type meets the requirements specified under the rules. The unmanned aircraft system will be an aircraft that can be operated autonomously or can be operated remotely without a pilot on board. A particular type of unmanned aircraft system will be of the same basic design including all modifications thereto, except those modifications which result in a change in handling or flight characteristics. A Unique Identification Number will be issued for registering an unmanned aircraft system in India. There shall be an "Unmanned Aircraft System Traffic Management System" for providing traffic management for safe and expeditious flow of unmanned aircraft traffic and avoiding collision between manned and unmanned aircraft through the collaborative integration of persons, information, technology, facilities and services.

5.2.4 Registration Procedure

The application and registration procedure have been clearly illustrated by the Ministry of Civil Aviation vide its Drone Rules, 2021 and Drone (Amendment) Rules, 2022. Any person who intends to register and obtain a unique identification number for his unmanned aircraft system shall make an application in Form D-2 on the digital sky platform along with the fee and provide requisite details including the unique number of the type certificate to which such unmanned aircraft system conforms to. The digital sky platform shall verify the details and issue a unique identification number to the applicant. The unique identification number of an unmanned aircraft system shall be linked to the unique serial number provided by the manufacturer and the unique serial numbers of its flight control module and remote pilot station. No person shall replace the flight control module or remote pilot station of an unmanned aircraft system whose serial number is linked to such unmanned aircraft system's unique identification number, without first updating, on the digital sky platform, the unique serial number of the new flight control module or remote pilot station, within a period of 7 days from the date of such replacement or before operating such unmanned aircraft system, whichever is earlier.

5.2.5 Financial Assistance for Agricultural Services through Drones

The Mechanization & Technology Division of the Ministry of Agriculture & Farmers Welfare (MoAFW), in January 2022, brought drone services for the agriculture sector under the Sub-Mission on Agricultural Mechanization (SMAM). MoAFW has clearly stated in one of its important office memorandums that the use of drone technology has the potential to provide a sustainable solution in the context of enhancing the productivity as well as efficiency of the agriculture sector. Agriculture drone empowers the farmer to adapt to specific environments and make relevant choices in regulating crop health and crop treatments. Under SMAM, the following provisions are made additionally by MoAFW under agricultural services through drone application.

5.2.5.1 Drone Technology Demonstrations

5.2.5.1.1 Objectives

- To conduct large-scale demonstrations on cropping system-based approach to promote the use of drones in agriculture.
- To introduce improved/newly developed drone-based agricultural practices in crop production/post-harvest management system.
- Conduct demonstrations with the participation of Custom Hiring Centres (CHCs) and with the partnership of drone manufacturers.

5.2.5.1.2 Financial Assistance for Drone Demonstrations

Financial assistance for the purchase of drone for demonstration will be applicable till 31st March 2023, and also the grants to meet the contingent expenditure for demonstration shall be applicable till 31st March 2023 and only once in any field.

- Depending on the requirements, FMTTIs, Indian Council of Agricultural Research (ICAR) institutions, KVKs and SAUs will be provided financial assistance (100% grant in aid) for the purchase of agricultural drones and their attachments (actual cost of the agricultural drone and its attachments or Rs. 10.00 lakhs, whichever is lower) for conducting demonstrations on the farmers' fields.

- FPOs will be provided 75% grant-in-aid for the purchase of agricultural drones and their attachments for conducting demonstrations on the farmers' fields.
- The proposals for the purchase of drones and their attachments will be submitted by the implementing agencies and will be considered by the Executive Committee of the Scheme for allocation of funds.
- The implementing agencies that do not want to purchase the agricultural drones for demonstration will hire the drones to be demonstrated from the Hi-tech hubs/Manufacturers/Start-ups. In such cases, the Department will provide 100% assistance @ Rs. 6,000 per hectare to meet contingency expenditures such as charges towards hiring of Drones/Drone Pilots, Expenditure towards hands-on training and Miscellaneous expenditure such as expenditures towards transport, labour, publicity and printing of technical literature.
- The rate of grant to meet the contingent expenditure will be Rs. 3,000/- per hectare in the case of the implementing agencies that will be provided grants for the purchase of agricultural drones for demonstration.
- The FPOs will submit their proposals for demonstration through the State Governments. The proposals of the ICAR institutions, KVKs and SAUs may be routed through the Division of Agricultural Engineering, ICAR.

5.2.5.1.3 Implementing Agencies

Farm Machinery Training and Testing Institutes (FMTTIs); ICAR institutions, KVKs, FPOs and SAUs.

5.2.5.2 Financial Assistance for Establishment/Upgrading the Custom Hiring Centres/Hi-tech Hub

Financial assistance @ 40% of the basic cost of the agricultural drone and its attachments or Rs. 4.00 lakhs, whichever is less, will be available to the existing CHCs/ Hi-tech Hubs established by the Cooperative Societies of Farmers, FPOs and Rural entrepreneurs:

- For upgrading their CHCs/Hi-tech Hubs for providing agricultural services through drone application.
- The new CHCs or the Hi-tech Hubs that will be established by the Cooperative Societies of Farmers, FPOs and Rural entrepreneurs with financial assistance from SMAM, RKVY or any other schemes may also include drone as one of the machines along with other agricultural machines in the projects of CHCs/Hi-tech Hubs.
- These CHCs/Hi-tech Hubs will also be eligible for availing finance under the Agriculture Infrastructure Fund (AIF).
- The CHCs/Hi-tech Hubs upgraded/established for providing agricultural services through drone application will be supervised by the District

Agriculture Officer for its operation so that hiring charges are reasonable and affordable to farmers.

- Financial Assistance for Agriculture Graduate for establishment/Upgrading of CHCs/Hi-tech Hubs will be @ 50% of the basic cost of the agricultural drone and its attachments or Rs. 5.00 lakhs, whichever is less.

5.2.5.3 Implementing Agencies

State Governments are primarily the implementing agencies.

5.2.6 Compliances to Be Ensured by the Demonstrating Institutions and Drone Service Providers (for Agricultural Services)

- "Drone Rules 2021" vide GSR No. 589(E) dated 25th August 2021 and The Drone (Amendment) Rules, vide G.S.R. No. 108(E) dated 11th February 2022 published by the Ministry of Civil Aviation to regulate the use and operation of drones in India, which shall be complied for all drone applications.
- "Standard Operating Procedures (SOPs)" published by the Ministry of Agriculture & Farmers Welfare for use of drone application with pesticides for crop protection in agricultural, forestry, non-cropped areas, etc., and for spraying soil and crop nutrients, which shall be adhered to by all the demonstrating institutions and all the drone service providers for agricultural services

5.3 REGULATORY FRAMEWORK FOR DRONES FOR SPRAY APPLICATIONS OF CROP PROTECTION CHEMICALS

The SOPs for regulating spray applications of crop protection chemicals through drones have been published by The Ministry of Agriculture & Farmers Welfare, which cover important aspects, namely, statutory provisions, flying permissions, area distance restrictions, weight classification, overcrowded areas restriction, drone registration, safety insurance, piloting certification, operation plan and weather conditions.

5.3.1 Registration of Drone Application for Crop Protection Chemicals

Spray application of chemicals, especially for crop protection purposes, has primarily been done either through knapsack sprayers manually or through tractor-mounted sprayers in conventional farming systems, and in all kinds of spray applications, significant water volume is required, and farmers usually struggle in arranging labours for getting the spray done within the stipulated time of pest-attack in their crops. With the advent of drone technology and Government's focus on bringing faster approvals of pesticide applications through drones, there could be a transformation in every aspect associated with spray applications. Substantial reduction in time as well as quantity of water required for sprays; virtually, no exposure of spraying chemicals to spray operators (labours); better spray applications in cost-effective manner; no wastage of spray solutions; peace of mind for farmers and many more such aspects could be possible with this new drone technology. In fact, custom hiring and use of cooperatives while using drones will be encouraged by the Government. Such a great move of the Indian Government will certainly motivate business start-ups to come forward; however, considering the existing legislative framework, the use of drones in the crop protection sector is being regulated by the Ministry of Agriculture & Farmers Welfare.

Registration Committee (RC) vide its 438th meeting while granting provisional/interim approval for 2 years from the date of approval granted by the Central Government recommended pesticide formulations of insecticides, fungicides, plant growth regulators (PGRs), biopesticides and botanical pesticides to be used with drones. Subsequently, the Registration Committee vide its 439th meeting updated this list and finally published 475 provisionally approved pesticides to be used with drones (337 insecticides/fungicides/biopesticides/others + 138 mixtures of insecticides/fungicides = total 475) keeping the other conditions same as per previous RC decision (Table 5.1). However, this interim approval is subject to the intimation of the action plan for regulatory data generation with drone technology provided by the applicant to the Central Insecticide Board & Registration Committee (CIB&RC). It is pertinent to mention that registered pesticides approved for use as public health/household insecticides; stored grain insecticides/fumigants; rodenticides; herbicides, seed treatment products and nematicides are not considered feasible with drones as per the Government's decision.

TABLE 5.1
The Pesticide Formulations Registered by Central Insecticide Board & Registration Committee (CIB&RC) for Use through Drones (*ad-interim* Approval for 2 Years)

Sr. No.	Registered Formulations of Insecticides and Fungicides
1.	Abamectin 1.9% EC
2.	Acephate 75% SP
3.	Acephate 95% SG
4.	Acephate 97% DF
5.	Acteamiprid 20% SP
6.	Afidopyropen 50 g/L DC
7.	Alpha cypermethrin 5% WP
8.	Alpha cypermethrin 10% EC
9.	Alpha cypermethrin 10% SC
10.	Alphanaphthyl acetic acid 4.5% SL
11.	Aureofungin 46.15% SP
12.	Azadirachtin (neemproducts) 0.03% EC
13.	Azadirachtin (neemproducts) 0.1% EC

(*Continued*)

TABLE 5.1 (*Continued*)
The Pesticide Formulations Registered by Central Insecticide Board & Registration Committee (CIB&RC) for Use through Drones (*ad-interim* Approval for 2 Years)

Sr. No.	Registered Formulations of Insecticides and Fungicides
14.	Azadirachtin (neemproducts) 0.15% EC
15.	Azadirachtin (neemproducts) 0.3% EC
16.	Azadirachtin (neemproducts) 1%EC
17.	Azadirachtin (neemproducts) 5%EC
18.	Azoxystrobin 23% SC
19.	Bendiocarb 8% WP
20.	Benfuracarb 40% EC
21.	Benzpyrimoxam 10% SC
22.	Beta-cyfluthrin 2.45% SC
23.	Bifenazate 22.6% SC
24.	Bifenazate 50% WP
25.	Bifenthrin 8% SC
26.	Bifenthrin 8.8% CS
27.	Bifenthrin 10% EC
28.	Bifenthrin 10% WP
29.	Bitertano l25% WP
30.	Buprimate 26.7% EC
31.	Buprofezin 25% SC
32.	Captan 50% WP
33.	Captan 50% WG
34.	Captan 75% WS
35.	Captan 75% WP
36.	Carbendazim 46.27% SC
37.	Carbendazim 50% WP
38.	Carbosulfan 25% EC
39.	Carboxin 75% WP
40.	Carpropamid 27.8% SC
41.	Cartap Hydrochloride 50% SP
42.	Cartap Hydrochloride 75% SG
43.	Chlorantraniliprole 18.5% SC
44.	Chlorantraniliprole 35% WG
45.	Chlorfenapyr 10% SC
46.	Chlorfluazuron 5.4% EC (5%w/v)
47.	Chlormequat Chloride 50% SL
48.	Chlorothalonil 75% WP
49.	Chlorpyriphos 20% EC
50.	Chlorpyriphos 20% CS
51.	Chlorpyriphos 50% EC
52.	Chlorpyriphos 75% WG

(*Continued*)

TABLE 5.1 (*Continued*)
The Pesticide Formulations Registered by Central Insecticide Board & Registration Committee (CIB&RC) for Use through Drones (*ad-interim* Approval for 2 Years)

Sr. No.	Registered Formulations of Insecticides and Fungicides
53.	Chlorpyriphos Methyl 40% EC
54.	Chromafenozide 80% WP
55.	Clothianidin 50% WG
56.	Copper hydroxide 46.1% WG
57.	Copper hydroxide 77% WP
58.	Copper oxychloride 50% WP
59.	Copper oxychloride 50% WG
60.	Copper oxychloride 56% OP
61.	Copper sulphate 2.62% SC
62.	Cyantraniliprole 10.26% OD
63.	Cyazofamid 34.5% SC
64.	Cyenopyrafen 30.0% w/w SC
65.	Cyflufenamide 5% EW
66.	Cyflumetofen 20% SC
67.	Cyfluthrin 5% EW
68.	Cyfluthrin 10% WP
69.	Cymoxanil 50% WP
70.	Cypermethrin 0.1% Aqueous
71.	Cypermethrin 10% EC
72.	Cypermethrin 25% EC
73.	Cyphenothrin 5% EC
74.	Decamethrin (Deltamethrin) 1.25% ULV
75.	Decamethrin (Deltamethrin) 1.8% EC
76.	Decamethrin (Deltamethrin) 2% EW
77.	Decamethrin (Deltamethrin) 2.8% EC
78.	Decamethrin (Deltamethrin) 11% EC
79.	Diafenthiuron 47.8% SC
80.	Diafenthiuron 50% WP
81.	Dichloropropene and Dichloropropane mixture (DDMixture)
82.	Dicofol 18.5% EC
83.	Difenoconazole 3% WS
84.	Difenoconazole 25% EC
85.	Diflubenzuron 25% WP
86.	Dimethoate 30% EC
87.	Dimethomorph 50% WP
88.	Dinocap 48% EC
89.	Dinotefuran 20% SG
90.	Dithianon 75% WP
91.	Dodine 40% SC

(*Continued*)

TABLE 5.1 (*Continued*)
The Pesticide Formulations Registered by Central Insecticide Board & Registration Committee (CIB&RC) for Use through Drones (*ad-interim* Approval for 2 Years)

Sr. No.	Registered Formulations of Insecticides and Fungicides
92.	Dodine 65% WP
93.	Edifenphos 50% EC
94.	Emamectin Benzoate 1.9% EC
95.	Emamectin Benzoate 5% SG
96.	Ethephon 39% SL
97.	Ethion 50% EC
98.	Ethofenprox (Etofenprox) 10% EC
99.	Etoxazole 10% SC
100.	Fenazaquin 10% EC
101.	Fenazaquin 18.3% SC
102.	Fenitrothion R40% WP
103.	Fenitrothion 20% OL (banned in agriculture except for locust in scheduled dessert area and public health)
104.	Fenobucarb (BPMC) 50% EC
105.	Fenpropathrin 10% EC
106.	Fenpropathrin 30% EC
107.	Fenpyroximate 5% EC
108.	Fenpyroximate 5% SC
109.	Fenvalerate 0.4% DP
110.	Fenvalerate 20% EC
111.	Fipronil 0.6% WG
112.	Fipronil 2.92% EC
113.	Fipronil 5% SC
114.	Fipronil 18.87% SC
115.	Fipronil 80% WG
116.	Flonicamide 50% WG
117.	Flubendiamide 20% WG
118.	Flubendiamide 39.35% SC
119.	Flufenoxuro n10% DC
120.	Flumite 20% SC
121.	Fluopyram 34.48% SC
122.	Flupyradiflurone1 7.9% w/wSL
123.	Flupyrimin 10% SC
124.	Flusilazole 40% EC
125.	Fluvalinate 25% EC
126.	Fluxametamide 10% EC
127.	Forchlorfenuron(CPPU) 0.1% L
128.	Forchlorfenuron 0.12% EC
129.	Fosetyl-Al 80% WP

(*Continued*)

TABLE 5.1 ***(Continued)***
The Pesticide Formulations Registered by Central Insecticide Board & Registration Committee (CIB&RC) for Use through Drones (*ad-interim* Approval for 2 Years)

Sr. No.	Registered Formulations of Insecticides and Fungicides
130.	Gibberellic Acid 0.001% L
131.	Gibberellic Acid 0.186% SP
132.	Gibberellic Acid 0.45% SL
133.	Gibberellic Acid 40% WSG
134.	Hexaconazole 2% SC
135.	Hexaconazole 5% SC
136.	Hexaconazole 5% EC
137.	Hexaconazole 75% WG
138.	Hexythiazox 5.45% EC
139.	Hydrogen Cyanamide 49% AS
140.	Hydrogen Cyanamide 50% SL
141.	Imidacloprid 17.1% SL
142.	Imidacloprid 17.8% SL
143.	Imidacloprid 30.5% SC
144.	Imidacloprid 70% WG
145.	Indoxacarb 14.5% SC
146.	Indoxacarb 15.8% EC
147.	Iprobenfos(Kitazin) 48% EC
148.	Iprodione 50% WP
149.	Isoprothiolane 40% EC
150.	Kasugamycin 3% SL
151.	Kresoxim-methyl 44.3% (500g/L) SC
152.	Lambda-cyhalothrin 2.43% CS
153.	Lambda-cyhalothrin 2.5% EC
154.	Lambda-cyhalothrin 4.9% CS
155.	Lambda-cyhalothrin 5% EC
156.	Lambda-cyhalothrin 9.7% CS
157.	Lambda-cyhalothrin 10% WP
158.	Lambda-cyhalothrin 22.8% CS
159.	LimeSulphur 22% SC
160.	Lufenuron 5.4% EC
161.	Malathion 25% WP
162.	Malathion 50% EC
163.	Malathion 96% ULV
164.	Mancozeb 35% SC
165.	Mancozeb 75% WG
166.	Mancozeb 75% WP
167.	Mandipropamid 23.4% SC
168.	MepiquatChloride 5% AS

(Continued)

TABLE 5.1 (*Continued*)
The Pesticide Formulations Registered by Central Insecticide Board & Registration Committee (CIB&RC) for Use through Drones (*ad-interim* Approval for 2 Years)

Sr. No.	Registered Formulations of Insecticides and Fungicides
169.	Meptyldinocap 35.7% EC
170.	Metaflumizone 22% SC
171.	Metalaxyl 35% WS
172.	Metalaxyl–M 31.8% ES
173.	Methoxyfenazide 21.8% SC
174.	Metiram 70% WG
175.	Metrafenone 500 g/L SC
176.	Milbemectin 1% EC
177.	Monocrotophos R15% w/w SG
178.	Monocrotophos R36% SL
179.	Myclobutanil1 0% WP
180.	Novaluron 10% EC
181.	Novaluron 8.8% SC
182.	Oxathiapipron 10.1% OD
183.	Oxycarboxin 20% EC
184.	Oxydemeton-methy 25% EC
185.	Paclobutrazol 40% SC
186.	Paclobutrazol 23% SC
187.	Penconazole 10% EC
188.	Pencycuron 22.9% SC
189.	Permethrin 5% SG
190.	Permethrin 25% EC
191.	Phenthoate 50% EC
192.	Phosalone 35% EC
193.	Picoxystrobin 22.52% SC
194.	Polyoxin D Zinc salt 5% SC
195.	Prochloraz 39.6% EC
196.	Profenophos 50% EC
197.	Prohexadione calcium 10% WG
198.	Propergite 57% EC
199.	Propetamphos 20% EC
200.	Propiconazole 25% EC
201.	Propineb 70% WP
202.	Pymetrozine 50% WG
203.	Pyraclostrobin 100 g/L CS
204.	Pyraclostrobin 20% WG
205.	Pyrethrum 2.0% EC
206.	Pyridaben 20% WP
207.	Pyridalyl 10% EC

(*Continued*)

TABLE 5.1 (*Continued*)
The Pesticide Formulations Registered by Central Insecticide Board & Registration Committee (CIB&RC) for Use through Drones (*ad-interim* Approval for 2 Years)

Sr. No.	Registered Formulations of Insecticides and Fungicides
208.	Pyriproxifen 10% EC
209.	Quinalphos 20% AF
210.	Quinalphos 25% EC
211.	Sodium para nitro phenolate 0.3% SL
212.	Sodium para nitro phenolate 1.8% SL
213.	Spinetoram 11.7% SC
214.	Spinosad2.5% SC
215.	Spinosad 45% SC
216.	Spiromesifen 22.9% SC
217.	Spirotetramat 15.31% OD
218.	Sulfoxaflor 21.8% SC
219.	Sulphur 40% SC
220.	Sulphur 52% Flowable
221.	Sulphur 55.16% SC(800 gm/L)
222.	Sulphur 80% WG
223.	Sulphur 80% WP
224.	Tebuconazole 2% DS
225.	Tebuconazole 25% WG
226.	Tebuconazole 25.9% EC
227.	Tebuconazole 38.39% SC
228.	Tetraconazole 11.6% w/w (12.5% w/v) SL
229.	Tetraconazole 3.8% EW
230.	Tetraniliprole 18.18% SC
231.	Thiacloprid 21.7% SC
232.	Thifluzamide 24% SC
233.	Thiocyclam hydrogen oxalate 50% SP
234.	Thiodicarb 75% WP
235.	Thiamethoxam 25% WG
236.	Thiomethoxam 70% WS
237.	Thiomethoxam 75% SG
238.	Thiophanate-methyl 70% WG
239.	Thiophanate-methyl 70% WP
240.	Thiram 75% WS
241.	Tolfenpyrad 15% EC
242.	Triadimefon 25% WP
243.	Tricontanol 0.05% EC
244.	Tricontanol 0.1% EW
245.	Tricyclazole 70% WG
246.	Tricyclazole 75% WP

(Continued)

TABLE 5.1 (*Continued*)
The Pesticide Formulations Registered by Central Insecticide Board & Registration Committee (CIB&RC) for Use through Drones (*ad-interim* Approval for 2 Years)

Sr. No.	Registered Formulations of Insecticides and Fungicides
247.	Triflumezopyrim 10% SC
248.	Triflumizole 42.14% SC
249.	Validamycin 3% L
250.	Zineb 75% WP
251.	Ziram 27% SC
252.	Ziram 80% WP
253.	*Ampelomyces quisqualis* 2.0% A.S. (IPL/AQ/06)
254.	*Ampelomyces quisqualis* 2.0% WP (Aq-1MTCC-5683)
255.	*Bacillus subtilis* 1.0% WP (CPBT15TCC-5527)
256.	*Bacillus subtilis* 1.5% LF (Bs-1, MTCC25072)
257.	*Bacillus thuringiensis var. galleriae* 1.3% FC
258.	*Bacillus thuringiensis var. sphaericus* 1.3% FC
259.	*Bacillus thuringiensis kurstaki* 0.5% WP Serotype 3a, 3b, 3c (DORBt-1, B-01118)
260.	*Bacillus thuringiensis kurstaki* 5% WP
261.	*Bacillus thuringiensis kurstaki* 2.5% AS Serotype 3a, 3b, 3c
262.	*Bacillus thuringiensis kurstaki* 3.5% ES Serotype 3a, 3b, 3c
263.	*Bacillus thuringiensis var.israelensis* 0.5% AS
264.	*Bacillus subtilis* 1.5% A.S. (BIL, BS-168)
265.	*Bacillus subtilis* 1.5% A.S. (KTSBMTCC5786)
266.	*Bacillus subtilis* 1.15% A.S
267.	*Bacillus subtilis* 2.0% A.S. (IPL/BS/09, MTCC-5728)
268.	*Bacillus thuringiensis var. israelensis* 5% AS, B-17 serotype H-14 (VCRC-B-17, MTCC-5596)
269.	*Bacillus thuringiensis var. israelensis* 0.5% WP
270.	*Bacillus thuringiensis var. israelensis* 5% WP
271.	*Bacillus thuringiensis var. israelensis* 12% AS
272.	*Bacillus thuringiensis var. krustaki,* serotype H-39, 3B, starin Z52
273.	*Beauveria basiana* 1.15% WP (Bb-1, MTCCNo. 5171)
274.	*Beauveria basiana* 1.15% WP (BB-5372)
275.	*Beauveria basiana* 1.15% WP (ICAR-Umaim, NAIMCC-03045)
276.	*Beauveria basiana* 1.50% LF (Bb-1, MTCCNo.5171)
277.	*Beauveria bassiana* 1.00% WP (NBRI-9947)
278.	*Beauveria bassiana* 1.00% WP (SVBPU/CSP/BB-10/ ITCC-7520)
279.	*Beauveria bassiana* 2.0% A.S. (IPL/BB/MI/01, 6897)
280.	*Beauveria bassiana* 5.0% SC (NBAI, ITCC7102)
281.	*Beauveria bassiana* 5.0% WP (IARI, 7353)

(*Continued*)

TABLE 5.1 (*Continued*)
The Pesticide Formulations Registered by Central Insecticide Board & Registration Committee (CIB&RC) for Use through Drones (*ad-interim* Approval for 2 Years)

Sr. No.	Registered Formulations of Insecticides and Fungicides
282.	*Beauveria bassiana* 1.0% W.P. (SVBPU/CSP/Bb-10, ITCC-7520)
283.	*Beauveria bassiana* 1.15% WP (BB-ICAR-RJP, MCC-1022)
284.	*Hirsutella thompsonii* 2.0% A.S. (IPL/HT/01)
285.	*Metarhizium anisopliae* 2.0% SC (KSCL/Ma-59 ITCC-7058)
286.	*Metarhizium anisopliae* 1.15% WP (Ma-1MTCC5173)
287.	*Metarhizium anisopliae* 2.0% A.S (IPL/KC/44 ITCC-6895)
288.	*Metarhizium anisopliae* 1.15% WP (Ma-1MTCC-5173)
289.	*Metarhizium anisopliae* 1.50% LF(Ma-1MTCC5173)
290.	*Metarhizium anisopliae* 5.0% SC(NBAII,7103)
291.	*Metarhizium anisopliae* 1.00% WP(CPB/PSP-T26MTCC5699)
292.	*Metarhizium anisopliae* 1.00% WP (IPL/KC/44ITCC-6895)
293.	*Metarhizium anisopliae* 1.15% WP (NAIMCC-F-03037)
294.	*NPV of Helicoverpa armigera* 0.43% AS
295.	*NPV of Helicoverpa armigera* 0.5% AS
296.	*NPV of Helicoverpa armigera* 1% WP
297.	*NPV of Spodoptera litura* 2.0% AS
298.	*NPV of Spodoptera litura* 0.5% AS
299.	*Paecilomyces lilacinus* 1.0% S.P. (IIHR-PL-2ITCC-6887)
300.	*Paecilomyces lilacinus* 1.0% W.P. (IIHR-PL-2 ITCC-6887)
301.	*Paecilomyces lilacinus* 1.5% LF
302.	*Pseudomonas fluorescens* 0.5% W.P. (TNAUPF-1ITCCBE-0005)
303.	*Pseudomonas fluorescens* 1.75% WP (PF-1MTCC No. 5671)
304.	*Pseudomonas flourescens* 1.5% LF (PF-1, MTCC No. 5671)
305.	*Pseudomonas fluorescens* 0.5% W.P. (BIL-331MTCC-5866)
306.	*Pseudomonas fluorescens* 1.0% W.P. (IPL/PS-01, MTCC-5727)
307.	*Pseudomonas fluorescens* 1.0% W.P. (IIHR-PF-2, ITCCB-0034)
308.	*Pseudomonas fluorescens* 1.5% A.S. (AS-AAIF, MCC2539)
309.	*Pseudomonas fluorescens* 1.5% WP
310.	*Pseudomonas fluorescens* 2.0% A.S. (IPL/PS-01)
311.	*Trichoderma harzianum* 1.0% WP (Th3, ITCC-5593)
312.	*Trichoderma ressei* 3.0% WP (CSR-T-3, NAIMCC-SF-0030)
313.	*Trichoderma viride* 2% W.P. (BHU, NAIMCC-F-02976)
314.	*Trichoderma viride* 0.50% W.P.
315.	*Trichoderma viride* 1.0% AS (TV-AAU-RJP, MCC-1013)
316.	*Trichoderma viride* 1.0% W.P. (Tv-1, ITCC-6914)
317.	*Trichoderma viride* 1.00% W.P. (BIL-198)
318.	*Trichoderma viride* 1.00% W.P. (IPL/VT/101)
319.	*Trichoderma viride* 1.00% W.P. (KAU, MTCC-5694)

(*Continued*)

TABLE 5.1 (*Continued*)
The Pesticide Formulations Registered by Central Insecticide Board & Registration Committee (CIB&RC) for Use through Drones (*ad-interim* Approval for 2 Years)

Sr. No.	Registered Formulations of Insecticides and Fungicides
320.	*Trichoderma viride* 1.00% W.P. (T-14)
321.	*Trichoderma viride* 1.5% LF (Tv-1, MTCC-5170)
322.	*Trichoderma viride* 5% LF (NAIM-F-03034)
323.	*Trichoderma viride* 1.50% W.P .(IIHR-Tv-5, ITCC-6889)
324.	*Trichoderma viride* 5% WP
325.	*Trichoderma harzianum* 5.0% WP
326.	*Trichoderma harzianum* 5.0% SC (ITCC-7111)
327.	*Trichoderma harzianum* 0.5% WS (T-39)
328.	*Trichoderma harzianum* 1.0% W.P. (IIHR-Th-2, ITCC-6888)
329.	*Trichoderma harzianum* 2.0% A.S. (IPL/VT/102 6893)
330.	*Trichoderma harzianum* 2.0% WP (NBRI-1055)
331.	*Verticilium lecanii* 1.50% LF (Vl-1, MTCC No. 5172)
332.	*Verticillium chlamydosporium* 1.0% W.P. (IIHR-VC-3, ITCC6898)
333.	*Verticillium lecanii* 2.0% A.S. (IPL/VL/05)
334.	*Verticillium lecanii* 5.0% SC (NBAII2638)
335.	*Verticillium lecanii* 1.15% WP (AAI,NAIMCC-F-03007)
336.	*Verticillium lecanii* 1.15% WP (AS-MEGH-VL, MCC-1028)
337.	*Verticillium lecanii* 1.15% WP (Vl-1, MTCC No. 5172)

Registered Formulations of Combination of Pesticides

Sr. No.	Formulations Registered
	A. Insecticides
1.	Acephate 25% + Fenvalerate 3% EC
2.	Acephate 50% + Bifenthrin 10% WDG
3.	Acephate 50% + Fipronil 5% WDG
4.	Acephate 50% + Imidacloprid 1.8% SP
5.	Acetamiprid 0.4% + Chlorpyriphos 20% EC
6.	Acetamiprid 1.1% + Cypermethrin 5.5% EC
7.	Betacyfluthrin 8.49% + Imidacloprid 19.81% OD
8.	Bifenthrin 3% + Chlorpyriphos 30% EC
9.	Buprofezin 20% + Acetamiprid 2% WP
10.	Buprofezin 15% + Acephate 35% WP
11.	Buprofezin 20% + Acephate 50% WP
12.	Buprofezin 22.0% + Fipronil 3.0% SC
13.	Buprofezin 23.1% + Fipronil 3.85% SC
14.	Buprofezin 9.0% + Acephate 24.0% WP
15.	Cartaphydrochloride 50% + Buprofezin 10% WP

(*Continued*)

TABLE 5.1 (*Continued*)
The Pesticide Formulations Registered by Central Insecticide Board & Registration Committee (CIB&RC) for Use through Drones (*ad-interim* Approval for 2 Years)

Registered Formulations of Combination of Pesticides

Sr. No.	Formulations Registered
16.	Chlorantraniliprole 4.3% + Abamectin 1.7% SC
17.	Chlorantraniliprole 8.8% + Thiamethoxam 17.5% SC
18.	Chlorantraniliprole 9.3% + Lambdacyhalothrin 4.6% ZC
19.	Chloropyriphos 16% + Alphacypermethrin 1% EC
20.	Chlorpyriphos 50% + Cypermethrin 5% WG
21.	Chlorpyriphos 50% + Cypermethrin 5% EC
22.	Cyclanilide 2.10% + MepiquatChloride 8.40% SC
23.	Cypermethrin 10% + Indoxacarb 10% SC
24.	Cypermethrin 3% + Quinalphos 20% EC
25.	Deltamethrin 0.72% + Buprofezin 5.65% EC
26.	Diafenthiuron 40.1% + Acetamiprid 3.9% WP
27.	Diafenthiuron 30% + Pyriproxifen 8% SE
28.	Difenthiuron 47.0% +Bifenthrin 9.4% SC
29.	Dinotefuran 15% + Pymetrozine 45% WG
30.	Dinotefuran 4% + Acephate 50% SG
31.	Emamectin benzoate 1.5% + Profenofos 35% WDG
32.	Emamectin Benzoate 1.5% + Fipronil 3.5% SC
33.	Emamectin Benzoate 5% + Lufenuron 40% WG
34.	Emamectin Benzoate 3% + Thiamethoxam 12% WG
35.	Ethion40% + Cypermethrin 5% EC
36.	Ethiprole 40% + Imidacloprid 40% WG
37.	Fenazaquin 10% + Bifenthrin 4% EC
38.	Fenobucarb 20% + Buprofezin 5% SE
39.	Fipronil 40% + Imidacloprid 40% WG
40.	Fipronil 5% + Buprofezin 20%
41.	Fipronil 7% + Hexythiazox 2% w/w SC
42.	Fipronil 4% + Acetamiprid 4% w/w SC
43.	Fipronil 4% + Thiamethoxam 4% w/w SC
44.	Flubendiamide 19.92% + Thiacloprid 19.92% SC
45.	Flubendiamide 3.50% + Hexaconazole 5% WG
46.	Flubendiamide 4% + Buprofezin 20% SC
47.	Flubendiamide 7.5% + Kresoxim methyl 37.5% SC
48.	Flubendiamide 8.33% + Deltamethrin 5.56% SC
49.	Hexythiazox 3.5 + Diafenthiuron 42% WDG
50.	Imidacloprid 21% + Beta-cyfluthrin 10.5% SC
51.	Imidacloprid 6.0% + Lambdacyhalothrin 4.0% SL

(*Continued*)

TABLE 5.1 (*Continued*)
The Pesticide Formulations Registered by Central Insecticide Board & Registration Committee (CIB&RC) for Use through Drones (*ad-interim* Approval for 2 Years)

Registered Formulations of Combination of Pesticides

Sr. No.	Formulations Registered
52.	Indoxacarb 14.5% + Acetamiprid 7.7% SC
53.	Indoxacarb 5% + Fipronil 5% SC
54.	Isoprothiolane 28% + Fipronil 5% EC
55.	Novaluron 5.25% + Emamectinbenzoate 0.9% w/w SC
56.	Novaluron 5.25% + Indoxacarb 4.5% SC
57.	Phenthoate 45% + Cypermethrin 6% EC
58.	Profenofos 40% + Cypermethrin 4% EC
59.	Profenofos 40% + Fenpyroximate 2.50% EC
60.	Propargite 42% + Hexythiazox 2% EC
61.	Propargite 50% + Bifenthrin 5% SE
62.	Pyriproxyfen 10% + Bifenthrin 10% EC
63.	Pyriproxyfen 5% + Fenpropathrin 15% EC
64.	Pyriproxyfen 5.0% + Diafenthiuron 25% SE
65.	Spirotetramat 11.01% + Imidacloprid 11.01% SC
66.	Thiamethoxam 12.6% + Lambda-cyhalothrin 9.5% ZC
	B. Fungicides
1.	Ametoctradin 27% + Dimethomorph 20.27% SC
2.	Azoxysrobin 120 g/L + Tebuconazole 240 g/L SC
3.	Azoxystrobin 11.0% + Tebuconazole 18.3% SC
4.	Azoxystrobin 12.5% + Tebuconazole 12.5% SC
5.	Azoxystrobin 18.2% + Cyproconazole 7.3% SC
6.	Azoxystrobin 18.2% + Difenoconazole 11.4% SC
7.	Azoxystrobin 4.8% + Chlorothalonil 40.0% SC
8.	Azoxystrobin 7.1% + Propiconazole 11.9% SE
9.	Azoxystrobin 8.3% + Mancozeb 66.7% WG
10.	Azoxystrobin 11.5% + Mancozeb 30.0% WP
11.	Azoxystrobin 16.7% + Tricyclazole 33.3% SC
12.	Benalaxyl 8.0% + Mancozeb 65% WP
13.	Benalaxyl-M 4.0% + Mancozeb 65.0% WP
14.	Boscalid 25.2% + Pyraclostrobin 12.8% WG
15.	Captan 70% + Hexaconazole 5% WP
16.	Carbendazim 12% + Mancozeb 63% WP
17.	Carbendazim 12% + Mancozeb 63% WS
18.	Carbendazim 25% + Flusilazole 12.5% SE
19.	Carbendazim 25% + Mancozeb 50% WS
20.	Carfentrazone Ethyl 20% + Sulfosulfuron 25% WG
21.	Chlorothalonil 40% + difenoconazole 4% w/w SC

(*Continued*)

TABLE 5.1 (*Continued*)
The Pesticide Formulations Registered by Central Insecticide Board & Registration Committee (CIB&RC) for Use through Drones (*ad-interim* Approval for 2 Years)

Registered Formulations of Combination of Pesticides

Sr. No.	Formulations Registered
22.	Copper Sulphate 47.15% + Mancozeb 30% WDG
23.	Cymoxanil 8% + Mancozeb 64% WP
24.	Difenconazole 10% + Mancozeb 50% WDG
25.	Dimethomorph 12% + Pyraclostrobin 6.7% WG
26.	Famoxadone 16.6% + Cymoxanil 22.1% SC
27.	Fenamidone 10% + Mancozeb 50% WDG
28.	Fenamidone 4.44% + Fosetyl-Al 66.66% WDG
29.	Fluopicolide 5.56% + Propamocarbhydrochloride 55.6% SC
30.	Fluopyram 17.7% + Tebuconazole 17.7% SC
31.	Fluxapyroxad 167 g/L + Pyraclostrobin 333 g/L SC
32.	Fluxapyroxad 250 g/L + Pyraclostrobin 250 g/L SC
33.	Fluxapyroxad 62.5 g/L + Epoxiconazole 62.5g/L EC
34.	Fluxapyroxad 75g/L + Difenoconazole 50g/L SC
35.	Hexaconazole 4% + Carbendazim 16% w/w SC
36.	Hexaconazole 4% + Zineb 68% WP
37.	Hexaconazole 5% + Validamycin 2.5% SC
38.	Iprovalicarb 5.5% + Propineb 61.25% WP
39.	Iprodione 25% + Carbendazim 25% WP
40.	Kasugamycin 5% + Copper Oxychloride 45% WP
41.	Kasugamycin 6% + Thifluzamide 26% w/v SC
42.	Kresoxim methyl 15% + Chlorothalonil 56% WG
43.	Kresoximmethyl 18% + Mancozeb 54% w/w WP
44.	Kresoximmethyl 40% + Hexaconazole 8% w/w WG
45.	Mancozeb 40% + Azoxytrobin 7.0% w/w OS
46.	Mancozeb 50% + Thiophanatemethyl 25% WG
47.	Mandipropamid 5.0% + Mancozeb 60.0% WG
48.	Metalaxyl–M 3.3% + Chlorothalonil 33.1% SC
49.	Metalaxyl-M 8% + Mancozeb 64% WP
50.	Metalaxyl-M 4% + Mancozeb 64% WP
51.	Metiram 44% + Dimethomorph 9% WG
52.	Metiram 55% + Pyraclostrobin 5% WG
53.	Picoxystrobin 6.78% + Tricyclazole 20.33% SC
54.	Picoxystrobin 7.05% + Propiconazole 11.7% SC
55.	Prochloraz 24.4% + Tebuconazole 12.1% EW
56.	Prochloraz 34.8% + Propiconazole 7.8% EC
57.	Prochloraz 5.7% + Tebuconazole 1.4% ES
58.	Prochloraz 23.5% + Tricyclazole 20% SE

(*Continued*)

TABLE 5.1 (*Continued*)
The Pesticide Formulations Registered by Central Insecticide Board & Registration Committee (CIB&RC) for Use through Drones (*ad-interim* Approval for 2 Years)

Registered Formulations of Combination of Pesticides

Sr. No.	Formulations Registered
59.	Propiconazole 10.7% + Tricyclazole 34.2% SE
60.	Propiconazole 13.9% + Difenoconazole 13.9% EC
61.	Propineb 54.2% + Tricyclazole 15.0% WP
62.	Pyroclostrobin 133g/L + Epoxiconazole 50 g/L (w/v) SE
63.	Streptomycin + Tetracycline (90 + 10)
64.	Tebuconazole 10% + Sulphur 65% WG
65.	Tebuconazole 15% + Zineb 57% WDG
66.	Tebuconazole 50% + Trifloxystrobin 25% WG
67.	Tebuconazole 6.7% + Captan 26.9% SC
68.	Tricyclazole 18% + Mancozeb 62% WP
69.	Tricyclazole 18.0 + Tebuconazole 14.4% SC
70.	Tricyclazole 20.4% + Azoxystrobin 6.8% SC
71.	Tricyclazole 45% + Hexaconazole 10% WG
72.	Flubendiamide 3.5% + Hexaconazole 5% WG

Source: MoM of 439th Registration Committee (RC) Meeting, CIB&RC, DPPQS, DoA&FW, MoA&FW, 2022. OM dated 18th April 2022 further extended for one year till 18th April 2025 by OM dated 25th April 2024.

RC, while granting interim approvals of pesticides with drones, clarified that it shall be the sole liability/responsibility of the applicant/registrant/drone operator to adhere to the requirements as prescribed in the SOP for the use of drone application with pesticides for crop protection and for spraying soil and crop nutrients in agricultural, forestry, non-cropped areas, etc. RC also made it mandatory to follow guidelines and safety measures prescribed by the CIB&RC and envisaged under Rule 43 of the Insecticide Rules, 1971 ensuring that no damage/crop loss is caused/occurred which is detrimental to the interest of the farmers. It was further specified by the RC that during the interim period, the registrants should generate the requisite data on priority as per RC-approved guidelines for use with drone and get their pesticides endorsed with the approval of RC. RC recommended that for proper use of approved pesticides with drones, State Agricultural Universities (SAUs), State Agriculture Departments, KVKs and CIPMCs should organise field trainings for the applicant/registrant/drone operator on the safe use of drones.

The aerial spraying operations are regulated as per statutory provisions under the Insecticide Act, 1968, and rules framed therein. Some of the provisions applicable for drones w.r.t. spray application of crop protection chemicals are (i) marking of the area shall be the responsibility of the operators; (ii) the operators shall use only approved insecticides and their formulations at approved concentrations and height; (iii) washing decontamination and first aid facilities shall be provided by the operators; (iv) all aerial operations shall be notified to be public not less than 24 hours in advance through competent authorities (written intimation to be sent to the Executive Officer of the respective Gram Panchayat and Panchayat Samiti, as well as the concerned Agriculture Officer of the area, at least 24 hours in advance; (v) animals and persons not connected with the operations shall be prevented from entering such areas for a specific period; the pilots shall undergo specialisation training including clinical effects of the insecticides.

In addition to the above, the precautions and prerequisites shall also be adhered to before application, during application and post application. As per SOP, some critical parameters for drone-based application of crop protection chemicals are summarised below:

a. **Statutory requirements for drones:** Among drone-related parameters, only DGCA-approved drone shall be permitted to carry out agricultural sprays. The other drone parameters can be referred from the SOP published by The Ministry of Agriculture & Farmers Welfare.
b. **Statutory requirements for crop protection chemicals:** The insecticides, fungicides, herbicides, plant growth regulators, etc., approved by the CIB&RC shall be allowed. Their compatibility with the drone spray system shall be established prior to the desired dilution. The minimum dilution shall be decided based on the fulfilment of the above-mentioned requirements and ensuring satisfactory coverage of sprayed inputs horizontally as well as vertically. The dilution with either clean water or with suitable ingredients shall be done as has been approved by CIB&RC.
c. **Environment limitations:** The drone-based spray may be permitted under conducive weather conditions to get the best results in terms of appropriate wind speed, temperature and relative humidity, etc.
d. **Pilot trainings:** Only DGCA-certified pilots shall be permitted to fly the agri-drones. The training modules devised by the National Institute of Plant Health & Management (NIPHM), Hyderabad, shall be made mandatory for these pilots.
e. **Drift management parameters:** Spray height above crop canopy, speed of the drones, suitable nozzles and droplet size, buffer zone, spray schedule and other guidelines to minimise the spray drift apart from wind limitations shall be suitably adopted as prescribed by CIB&RC from time to time.
f. **Safeguarding the non-targets:** Keeping buffer zones between adjacent farms, maintaining distance by drone operator from the drone, water bodies, residential areas, fodder crops, public utilities, dairy, poultry, etc., not allowing human or animals in the farms during or immediately after the spray are some of the important parameters which need to be followed as per approval of CIB&RC and DGCA guidelines.
g. **Submission of spray monitoring data:** Spray monitoring data duly filled by the operator/service provider shall be submitted within 7 days of application of pesticide through drone in the prescribed format.

5.4 PROSPECTS FOR INDIAN FARMERS

Drones become increasingly useful to farmers as agri-inputs, namely, crop protection chemicals (insecticides, fungicides, herbicides, plant growth regulators, etc.), biostimulants, fertilisers, seeds, spray applications of nutrients, etc. are being optimised through drones. Farmers will not only be getting rid of insects-pests-weeds diseases in a timely manner but also a distinctive cost saving will increase their profitability. Drones are going to be instrumental in bringing the accuracy in spray applications.

Robust crop surveillance through drones will bring new dimensions in ensuring crop health, soil health, improved farm-irrigation patterns and monitoring of water stress in plants, etc. Additionally, drones could be immensely deployed in resolving matters related to climate change, monitoring of flood-affected areas, etc. All such factors can be precisely studied with drone technology and prevention or execution of corrective measures could be appropriately followed.

Drone services could be utilised, particularly for obtaining precise data on soil fertility and the status of minerals in soils. This will enable farmers tweaking their decisions on taking appropriate rotational crops, using the right quantity and quality of fertilisers in their fields. All this put together will fetch higher yields and higher returns on investment.

Eventually, farming could turn out to be a trustworthy profession with ease of operations and peace of mind. The younger rural generations moving out of villages would probably be encouraged to stay back with their farms and become entrepreneurs. We shall be able to achieve our GDP to the next level providing employment opportunities to our youth.

Marginal farmers could be considerably benefitted with drone technology as all drone-related operations are expected to be optimised in cost-effective manner in the near future once the technology is popularised and successful.

5.5 FUTURE SCENARIO, CHALLENGES AND POLICY FRAMEWORK

The introduction of drone technology in India is going to transform the traditional/manual application technologies. With a lot of positive initiatives of the Government of India, the future of the drone industry looks to be bright. Farming in India shall be revolutionised more towards modern agriculture, especially in view of the current labour crisis. The spraying of crop protection chemicals will become more precise with almost no exposure to spray operators. The agriculture in hilly terrains particularly fruit crops as well as crops of challenging and complex topography shall be transformed significantly. As a matter of fact, drones shall contribute significantly to the Government of India's vision of "Doubling Farmers' Income" along with strengthening the food security mission.

Success comes with challenges, and drones are no exception to it. Spray drift to adjacent fields, water bodies, poultry fields and residential areas; off-label use of drones; possible signal interferences due to high-tension electric wires; absence of relevant flying permissions and many other such aspects could possibly be some of the stumbling blocks and need to be carefully examined to make this glorious journey of drone technology a true success.

The Government of India's regulatory policies in transforming the application technologies in the farming sector through the advent of drones have been phenomenal. The public–private partnerships could be strengthened by further incentivising drone manufacturers. More specifically, the area of artificial intelligence and machine learning in deploying drones in Indian farms would open up employment avenues for the country's youth to compensate for the labour deficit. Integrated Pest Management (IPM) practices by various institutes of ICAR would require to be rearticulated with the advent of drones as surveillance of diseases, pests, weeds; soil mapping; irrigation and everything related to agriculture is going to be digitalised in near future.

Let's unfold the future of crop protection with the power of drone technology!

REFERENCES

Department of Agriculture & Farmers' Welfare, Ministry of Agriculture & Farmers' Welfare, Government of India. 2022. Standard Operating Procedure (SOP) for Use of Drone Application with Pesticides for Crop Protection and For Spraying Soil and Crop Nutrients in Agricultural, Forestry, Non-Cropped Areas, etc, New Delhi, India, pp. 1–36.

Department of Agriculture & Farmers' Welfare, Ministry of Agriculture & Farmers' Welfare. 2022. *MoM of 439th Registration Committee (RC) Meeting, Central Insecticide Board & Registration Committee, Directorate of Plant Protection, Quarantine & Storage*, 25th April, 2022, New Delhi, India, pp. 267–283.

Hogan, S., Kelly, M., Stark, B., Chen, Y. 2017. Unmanned aerial systems for agriculture and natural resources. *California Agriculture Berkeley*, 71(1), 5–14.

ICAO, Doc. 2015. 10019, "Manual on Remotely Piloted Aircraft Systems (RPAS)". https://store.icao.int/products/manual-on-remotely-piloted-aircraft-systems-rpas-doc-10019. ICAO.2011.Cir '328 AN/190', Unmanned Aircraft Systems (UAS) Circular.

Malveaux, C., Hall, S. G., Price, R. 2014. *Using Drones in Agriculture: Unmanned Aerial Systems for Agricultural Remote Sensing Applications*. American Society of Agricultural and Biological Engineers, Quebec Canada.

Matthew, A., Selorm, A., Albert, A. N. 2021. Multifaceted applicability of drones: A review. *Technological Forecasting & Social Change*, 167, 120677.

Mechanization & Technology Division, Department of Agriculture & Farmers' Welfare, Ministry of Agriculture & Farmers' Welfare, Government of India. 2022. *Providing Agricultural Services through Drone Application under Sub-Missionon Agricultural Mechanization (SMAM): Additional Guidelines*, 17th January, 2022, New Delhi, India, pp. 1–3.

Ministry of Civil Aviation, Government of India. 2021. The Drone Rules, G.S.R. 589(E). 25th August, 2021, pp. 1–43.

Ministry of Civil Aviation, Government of India. 2022. The Drone (Amendment) Rules, G.S.R., 108(E). 11th February, 2022, pp. 1–3.

Nelson, J., Gorichanaz, T. 2019. Trust as an ethical value in emerging technology governance: The case of drone regulation. *Technology in Society*, 59, 101131. https://doi.org/10.1016/j.techsoc.2019.04.007.

6 Sero-Diagnosis

A Novel Tool in Precise Detection of Pathogens, Bioinoculants, and Induced Immunity in Plants

Bishwanath Chakraborty and Usha Chakraborty

6.1 INTRODUCTION

A variety of macromolecules such as all proteins, many polysaccharides, lipoproteins, nucleoproteins, numerous synthetic polypeptides, and many small molecules if they are suitable linked to proteins or to synthetic polypeptides can behave as an antigens. The actual significance of the term "antigen" lies in its capacity to stimulate the formation of the corresponding antibodies, i.e., immunogenicity, and its ability to react specifically with these antibodies. However, those portions of antigen molecules that have the ability only to react with antibody but unable to stimulate antibody formation can also be considered as an antigen. Substances known as haptens are not immunogenic but do react specifically with the appropriate antibodies. Immunogenicity is not an inherent property of a macromolecule, rather, it is dependent on the system and condition employed. The dynamic role of antigens in plant–parasitic interactions has been illustrated (Devay et al., 1967), and common antigenic relationship between host and parasite was first reviewed by Devay and Adler (1976). "Antigenic disparity" is the term for the qualitative or quantitative alteration in recognition molecules that causes cells in resistance-induced tissues to see a compatible pathogen as incompatible (Chakraborty, 1988). Such alteration in antigenic patterns followed by enhanced production of glyceollin (phytoalexin) after induction of resistance in soybean plants against *Macrophomina phaseolina* has been elucidated by Chakraborty and Purkayastha (1987).

Detection of phytopathogens and disease diagnosis using conventional methods including isolation, characterization, and identification of the pathogen following completion of Koch's postulate are time-consuming. As an alternative to conventional methods, serological diagnostic methods for early detection of phytopathogens have been introduced. Following the development of monoclonal antibodies, standardization of many serological assays has revolutionized detection of phytopathogens (Chakraborty and Chakraborty, 2002; Lopez et al., 2003; Gawande et al., 2006; Chakraborty and Chakraborty, 2012). Subsequent to this, attempts have been

DOI: 10.1201/9781003531944-6

made for constant advancement of our knowledge in the field of molecular biology, sequencing, nanotechnology, and computer sciences for specific pathogen detection (Chakraborty and Chakraborty, 2021; Umesha and Raghava, 2021).

6.1.1 Immunodetection of Phytopathogens

The goal of detection is to confirm the existence of a certain target organism in a sample, with a focus on asymptomatic individuals. When a plant exhibits symptoms of a disease, the diagnosis identifies the type and origin of the illness. High levels of specificity, sensitivity, and quickness are necessary for routine illness detection accuracy. In this sense, specificity is the capacity to identify the target organism without producing false positives or negatives. The least number of pathogens that can be consistently detected in an assay or sample is what is meant by sensitivity. To reduce the danger of illness and to halt the spread of inoculum, the plant material may be intended to zero tolerance, as is the case with many seed-borne pathogens or quality pathogens, or to controlled zero tolerance, as is the case with quarantine pathogens. In the case of quarantining viruses and bacteria, high sensitivity and, of course, specificity procedures are especially necessary. Such methods must be used in sanitary and eradication programs when, in addition, the earliest feasible discovery is required. In contrast to what has been possible with conventional methods, fungi, bacteria, and viruses have been detected using serological techniques such as the enzyme-linked immunosorbent assay (ELISA), dot immunobinding assay (DIBA), Western blot analysis, indirect immunofluorescence (IF), immunocytochemical, and immunogold localization of pathogens in plant tissues.

6.1.1.1 Fungi

Detection of fungal pathogens through conventional techniques has been a time-consuming process, and in many cases, especially in root diseases, by the time symptoms appear at aboveground level and detection can be done, it becomes too late. Since growth of many plant pathogenic fungi take time, confirmation of the pathogen cannot be done early. Besides, detection of a pathogen does not necessarily prove that it caused the disease at hand. For example, one can usually culture, or induce by incubation, several pathogens from a plant that are not actively causing the observed symptoms. The diagnostician must identify the various causal factors and determine their relative importance. This task can be difficult, as different factors may cause similar symptoms and different symptoms may be caused by the same factor. Just as a positive test result does not prove that the pathogen detected caused the disease, a negative result does not definitely rule out the presence of a particular pathogen. The test itself may fail, or improper selection of plant tissues can lead to erroneous conclusions. Attempts to use culture techniques to recover a vascular pathogen, such as *Fusarium oxysporum* f.sp. *lycopersici*, from wilted tomato foliage rather than vascular tissues would give erroneous negative results. False-negative results can occur with any assay, and the diagnostician must be aware of the likelihood of their occurrence.

Obviously, the choice of diagnostic test(s) can greatly influence the diagnosis. Many factors influence the interpretation of test results and the value of the final diagnosis.

The quality and case history of the sample provided, available resources, and expertise of the diagnostician all play a role in the accuracy of the diagnosis. Sample quality is of paramount importance and is probably the most common limiting factor to accurate diagnosis. The best diagnosis would be one for which the affected plant(s) could be observed both in the field and in the laboratory. Unfortunately, expense, time, and logistics usually preclude field visits by the diagnostician. Case history information provided with a specimen can be more important to the diagnosis than the specimen itself. This is particularly true when biotic agents play a minor role in the disease. Indeed, many plant specimens submitted to diagnostics labs are afflicted with abiotic problems, such as nutritional imbalances, chemical injury, temperature or moisture extremes, or air pollution. Most abiotic problems must be diagnosed from case history information and symptom directions in the literature. Inadequate or incorrect information often leads to an inaccurate diagnosis.

The competent diagnostician must be able to synthesize information about many different aspects of plant health and know the plant in health as well as disease. The diagnostician has historically been a generalist rather than a specialist, well versed not only in pathology, but also in agronomy, horticulture, entomology, soil science, and weed science. To understand and interpret modern diagnostic techniques, he or she must also have working knowledge of immunology and molecular biology. In addition to familiarity with pathology, the diagnostician must develop expertise in production practices, response of plants to environmental factors, and characteristics of specific cultivars or hybrids of many different crops. To this extent, the individual must become a crop specialist. The difficulties inherent in becoming both a specialist and a generalist in a field that deals with hundreds of plant species and exponentially more pathogens or abiotic agents are obvious.

Plant pathogenic fungi most readily detected using various immuno-techniques (Chakraborty, 2021) have the following extra advantage:

a. Identification by morphological characteristics.
b. Species identification.
c. Prior to development of aboveground symptoms detection of root pathogens.
d. Large number of samples can be processed for a specific disease within a short span of time.
e. On-site rapid detection is possible for developing disease management strategies for high-value crops.
f. Regulations governing the use of pesticides, which require demonstration of presence of a particular pathogen.
g. Plant material is subject to quarantine regulation.

ELISA is one of the most commonly used immunoassays for detection of phytopathogens. To boost the specificity and sensitivity for fungi in ELISA, various modifications and amplifications are applied. Addition of various blocking agents that saturate non-specific binding regions can minimize or eliminate background responses. These include skimmed milk, gelatin, albumin from chicken protein, bovine serum albumin, and serum from non-immunized animals for fungi. Only when enough specific antibodies are available for the given application may modifications and amplifications improve ELISA findings. Therefore, antiserum purity is unquestionably the most crucial factor in the ELISA method. The majority of the serological tests for fungi that are now on the market use either direct or indirect ELISA using polyclonal or monoclonal antibodies. Different ELISAs for fungi have been developed, some of which are better suited for diagnostic labs than others. Multiwell, flow-through, dipstick, dot blot, and tissue print tests are among the formats that are offered. Commercial kits for multiwell ELISA have pre-coated wells with antibodies that are specific to the target species. When tissue homogenate is added to the wells, antibodies can bind to antigens, which are then identified by an antibody–enzyme conjugate. Numerous plant pathogenic fungi have been effectively detected using multiwell ELISA. The multiwell ELISA has a high sensitivity. Rapidity is the key benefit of multiwell ELISA over traditional methods. An absorbent plastic device with antibodies bound on the surface is used for flow-through ELISA diagnostic tests. Drop by drop, the sample is transferred to the top surface, passing by immobilized antibodies that capture the antigen. Enzyme conjugate detects antigen. The flow through kits are especially well suited for field diagnosis because they are quick, simple to use, and self-explanatory. To enable antigen–antibody binding, the membrane on the dipstick, which has been pre-coated with particular antibodies, is soaked in buffered crude plant extracts. The bound antigens are discovered using enzyme conjugate after the dipsticks have been rinsed.

In contrast to double antibody sandwich (DAS)-ELISA, the dot-blot format spots the antigen or plant sap directly onto nitrocellulose or polyvinylidene membranes without first subjecting them to an antibody pretreatment. A cross section of the plant tissue is pressed directly onto a membrane in a tissue-print ELISA, doing away with the requirement for an extraction phase. The antigen is transported from the plant tissue to the membrane if the target pathogen is present. Antibody–enzyme conjugates are used to identify membrane-bound antigen. Tissue-print ELISA is quicker than dot-blot, easier to use, and requires less equipment. To increase the likelihood of detection, prints can be quickly made of numerous different plant tissue sections. Similar to dot blots, strong background color from healthy plant sap can occur, although background color can be reduced by changing the ELISA method. Different immunological formats used for detection of plant pathogenic fungi in various crops have been listed in Table 6.1. Some specific examples are discussed below.

6.1.1.1.1 Foliar Pathogens

6.1.1.1.1.1 Exobasidium sp. In order to create field management measures, the causal agent of the blister blight of tea, *Exobasidium vexans*, immunoassays such as plate-trapped antigen-coated enzyme-linked immunosorbent assay (PTA-ELISA), dot immunobinding assay (DIBA), and indirect IF, was discovered. Significance of PTA-ELISA format was also confirmed for screening resistance of 30 tea varieties toward fungal pathogens using IgG of *E. vaxans* reacted with healthy leaf antigens and compared with blister blight disease reactions of these tea varieties in the field (Figure 6.1). DIBA of these 30 healthy tea leaf antigens when reacted with the homologous IgG, yielded color reactions, and based on color intensity developed on nitrocellulose paper due to cross reactive antigens (CRA) shared by host and pathogen, these tea varieties were categorized/screened accordingly for their possible disease

TABLE 6.1
Serological Formats Used for Detection of Phytopathogenic Fungi

Crops		Fungal Pathogens	Serological Formats	References
Cereals	Rice	*Drechslera oryzae*	PTA-ELISA, DIBA, IIF	Khati and Chakraborty (2019)
	Wheat	*Bipolaris sorokiniana*	PTA-ELISA, DIBA, IIF	Chakraborty et al. (2016a)
Pulses and Oil seeds	Soybean	*Macrophomina phaseolina*	PTA-ELISA, IIF	Chakraborty and Purkayastha (1983)
		Fusarium oxysporum	PTA-ELISA, IIF	Chakraborty et al. (1997)
		Fusarium graminearum	PTA-ELISA, IIF	
		Sclerotium rolfsii	PTA-ELISA	Chakraborty and Chakraborty (2002)
		Phytophthora infestans	DAS-ELISA	
	Sunflower	*Sclerotinia sclerotiorum*	DAS- ELISA	
Vegetables	Potato	*Phoma exigua*	PTA-ELISA, DAS-ELISA	
		Phytophthora infestans	PTA-ELISA	
	Carrot	*Pythium violae*	Competition ELISA	
	Cabbage	*Plasmodiophora brassicae*	PTA-ELISA	
Fruits	Grapevines	*Botrytis cinerea*	PTA-ELISA	
		Eutypa armeniaceae	DIF, IIF	
	Strawberry	*Phytophthora fragariae*	PTA-ELISA	
	Mandarin	*Macrophomina phaseolina*	PTA-ELISA, DIBA, IIF	
		Fusarium oxysporum	PTA-ELISA, DIBA, IIF	
		Fusarium solani	PTA-ELISA, DIBA, IIF	
Fiber	Cotton	*Verticillium alboatrum*	PTA-ELISA	
		Fusarium vasinfectum	PTA-ELISA, DIF	
Forage	Sugarcane	*Colletotrichum falcatum*	PTA-ELISA	Viswanathan et al. (1998)
Plantation	Tea	*Bipolaris carbonum*	PTA-ELISA, IIF	Chakraborty and Saha (1994)
		Pestalotiopsis theae	PTA-ELISA, DAS-ELISA, IIF	Chakraborty et al. (1995)
		Glomerella cingulata	PTA-ELISA, DAS-ELISA, IIF	Chakraborty et al. (2008)

(*Continued*)

TABLE 6.1 (*Continued*)
Serological Formats Used for Detection of Phytopathogenic Fungi

Crops		Fungal Pathogens	Serological Formats	References
		Exobasidium vexans	PTA-ELISA, DIBA, IIF	Chakraborty and Sharma (2007)
		Corticium invisum	PTA-ELISA, DIBA, IIF	Chakraborty and Das Biswas (2008)
		Fomes lamaoensis	PTA-ELISA, IIF	Chakraborty et al. (2002a)
		Ustulina zonata	PTA-ELISA, IIF	Chakraborty et al. (2002b)
		Poria hypobrunnea	PTA-ELISA, DIBA	Chakraborty and Chakraborty (2018)
		Rosellinia arcuata	PTA-ELISA, DIBA	
		Sclerotium rolfsii	PTA-ELISA, DIBA, IIF	Bhagat and Chakraborty (2020)
		Sphaerostilbe repens	PTA-ELISA, DIBA, IIF	Chakraborty et al. (2022)
		Alternaria alternata	PTA-ELISA, DIBA, IIF	Das Biswas and Chakraborty (2020)
Sericulture	Som	*Pestalotiopsis dessiminata*	PTA-ELISA, DIBA, IIF	Acharya et al. (2015)
		Colletotrichum gleosporioides	PTA-ELISA, DIBA, IIF	
		Curvularia lunata	PTA-ELISA, DIBA, IIF	Acharya et al. (2021)

reactions upon inoculation of the plant. For the first time, Chakraborty and Sharma (2007) demonstrated immunogold localization of CRA shared by *Camellia sinensis* and *E. vexans*. Using IF study, apple green fluorescence of germinated basidiospores with appressoria formation *in vitro* (Figure 6.2f) and disease development in tea leaf tissue using FITC conjugates (Figure 6.3b) as well as immunogold localization of pathogen in blister blight infected tea leaf tissue (Figure 6.4a) have been documented. Another simple protocol was developed using IgG of *E. vexans,* which reacted with fast blue BB salt and could be observed under bright field microscope (Figure 6.3c).

6.1.1.1.1.2 Bipolaris sp. Using PTA-ELISA format and indirect IF, CRA shared between *Bipolaris carbonum* and *Camellia sinensis* and its cellular location in epidermal and mesophyll tissues of tea varieties (Chakraborty and Saha, 1994) and sero-diagnosis of *Bipolaris sorokiniana* causing spot blotch disease of wheat using PTA-ELISA, DIBA, Western blot analysis, IF, and immunogold localization of pathogen in leaf tissue (Figure 6.4b) has also been demonstrated (Chakraborty et al., 2016a).

6.1.1.1.1.3 Colletotrichum sp. Infection of sugarcane stalks by *Colletotrichum falcutum* could be very rapidly detected well before symptom development. ELISA technique has been found effective for determining high pathogen colonization on root eyes and buds. Based on the pathogen colonization in the cane stalks, this

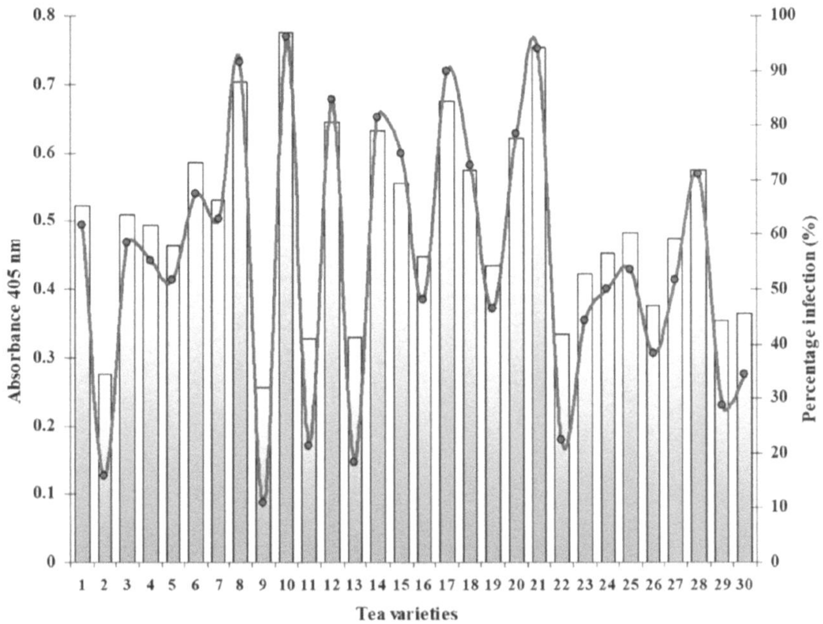

FIGURE 6.1 Screening of tea varieties for resistance toward *Exobasidium vexans* using PTA-ELISA format and comparison with their disease reaction following challenge inoculation with the pathogen. Tea varieties: (1) AV-2 (2) BS-7A/1/76 (3) CP-1 (4) HV-39 (5) P-1258 (6) RR-17 (7) TS-449 (8) T-135 (9) T-17/1/54 (10) T-78 (11) BSS-1 (12) BSS-2 (13) BSS-3 (14) UP-2 (15) UP-3 (16) UP-8 (17) UP-9 (18) UP-17 (19) UP-26 (20) TV-9 (21) TV-18 (22) TV-20 (23) TV-22 (24) TV-23 (25) TV-25 (26) TV-26 (27) TV-27 (28) TV-28 (29) TV-29 (30) TV-30.

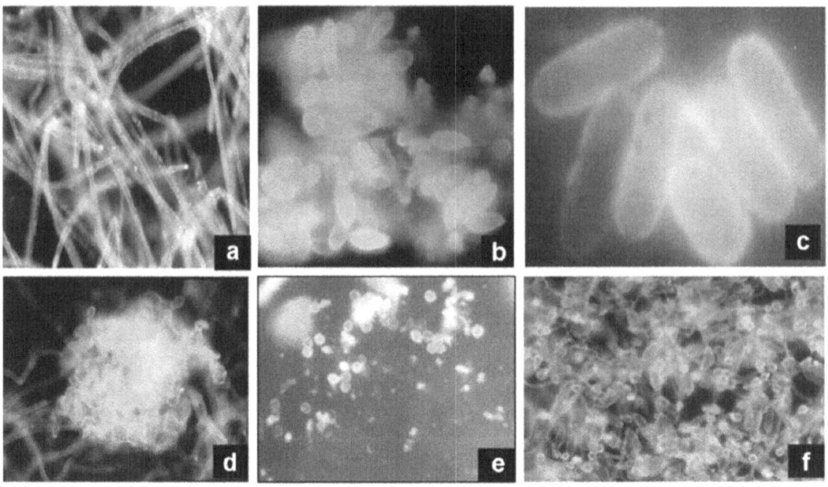

FIGURE 6.2 Indirect immunofluorescence of (a) young hyphae and (b) conidia of *Sprerostilbe repens*; (c) conidia of *Glomerella cingulata*; (d) sclerotium of *Corticium invisum*; (e) chlamydopsores of *Fusarium solani*; (f) Germinated basidiospores of *Exobasidium vexans* showing appressoria; reacted with IgG raised against respective pathogens and labeled with FITC coagulates.

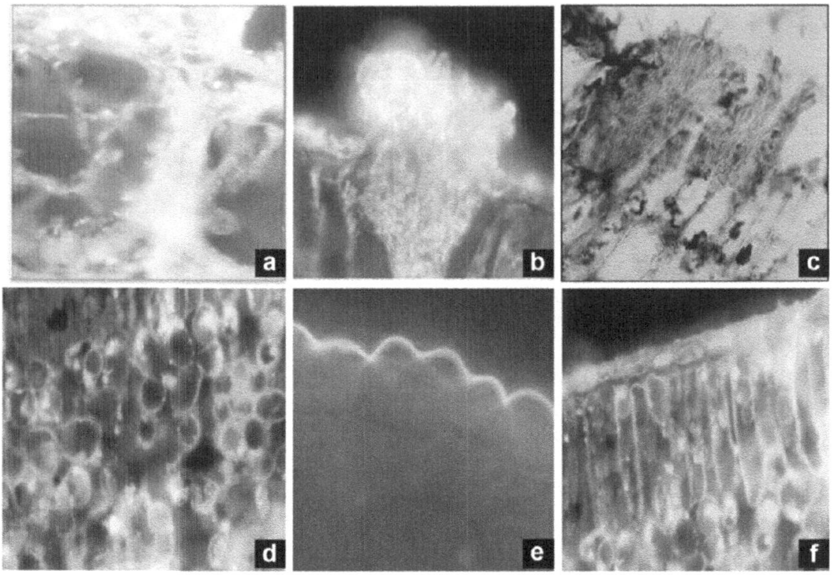

FIGURE 6.3 Transverse sections of tea root (a) and leaf (b–f) tissues showing disease developments using indirect immunofluorescence and labeled with FITC conjugates (a, b, d–f) and fast blue BB salt (c). (a) Brown rot infected root treated with IgG of *Phellinus noxius*; (b and c) Blister blight infected leaf treated with IgG of *Exobasidium vexans*; (d and f) Black rot infected leaf treated with IgG of *Corticium invisum* and (e) Healthy tea leaf tissue showing autofluorescence on cuticle.

technique was also utilized to differentiate host as resistant, moderately resistant, and susceptible against the pathogen (Viswanathan et al., 1998). Using IgG raised against mycelia, cell wall, and conidia (Figure 6.2c) of the pathogen (*Glomerella cingulata*) causing brown blight disease of tea, serological cross-reactivity between host and pathogen as observed in PTA-ELISA and DAS-ELISA tests, revealed that PAb-raised against cell wall preparations gave better recognition than that against mycelia preparation (Chakraborty et al., 2002c). Immunogold localization of *Colletotrichum gloeosporioides* in leaf tissue of som plant (*Persea bombycina*) causing anthracnose disease has been developed (Figure 6.4c).

6.1.1.1.1.4 Pestalotiopsis sp. IgG-raised mycelia and cell wall antigens of *Pestalotiopsis theae* causing gray blight disease of tea were used after purification by ammonium sulfate fractionation and chromatography on DEAE-Sephadex for PTA-ELISA, DAS-ELISA, and Competition ELISA tests (Chakraborty et al., 1995). Pathogen was detected in tea leaves as early as 12h after inoculation with *P. theae*. IgG reacted with two other isolates of *P. theae* but not with the mycelia antigens of *Glomerella cingulata* and *Corticium invisum* or with tea leaves inoculated with these pathogens (Chakraborty et al., 1996). The development of management strategies has been strengthened by the early identification of *Pestalotiopsis disseminata* causing gray blight disease in som plant (Acharya et al., 2015). Immunogold localization of pathogen (*P. disseminata*) in leaf tissue has also been demonstrated (Figure 6.4d).

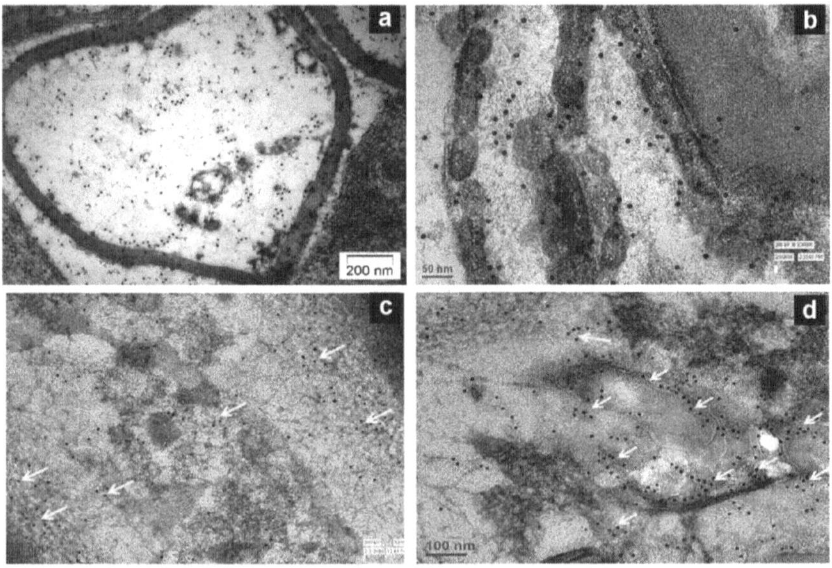

FIGURE 6.4 Transmission electron micrographs of infected leaves reacted with PAb and labeled with gold conjugates. (a) Blister blight infected tea leaf probed with IgG of *Exobasidium vexans*. (b) Spot blotch infected wheat leaf probed with IgG of *Bipolaris sorokiniana*. (c) Brown blight infected some leaves probed with IgG of *Colletotrichum gloeosporioides*. (d) Gray blight infected some leaves probed with IgG of *Pestaloitiopsis dessiminata*.

6.1.1.1.1.5 Corticium sp. Black rot of tea is caused by *Corticium invisum*, one of the important fungal diseases prevalent in tea gardens, for which serological detection kit was developed. Strong apple green fluorescence was evident in immmuofluorescence study with young sclerotium of *C. invisum* formed from hyphal strands (Figure 6.2d). The pathogen could be detected in leaf tissue as early as 12 hours following inoculation with *C. invisum* using PTA-ELISA and DIBA. Immunoenzymatic staining of leaf tissue using PAb of the pathogen and fast blue BB salt could detect the pathogen in mesophyll tissue (Chakraborty and Das Biswas, 2008). Cellular localization of the pathogen was also evident after 42 hours of inoculation as bright fluorescence mainly in mesophyll tissues, spongy parenchyma (Figure 6.3d), palisade parenchyma (Figure 6.3f) reacting with PAb of *C. invisum* and labeled with FITC. Healthy tea leaf tissue exhibited autofluorescence as evident in cuticle (Figure 6.3e).

6.1.1.1.1.6 Alternaria sp. Antigenic characteristic of *Alternaria dauci*, pathogenic to carrot and parsley, was examined by electro-blot-immunoassay (Western blot). Similarities and differences between isolates of the same and distinct species were revealed by antigenic features, demonstrating their complex relationships, which accurately reflect their diversity in nature. Immunodetection of *Alternaria alternata*, causing Alternaria blight, very common in nursery grown tea plant using PTA-ELISA, DIBA, and cellular localization of the *A. alternata* using IgG of pathogen labeled with FITC in mesophyll tissue has been illustrated (Das Biswas and Chakraborty, 2020).

6.1.1.1.1.7 Pythium sp. Monoclonal antibody, the product of the hybridoma cell line PA5, was developed for *P. aphanidermatum* to detect this pathogen in turfgrass involved with Pythium blight. Besides, competition ELISA tests using polyclonal antibodies against *P. violae* and *P. sulcatum* confirmed the predominance of *P. violae* in cavities developed on field-grown carrots.

6.1.1.1.2 Root Pathogens

6.1.1.1.2.1 Phytophthora sp. Multiwall ELISA kit was developed for detection of *Phytophthora* in plant tissue and also in soil where soybean (*Glycine max*) was damaged by *P. megasperma* f.sp. *glycinea*. PAb raised against *P. infestans* reacted in ELISA with mycelial antigens of two *Phytophthora* species but not with those of 10 other microorganisms found on potato. Immunoassay kits and culture plate results for detection of *P. cinnamomi* were most consistent beginning 1 week after inoculation. A dipstick immunoassay that was specific for *P. cinnamomi* was developed for use in soils. *P. fragariae* was detected by ELISA in strawberry roots before symptoms could be detected in microscopical observations (Mohan, 1988).

6.1.1.1.2.2 Fusarium sp. Serological assays were developed for detection of isomarticin, a naphthazarin toxin produced by *Fusarium solani* (Phelps et al., 1990). A carbodimide procedure was used to couple the hapten isomarcitin to BSA for immunogen and to alkaline phosphatase for the enzyme-linked tracer. By using a competitive ELISA technique, naphthazarin toxins of *F. solani* were also found in the xylem fluid of roots rotted by the organism. Using PTA-ELISA format, *F. oxysporum* in soybean root tissue was demonstrated (Chakraborty et al., 1997). Indirect IF assay was also developed for detection of chlamydosores of *F. solani* directly from soil (Figure 6.2e).

6.1.1.1.2.3 Macrophomina sp. Major cross-reactive antigens (CRAs) shared between *Macrophomina phaseolina* and soybean (*Glycine max*) root tissue have been elucidated (Chakraborty and Purkayastha, 1983). Quick and accurate detection of *M. phaseolina* from soil causing root rot disease of mandarin (*Citrus reticulata*) was carried out using PAb raised against pathogen using immunological formats such as immunodiffusion, PTA-ELISA, DIBA, Western blot analysis, and indirect IF (Chakraborty et al., 2012).

6.1.1.1.2.4 Phellinus sp. IgG raised against mycelial antigen of *Phellinus noxius* causal agent of brown root rot disease of tea and analyzed using PTA-ELISA formats. Such format was very useful in detecting the pathogen in infested soil, which enables disease prevention at an early stage (Chakraborty et al., 2002a). Histopathological study of brown rot infected tea root tissue from collar region using IgG of *P. noxius* revealed hyphal penetration in the vascular tissue using indirect IF assay (Figure 6.3a).

6.1.1.1.2.5 Sclerotinia sp. Jamaux and Spire (1994) developed a serological test that allows the early detection of infection of young petals of rapeseed caused by *Sclerotinia sclerotiorum*.

6.1.1.1.2.6 Sclerotium sp. Serological detection of *Sclerotium rolfsii* causing sclerotial blight disease of tea one of the important fungal diseases, which appear in the nursery grown tea seedling, has been reported by Bhagat and Chakraborty (2020). IgG-based immunotechniques have been developed for detection of pathogen (*S. rolfsii*) in soil and tea root tissue, and their cellular localization has also been determined following indirect IF.

6.1.1.1.2.7 Sphaerostilbe sp. Effectiveness of raising antibody against the mycelia antigen of *Sphaerostilbe repens*, causing violet root rot disease of tea, was confirmed by immunodiffusion test and DIBA. Immunodetection of pathogen in soil samples was developed using PTA-ELISA format (Das et al., 2022). Treatment of mycelia and conidia of *S. repens* with homologous PAb and reacted with FITC showed a strong apple green fluorescence that was more intense on young hyphal tips (Figure 6.2a), while conidia showed a general fluorescence throughout the surface (Figure 6.2b)

ELISA is now widely used as a standard diagnostic technique in laboratories all around the world. The technique's versatility, sensitivity, speed, and accuracy of results, as well as its impact, have been most noticeable in the large-scale diagnosis of diseases of perennial and vegetatively propagated crops such as trees, bulbs, and ornamentals. Perhaps most significantly, potatoes and foliar and root diseases of tea (*Camellia sinensis*) have also benefited. Studies on host–pathogen interactions, epidemiological investigations, and characterization studies have all made use of the ELISA technique's potential for obtaining quantitative data.

Thin tissue sections of plants or soil samples put on microscope slides are fixed by a brief drying period or by chemical treatment. Each sample's (antigen's) best fixing technique must be identified empirically. For *in situ* investigations of phytopathogenic fungi, indirect IF has been preferred up till now. Direct IF has not been shown to be as effective for soil or plant material detection. Because the cells may be seen directly, IF is particularly well suited for detecting and localizing antigens in plant tissue and soil. Thus, the pathogen could be identified by IF in plants that had latent phytopathogenic fungal infections. By measuring and counting hyphae as well as dormant spores, quantitative assessments in addition to pathogen identification are achievable.

The selection and preparation of an appropriate antigen and the generation of antibodies are of utmost importance for immunological studies. The choice and specific modification of the test procedure will be crucial for the majority of host–pathogen combinations. The relevant structures or metabolic products for the pathogen under research will determine the choice of a fungal antigen. Understanding the pathogen's biology and epidemiology is necessary for this. The antigen needs to be stable, present in a sufficient quantity, and have the greatest immunogenicity achievable. Young hyphal tips are often where fungal antigenic determinants are concentrated, and metabolites that fall under the category of surface components may result in the production of highly specific antibodies. Time-saving and simple procedures are crucial for everyday use with high sample sizes. Immunological techniques such as ELISA, dot blot, and dip stick test have become more and more crucial in the plant protection services where processing a huge number of samples is required.

6.1.1.1.3 Seed-Borne Pathogens

The environment is typically quite conducive to the development of disease in seedling establishing systems. Therefore, it is crucial to prevent the introduction of any potentially harmful infections onto seeds. The most efficient way to do this is by exclusion, which involves identifying contaminated seed lots using assays for seed detection before discarding or treating them. The widely used conventional seed detection tests, such as eye inspection, selective media, serological assays, and the seedling grow-out assay, all have drawbacks ranging from inefficiency to a lack of specificity and sensitivity.

In serological seed testing, specific antigens on the surfaces of plant diseases are targeted by antibodies (polyclonal or monoclonal). The enzymatic breakdown of substrates or the presence of fluorescent tags can be used to identify antibodies once they bind tightly and specifically to their antigens. The commonly used ELISA and IF microscopy are two formats for serology-based seed tests. Serological assays are relevant to both biotrophic and necrotrophic seed-borne diseases since they do not require pure isolations of the pathogen.

The lack of species-specific antibodies is a shortcoming of serology, which has been frequently utilized to detect diseases such as bacteria, fungi, and viruses that are transmitted by seeds. The quality of the antibody and the testing format also have a substantial impact on the detection thresholds of serology-based assays. Finally, it is feasible to detect nonviable infections using assays based on serology, which can lead to incorrect (false positive) interpretation. Seed detection tests must be based on novel technology, much like in other fields where pathogen detection is important.

However, it is crucial to thoroughly assess these assays' applicability, precision, and accuracy, high-throughput testing of naturally infested seeds, before using them. Numerous reports of novel seed detection assays have been published in the scientific literature; however, they must first undergo rigorous multi-laboratory validation testing so that their reproducibility and repeatability can be effectively assessed (Walcott, 2003).

6.1.1.1.4 Air-Borne Pathogens

By microscopic analysis of surfaces where spores have been touched, airborne spores of fungal plant diseases are frequently found and counted. However, the use of nucleic acid or antibody probes in fungal diagnostics offers the possibility of creating quick systems for identifying and measuring airborne spores of fungi that cause plant diseases. The ascospore of *Venturia inaequalis*, the cause of apple scab, is passively deposited on a membrane surface in an immunoassay system created by Spore View (Chaparral Diagnostics, Burlington, Vt.). Similar to these experiments, efforts to create an immunoassay based on antibodies to detect *Sclerotinia sclerotiorum*, a significant fungus pathogen of oilseed rape (*Brassica napus*), early have solely depended on passive ascospore deposition on rapeseed petals and subsequent mycelial growth.

6.1.1.1.4.1 Microtiter Immunospore Trapping Method Horticulture Research International, Warwickshire, has created a new microtiter immunospore trapping device (MTIST device), which employs a suction system to directly impact air spores in microtiter wells. This device can be used for the quick detection and quantification of ascospores of *M. brassicicola* and conidia of *Botrytis cinerea* by an ELISA. Target organisms in a microflora can be quickly distinguished, identified, and precisely quantified using the MTIST instrument.

The MTIST device should be helpful for monitoring airborne particles and microbes in a variety of situations. It is a portable, reliable, and affordable system that may be used to carry out several tests within a single sample time. For *M. brassicicola* ascospores, correlation coefficients (r^2) of 0.943 and 0.9514 were found for the number of ascospores per microtiter well that were affected by the MTIST device and the ELISA-determined absorbance values, respectively. The adoption of a mixed fungal spore population has no impact on these results. There was a correlation between the amount of illness present on exposed *Brassica oleracea* plants and the quantity of *M. brassicicola* ascospores captured by the MTIST device per liter of air analyzed (Brussels sprouts).

A correlation coefficient of 0.8797 was found between the absorbance values produced by the ELISA and the observed number of conidia per microtiter well when the MTIST device was employed to trap *B. cinerea* conidia. In controlled plant growth chambers with constrained airflow, the MTIST device's relative collection efficiency was 1.7 times higher than that of a Burkard 7-day volumetric spore trap for the collection of *M. brassicicola* ascospores (Kennedy et al., 2000). For the purpose of early detection of the fungus responsible for blister blight disease of tea (*Camellia sinensis*) and the development of management measures, the MTIST method (Figure 6.5a and b) was used to trap the basidiospores (Figure 6.5c) of *Exobasidium vexans* (Chakraborty and Chakraborty, 2018).

FIGURE 6.5 (a) Microtiter immunospore trapping (MTIST) device. (b) Microtiter wells. (c) Basidiospores of *Exobasidium vexans* trapped from air in tea plantation, after reaction with IgG of the pathogen and labeling with FITC showing immunofluorescence.

6.1.1.2 Bacteria

Numerous contributions from basic and practical research focused on the quick detection and identification of plant pathogenic bacteria have benefited in the diagnosis of bacterial illnesses. There are numerous selective and semi-selective media available. Serology keeps improving in usefulness as assay times get shorter and sensitivity rises. Numerous test formats, such as DAS-ELISA, dot blot, and IF, have been created for bacteria that are comparable to those created for fungi and viruses. Numerous businesses have created a variety of kits based on slide agglutination, IF, and ELISA for the quick and/or *in-situ* detection of bacteria utilizing polyclonal or monoclonal antibodies. Latent bacterial infections can be found using methods based on IF because of its sensitivity.

Only plants that have already begun to display symptoms should be diagnosed using agglutination or ELISA assays. Devices for the identification of bacteria have also been developed using tissue prints, ELISA, and lateral flow. However, these techniques have considerable sensitivity for detecting viruses but relatively low sensitivity for detecting bacteria. It has been demonstrated that flow cytometry, a method for quickly identifying bacteria using fluorescent dyes attached to certain antibodies and detected electronically using a fluorescence-activated cell sorter, has the potential to be used for routine seed health assessment (Lopez et al., 2003). New technologies, such as nucleic acid probes and polymerase chain reaction (PCR), have increased the boundaries of specificity. Fatty acid analysis has also demonstrated to be a viable method of identification (Umesha, 2021).

6.1.1.3 Viruses

Before the development of sero-diagnosis, many diagnostics laboratories used time-consuming procedures such as transmission to indicator hosts for the regular detection of viruses. Earlier serological methods required more time and were less accurate than those in use today. Modern, quicker, and more sensitive methods have largely taken the place of traditional methods for viral detection; however, some of the traditional methods are still used in diagnostic laboratories. Utilizing starting and planting materials that are virus-free is necessary for effective management of plant viruses. Citrus tristeza virus (CTV), one of the most significant citrus virus infections, necessitates sensitive, focused, and trustworthy pathogen-detection techniques. Grafting chosen citrus scions that are virus-free onto rootstocks that are CTV-tolerant or -resistant is how CTV is controlled. Programs for certification and quarantine are crucial for preventing the admission and spread of dangerous CTV strains.

Using CTV-specific monoclonal antibodies, Vidal et al. (2012) compared the tissue-print (TP) real-time reverse-transcriptase polymerase chain reaction (RT-PCR) with the validated TP ELISA. In comparison to TP real-time RT-PCR, they estimated TP-ELISA to have a sensitivity of 0.9820, specificity of 0.8519, and positive and negative likelihood ratios of 6.63 and 0.021, respectively. In contrast, they estimated TP-ELISA to have a sensitivity of 0.8015, specificity of 0.9963, and a positive and negative likelihood ratio of 216.42 and 0.199, respectively. The utility of the molecular approach for routine CTV detection is validated by these diagnostic characteristics, which demonstrate that TP-ELISA had the highest specificity and

TP real-time RT-PCR had the highest sensitivity. For many years, the go-to regular approach for virus-free starting and planting material was the ELISA test (Boonham et al., 2014).

Given its ability to cut down on handling time and the amount of consumables and reagents required, multiplex testing, or testing for several targets in a single sample, can indeed provide considerable time and cost savings, especially when testing larger numbers of samples. Due to the diverse, partially overlapping fluorescence signals produced by the various reporter and quencher dyes utilized for detection, TaqMan's multiplex capabilities are restricted to four targets. The number of tests that may be accurately detected in a single tube is thus constrained. As a result, numerous TaqMan assays must be run in parallel for genuine multiplex testing, increasing costs and decreasing time savings.

In the Luminex approach, targets are selectively captured on a wide variety of polystyrene, paramagnetic beads while a general fluorochrome is used for detection. True multiplex capabilities are provided, enabling simultaneous detection of up to 500 targets using either molecular or serological technologies. Both the molecular-based xTAG technology and the serology-based Luminex xMAP technology offer a rapid and cutting-edge platform that uses multiplex detection while maintaining the conventional 96-well format and methodology (Rene et al., 2015).

6.1.1.3.1 Luminex xMAP Serological Assay

The majority of current DAS-ELISA tests use certain antisera that enable the identification of a particular disease. Separate assays are typically needed to detect additional diseases. The universal bead immune assay (Joos et al., 2000; Vignali, 2000) is the foundation of the xMAP technology, which was created by Luminex (Austin, TX, USA). It enhances the multiplex immunological detection power of ELISA and permits the simultaneous identification of numerous diseases. The system's foundation is an array with a theoretical maximum capacity of up to 500 separate individual bead addresses made of 6.5 μM polystyrene beads that have been internally dyed with two fluorochromes.

Additionally paramagnetic, these beads allow for extremely rigorous cleaning processes utilizing a magnet. This greatly eliminates background issues in complicated or challenging matrices. A pathogen-specific antibody is covalently bonded with beads that have a specific bead address or a special combination of internal dyes. Then, various beads are put together to make unique "bead sets" for the detection of certain mixtures of plant diseases. The bead-based immune assay's technique is the same as that of the standardized (DAS)-ELISA and may be easily incorporated into current processes. The standard 96-well microtiter plate is used to hold plant samples that have been processed in the standard ELISA buffer.

The pathogens in the sample are caught by the various antibodies bound to the individual beads in the antibody coated bead set. The beads are then incubated with the proper combination of biotinylated secondary antibodies following a brief washing step in which a magnet is used to collect the beads from the bottom of the plate. A fluorescent reporter (Streptavidin-R-Phycoerythrin) is introduced after the magnetic wash stage. The samples are examined after a brief incubation, during which the fluorescent reporter binds to the biotinylated secondary antibodies.

Either a tiny flow cytometer with a laser or an LED base detects the various beads and the fluorescent reporter that is attached to them. The analyzer queries a user-defined number of unique beads of each particular address for both their internal color (the "bead-address") and amount of the fluorescent reporter attached to the secondary antibody.

The fluorochrome and the virus-specific bead address provide a positive signal in the presence of a certain infection. The median level of fluorescence recorded on the virus- or pathogen-specific beads (Median Fluorescent Intensity = MFI) correlates with the amount of viral or pathogen present in the sample, much like a typical DAS-ELISA. While the Luminex xMAP technique typically takes no longer than 3–4 hours, including sample preparation, typical DAS-ELISA protocols may take 1–2 days. Pre-coating of ELISA plates is avoided by using pre-defined bead combinations that are stable for a long time, and reduced incubation times are made possible by enhanced fluid dynamics (Bergervoet et al., 2008). Improved detection limits and a significantly higher dynamic range are further benefits (Rao et al., 2004).

6.1.2 Immunodetection of Bioinoculants

Soil and rhizosphere microbiome such as *Trichoderma* sp., AM fungi, and plant growth promoting rhizobacteria (PGPR) are frequently associated with many nutrient and biological processes at root level, and these are being used as bioinoculants (Chakraborty and Chakraborty, 2019). The sustainability of the bioinoculants applied in the soil has to be ascertained prior to recommending their application. Immunodetection of these microorganisms using their specific antisera can be used to ascertain their presence after varying periods of time.

6.1.2.1 Trichoderma

Trichoderma are abundant in the environment and have been extensively studied for their variety of mechanisms, which include antagonistic interactions, stimulation of plant growth, production of various enzymes, and induction of defense responses in plants that play a significant role in biotic and abiotic stress tolerance (Chakraborty et al., 2011). Because of its many positive properties, *Trichoderma* is regarded as nature's gift to humans. If used properly, *Trichoderma* can result in sustainable, environmentally friendly agriculture, displacing the usage of chemical fertilizers and fungicides/pesticides. *Trichoderma*-based products have been particularly acknowledged as effective biological control agents for managing soil-borne diseases and fostering resilience to biotic stresses (Chakraborty, 2020). However, some *Trichoderma* strains have also been shown to exhibit biostimulant activity, promote plant development, increase yield and nutritional quality, as well as mitigate the negative effects of abiotic stressors, in addition to their usefulness as a biocontrol agent. Polyclonal antibody-based immunodiagnostic kits have been developed for detection of *Trichoderma* sp. in soil. Indirect IF test using IgG raised against *Trichoderma harzianum* labeled with FITC confirmed the identity of this fungus as bright apple green fluorescence (Figure 6.6e–j). Sustainability of the introduced *Trichoderma* sp. in the soil for plant protection can easily be determined specifically with the help of such immunological assay format.

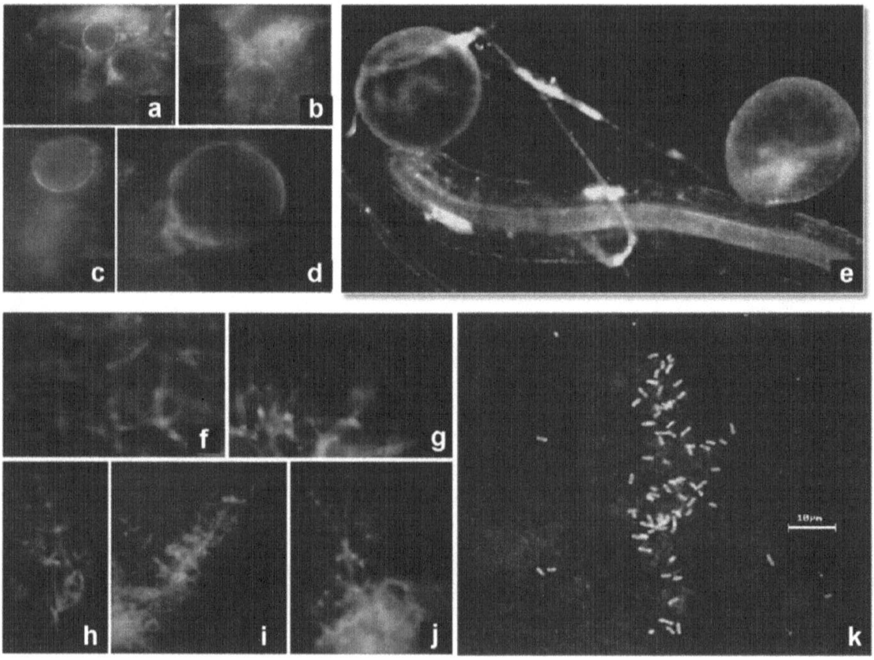

FIGURE 6.6 Indirect immunofluorescence of bioinoculants (AMF, BCA, PGPR) using IgG of (a–d) *Rhizophagus fasciculatus*; (e) *Gigaspora gigantea*; (f–j) *Trichoderma harzianum* (k) *Bacillus megaterium* and labeling with FITC conjugates.

6.1.2.2 AM Fungi

In response to an arbuscular mycorrhizal fungus (AMF) infection, plants may generate a stronger defense mechanism. Mycorrhiza-induced resistance (MIR) offers systemic defense against a variety of intruders and shares traits with induced systemic resistance (ISR) following root colonization by non-pathogenic rhizobacteria and systemic acquired resistance (SAR) following pathogen infection. MIR is typically thought to be caused only by the fungus stimulating the plant immune system. The "mycorrhizosphere effect," a phenomenon that is known to enhance biological activity in the rhizosphere, is a result of AMF infection. The attraction and selection of particular bacterial strains, such as PGPR, which can promote plant growth and control pests and diseases, are examples of this effect. Some of these bacteria that live in the mycorrhizosphere can function as "mycorrhiza-helper bacteria" and increase the effectiveness of the mycorrhizal symbiosis (Chakraborty, 2019).

Immunological formats using IgG raised against two dominant AM fungi (*Rhizophagus fasciculatus* and *Gigaspora gigantea*) were developed by Chakraborty and Allay (2022) for their detection in soil labeled with FITC conjugates following an indirect immunofluorescence test (Figure 6.6a–e). Bright apple green fluorescence was visible throughout the spore wall of AM fungus spores. Assisting hyphae also fluoresced apple green. In the root rhizosphere, spores with their hyphae were more noticeable. The presence of tiny arbuscule branches within the root cells was

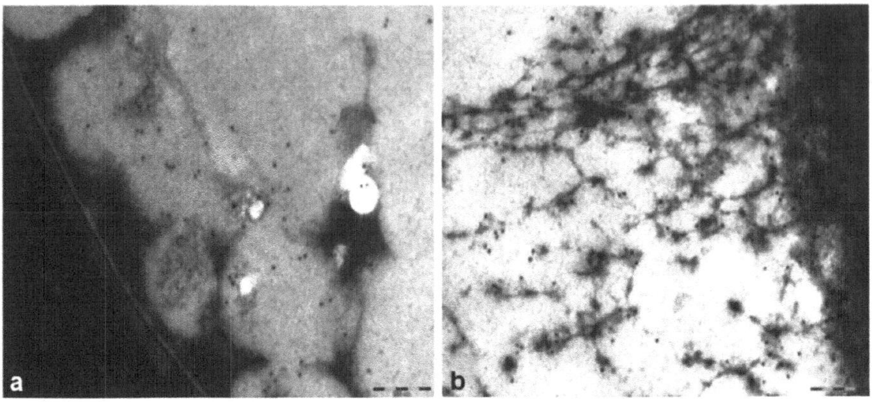

FIGURE 6.7 (a and b) Transmission electron micrographs of mandarin root colonized with *Rhizophagus fasciculatus*, reacted with IgG of *R. fasciculatus* and labeled with gold conjugates.

confirmed by ultrathin slices of an AMF colonized root stained with toluidine blue. *Rhizophagus fasciculatus* was immunogold localized in mandarin roots (Figure 6.7a and b), and the results showed that the gold particles were primarily concentrated close to the cell wall.

6.1.2.3 Plant Growth Promoting Rhizobacteria

PGPR, which are beneficial bacteria isolated from the root microbiome, have been studied for their remarkable impact on plant growth promotion. They do this by improving nutrient availability, modifying the host's hormonal balance, and inducing immunity against phytopathogens (Chakraborty et al., 2014; Backer et al., 2018). The ability to effectively compete with pathogens for nutrients and space, a potent direct antagonistic activity based on the secretion of small antimicrobials or hydrolytic enzymes, and the capacity to promote ISR are just a few of the traits that contribute to the biocontrol potential of PGPR (Köhl et al., 2019). Using IgG raised against *Bacillus megaterium*, dot immunobinding assay format was developed on colony blot transfer of PGPR isolated from rhizosphere. Sustainability of the introduced PGPR in the soil for improvement of plant health can easily be determined specifically with the help of such developed diagnostic kits. PGPR can also be detected through indirect IF assay by their bright apple green fluorescence when treated with specific homologous antibodies (Figure 6.6f).

6.1.3 Plant Immunity Induced by Bioinoculants

ISR is a systemically expressed resistance state that makes the host less vulnerable to recurrent infection. Because it is effective against a wide range of microbial diseases, nematodes, and insects, ISR is of great importance from an agronomical perspective (Pieterse et al., 2014; Gupta and Bar, 2020). ISR shares phenotypic similarities with SAR, which is mounted upon pathogen awareness and results from a sophisticated immune-related process. When various potential agriculturally important microbes (Chakraborty, 2013, 2016) are used as bioinoculants, their chitin, β-glucan, peptidoglucan (PGN), and lipopolysaccharides (LPS), which have been shown to immediately activate an immune response in plants upon recognition, including the accumulation of defense enzymes (chitinase, β, 1,3-glucanase, peroxidase), microbe-associated molecular patterns (MAMPs) are detected by plants leading to synthesis of pathogenesis-related proteins (PRPs). Accumulation of such protective PRPs induced in several plant species triggered by biotic and abiotic elicitors has been intensively studied (Chakraborty and Sharma, 2008). Furthermore, R proteins are typically NB-LRR (nucleotide-binding leucine rich repeat) proteins that interact with one another to create effector-triggered immune responses, which stimulate production of antimicrobial enzymes and secondary metabolites, known as phytoalexins (Chakraborty, 2018). Induced immunity in plants using bioinoculants and biostimulants has been elucidated (Chakraborty and Chakraborty, 2019).

6.1.3.1 Trichoderma sp.

Numerous *Trichoderma* species, including *T. asperellum*, *T. harzianum*, *T. virens*, *T. viride*, and *T. atroviride*, can produce immunity and trigger systemic resistance in connection with plant roots. MAMPs produced by *Trichoderma* have been shown to have a demonstrable biological effect on plants. The role of proteins, secondary metabolites (SM), microbial volatile organic compounds (MVOCs), peptaibols, and microbial phytohormones (MP) in the complex interactions between plant roots and *Trichoderma* has also been clarified (Monti et al., 2020). There have been reports of *T. harzianum* and *T. asperellum* inducing both systemic and local resistance in a variety of crops, including rice, maize, beans, lettuce, cotton, and tobacco, as well as in other cereals, pulses, plantation, and horticultural crops (Chakraborty et al., 2020). *Trichoderma* strains can significantly alter plant metabolism and genome, leading to accumulation of antimicrobial chemicals in plants. *Trichoderma* produces an elicitor called Sm1/Ep11, a member of the cerato-platanin family, which induces immunity by increasing the expression of defense-related genes (Djonovic et al., 2007). Sharma et al. (2011) performed a thorough analysis of *Trichoderma* species' biocontrol genes. Full-length cDNA and genomic DNA sequences of hyb2, a minor gene, cloned by Gulijimila et al. (2012) from *T. asperellum* T4 are hydrophobic proteins. Hyb2's chitinase activity using gene recombination technique has been improved further (Cong et al., 2012).

6.1.3.2 AM Fungi

AM fungi induce systemic resistance to control plant pests and diseases. While MIR has been linked to SAR-like priming of salicylic acid (SA)-dependent genes, it more frequently occurs in conjunction with priming of jasmonic acid (JA)-dependent defenses and cell wall defenses. The induced resistance shares characteristics with both pathogen-induced SAR and rhizobacterial ISR. As a result, MIR offers defense against a variety of adversaries, including nematodes, herbivorous arthropods, biotrophic pathogens, and necrotrophic diseases. It has been suggested that MIR is caused by the active inhibition of defense pathway components that depend on SA, which results in systemic priming of JA-dependent defenses. It has been shown that mandarin plants can develop immunity to Fusarium root rot disease when

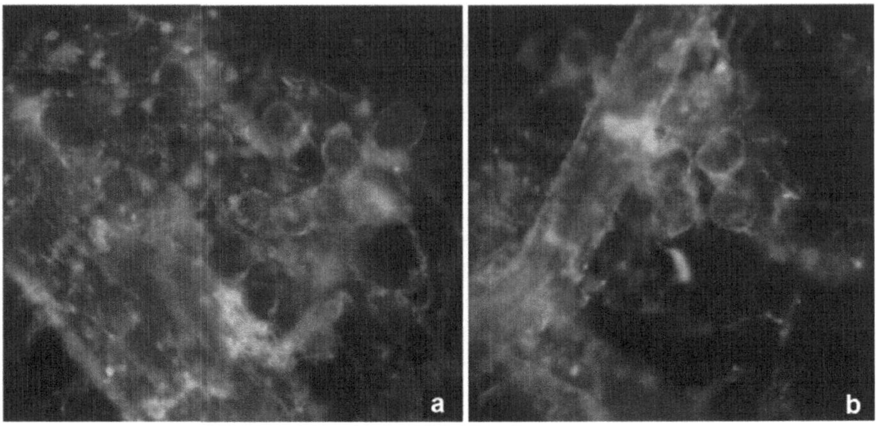

FIGURE 6.8 (a and b) Mandarin rhizosphere showing successful root colonization with *Rhizophagus fasciculatus* reacted with homologous IgG and labeled with FITC following induced immunity.

co-inoculated with bioinoculants (*Glomus mosseae* and *Trichoderma hamatum*) (Allay and Chakraborty, 2010). Strong bright apple green fluorescence was evident in mandarin rhizosphere following successful colonization of roots with *Rhizophagus fasciculatus* using indirect IF tests (Figure 6.8a and b). The colonization of mandarin roots with *Gigaspora gigantea* (AMF) and treatment with *T. asperellum* (BCA) represent additional attempts to induce resistance. Induction of resistance in tea against *Ustulina zonata* using AMF and PGPR (Bhutia et al., 2012) and by dual application of *Rhizophagus fasciculatus* and *Bacillus pumilus* against *Sclerotium rolfsii* has been illustrated (Chakraborty et al., 2016b).

Through genetic modification of the associated biosynthetic pathways in the host plant, it is possible to confirm the involvement of putative plant metabolites as regulators of mycorrhizosphere bacterial activities that induce resistance. To determine the precise contribution of rhizobacteria to MIR, additional comparative research on the signaling systems controlling MIR and ISR will be required.

6.1.3.3 Plant Growth Promoting Rhizobacteria

A variety of plasma membrane-anchored immune sensors known as pattern recognition receptors (PRRs) are at the forefront, detecting molecular properties of invading microorganisms with great specificity and sensitivity in the apoplast. These PRRs bind certain structural motifs (epitopes) of bigger (macro) molecules while maintaining crucial activities for microbial fitness. The PGPR triggers similar early immunological responses such an oxidative burst, ion fluxes, and a phosphorylation cascade. PGPR frequently activates the host's defense mechanisms, but only after a pathogen challenge. There have been reports of genes involved for creating protective proteins and enzymes being upregulated. Prsic and Ongena (2020) have updated their summary of the PGPR determinants that, in the majority of cases, are small-size chemicals released by the colonizing bacteria and are responsible for ISR elicitation. However, they have only discussed about substances that have been tested in pure (proven) form at biologically relevant concentrations and/or via specifically suppressed mutants, and for which the ability to stimulate systemic resistance was clearly demonstrated to be independent from direct antimicrobial effect. They have highlighted their chemical diversity and structure- and dose-dependent activity.

Rhizobacterial cyclic lipopeptides (CLPs) are multipurpose secondary metabolites engaged in biocontrol principally because of their antibacterial activity and developmental processes such as motility and biofilm formation (Geudens and Martins, 2018). However, several CLPs produced by helpful bacilli and pseudomonads have also started to be recognized as significant plant immunity inducers. As a pure compound, they caused ISR in bean, tomato, tobacco, melon, and peanut against *Sclerotium rolfsii* and *Botrytis cinerea* when administered to the roots (Rodriguez et al., 2018). Rhamnolipids (RLs) are amphiphilic glycolipids made by a variety of species, including pathogenic isolates as well as some *Pseudomonas* and *Burkholderia* species that are helpful to plants (Abdel-Mawgoud et al., 2010). They are necessary for the establishment of bacterial biofilms and surface motility (Chrzanowski et al., 2012). Use of a wide range of doses (10–300 mM) of mono- and di-RLs as pure compounds has been found to trigger plant defensive responses and to promote resistance against several diseases in grapevine, *Arabidopsis*, and *Brassica napus* (Monnier et al., 2020). In addition to acting as plant immunity inducers, PGPR siderophores are known to inhibit pathogen populations by reducing the quantity of iron in soil, leading to competition for this crucial element (Kramer et al., 2020). In particular, *P. fluorescens* CHA0 on tobacco infected with Tobacco necrosis virus and *P. fluorescens* WCS374 on radish protected against Fusarium wilt demonstrated their critical involvement in systemic resistance induction. Additionally, it was said that these chromopeptides were inducers of resistance in a variety of plant species, including bean, tomato, *Arabidopsis*, tobacco, eucalyptus, and rice against a variety of microbial diseases. *P. aeruginosa* 7NSK2 also produces the chelating chemical pyochelin in addition to pyoverdine as well as its precursor SA, which was found to be critical for mounting the plant immunity when tested on the pathosystem grapevine/*B. cinerea*. Although soluble chemicals make up the majority of the well-characterized PGPR elicitors, several volatile organic compounds (VOCs) have also been shown to cause systemic resistance in the host plant.

Bacillus megaterium (Chakraborty et al., 2006) and *Ochrobactrum anthropi* (Chakraborty et al., 2009) and their talc-based formulations have been evaluated for growth promotion and induced resistance in tea plants against brown root rot disease. Induced defense response in wheat plants against *Bipolaris sorokiniana* following application of PGPR (*Bacillus safensis* and *Ochrobacterum pseudogrignonense*) has been demonstrated by Sarkar et al. (2018). Biopriming by PGPR (*Bacillus altitudinis* and *Bacillus pumilus*) application led to enhanced production of PR proteins such as chitinase, β,1,3-glucanase, and peroxidase and thereby induced resistance in *Brassica juncea* against *Thanatephorus cucumeris* causing seedling blight disease (Sunar et al., 2020).

It is unclear how PGPR elicitors are interpreted at the plasma membrane level, the early stages of downstream signaling still show a distinct convergence between PGPR-triggered immunity and pathogen-induced PTI. Similar early immunological responses seen upon pathogen recognition, such as an oxidative burst, ion fluxes, and a phosphorylation cascade, are induced by the most well-studied PGPR elicitors. Again, in contrast to PTI, PGPR elicitor priming does not require fitness costs, substantial transcriptional reprogramming, or fitness costs until the pathogen is discovered, but it nonetheless gets the plant ready to mount an effective defense. Generally speaking, the defense mechanisms that prevent pathogen entry and are elicited by helpful microorganisms or their elicitors match those seen upon pathogen MAMP sensing. Notably, it entails the upregulation of genes related to hormone signaling and the manufacture of antimicrobial enzymes, proteins, or metabolites. In plants treated with elicitors following pathogen infection, higher expression of genes associated to pathogenesis, including lipoxygenase, plant defensin factor, and phenylalanine ammonia lyase, is frequently observed.

6.1.3.4 Immunolocalization of Defense Enzymes in Plant Tissue Following Induced Immunity by Bioinoculants

Following the induced immunity triggered by bioinoculants (BCA, AMF, PGPR) in tea, wheat, and mandarin plants, enhancement of chitinase and glucanase was seen in the epidermal cell and cortical tissue in roots and mesophyll tissues in leaves (Chakraborty, 2020). Transmission electron microscopy was used to show the immunogold localization of chitinase (Figure 6.9a) and glucanase (Figure 6.9b) in tea leaf tissue after application of *Trichoderma asperellum* (BCA) and *Bacillus megaterium* (PGPR) and challenge inoculation with *Phellinus noxius* and in wheat leaf tissue

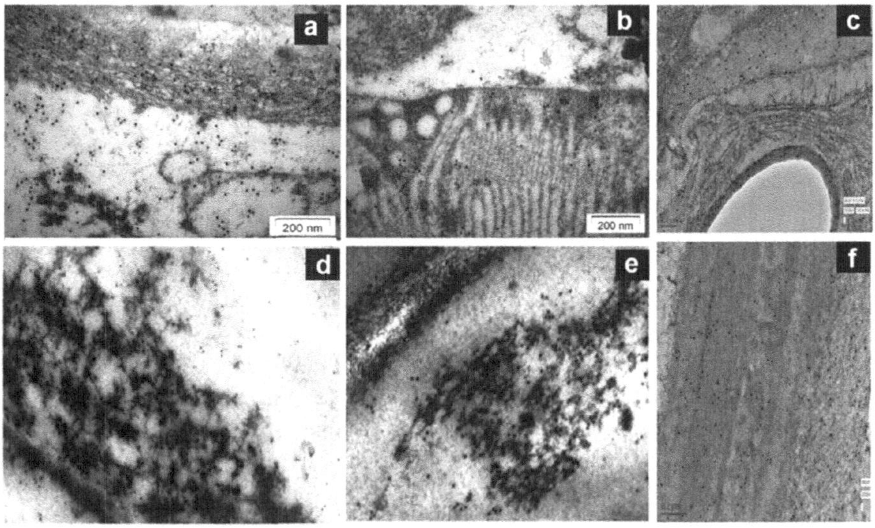

FIGURE 6.9 Immunogold localization of defense enzymes (chitinase and glucanase) in tea leaf tissue (a and b) following joint application of *Trichoderma asperellum* and *Bacillus megaterium* and challenge inoculation with *Phellinus noxius*, wheat leaf tissue (c and f) following application of *Bacillus methylotropicus* (PGPR) and challenge inoculation with *Bipolaris sorokiniana* and mandarin root tissue (d and e) following successful colonization with *Rhizophagus fasciculatus* and challenge inoculation with *F. solani*, probed with PAb of chitinase (a, c, d) and glucanase (b, e, f) and labelled with gold conjugates.

following application of *Bacillus methylotropicus* (PGPR) and challenge inoculation with *Bipolaris sorokiniana* using PAb of chitinase (Figure 6.9c) and glucanase (Figure 6.9f) gold labeled conjugates (Chakraborty et al., 2016). Immunogold localization of chitinase and glucanase in mandarin roots following successful colonization with *Rhizophagus fasciculatus* and challenge inoculation with *F. solani* was obvious as substantial deposition of gold particles close to the cell wall of infected roots, as shown by transmission electron microscopy (Figure 6.9d and e). Additionally, some gold particles were observed in the cell wall regions where fungal hyphae were developing between individual cells. Elicitation of chitinase was visible as extensive gold labeling over both cell walls and wall appositions when plants were treated with *T. asperellum* followed by challenge inoculation with *F. solani*, activating the defense response in plants. Increased nutrient availability, host immunity, or pathogen inhibition directly in the soil have all been linked to better health in mandarin plants and a reduction in disease incidence (Chakraborty et al., 2019). When *Rhizophagus fasciculatus* and *Gigaspora gigantea* were tested separately and together for their ability to prevent *Fusarium solani* causing root rot of mandarin seedlings in field conditions, joint inoculation with AMF could effectively reduce disease incidence. Using PAb raised against chitinase and FITC conjugates, it was shown that defense enzymes, primarily chitinase, are localized within cells. Strong bright apple green fluorescence was seen in the epidermal cells and uniformly in the cortical tissues in roots during indirect immunofluorescence tests (Chakraborty and Allay, 2022). Multifold increase in the activities of chitinase, β-1,3-glucanase, peroxidase, and phenylalanine ammonia lyase in roots as well as in leaf of mandarin plants was observed after application of AM fungi followed by inoculation with *F. solani,* which was correlated with disease reduction.

6.1.4 Prospects

One strategy could be the development of compounds that can be aimed at inhibiting the ligand-receptor binding site in order to impede the specificity of the host–pathogen relationship. This would require knowledge of the protein structures of the components engaging in the pathosystem. The use of anti-IT antibodies in plant–pathogen interactions will also enable us to examine and comprehend the intricate network of signal transduction pathways that function after the initial recognition event. Anti-idiotypic (Anti-IT) antibodies can mimic phytopathogens' capacity to cause the accumulation of transcripts linked to plant defense responses.

Advanced molecular methods, such as proteomics and transcriptomics, can be used to completely understand how plants and pathogens respond during the development of disease. This understanding can then be helpful for creating disease management strategies, including biocontrol. The expression of virulence genes can be determined by genome and transcriptome studies, allowing for the screening of virulence suppressive factors. Today, it is conceivable to filter the genome libraries of some advantageous species in an effort to find novel PGP features. Studying the expression of several Trichoderma genes was challenging because their activity might only be related to defense against other microorganisms or multicellular creatures. Growing fungus in environments that encourage competition and permit the regular metabolic pathways to function can start certain strategies to activate silent gene clusters. Before they are used commercially as biocontrol agents, the roles of Trichoderma gene products can be clarified by thorough metabolomic and genomic research, which can also lead to the discovery of a large number of novel compounds and the unravelling of their functions.

6.1.5 Conclusion

Immunological diagnostic techniques have drawn more attention in recent years as an alternative to or addition to traditional techniques. False positives brought on by antibody interactions with plant debris or unrelated species are a significant downside of serological assays, nevertheless. It has been demonstrated that serological assays may accurately detect plant pathogens (fungi, bacteria, and viruses) and bioinoculants (BCA, AMF, and PGPR). To analyze a large number of samples quickly, molecular approaches are being utilized increasingly frequently for pathogen detection in plant tissues due to their higher specificity and sensitivity compared to serological techniques. Before symptoms occur, early and accurate detection of phytopathogens in plant tissue and/or soil can open the door to creating plant disease management techniques.

With the aid of established sero-diagnostic kits, the viability of newly introduced agriculturally significant bacteria in the soil for the enhancement of plant health can be particularly determined. *Trichoderma* has been shown to have several advantageous qualities and is therefore a blessing to agriculture. In order to determine the cellular location of defense enzymes following induced immunity in plants employing bioinoculants, more recent research studies have replaced earlier ones with more sophisticated ones using immuno-techniques. In root and leaf tissues primed with PGPR and/or AM fungus, immunogold localization of chitinase and glucanase is also observed in eliciting immunity in plants. Such research can be beneficial for creating illness prevention plans. Additionally, genome and transcriptome analysis can pinpoint how virulence genes are expressed, enabling the identification of virulence suppressive agents whose activity may only be related to defense against other microorganisms or multicellular organisms. Commercial application of these eco-friendly bioinoculants should be pursued more actively, not only as a biocontrol agents (BCAs), but also as growth promoters and immune system stimulators.

REFERENCES

Abdel-Mawgoud, A.M.; Lepine, F.; Deziel, E. Rhamnolipids: Diversity of structures, microbial origins and roles. *Applied Microbiology and Biotechnology*, **2010,** 86, 1323–1336.

Acharya, A.; Ghosh, S.; Chakraborty, B.N. Serological detection of *Pestalotiopsis disseminata* in *Persea bombycina* causing grey blight disease. *International Journal of Advanced Research*, **2015**, 5(4), 327–333.

Acharya, A.; Mulka, N.A.; Chakraborty, B.N. Development of polyclonal antibody based detection of *Curvularia lunata* causing leaf blight of *Persea bombycina*. *Journal of Mycopathological Research*, **2021**, 58(4), 259–265.

Allay, S.; Chakraborty, B.N. Activation of defense response of mandarin plants against Fusarium root rot disease using *Glomus mosseae* and *Trichoderma hamatum*. *Journal of Mycology and Plant Pathology*, **2010**, 40(4), 499–511.

Backer, R.; Rokem, J.S.; Ilangumaran, G.; Lamont, J.; Praslickova, D.; Ricci, E. et al. Plant growth-promoting rhizobacteria: Context, mechanisms of action, and roadmap to commercialization of biostimulants for sustainable agriculture. *Frontiers in Plant Science*, **2018**, 9, 1473.

Bergervoet, J.H.W.; Peters, J.; van Beckhoven, J.R.C.M.; van den Bovenkamp, G.W.; Jacobson, J.W.; van der Wolf, J.M. Mulitplex microsphere immunodetection of potato virus Y, X and PLRV. *Journal of Virological Methods*, **2008**, 149, 63–68.

Bhagat, I.; Chakraborty, B.N. Serological detection of *Sclerotium rolfsii* in tea root tissue. *Journal of Mycopathological Research*, **2020**, 57(4), 237–242.

Bhutia, L.P.; Chakraborty, B.N.; Chakraborty, U. Management of charcoal stump rot disease using AMF and PGPR in Temi Tea Estate, Sikkim. *Journal of Mycology and Plant Pathology*, **2012,** 42(1), 1–12.

Boonham, N.; Kreuze, J.; Winter, S.; van der Vlugt, R.; Bergervoet, J.; Tomlinson, J.; Mumford, R. Methods in virus diagnostics: From ELISA to next generation sequencing. *Virus Research*, **2014**, 186, 20–31.

Chakraborty, B.N. Antigenic disparity. In: *Experimental and Conceptual Plant Pathology*; Singh, R.S., Singh, U.S., Hess, W.M., Weber, D.J., Eds.; Oxford & IBH Publishing Co., New Delhi, **1988**; pp. 477–484.

Chakraborty, B.N. Arbuscular mycorrhizal fungi and mycorrhiza helper bacteria: Molecular detection and their role in symbiosis and crop protection. *Journal of the Botanical Society of Bengal*, **2019**, 73, 1–20.

Chakraborty, B.N. Immunotechniques. In: *Instrumentations Manual in Biology*; Bhandari, J.B., Gurung, C., Eds.; Narosa Publishing House, New Delhi, **2021**; pp. 30–54.

Chakraborty, B.N. Induced immunity developed in plants against phytopathogens using bioinoculants. *Journal of Mycology and Plant Pathology*, **2020,** 50(1), 1–14.

Chakraborty, B.N. Microbial resources and their importance in agricultural biotechnology. In: *Eco-Conservation and Sutainable Living*; Gurung, C., Bhandari, J.B., Eds.; Narosa Publishing House, New Delhi, **2013**; pp. 11–30.

Chakraborty, B.N. Molecular recognition of fungal pathogens and activation of plant immune response. *Indian Phytopathology*, **2018**, 71(4), 471–483.

Chakraborty, B.N. Scoping the potential uses of beneficial microorganisms for biopesticide industry and entrepreneurship development in crop protection. In: *Perspectives of Plant Pathology in Genomic Era*; Chowdappa, P., Sharma, P., Singh, D., Misra, A.K., Eds.; Today and Tomorrow Publishers, New Delhi, **2016**; pp. 335–355.

Chakraborty, B.N.; Allay, S. Immunodetection of *Rhizophagus fasciculatus* and *Gigaspora gigantea* in soil and root tissues in *Citrus reticulata*, their exploitation as bioinoculants and cellular localization of defense enzymes following induced immunity developed against *Fusarium solani*. *Kavaka*, **2022**, 58(3), 1–10.

Chakraborty, B.N.; Basu, P.; Das, R.; Saha, A.; Chakraborty, U. Detection of cross reactive antigens between *Pestalotiopsis theae* and tea leaves and their cellular location. *Annals of Applied Biology*, **1995**, 127, 11–21.

Chakraborty, B.N.; Chakraborty, U. Immunodetection of plant pathogenic fungi. In: *Frontiers of Fungal Dversity in India*; Rao, G.P., Manoharachari, C., Bhat, J., Rajak, R.C., Lakhanpal, T.N., Eds.; International Book Distributing Co., Lucknow, **2002**; pp. 23–42.

Chakraborty, B.N.; Chakraborty, U. Induced immunity in plants using bioinoculants and biostimulants. In: *Microbial Antagonist: Role in Biological Control of Plant Diseases, Part-II Bacterial Antagonsists and Bacteriophages*; Pandey, R.N., Chakraborty, B.N., Singh, D., Sharma, P., Eds; Indian Phytopathological Society and Today & Tomorrow's Publishers, New Delhi, **2019**; pp. 693–712.

Chakraborty, B.N.; Chakraborty, U. Molecular detection of fungal pathogens and induction of phytoimmunity using bioinoculants. *Indian Phytopathology*, **2021**, 74(2), 307–322.

Chakraborty, B.N.; Chakraborty, U. Serological diagnosis of plant diseases. In: *Glimpses of Phytopathology*; Chourasia, H.K., Roy, A.K., Kumari, U., Eds.; A.B. Publication, Delhi, **2012**; pp. 1–9.

Chakraborty, B.N.; Chakraborty, U. Tea diseases: Early detection and their management strategies. In: *Recent Approaches for Plant Diseases*; Das, S., Dutta, S., Chakraborty, B.N., Singh, D., Eds.; Indian Phytopathological Society and Today and Tomorrow's Publishers, New Delhi, **2018**; pp. 175–208.

Chakraborty, B.N.; Chakraborty, U.; Allay, S. Wilt root rot complex in mandarin plants and activation of defense against pathogen. In: *Wilt Diseases of Crops*; Bhattacharya, A., Chakraborty, B.N., Pandey, R.N., Singh, D., Dubey, S.C. Eds.; Indian Phytopathological Society and Today and Tomorrow's Publishers, New Delhi, **2019**; pp. 293–321.

Chakraborty, B.N.; Chakraborty, U.; Chakraborty, A.P.; Sashankar, P. Serological and molecular detection of *Bipolaris sorokiniana* Sacc. causing spot blotch disease of wheat. *Journal of Mycopathological Research*, **2016a**, 54(1), 117–125.

Chakraborty, B.N.; Chakraborty, U.; Das, R.; Som, R. Development of immunoassays for early detection of *Glomerella cingulata* in tea leaf tissues. *Journal of Mycology and Plant Pathology*, **2008,** 38, 478–484.

Chakraborty, B.N.; Chakraborty, U.; De, U.; Chakraborty, A.P. Induction of resistance in *Camellia sinensis* against *Sclerotium rolfsii* by dual application of *Rhizophagus fasciculatus* and *Bacillus pumilus*. *Archives of Phytopathology and Plant Protection*, **2016b**, 48, 825–840.

Chakraborty, B.N.; Chakraborty, U.; Deb, D.; Sengupta, D.; Das, J. Immuno-enzymatic detection of *Fomes lamaoensis* in soil and root tissue. *Journal of Plantation Crops*, **2002a**, 15, 497–503.

Chakraborty, B.N.; Chakraborty, U.; Dey, P.L. Potential application of *Trichoderma* as biocontrol agents, their molecular characterization and diversity analysis. In: *Plant Diseases and Their Biological Control*; Singh, A. Ed.; Aavishkar Publishers, Jaipur, **2011**; pp. 186–216.

Chakraborty, B.N.; Chakraborty, U.; Rai, K.; Sunar, K.; Dey, P.L. Serological and molecular detection of *Macrophomina phaseolina* causing root rot of *Citrus reticulata*. *NBU Journal of Plant Sciences*, **2012,** 6, 77–86.

Chakraborty, B.N.; Chakraborty, U.; Sengupta, D.; Deb, D.; Das, J. Development of immunodiagnostic kits for detection of *Ustulina zonata* in soil and tea root tissues. *Journal of Basic and Applied Mycology*, **2002b**, 1, 58–61.

Chakraborty, B.N.; Chakraborty, U.; Sunar, K. Induced immunity developed by *Trichoderma* spp. in plants. In: Trichoderma – *Host Pathogen Interactions and Applications*; Sharma, A.K., Sharma, P., Eds.; Springer, Berlin, **2020**; pp. 125–147.

Chakraborty, B.N.; Chakraborty, U.; Sunar, K. Induced immunity developed by *Trichoderma* species in plants. In: Trichoderma – *Host Pathogen Interactions and Applications*; Sharma, A.K., Sharma, P., Eds.; Springer Nature, Singapore, **2020**; pp. 125–147.

Chakraborty, B.N.; Das Biswas, R. Serological Detection of *Corticium invisum* in Tea Leaf Tissue. *Journal of Mycology and Plant Pathology*, **2008**, 38(3), 458–465.

Chakraborty, B.N.; Purkayastha, R.P. Alteration in glyceollin synthesis and antigenic patterns after chemical induction of resistance in soybean to *Macrophomina phaseolina*. *Canadian Journal of Microbiology*, **1987**, 33, 835–840.

Chakraborty, B.N.; Purkayastha, R.P. Serological relationship between *Macrophomina phaseolina* and soybean cultivars. *Physiological Plant Pathology*, **1983**, 23,197–205.

Chakraborty, B.N.; Saha, A. Detection and cellular location of cross-reactive antigens shared by *Camellia sinensis*and *Bipolaris carbonum*. *Physiological and Molecular Plant Pathology*, **1994**, 44, 403–416.

Chakraborty, B.N.; Sarkar, B.; Chakraborty, U. Detection of cross-reactive antigens shared by *Fusarium oxysporum* and *Glycine max* by indirect ELISA and their cellular location in root tissues. *Folia Microbiologica*, **1997**, 42, 607–612.

Chakraborty, B.N.; Sharma, M. Pathogenesis-related proteins in plant defence. In: *Review of Plant Pathology*; Reddy, S.M., Gour, H.N, Eds.; Scientific Publishers, Jodhpur, **2008**; pp. 105–138.

Chakraborty, B.N.; Sharma, M. Serological detection and immunogold localization of cross-reactive antigens shared by *Camellia sinensis* and *Exobasidium vexans*. *Journal of Applied Microbiology*, **2007**, 103, 1669–1680.

Chakraborty, U.; Basu, P.; Das, R.; Saha, A.; Chakraborty, B.N. Evaluation of antiserum raised against *Pestalotiopsis theae* for detection of grey blight disease of tea by ELISA. *Folia Microbiologica*, **1996**, 41, 413–418.

Chakraborty, U.; Chakraborty, B.N.; Basnet, M. Plant growth promotion and induction of resistance in *Camellia sinensis* by *Bacillus megaterium*. *Journal of Basic Microbiology*, **2006**, 46(3), 186–195.

Chakraborty, U., Chakraborty, B.N., Basnet M., Chakraborty, A.P. Evaluation of *Ochrobactrum anthropi* TRS-2 and its talc based formulation for enhancement of growth of teaplants and management of brown root rot disease. *Journal of Applied Microbiology*, **2009**, 107(2), 625–634.

Chakraborty, U.; Chakraborty, B.N.; Chakraborty, A.P.; Sunar, K.; Dey, P.L. Plant growth promoting rhizobacteria: Diversity, mechanisms of action and perspectives in agriculture. In: *Review of Plant Pathology*; Chakraborty, B.N., Chakraborty, U., Eds.; Scientific Publishers (India), Jodhpur, Vol. 6, **2014**; pp. 215–268.

Chakraborty, U.; Das, R.; Basu, P.; Guha, S.; Chakraborty, B.N. Serological cross reactivity between *Glomerella cingulata* and *Camellia sinensis*. *Indian Phytopathology*, **2002c**, *55*(1), 1–7.

Chrzanowski, Ł.; Ławniczak, Ł.; Czaczyk, K. Why do microorganisms produce rhamnolipids? *World Journal of Microbiology and Biotechnology*, **2012**, 28, 401–419.

Cong, D.; Li, Y.; Xian, H. Purification, renaturation and characterization of chitinase gene from *Trichoderma asperellum*. *Chinese Agricultural Science Bulletin*, **2012**, 28, 34–38.

Das Biswas, R.; Chakraborty, B.N. Immunodetection of *Alternaria alternata* and evaluation of antifungal compounds in tea leaf tissues following challenge inoculation with pathogen. *Journal of Mycology and Plant Pathology*, **2020,** 50(3), 236–248.

Das, J.; Chakraborty, U.; Chakraborty, B.N. Immunodetection of *Sphaerostilbe repens* causing Violet root rot disease of *Camellia sinensis*. *Journal of Mycopathological Research*, **2022,** 60(1), 41–53.

Devay, J.E.; Adler, H.E. Antigens common to hosts and parasites. *Annual Review of Microbiology*, **1976,** 30, 147–168.

Devay, J.E.; Schnathorst, W.C.; Foda, M.S. Common antigens and host-parasite interactions. In: *The Dynamic Role of Molecular Constituents in Plant Parasite Interactions*; Mirocha, C.J., Uritani, I., Eds; Bruce Publishing Corporation, Minneapolis, MN, **1967**; pp. 313–328.

Djonovic, S.; Vargas, W.A.; Kolomiets, M.V.; Horndeski, M.; Wiest, A.; Kenerley, C.M. A proteinaceous elicitor *Sm1* from the beneficial fungus *Trichoderma virens* is required for induced systemic resistance in maize. *Plant Physiology*, **2007**, 145, 875–899.

Gawande, S.J.; Chimote, V.P.; Shukla, A.; Gadewar, A. Serological and molecular diagnosis of plant pathogens: A plant protection approach. In: *Plant Protection in New Millennium, Vol 1*; Gadewar, A.V., Singh, B.P., Eds.; Satish Serial Publishing House, Delhi, **2006**; pp. 563–588.

Geudens, N.; Martins, J.C. Cyclic lipodepsipeptides from *Pseudomonas* spp. - biological swiss-army knives. *Frontiers in Microbiology*, **2018**, 9, 1867.

Gulijimila, M.; Fan, H.J.; Liu, Z.H.; Wang, N.; Dou, K.; Huang, Y.; Wang, Z.Y. Cloning and sequence analysis of small molecular hydrophobin protein hyb2 gene from *Trichoderma asperellum* T4. *Chinese Agricultural Science Bulletin*, **2012**, 28, 85–91.

Gupta, R.; Bar, M. Plant immunity, priming, and systemic resistance as mechanisms for *Trichoderma* spp. biocontrol. In: Trichoderma – *Host Pathogen Interactions and Applications*; Sharma, A.K., Sharma, P., Eds.; Springer, Berlin, **2020**; pp. 81–110.

Jamaux, I.; Spire, D. Development of a polyclonal antibody-based immunoassay for the detection of *Sclerotinia sclerotiorum* in rapeseed petals. *Plant Pathology*, **1994**, 43, 847–862.

Joos, T.O.; Schrenk, M.; Hopfl, P.; Kroger, K.; Chowdhury, U.; Stoll, D.; Schorner, D.; Durr, M.; Herick, K.; Rupp, S.; Sohn, K.; Hammerle, H. A microarray enzyme-linked immunosorbent assay for autoimmune diagnostics. *Electrophoresis*, **2000,** 21(13), 2641–2650.

Kennedy, R.; Wakeham, A.J.; Byrne, K.G.; Meyer, U.M.; Dewey, F.M. A new method to monitor airborne inoculums of the fungal plant pathogens *Mycosphaerella brassicicola* and *Botrytis cinerea. Applied and Environmental Microbiology*, **2000**, 66(7), 2996–3003.

Khati, S.; Chakraborty, B.N. Immunodetection of *Drechslera oryzae* causing brown spot of rice, its root colonization with arbuscular mycorrhizal fungi and their use for induction of resistance. *Journal of The Botanical Society of Bengal*, **2019**, 73(2), 43–52.

Kramer, J.; Özkaya, Ö.; Kümmerli, R. Bacterial siderophores in community and host interactions. *Nature Reviews Microbiology*, **2020**, 18, 152–163.

Lopez, M.M.; Bertolini, E.; Olmos, A.; Caruso, P.; Gorris, M.T.; Llop, P.; Penyalver, R.; Cambra, M. Innovative tools for detection of plant pathogenic viruses and bacteria. *International Microbiology*, **2003**, 6, 233–243.

Mohan, S.B. Evaluation of antisera raised against Phytophthora fragariae for detecting the red core disease of strawberries by enzyme-linked immunosorbent assay (ELISA). *Plant Pathology*, **1988**, 37, 206–216.

Monnier, N.; Cordier, M.; Dahi, A.; Santoni, V.; Guenin, S.; Clement, C.; et al. Semi-purified rhamnolipid mixes protect *Brassica napus* against *Leptosphaeria maculans* early infections. *Phytopathology*, **2020**, 110, 834–842.

Monti, M.M.; Pedata, P.A.; Gualtieri, L.; Ruocco, M. The vocabulary of *Trichoderma*-plant interactions. In: Trichoderma – *Host Pathogen Interactions and Applications*; Sharma, A.K., Sharma, P., Eds.; Springer Nature, Singapore, **2020**; pp. 19–34.

Phelps, D.C.; Nemec, S.; Baker, R.; Mansell, R. Immunoassay for naphthazarin phytotoxin produced by *Fusarium solani. Phytopathology*, **1990,** 80(3), 298–302.

Pieterse, C.M.; Van der Does, D.; Zamioudis, C.; Leon-Reyes, A.; Van Wees, S.C. Hormonal modulation of plant immunity. *Annual Review of Cell and Developmental Biology*, **2012,** 28, 489–521.

Prsic, J.; Ongena, M. Elicitors of plant immunity triggered by beneficial bacteria. *Frontiers in Plant Science*, **2020**, 11, 594530.

Rao, R.S.; Visuri, S.R.; McBride, M.T.; Albala, J.S.; Matthews, D.L.; Coleman, M.A. Comparison of multiplexed techniques for detection of bacterial and viral proteins. *Journal of Proteome Research*, **2004**, 3(4), 736–742.

Rene, A.A.; van der, V.; van Raaij H.; de Weerdt, M.; Bergervoet, J.H.W. Multiplex detection of plant pathogens through the luminex magplex bead system. In: *Plant Pathology: Techniques and Protocols, Methods in Molecular Biology*; Lacomme, C., Ed.; Springer Science +Business Media, New York, **2015**; vol. 1302, pp. 283–299.

Rodriguez, J.; Tonelli, M.L.; Figueredo, M.S.; Ibáñez, F.; Fabra, A. The lipopeptide surfactin triggers induced systemic resistance and priming state responses in *Arachis hypogaea* L. *European Journal of Plant Pathology*, **2018**, 152, 845–851.

Sarkar, J.; Chakraborty, U.; Chakraborty, B.N. Induced defense response in wheat plants against *Bipolaris sorokiniana* following application of *Bacillus safensis* and *Ochrobacterum pseudogrignonense. Indian Phytopathology*, **2018,** 74(1), 49–56.

Sharma, P.; Kumar, P.V.; Ramesh, R.; Saravanan, K.; Deep, S.; Shrama, M.; Mahesh, S.; Singh, D. Biocontrol genes from *Trichoderma* species: A review. *African Journal of Biotechnology*, **2011**, 10(86), 19898–19907.

Sunar, K.; Chakraborty, U.; Chakraborty, B.N. *Bacillus altitudinis* and *Bacillus pumilus* induce resistance in *Brassica juncea* L. against *Thanatephorus cucumeris* causing Seedling blight. *Journal of Mycopathological Research*, **2020,** 58(3), 167–179.

Umesha, S.; Raghava, S. Advanced molecular diagnostics for detection of plant pathogenic bacteria. *Indian Phytopathology*, **2021**, 74(2), 431–436.

Vidal, E.; Yokomi, R.K.; Moreno, A.; Bertolini, E.; Cambra, M. Calculation of diagnostic parameters of advanced serological and molecular tissue-print methods for detection of Citrus tristeza virus: A model for other plant pathogens. *Phytopathology*, **2012**, 102, 114–121.

Vignali, D.A.A. Multiplexed particle based flow cytometric assays. *Journal of Immunological Methods*, **2000**, 243(1–2), 243–255.

Viswanathan, R.; Samiyappan, R.; Padmanaban, P. Specific detection of *Colletotrichum falcatum* in sugarcane by serological technique. *Sugarcane*, **1998**, 3, 18–23.

Walcott, R.R. Detection of seed borne pathogens. *Horticultural Technology*, **2003**, 13(1), 40–47.

7 Challenges and Opportunities for Sustainable Bio-Intensive Integrated Disease Management

Sunil Chandra Dubey, Aravindaram Kandan, Raman Ramesh, and O. P. Sharma

7.1 INTRODUCTION

In the world of increasing population, there has been a continuous increase in demand for the food and scarcity in supply. Indian agriculture has made tremendous progress as an import dependent on an exporting country. However, plant diseases are considered a major biotic stress in crop production since average crop losses due to pests and diseases in the agricultural fields were estimated to be 17%–30% for major crops (Savary et al. 2019). Different strategies and measures have been followed for the management of diseases such as use of agrochemicals, resistance breeding, and biological control, in which each has its own advantages and disadvantages. The cultivation of resistant varieties is one of the most practical and cost-effective options available for the management of the diseases, but the performance of resistant varieties is not satisfactory across the locations due to the presence of the high level of pathogenic variability among the majority of the pathogens (Dubey 2016). Thus, the pesticides continue to be the predominant and much relied weapon used against pests by farmers since long. Their overuse had witnessed a steady rise in consumption over the years. In the past, pesticides use in India had been increasing @ 2%–5% per annum; however, at present, it accounts for 2% of the total pesticides used in the world. Development of resistant varieties is time-consuming and cannot provide resistance against broad spectrum of pathogens. On the other hand, use of agrochemicals adversely affects the soil beneficial microorganisms, degrading the soil quality, contaminating the land and water resources, and most importantly, accumulating in the plant tissues posing threat to human health. The use of chemical pesticides has led to dramatic improvements in the production of crop plants. However, with passage of time, their impact has waned, and the pests have become an increasingly serious constraint. Diverse consumption of food has also changed, and people have

DOI: 10.1201/9781003531944-7

become aware of green or lesser pesticide foods. However, like other eco-conscious human beings, we are equally committed to sustainable agriculture incorporating Integrated Pest Management (IPM) as defined by the FAO, as an economically viable, ecologically safe, and socially acceptable approach to crop protection.

Owing to the awareness in food safety, degradation of soil, and resistance of pathogens to chemicals, the trend has been shifted toward sustainable disease management involving bio-intensive control of plant diseases. Biological control being eco-friendly and less expensive is considered a better alternative for control of plant pathogens.

7.2 SUSTAINABLE INTEGRATED DISEASE MANAGEMENT

The concept of integrated disease management refers to the practice of using a combination of various preventive and remedial measures to minimize the risk of disease incidence and damage. Hence, the idea of sustainable disease management involves the strategical management of the host, pathogen, and the environment, the factors of a disease triangle. Preventive measures include cultural practices through modification of environment, which does not favor the growth of pathogen such as decontamination of the farm equipment, crop rotation, pruning, use of resistant planting materials (Dufault et al. 2019). Control measures involve use of biological control and modified nanoformulations of organic amendments (Mahapatra et al. 2022) over conventional agrochemicals. The effective management should be targeted to exclusion of the pathogen as well as by reduction in the amount and efficiency of the initial inoculum. Use of pathogen-free seeds, field sanitation by removal of plant debris, soil solarization, disinfestation of infested soil, cultivation of resistant cultivars, sowing date, and seed treatment with chemicals/biocontrol agents are some of the practical approaches available for the management of majority of the diseases (Dubey 2016).

7.2.1 Biological Control

As the word implies, the term biological control refers to the use of microbial antagonists, modified organisms, or modified genes or their derived products to suppress the disease as well as to promote the crop growth, desirable microorganisms, and predators (Heydari and Pessarakli 2010). An organism that suppresses the pathogen is referred to as a biological control agent (BCA). A BCA colonizes the rhizosphere/the site requiring protection and leaves no toxic residues as opposed to chemicals. The first requirement for biological control is the identification and deployment of highly effective species.

7.2.1.1 Advantages of Biocontrol Agents

1. They avoid environmental pollution of soil, air, and water as is being experienced in chemical control.
2. They avoid adverse effect on the beneficial microbes including antagonists in the soil, whereas chemicals are lethal.
3. They are less expensive compared to chemical control.

4. Continuous uses of bioagents avoid the development of resistant strains of pathogens.
5. Provide sustainable management of plant diseases.
6. Improve immunity of the plants and enhance plant growth and yield attributing factors including yield.

7.2.1.2 Mechanisms of Disease Control by Biocontrol agents

7.2.1.2.1 Antibiosis

Certain secondary metabolites produced by the microorganisms that are synthesized and released into the environments could be harmful or deleterious to other microorganisms. Antibiosis is considered the most potential mode of action against pathogens as antibiotic-producing microbes have competitive advantage in resource-limited environments (Raaijmakers and Mazzola 2012). Production of antimicrobial metabolites, mostly with broad-spectrum activity, has been reported for biocontrol bacteria belonging to *Agrobacterium* spp., *Bacillus* spp., *Pantoea* spp., *Pseudomonas* spp., *Serratia* spp., *Streptomyce* spp., and many other genera (Kohl et al. 2019). Plant-associated fluorescent and non-fluorescent pseudomonads, *Burkholderia* sp., *Streptomyces* sp., are well known for the production of antibiotics such as phenazine-1-carboxylic acid, 2,4-diacetylphloroglucinol (DAPG), pyrrolnitrin, rhamnolipids, ecomycin, butyrolactones, sulphonamides, pyoluteorin, pseudomonic acid (murporicin), and azomycin against a wide variety of plant pathogens (Dilantha-Fernando et al. 2006; Weller 2007). 2,4-diacetylphloroglucinol (DAPG), hydrogen cyanide (HCN), and siderophores are diffusible and volatile antimicrobials produced by endophytes and reported to inhibit pathogenic *Ralstonia solanacearum* strains (Zhou et al. 2012). Use of volatile compound ammonia as soil amendments has been reported to control bacterial wilt (Yadessa et al. 2010). Endophytic and rhizobacterial strains of *Burkholderia* sp., *Bacillus* sp., and DAPG-producing *Pseudomonas* sp. from brinjal have been reported to be inhibitory to *R. solanacearum* (Ramesh et al. 2009; Ramesh and Phadke 2012). Strains of genera *Enterobacter* sp., *Pseudomonas* sp., *Bacillus* sp., and *Staphylococcus* sp isolated from xylem of brinjal and chilli produced antimicrobials, HCN, siderophores, and ammonia against *R. solanacearum*, prevented bacterial wilt, and also promoted brinjal growth (Achari and Ramesh 2014; Achari 2015).

Antibiotics produced by microorganisms have the ability in overwhelming plant pathogens resulting in reduction in diseases caused by them. The volatile and non-volatile toxic metabolites produced by *Trichoderma* species inhibited the growth of several plant pathogenic fungi as *R. solani* (Dubey and Patel 2001) and *R. bataticola* (Dubey 2002). The secondary metabolites from the culture filtrate and mycelial mats of *Trichoderma* species were found to be inhibitory to mycelial growth of *R. bataticola* and *Fusarium oxysporum* f. sp. *ciceris* causing dry root rot and wilt, respectively, in chickpea. Gas chromatography-mass spectrometry (GC-MS/MS) was used to chemically characterize some of these compounds from these metabolites. *Trichoderma* species showed the activities of enzymes related to defense, namely chitinase and β-1, 3-glucanase. Activity of Chitinase was higher in *T. virens*, whereas *T. viride* showed higher activity of β-1,3-glucanase (Dubey et al. 2011c).

7.2.1.2.2 Competition

Necrotrophic pathogens during their saprophytic life cycle germinate, survive, reproduce independent of the host, which depends on the exogenous nutrition available in the environment, which makes the pathogens vulnerable for nutrient competition needed for its survival. Saprophytic microbial agents, which are potential in degrading the dead organic matter, and their ability to produce siderophores help them scavenge the essential nutrients and colonize first in the environment eliminating the pathogens (Spadaro and Droby 2016). Examples are the application of *Microsphaeropsis ochracea* controlling *Venturia inaequalis* in apple (Carisse et al. 2000), and *Trichoderma asperellum* producing iron-binding siderophores controls Fusarium wilt (Segarra et al. 2010). Xylem-residing bacteria from solanaceous vegetables can recolonize brinjal and prevent bacterial wilt infection through niche exclusion (Achari 2015). Xue et al. (2009) reported that soil drenching with antagonistic rhizosphere colonizing strains of *Acinetobacter* sp. and *Enterobacter* sp. reduced bacterial wilt incidence in tomato plants. Xylem-colonizing strain of *Bacillus subtilis* reduced bacterial wilt incidence in mulberry through production of antimicrobials and niche exclusion as seen from its high endophytic and rhizosphere population densities (Ji et al. 2008). Wei et al. (2013) have reported recolonization of tomato xylem by endophytic strain of *Ralstonia pickettii* and prevention of bacterial wilt by niche exclusion. Hu et al. (2016) have demonstrated the importance of diversity of consortium consisting eight strains of antagonistic *Pseudomonas* sp. in the plant rhizosphere in creating an intensified competition for resources and prevention of bacterial wilt in tomato.

7.2.1.2.3 Hyper parasitism

This is a kind of interaction often reported in fungal biocontrol agents in which they acquire nutrients from the pathogens, which are also the parasite of the plant and, hence, the term hyper parasitism. Common mechanism of parasitism is the excretion of cell-wall-degrading enzymes combined in some cases with excretion of secondary metabolites in close contact with the host cell, leading to openings in the cell wall and subsequent disorganization of the cytoplasm. Cell wall degradation is typically caused by a range of chitinases, β-1,3-glucanases, and proteases or, in case of hyperparasites of oomycota, cellulases. The most studied mycoparasites belong to the genera *Trichoderma* spp. and *Clonostachys* spp. Antagonistic isolates of these genera vary in host range, and individual strains mostly have a range of plant pathogenic hosts. They produce structures for attachment and infection and kill their hosts by CWDEs, often in combination with antimicrobial secondary metabolites (Mukherjee et al. 2012; Karlsson et al. 2017; Nygren et al. 2018).

Generally, three types of interactions occur between plants and the rhizosphere microbiota, namely positive, negative, and neutral (Whipps 2001). Plant growth-promoting rhizobacteria (PGPR) can be defined as an indispensable part of rhizosphere microbiota that imparts a positive effect on the plant's growth and health either directly or indirectly (Vejan et al. 2016). PGPR plays a direct role in plant growth promotion by solubilization of nutrients such as phosphate, which involves basic mechanisms such as the expulsion of complex and mineral-dissolving compounds such as organic acid anions, protons, OH− ions, and CO_2, production of extracellular enzymes, and biological phosphate mineralization (Sharma et al. 2013). Biological nitrogen fixation, production of siderophores, which are low-molecular-weight iron-chelating compounds (Schalk et al. 2011), production of phytohormones such as auxins, gibberellins, cytokinins which exert profound effect on root and shoot elongation and subsequently increase the nutrient and water uptake also add up to the direct mechanisms involved in plant growth regulation (Arora et al. 2013). Indirect role of PGPR includes antagonistic activity against pathogenic microorganisms, eliciting induced systemic resistance against a broad spectrum of plant pathogens, and stimulating ethylene (ET) and jasmonic acid pathway (Glick 2012). A recent study reports that rhizobacterial strains *Alcaligenes faecalis* and *Bacillus amyloliquefaciens* induced resistance in rice against blast and sheath blight (Kakar et al. 2018).

The hyphae of *Trichoderma* coiled around the mycelium of *R. solani* and killed them. They entered inside the hyphae of *R. solani* and sporulated abundantly around the point of entry (Dubey and Patel 2001). It was further observed that sclerotia of *R. solani* were colonized and lost their viability at 14 days. The hyphae of *Trichoderma* were observed inside the sclerotial cells. Dubey (1998) and Kumar and Dubey (2001) observed similar interactions with other plant pathogenic fungi.

7.2.1.2.4 Induced systemic resistance

Plants possess different range of protective mechanisms against disease-causing pathogens and insect pests, which are usually triggered by biotic factors. Induced systemic resistance is a "state of enhanced defence capacity" elicited by specific environmental stimuli whereby plant's innate defense mechanisms are activated and protected against a broad spectrum of pathogens (Choudhary et al. 2007). Induced systemic resistance is mediated by a network of interconnected signaling pathways. Induced defense mechanisms involve the production of reactive oxygen species, phytoalexins, phenolic compounds, or pathogenesis-related proteins or the formation of physical barriers such as modifications of cell walls and cuticles by the induced plant (Wiesel et al. 2014). Hoffland et al. (1995) provided evidence that in radish *P. fluorescens* WCS417r-ISR against *F. oxysporum* developed without accumulation of the pathogenesis-related (PR) proteins that are characteristic for systemic acquired resistance (SAR). Similarly, *P. fluorescens* WCS417r-ISR in *Arabidopsis* was shown to develop without PR gene activation in systemic leaf tissue. *Trichoderma virens* has been reported to induce resistance in maize (Djonovic et al. 2007), and various *Bacillus* species have been studied for their capability in inducing systemic resistance against broad range of pathogens (Kloepper et al. 2004). The role of phytohormones in basal resistance and *Trichoderma* induced systemic resistance to *Botrytis cinerea* in *Arabidopsis thaliana* (Korolev et al. 2008). *Trichoderma* species induce defense mechanism into the plants, which leads to reduction in disease development. The host defenses may be local and/or systemic in nature. Induction of host SAR, which is mediated by salicylic acid (SA) pathway following pathogen infection, typically leads to the expression of (PR proteins responsible for induced localized cell death. Induced systemic resistance (ISR) is arbitrated by jasmonic acid (JA) and/or ET pathway in response to some non-pathogenic rhizobacteria (Dubey and Tripathi 2019). Majority of endophytic and rhizosphere bacteria activate ISR against all types of plant pathogens, including viruses, fungi, bacteria, and nematodes via an SA-independent pathway involving jasmonate (JA) and ET signals. However, acetoin, a volatile metabolite produced by plant-associated bacteria, can elicit ISR via SA and ET signaling pathways. Acetoin-producing *Bacillus* sp. and *Enterobacter* sp. are useful as consortia for induction of ISR in plants against phytopathogens (Jetiyanon and Kloepper 2002; Yi et al. 2013). Acetoin production by bacterial strains isolated is also reported as one of the mechanisms for ISR in brinjal by xylem-residing antagonistic bacteria isolated from brinjal (Achari 2015). Fluorescent *Pseudomonas* sp. actively secretes SA in rhizosphere in nanogram quantities, which is converted to siderophore pyochelin, which can be effective in inducing resistance against bacterial wilt. Pyochelin and pyocyanin act synergistically to produce reactive oxygen species that cause cell damage and induce resistance against several invading plant pathogens (Dilantha-Fernando et al. 2006).

Association of *Trichoderma* with plants enhanced root development, productivity of crop, resistance to biotic and abiotic stresses, and nutrients uptake (Dubey and Tripathi 2019). Association of *Trichoderma* spp. with plants enhanced the growth of various crop plants (Dubey 2000, 2003a,b, 2006; Dubey et al. 2007). This response may be due to the suppression of deleterious root microflora including rot causing diseases, production of growth stimulating factors (hormones or growth factors) and increased nutrient uptake through solubilization and sequestration of nutrients, and/or enhanced root growth. Enhanced root development is also helpful for plants in tolerating biotic and abiotic stresses. Soil application of *Trichoderma* multiplied on various substrates significantly enhanced the shoot and root length of mungbean and urdbean crops (Dubey and Patel 2002; Dubey et al. 2009, 2011a, 2013).

7.2.1.3 Microbes Involved in Biocontrol of Plant Pathogens

7.2.1.3.1 Bacteria

Bacteria benefit from the plant nutrients provided by the roots, but plants can benefit from their rhizobacteria as well. Some bacteria serve as a first line of a plant's defense against pathogenic bacteria, fungi, and other parasites. Capability of endophytic and rhizosphere bacteria to colonize internal host tissues has made them valuable tool in agriculture to improve crop performance (Compant et al. 2005; Rosenblueth and Martínez-Romero 2006). One of the most commonly studied groups of organisms belongs to *Pseudomonas* species, which have been implicated as protective bacteria that suppress root-fungus disease. Mercado-Blanco and his coworkers (2004) isolated a bacterial endophyte *Pseudomonas fluorescens* PICF7 from the roots of olive trees and tested it for the suppression of *Verticillium* wilt, one of the most serious diseases affecting olive trees worldwide. Bacteria from the *Bacillus* group are microorganisms that inhabit a large number of different habitats. They are well known as producers of a wide array of antagonistic compounds of different structures, having between 5% and 8% of the total genome devoted to biosynthesis of secondary metabolites (Fira et al. 2018). *Bacillus amyloliquefaciens* FZB42 and GA1 were shown to possess biocontrol activity against the *Xanthomonas* strains by producing the antibiotic compounds difficidin and bacilysin (Arguelles-Arias et al. 2009). A bacterial endophyte from mangrove, identified as *Bacillus amyloliquefaciens* Bg-C31, was found to be effective in the biocontrol of *Capsicum* bacterial wilt caused by *Ralstonia solanacearum* in pot and field trials. Avirulent strains of *R. solanacearum, Pseudomonas* spp., *Bacillus* spp., and *Streptomyces* spp. are well-known biological control agents. Bacterial wilt suppression by endophytic bacteria is attributed to production of volatile and diffusible antimicrobial compounds, niche exclusion, and induced systemic resistance (Compant et al. 2005). These properties of endophytes make them valuable for biocontrol of bacterial wilt. Bacteria of the genera *Pseudomonas*, *Bacillus*, *Paenibacillus,* and *Sphingomonas* isolated from bacterial wilt resistant plants were reported to be highly antagonistic to *R. solanacearum* (Feng et al. 2013). Very high biocontrol effect against bacterial wilt and growth promotion in tomato was exhibited by *P. fluorescens* (Seleim et al. 2011). Endophytic strains of *Burkholderia* sp., *Bacillus* sp., and DAPG producing *Pseudomonas* sp. from brinjal were found inhibitory to *R. solanacearum* (Ramesh et al. 2009). These rhizobacteria possessing plant growth promoting ability in addition to antagonism against various plant pathogens make them promising for their use in agriculture. *Streptomyces lydicus* reduced root and seed rot severity in peas and resulted in significantly higher final emergence and significantly reduced disease incidence in spinach by *Pythium* and *Fusarium* (soil-borne fungi) (Hu et al. 2010).

7.2.1.3.2 Fungi

With fungi being abundant in soil, which is the natural environment for plant growth, various saprophytic fungi with specific interactions with specific pathogens can be utilized as an alternative of chemical fungicides. *Trichoderma* is identified as the genus with greatest potential comprising 25 biocontrol agents that have been used against a number of plant fungal diseases. In addition to *Trichoderma*, nine genera are recognized as significant comprising five or more known antagonistic species, namely, *Alternaria*, *Aspergillus*, *Candida*, *Fusarium*, *Penicillium*, *Pichia*, *Pythium*, *Talaromyces*, and *Verticillium* (Thambugala et al. 2020). *Trichoderma* species have been explored as biological control agents for over eight decades, now the several strains are commercially available. In India, a total of 971 various biopesticides are registered by Central Insecticides Board and Registration Committee (CIB & RC), and out of these, 300 are from *T. viride* and 55 from *T. harzianum*. *Trichoderma harzianum* Rifai, *Pochonia chlamydosporia,* and *Paecilomyces lilacinus* to suppress soybean root diseases have been reported in Northeast China. *T. harzianum* produces a butenolide metabolite called harzianolide that stimulates growth and defense mechanisms of tomato plants, resulting in a 16%–30% reduction of disease caused by *S. sclerotiorum* (Wei et al. 2015). *Paecilomyces farinosus, Penicillium chrysogenum, Trichoderma harzianum, T viride,* and *Verticillium lecanii* are reported to be antagonistic to powdery mildew (Kiss 2003). Some studies showed that yeast has the property to inhibit the potential pathogen and induce the stress tolerance ability of plant (Kamel et al. 2016). Arbuscular mycorrhiza, *Glomus versiforme,* reduced population of *R. solanacearum* in tomato rhizosphere and xylem by induction of ISR (Zhu and Yao 2004). *Pythium oligandrum,* a non-phytopathogenic oomycete, effectively colonized tomato rhizosphere and prevented occurrence of bacterial wilt (Takenaka et al. 2008). *Cunninghamella elegans* has been shown to be effective in prevention of bacterial wilt in tomato (Oliveira et al. 2017).

Trichoderma spp. have been proved as potential biocontrol agent against several fungal plant pathogens and exploited at field level for disease management of wilt in chickpea caused by *F. oxysporum* f. sp. *ciceris* (Kaur and Mukhapadhyay 1992; Dubey et al. 2007; Dubey et al. 2013; Dubey et al. 2015; Dubey et al. 2017); stem rot in pea caused by *F. oxysporum* f. sp. *pisi* (Kumar and Dubey 2001); web blight/wet root rot of urdbean (Dubey 2006; Dubey et al. 2018) and mungbean (Dubey 2007; Dubey et al. 2011a; Dubey and Singh 2013) and chickpea (Dubey et al. 2012) caused by *Rhizoctonia solani*; dry root rot of mungbean (Dubey et al. 2009) and chickpea (Dubey et al. 2011a) caused by *Rhizoctonia bataticola;* root rot and damping off complex in French bean caused by *R. solani* and *Pythium ultimum* (Dubey et al. 2017); and stem rot of chickpea caused by *Sclerotinia sclerotiorum* (Mandal et al. 2015).

7.2.1.3.3 Virus

Bacteriophages are emerging as a novel biocontrol agent, which could be effectively potential alternative than conventional fungal and bacterial biocontrol agents (Dy et al. 2018). Despite the promising early work, phage therapy did not prove to be a reliable and effective means of controlling bacterial plant diseases. More than 30 years after the earliest work, Goto in 1992 concluded that practical use of phages for control of bacterial plant disease in the field has not been successful. More than three decades later, several studies indicated the renewed interest in bacteriophages in plant disease control (Jones et al. 2007). Phage interactions are typical for a root microbiome, which suggests that phages could be useful tools in biocontrol. A major limiting factor in using phages for control of plant diseases was the probability of developing bacterial strains resistant to the phage. This risk was addressed early on by preparing a mixture of host range mutant phages (h-mutants) for disease control (Flaherty et al. 2001). Phages are utilized for controlling plant pathogens either in the rhizosphere or phyllosphere. Gill and Abedon (2003) identified several factors that can hinder success of disease control in the rhizosphere. A number of approaches such as applications of phages with avirulent pathogen strains (Tanaka et al. 1990), with biocontrol agents (Svircev et al. 2006), and with SAR inducers (Obradovic et al. 2005) have been used for incorporating phage applications into part of an integrated management strategy of plant bacterial diseases.

For a most successful tailored product, bacteriophages' life cycle and host interactions of the pathogen need to be thoroughly understood (Holtappels et al. 2020). The antimicrobial mechanisms are based on lysis of bacterial cells caused by virulent phages or lytic enzymes. The phages could also be manipulated genetically to deliver specific lethal genes to the target organisms or to sensitize them against traditional antimicrobials, which, however, raises the problems of spreading antibiotic resistance and responsible usage of genetically modified organisms (GMOs) in the environment. Overall, phages may represent an "outside of the box" solution in biocontrol, and such unconventional thinking is necessary to create new innovations in agriculture.

Fujiwara et al. (2011) have reported effectiveness of *R. solanacearum*-specific phage ϕRSL1, a jumbo phage belonging to *Myoviridae* in preventing bacterial wilt in tomato. Infective phage ϕRSL1 particles were detected in plant rhizosphere and bulk soil highlighting its usefulness as BCA. A new phage, ϕRS138 (*Siphoviridae*) has been isolated from tomato field soil lysed strains of *R. solanacearum* (Thi et al. 2016). Filamentous phages act as "double edged swords" by increasing or decreasing the expression of virulence factors in *R. solanacearum* (Yamada 2013). Nevertheless, lytic action of phages specifically against *R. solanacearum* and ability to survive in soil are an important feature for their use as BCA.

7.3 CHALLENGES IN EFFECTIVE UTILIZATION OF BIOLOGICAL CONTROL AGENTS

Major challenges related to the development and the use of fungal and bacterial endophytes as biocontrol agents despite the strong antagonism and ability of endophytes to benefit the host plant through inducing resistance and growth promotion need to be overcome when developing or using them as biocontrol agents. Endophytic bacteria and fungi face competition from the heterogeneous diversity of indigenous soil microflora and the varying degree of their population densities, usually affected by the environmental conditions. The early establishment of the endophytic community on root rhizosphere as well as in host tissues can help avoid the diminishing of the introduced endophyte (Ting 2014). Furthermore, potential human-pathogenic bacteria are also found among endophytes, which should be identified at the early stages (Hallmann and Berg 2006), which is a possible health concern. Identifying endophytes that are potentially pathogenic in early stages of the research is crucial, so they should not be considered for applications (Mercado-Blanco and Prieto 2012; Mercado-Blanco and Lugtenberg 2014) that some endophytes have excellent bioactivity in greenhouse trials, sometimes they are not used in the fields, possibly due to their poor viability during less shelf life, poor root colonization, toxicity to nontargeted organisms, or the fact that they are not amenable to large-scale applications because they produce very low yield in cultures (Yu et al. 2010; Ting 2014). Increased number of research reports on the use of biocontrol agents for plant disease management are coming nowadays, but their efficacy has not been documented through field trials. One of the important reasons for lack of consistent control is lack of appropriate formulation and delivery system. Though lot of scientific progress is made in understanding the nature of pathogen, biocontrol agents, plant–microbe interactions at the molecular level, which has yet to be translated into effective use in disease control. The low commercial contribution is highly associated with low commercial technology transfer such that its economic value is not yet realized by thc agricultural community, particularly in developing countries. The efficacy of many BCAs is usually strongly affected by biotic and abiotic factors, and its durability under continuous deleterious pathogen evolution is concerned, further constraining the application of the approach.

7.4 FUTURE OPPORTUNITIES

It is a well-known fact that plant pathogens always coexist with other soil and plant microflora; any augmentation of beneficial microflora would benefit the plants by suppressing the pathogen. Certain procedures used to study bacterial and fungal endophytes may improve our understanding of their antagonistic mechanisms and subsequently can lead to better control of plant diseases under field conditions. Adapting strategies such as genomics approaches, *in vivo* expression technology (IVET), and recombination *in vivo* expression technology (RIVET) (Leveau and Lindow 2001; Zhang et al. 2006) may provide an insight into genes that are required by bacteria to enter, compete, colonize the plant, suppress pathogens, and generally survive within the plant (Mercado-Blanco and Lugtenberg 2014). Root colonization is an important factor in evaluating the competence of potential microbial biocontrol agents since it is the delivery system of the beneficial microbes and their products (Malfanova et al. 2013). Studying the colonization process of endophytes through investigating their preferred entry site *via* fluorescence experiments can be very helpful in the evaluation of their competence as biocontrol agents. Especially, the availability of GFP gene mutants with different colors may allow various microbes labeled with GFP and derivatives to be visualized at the same time in the same preparation (Bloemberg et al. 2000; Malfanova et al. 2013). It is very important to understand that a particular mechanism of antagonism is not working in exclusion, and clear understanding may lead to new ideas for their successful deployment. Finally, it is very important to realize that biological control alone will not yield optimum dividend, and it should be integrated with cultural, host resistance and other management practices for improved efficacy.

REFERENCES

Achari, G. A. Interactions of *Ralstonia solanacearum*, a Bacterial wilt Pathogen with the Xylem Residing Bacteria in Solanaceous Vegetables, Ph.D. Thesis., **2015**, Goa University.

Achari, G. A.; Ramesh, R. Diversity, biocontrol, and plant growth promoting abilities of xylem residing bacteria from solanaceous crops. *Int. J. Microbiol.* **2014**, 296521, 1–14.

Arguelles-Arias, A.; Ongena, M.; Halimi, B. et al. *Bacillus amyloliquefaciens* GA1 as a source of potent antibiotics and other secondary metabolites for biocontrol of plant pathogens. *Microb. Cell Fact.* **2009**, 8, 1–12.

Arora, N. K.; Tewari, S.; Singh, R. Multifaceted plant-associated microbes and their mechanisms diminish the concept of direct and indirect PGPRs. In: Arora, N. (ed), *Plant Microbe Symbiosis: Fundamentals and Advances*, Springer: New Delhi, **2013**; pp. 411–449.

Bloemberg, G. V.; Wijfjes, A. H.; Lamers, G. E.; Stuurman, N.; Lugtenberg, B. J. Simultaneous imaging of *Pseudomonas fluorescens* WCS365 populations expressing three different auto- fluorescent proteins in the rhizosphere: New perspectives for studying microbial communities. *Mol. Plant-Microbe Interact.* **2000**, 13, 1170–1176.

Carisse, O.; Philion, V.; Rolland, D.; Bernier, J. Effect of fall application of fungal antagonists on spring ascospore production of the apple scab pathogen, *Venturia inaequalis. Phytopathology*, **2000**, 90, 31–37.

Choudhary, D. K.; Prakash, A.; Johri, B. N. Induced systemic resistance (ISR) in plants: Mechanism of action. *Indian J. Microbiol.* **2007**, 47, 289–297.

Compant, S.; Duffy, B.; Nowak, J.; Clement, C.; Ait Barka, E. Use of plant growth-promoting bacteria for biocontrol of plant diseases: Principles, mechanisms of action, and future prospects. *Appl. Env. Microbiol.* **2005**, 71, 4951–4959.

Dilantha-Fernando, W. G.; Nakkeeran, S.; Zhang, Y. Biosynthesis of antibiotics by PGPR and its relation in biocontrol of plant diseases. In: Siddiqui, Z. A. (ed), *PGPR: Biocontrol and Biofertilization*, Springer: Netherlands, **2006**; pp. 67–109.

Djonovic, S.; Vargas, W. A.; Kolomiets, M. V.; Horndeski, M.; Wiest, A.; Kenerley, C. M. A proteinaceous elicitor Sm1 from the beneficial fungus *Trichoderma virens* is required for induced systemic resistance in maize. *Plant Physiol.* **2007**, 145, 875–889.

Dubey, S. C. Evaluation of fungal antagonists of *Thanetiphorus cucumeris* causing web blight of horsegram. *J. Mycol. Pl. Pathol.* **1998**, 28, 15–17.

Dubey, S. C. Biological management of web blight of groundnut (*R. solani*). *J. Mycol. Pl. Pathol.* **2000**, 30, 89–90.

Dubey, S. C. Bio-agent based integrated management of collar rot of French bean. *Indian Phytopath.* **2002**, 55, 230–231.

Dubey, S. C. Integrated management of web blight of urd/mung bean by bio-seed treatment. *Indian Phytopath.* **2003a**, 56, 34–38.

Dubey, S. C. Integrated management of web blight of urd and mung bean. *Indian Phytopath.* **2003b**, 56, 413–417.

Dubey, S. C. Integrating bioagent with plant extract, oil cake and fungicide in various mode of application for the better management of web blight of urdbean. *Arch. Phytopathol. Plant. Protec.* **2006**, 39, 341–351.

Dubey, S. C. Integrating bioagent with botanical and fungicide in different mode of application for the better management of web blight and mungbean yield. *Indian J. Agril. Sci.* **2007**, 77, 162–165.

Dubey, S. C. Race profiling, genetic diversity, diagnostics and management of *Fusarium oxysporum* f. sp. *ciceris* causing wilt in chickpea. *Indian Phytopath.* **2016**, 69, 210–217.

Dubey, S. C.; Bhavani, R.; Singh, B. Development of Pusa 5SD for seed dressing and Pusa Biopellet 10G for soil application formulations of *Trichoderma harzianum* and their evaluation for integrated management of dry root rot of mungbean (*Vigna radiata*). *Biol. Control.* **2009**, 50, 231–242.

Dubey, S. C.; Bhavani, R.; Singh, B. Integration of soil application and seed treatment formulations of *Trichoderma* species for management of wet root rot of mungbean caused by *Rhizoctonia solani. Pest. Manage. Sci.* **2011a**, 67, 1163–1168.

Dubey, S. C.; Patel, B. Evaluation of fungal antagonist against *Tharetophours cucumeris* causing web blight of urd and mung bean. *Indian Phytopath.* **2001**, 54, 206–209.

Dubey, S. C.; Patel, B. Mass multiplication of antagonists and standardization of effective dose for management of web blight of urd and mung bean. *Indian Phytopath.* **2002**, 55, 338–341.

Dubey, S. C.; Singh, B. Integrated management of major diseases of mungbean by seed treatment and foliar application of insecticide, fungicides and bioagent. *Crop Prot.* **2013**, 47, 55–60.

Dubey, S. C.; Singh, B.; Gupta, O.; Saxena, D. R.; Sharma, O. P.; Kohire, O. D.; Anadani, V. P.; Singh, R. K.; Singh, S. K.; Tripathi, A. Management of wilt and root rots of chickpea (*Cicer arietinum*) using *Trichoderma harzianum* in India. *Indian J. Agril. Sci.* **2017**, 87, 1283–1287.

Dubey, S. C.; Singh, B.; Tripathi, A. Integrated management of wet root rot, yellow mosaic, and leaf crinkle diseases of urdbean by seed treatment and foliar spray of insecticide, fungicide, and biocontrol agent. *Crop Prot.* **2018**, 112, 269–273.

Dubey, S. C.; Singh, V.; Priyanka, K.; Upadhyay, B. K.; Singh, B. Combined application of fungal and bacterial bio-agents, together with fungicide and *Mesorhizobium* for integrated management of Fusarium wilt of chickpea. *BioControl.* **2015**, 60, 413–424.

Dubey, S. C.; Suresh, M.; Singh, B. Evaluation of *Trichoderma* species against *Fusarium oxysporum.* f. sp. *ciceris* for integrated management of chickpea wilt. *Biol. Control.* **2007**, 40, 118–127.

Dubey, S. C.; Tripathi, A. *Trichoderma* species as bio-agent in integrated disease management for sustainable agriculture. In: Parmar, B. S., Singh, S. B. and Walia, S. (eds), *Pesticides and Pests*, Cambridge Scholars Publishing: Cambridge, **2019**; pp. 264–273.

Dubey, S. C.; Tripathi, A.; Bhavani, R.; Singh, B. Evaluation of seed dressing and soil application formulations of *Trichoderma* species for integrated management of dry root rot of chickpea. *Biocontrol Sci. Techn.* **2011b**, 21, 93–100.

Dubey, S. C.; Tripathi, A.; Dureja, P.; Grower, A. Characterization of secondary metabolites and enzymes produced by *Trichoderma* species and their efficacy against plant pathogenic fungi. *Indian J. Agril. Sci.* **2011c**, 81, 455–461.

Dubey, S. C.; Tripathi, A.; Singh, B. Combination of soil application and seed treatment formulations of *Trichoderma* species for integrated management of wet root rot caused by *Rhizoctonia solani* in chickpea. *Indian J. Agril. Sci.* **2012**, 82, 357–364.

Dubey, S. C.; Tripathi, A.; Singh, B. Integrated management of Fusarium wilt by combined application of soil and seed dressing formulations of *Trichoderma* species to increase grain yield of chickpea. *Int. J. Pest. Manage.* **2013**, 59, 47–54.

Dubey, S. C.; Tripathi, A.; Tak, R.; Indira, D. S. Management of root rot and damping off complex in French bean by biocontrol agents. *Indian Phytopath.* **2017**, 70, 388–390.

Dufault, N.; Paret, M.; Small, I. Future of peanut pest management: A plant pathologist's perspective. *Peanut Sci.* **2019**, 46, 107–109.

Dy, R. L.; Rigano, L. A.; Fineran, P. C. Phage-based biocontrol strategies and their application in agriculture and aquaculture. *Biochem. Soc. Trans.* **2018**, 46, 1605–1613.

Feng, H.; Li, Y.; Liu, Q. Endophytic bacterial communities in tomato plants with differential resistance to *Ralstonia solanacearum*. *Afr. J. Microbiol.* **2013**, 7, 1311–1318.

Fira, D.; Dimkić, I.; Berić, T.; Lozo, J.; Stanković, S. Biological control of plant pathogens by Bacillus species. *J. Biotechnol.* **2018**, 285, 44–55.

Flaherty, J. E.; Harbaugh, B. K.; Jones, J. B.; Somodi, G. C.; Jackson, L. E. H-mutant bacteriophages as a potential biocontrol of bacterial blight of geranium. *HortScience*, **2001**, 36, 98–100.

Fujiwara, A.; Fujisawa, M.; Hamasaki, R.; Kawasaki, T.; Fujie, M.; Yamada, T. Biocontrol of *Ralstonia solanacearum* by treatment with lytic bacteriophages. *Appl. Environ. Microbiol.* **2011**, 77, 4155–4162.

Gill, J. J.; Abedon, S. T. Bacteriophage Ecology and Plants. *APSnet*, **2003**, https://www.apsnet.org/edcenter/apsnetfeatures/Pages/BacteriophageEcology.aspx

Glick, B. R. Plant growth-promoting bacteria: Mechanisms and applications. *Scientifica*, **2012**, 2012, 1–15.

Goto, M. *Fundamentals of Bacterial Plant Pathology*. Academic Press: San Diego, CA, **1992**.

Hallmann, J.; Berg, G. Spectrum and population dynamics of bacterial root endophytes. In: Schulz, B. J. E., Boyle, C. J. C., Sieber, T. N. (eds), *Microbial Root Endophytes*, Springer: Berlin, Heidelberg, **2006**; pp. 15–31.

Heydari, A.; Pessarakli, M. A review on biological control of fungal plant pathogens using microbial antagonists. *J. Biol. Sci.* **2010**, 10, 273–290.

Hoffland, E.; Pieterse, C. M. J.; Bik, L.; Van Pelt, J. A. Induced systemic resistance in radish is not associated with accumulation of pathogenesis-related proteins. *Physiol. Mol. Plant Pathol.* **1995**, 46, 309–320.

Holtappels, R.; Schader, S. I.; Oettel, O.; et al. Insufficient antigen presentation due to viral immune evasion explains lethal cytomegalovirus organ disease after allogeneic hematopoietic cell transplantation. *Front. Cell. Infect. Microbiol.* **2020**, 10, 157.

Hu, H. Q.; Li, X. S.; He, H. Characterization of an antimicrobial material from a newly isolated *Bacillus amyloliquefaciens* from mangrove for biocontrol of *Capsicum* bacterial wilt. *Biol. Control*, **2010**, 54, 359–365.

Hu, J.; Wei, Z.; Friman, V. P.; et al. Probiotic diversity enhances rhizosphere microbiome function and plant disease suppression. *mBio.* **2016**, 7, e01790–16.

Jetiyanon, K.; Kloepper, J. W. Mixtures of plant growth-promoting rhizobacteria for induction of systemic resistance against multiple plant diseases. *Biol. Control*, **2002**, 24, 285–291.

Ji, X.; Lu, G.; Gai, Y.; Zheng, C.; Mu, Z. Biological control against bacterial wilt and colonization of mulberry by an Endophytic *Bacillus subtilis* strain. *FEMS Microbiol. Ecol.* **2008**, 65, 565–573.

Jones, J. B.; Jackson, L. E.; Balogh, B.; et al. Bacteriophages for plant disease control. *Annu. Rev. Phytopathol.* **2007**, 45, 245–262.

Kakar, K. U.; Nawaz, Z.; Cui, Z.; Almoneafy, A. A.; Ullah, R.; Shu, Q. Y. Rhizosphere-associated Alcaligenes and *Bacillus* strains that induce resistance against blast and sheath blight diseases, enhance plant growth and improve mineral content in rice. *J. Appl. Microbiol.* **2018**, 124, 779–796.

Kamel, S. M.; Ebtsam, M. M.; Massoud, O. N. Potentiality of some yeast species as biocontrol agents against *Fusarium oxysporum* f. sp. *cucumerinum* the causal agent of cucumber wilt. *Egypt J. Biol. Pest*, **2016**, 26, 185–193.

Karlsson, M.; Atanasova, L.; Jensen, D. F.; Zeilinger, S. Necrotrophic mycoparasites and their genomes. *Microbiol. Spectr.* **2017**, 5, 5–2.

Kaur, N. P.; Mukhopadhayay, A. N. Integrated control of chickpea wilt complex by *Trichoderma* spp. and chemical methods in India. *Trop. Pest Manage.* **1992**, 38, 372–375.

Kiss, L. A review of fungal antagonists of powdery mildews and their potential as biocontrol agents. *Pest Manage. Sci.* **2003**, 59, 475–483.

Kloepper, J. W.; Ryu, C. M.; Zhang, S. Induced systemic resistance and promotion of plant growth by Bacillus spp. *Phytopathology*, **2004**, 94, 1259–1266.

Köhl, J.; Kolnaar, R.; Ravensberg, W. J. Mode of action of microbial biological control agents against plant diseases: Relevance beyond efficacy. *Front. Plant Sci.* **2019**, 10, 845.

Korolev, N.; David, D. R.; Elad, Y. The role of phytohormones in basal resistance and Trichoderma-induced systemic resistance to Botrytis cinerea in Arabidopsis thaliana. *BioControl*, **2008**, 53, 667–683.

Kumar, D.; Dubey, S. C. Management of collar rot of pea by the integration of biological and chemical methods. *Indian Phytopath.* **2001**, 57, 62–66.

Leveau, J. H.; Lindow, S. E. Predictive and interpretive simulation of green fluorescent protein expression in reporter bacteria. *J. Bacteriol. Res.* **2001**, 183, 6752–6762.

Mahapatra, D. M.; Satapathy, K. C.; Panda, B. Biofertilizers and nanofertilizers for sustainable agriculture: Phycoprospects and challenges. *Sci. Total Environ.* **2022**, 803, 149990.

Malfanova, N.; Lugtenberg, B. J.; Berg, G. Bacterial endophytes: Who and where, and what are they doing there. In: de Bruijn, F. J. (ed), *Molecular Microbial Ecology of the Rhizosphere*, Wiley-Blackwell: Hoboken, NJ, **2013**; pp. 393–403.

Mandal, A. K.; Dubey, S. C.; Tripathi, A. Bio-agent based integrated management strategy against stem rot for chickpea. *Indian Phytopathol.* **2015**, 68, 402–409.

Mercado-Blanco, J.; Lugtenberg, B. Biotechnological applications of bacterial endophytes. *Curr. Biotechnol.* **2014**, 3, 60–75.

Mercado-Blanco, J.; Prieto, P. Bacterial endophytes and root hairs. *Plant. Soil*, **2012**, 361, 301–306.

Mercado-Blanco, J.; Rodrıguez-Jurado, D.; Hervás, A.; Jiménez-Dıaz, R. M. Suppression of *Verticillium* wilt in olive planting stocks by root-associated fluorescent *Pseudomonas spp. Biol. Control*, **2004**, 30, 474–486.

Mukherjee, M.; Mukherjee, P. K.; Horwitz, B. A.; Zachow, C.; Berg, G.; Zeilinger, S. *Trichoderma* plant- pathogen interactions: Advances in genetics of biological control. *Indian J. Microbiol.* **2012**, 52, 522–529.

Nygren, K.; Dubey, M.; Zapparata, A.; Iqbal, M.; Tzelepis, G. D.; Durling, M. B. The mycoparasitic fungus *Clonostachys rosea* responds with both common and specific gene expression during interspecific interactions with fungal prey. *Evol. Appl.* **2018**, 11, 931–949.

Obradovic, A., Jones, J. B., Momol, M. T., et al. Integration of biological control agents and systemic acquired resistance inducers against bacterial spot on tomato. *Plant Dis.* **2005**, 89, 712–16

Oliveira, W. S.; Coelho, I. L.; Oliveira, J. R. S.; et al. Biological control of the bacterial wilt *Ralstonia solanacearum* by bioprotector with fungi chitosan from *Cunninghamella elegans* on tomatoes. *Afr. J. Agric. Res.* **2017**, 12, 42–49.

Raaijmakers, J. M.; Mazzola, M. Diversity and natural functions of antibiotics produced by beneficial and plant pathogenic bacteria. *Annu. Rev. Phytopathol.* **2012**, 50, 403–424.

Ramesh, R.; Joshi, A. A.; Ghanekar, M. P. Pseudomonads: Major antagonistic endophytic bacteria to suppress bacterial wilt pathogen, *Ralstonia solanacearum* in the eggplant (*Solanum melanogena* L.) *World J. Microb. Biotech.* **2009**, 25, 47–55.

Ramesh, R.; Phadke, G. S. Rhizosphere and endophytic bacteria for the suppression of eggplant wilt caused by *Ralstonia solanacearum. Crop Prot.* **2012**, 37, 35–41.

Rosenblueth, M.; Martínez-Romero, E. Bacterial endophytes and their interactions with hosts. *Mol. Plant Microbe. Inter.* **2006**, 19, 827–837.

Savary, S.; Willocquet, L.; Pethybridge, S. J. The global burden of pathogens and pests on major food crops. *Nat. Ecol.* **2019**, 3, 430–439.

Schalk, I. J.; Hannauer, M.; Braud, A. New roles for bacterial siderophores in metal transport and tolerance. *Environ. Microbiol.* **2011**, 13(11), 2844–2854.

Segarra, G.; Casanova, E.; Avilés, M.; Trillas, I. *Trichoderma asperellum* strain T34 controls Fusarium wilt disease in tomato plants in soilless culture through competition for iron. *Microb. Ecol.* **2010**, 59, 141–149.

Seleim, M. A. A.; Saeas, F. A.; Abd-El-Moneem, K. M. H.; Abo-ELyousr, K. A. M. Biological control of bacterial wilt of tomato by plant growth promoting rhizobacteria. *Plant Pathol. J.* **2011**, 10, 146–153.

Sharma, A.; Shankhdhar, D.; Shankhdhar, S. C. Enhancing grain iron content of rice by the application of plant growth promoting rhizobacteria. *Plant Soil Environ.* **2013**, 59, 89–94.

Spadaro, D.; Droby, S. Development of biocontrol products for postharvest diseases of fruit: The importance of elucidating the mechanisms of action of yeast antagonists. *Trends Food Sci. Technol.* **2016**, 47, 39–49.

Svircev, A. M.; Lehman, S. M.; Kim, W.; Barszcz, E.; Schneider, K. E.; Castle, A. J. Control of the fire blight pathogen with bacteriophages. *Proceeding of International Symposium on Biological Control of Bacterial Plant Diseases*, Berlin, Germany, **2006**; pp. 259–268.

Takenaka, S.; Sekiguchi, H.; Nakaho, K.; Tojo, M.; Masunaka, A.; Takahashi, H. Colonization of *Pythium oligandrum* in the tomato rhizosphere for biological control of bacterial wilt disease analysed by real-time PCR and confocal laser-scanning microscopy. *Phytopathology*, **2008**, 98, 187–195.

Tanaka, H.; Negishi, H.; Maeda, H. Control of tobacco bacterial wilt by an avirulent strain of *Pseudomonas solanacearum* M4S and its bacteriophage. *Ann. Phytopathol. Soc. Jpn.* **1990**, 56, 243–246.

Thambugala, K. M.; Daranagama, D. A.; Phillips, A. J.; Kannangara, S. D.; Promputtha, I. Fungi vs. fungi in biocontrol: An overview of fungal antagonists applied against fungal plant pathogens. *Front. Cell. Infect. Microbiol.* **2020**, 10, 284–293.

Thi, B. V. T.; Khanh, N. H. P.; Namikawa, R.; et al. Genomic characterization of *Ralstonia solanacearum* phage ϕRS138 of the family Siphoviridae. *Arch. Virol.* **2016**, 161, 483–486.

Ting, A. S. Y. Biosourcing endophytes as biocontrol agents of wilt diseases. In: Vijay, C. V. and Alan, C. G. (eds), *Advances in Endophytic Research*, Springer: New Delhi, **2014**; pp. 283–300.

Vejan, P.; Abdullah, R.; Khadiran, T.; Ismail, S.; Nasrulhaq Boyce, A. Role of plant growth promoting rhizobacteria in agricultural sustainability—a review. *Molecules*, **2016**, 21, 573–582.

Wei, W.; Xu, Y.; Li, S.; Zhu, L.; Song, J. Developing suppressive soil for root diseases of soybean with continuous long-term cropping of soybean in black soil of Northeast China. *Acta Agric. Scand. Sect. B Soil Plant Sci.* **2015**, 65, 279–285.

Wei, Z.; Huang, J.; Tan, S.; Mei, X.; Shen, Q.; Xu, Y. The congeneric strain Ralstonia pickettii QL-A6 of *Ralstonia solanacearum* as an effective biocontrol agent for bacterial wilt of tomato. *Biol. Control*, **2013**, 65, 278–285.

Weller, D. M. *Pseudomonas* biocontrol agents of soilborne pathogens: Looking back over 30 years. *Phytopathology*, **2007**, 97, 250–256.

Whipps, J. M. Microbial interactions and biocontrol in the rhizosphere. *J. Exp. Bot.* **2001**, 52, 487–511.

Wiesel, L.; Newton, A. C.; Elliott, I.; et al. Molecular effects of resistance elicitors from biological origin and their potential for crop protection. *Front. Plant Sci.* **2014**, 5, 655–665.

Xue, Q. Y.; Chen, Y.; Li, S. M.; et al. Evaluation of the strains of *Acinetobacter* and *Enterobacter* as potential biocontrol agents against *Ralstonia* wilt of tomato. *Biol. Control*, **2009**, 48, 252–258.

Yadessa, G. B.; van Bruggen, A. H. C.; Ocho, F. L. Effects of different soil amendments on bacterial wilt caused by *Ralstonia solanacearum* on the yield of tomato. *J. Plant Pathol.* **2010**, 92, 439–450.

Yamada, T. Filamentous phages of *Ralstonia solanacearum*: Double-edged swords for pathogenic bacteria. *Front. Microbiol.* **2013**, 4, 325–331.

Yi, H. S.; Yang, J. W.; Ryu, C. M. ISR meets SAR outside: Additive action of the endophyte *Bacillus pumilus* INR7 and the chemical inducer, benzothiadiazole, on induced resistance against bacterial spot in field-grown pepper. *Front. Plant Sci.* **2013**, 4, 122–133.

Yu, H.; Zhang, L.; Li, L.; et al. Recent developments and future prospects of antimicrobial metabolites produced by endophytes. *Microbiol. Res.* **2010**, 165, 437–449.

Zhang, X. X.; George, A.; Bailey, M. J.; Rainey, P. B. The histidine utilization (hut) genes of *Pseudomonas fluorescens* SBW25 are active on plant surfaces, but are not required for competitive colonization of sugar beet seedlings. *Microbiol.* **2006**, 152, 1867–1875.

Zhou, T.; Chen, D.; Li, C.; et al. Isolation and characterization of *Pseudomonas brassicacearum* J12 as an antagonist against *Ralstonia solanacearum* and identification of its antimicrobial components. *Microbiol. Res.* **2012**, 167, 388–394.

Zhu, H. H.; Yao, Q. Localized and systemic increase of phenols in tomato roots induced by *Glomus versiforme* inhibits *Ralstonia solanacearum*. *J. Phytopathol.* **2004**, 152, 537–542.

8 Plant Microbiome

Ecology, Functions, and Emerging Trends in Microbial Application

Aundy Kumar

8.1 INTRODUCTION

Microorganisms are fundamental to the functioning of the biosphere, yet we understand little about their role in ecosystem function especially the plant adaption in soil-associated biotic and abiotic environments. This is primarily due to a lack of understanding of microbial diversity and our inability to isolate and identify the microbiota using the currently available microbiological tools. However, the advent of NextGen Sequencing technologies has made it possible to identify **"all microbiota or microbiome"** sharing a particular ecological niche holistically (Vorholt 2012; Hardoim et al. 2015). The term "microbiome" was first time used decades ago in the 1950s in a scientific monthly magazine by **JL Mohr in 1952 and later by JM Whipps** during the 1980s, who described the microbiome as "a characteristic microbial community occupying a reasonably well-defined habitat which has distinct physio-chemical properties." The current definition of microbiome not only refers to the microorganisms involved but also encompasses their theater of activities (Berg et al. 2020). In general viruses, archaea, phytoplasmas, bacteria, actinomycetes, slime-molds, fungi, and other protozoan species are considered a component of plant microbiota (Yadav et al. 2018; Compant et al. 2019). Among them, prokaryotes are ubiquitous and can be found almost everywhere on the planet Earth. At the beginning of the 21st century, the research regarding the understanding of complex microbial assemblages and networks associated with higher organisms has got impetuous with the advent of nucleotide sequencing technologies. In agriculture and plant biology, manipulation of the microbiome by targeted application has the potential to contribute to sustainable crop productivity by reducing the incidence of plant biotic and abiotic stresses, enhancing resource utilization, reducing chemical inputs, and reducing emissions of greenhouse gases. Microbes interact with plants by their highly conserved molecules termed Microbe-Associated Molecular Patterns (MAMPs). Plant receptors termed Pattern Recognition Receptors (PRRs) are known to perceive MAMPs and trigger immunity in the plants often called MAMP Trigger Immunity (MTI). It is speculated that several MTI events are triggered in the plants by these microbiomes.

DOI: 10.1201/9781003531944-8

Although presumed, the role of these MTI events in suppressing harmful pathogens on the plant and the consequent plant survival in a natural ecosystem is not clearly understood.

8.2 PLANT MICROBIOMES: EXTENDED GENOMIC CAPABILITIES FOR PLANTS

Based on the available information, all microorganisms associated with a particular plant (part or organ or development stage) constitute the plant microbiota, and the associated-combined genome is known as the plant microbiome, or phytobiomes, that mostly provides metabolic flexibility and functional extension to the plant genomes (Vorholt 2012; Hardoim et al. 2015; Brader et al. 2017; Lemanceau et al. 2017; Fitzpatrick et al. 2020**;** Sessitsch et al. 2012). Plants, like other living forms, are home to a diverse group of commensals and mutualistic microbiota. Depending on the trophic relationships, the microbiota may be as free-living, symbionts, commensals, ammensals, or even pathogens, which are thought to play a pivotal role in the health and function of higher eukaryotic organisms including plants (Ezenwa et al. 2012). Considering the functional aspect of plant–microbial interactions and other factors of community assembly, the way plants interact with their microbial partners is useful for the plant as a meta-organism (Hardoim et al. 2015; Hacquard 2016).

8.3 ECOLOGY OF MICROBIOMES

Plants routinely attract and recruit their partner microorganisms from microbial hubs around them such as air (aerobiome), irrigation water (hydrobiome), and soil (soil biome) that together make up the environmental microbiome (Hardoim et al. 2015). Most of the microbiota found associated with roots is derived from the soil environment, which includes horizontally transferred soil microbiota. The soil microbiota is dominated by bacteria belonging to **Proteobacteria, Actinobacteria, Bacteroidetes, Acidobacteria, Verrucomicrobia,** and **Planctomycetes** (Fierer 2017). The rhizosphere, which is the tiny layer of soil under the significant influence of the root system, is considered to be one of the most complex biological systems present on earth (Hiltner 1904; Yadav et al. 2018). Nevertheless, in recent years, it has been observed that seeds are one of the important sources of vertically transmitted microorganisms that are acquired during the germination process (Hardoim et al. 2012; Kim et al. 2020). Similar to the below-ground plant part, above-ground parts also provide a unique habitat for the colonization of diverse kinds of microbes such as epiphytes and endophytes. However, distinct ecological differences have been observed between the epiphyte and endophyte; while the former survives in a nutrient-depleted environment, the latter ones in nutrient-rich habitat (Compant et al. 2019). Phyllosphere bacteria have also been recorded to have originated from the soil ecosystem and are influenced by plant and climatic parameters. However, the climatic factors or the environmental conditions have a more profound impact on phyllosphere microbiome assemblage (Zarraonaindia et al. 2015; Wallace et al. 2018; Kumar et al. 2021; Sahu et al. 2021).

8.4 MICROBIOME IMPACT ON FACTORS OF THE PLANT DISEASE TRIANGLE

Plant microbiomes and their activities can have a profound impact on the major disease inciting factors such as susceptibility of the host and virulence of the pathogen. Not only the pathogen and their inoculum potential are impacted by microbial interactions but also the metabolic capability of plants. Whereas the microbiome-associated conserved microbial molecular pattern may activate innate immunity, the pathogen activities may be affected by antimicrobial metabolic products (Sahu et al. 2021; Kumar et al. 2021). Microbial succession and energy transfers in the microbial network on the disease habitat might be a very natural phenomenon in microbial pathogen ecology. It is further speculated that the total metabolic activities of intra-microbial interactions can influence the plant microbial interaction outcomes. The contribution of the environmental microbiome to the disease triangle outcome is multipronged. Therefore, a shift from reductionist (pathogen- and plant-centric) to integrated (environmental microbial ecology–centric) approaches is essential to developing sustainable disease mitigation option in the future (Figure 8.1).

8.5 CORE MICROBIOME AND SATELLITE MICROBIOME

In general, the term "microbiome" refers to the collection of microorganism genomes found in a particular habitat or environmental niche (Stanley and van der Heijden 2017; del Carmen Orozco-Mosqueda et al. 2018; Kumar et al. 2021). Furthermore, the words microbiome and microbiota are interchangeable. Within the broader word "microbiome," a term known as "core plant microbiome" has become well-known in the field of microbiome science over time (Lundberg et al. 2012; Eyre et al. 2019; Kumar et al. 2021). The term core microbiome refers to a group of microbes that is

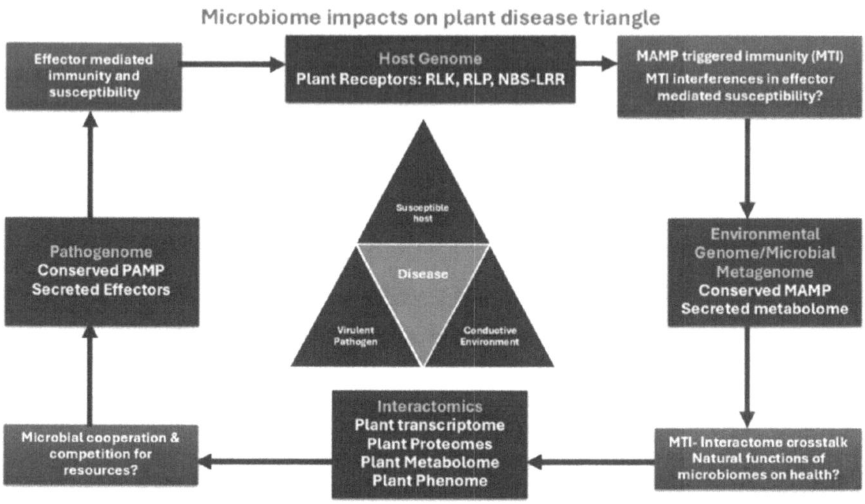

FIGURE 8.1 Predicted impacts of plant microbiome on disease-conditioning factors.

closely associated with a certain plant species/genotype, regardless of environmental/soil conditions, and that has been formed by the evolutionary process of selection (Busby et al. 2016; Toju et al. 2018). This category includes keystone microbial taxa that are extremely important for the health and functions of plants (Lemanceau et al. 2017). As per the literature, the core microbiome is assumed to comprise keystone microbial taxa that are thought to be the central regulator of microbial diversity in a particular ecological niche (Lemanceau et al. 2017). These keystone species are critical for plant fitness as well as performance. Their association with plants is formed by evolutionary mechanisms of selection (such as selection through human heterogeneity and natural diversification) and enrichment of microbial taxa containing vital functions genes (e.g., genes important for microbial functional processes such as nitrogen fixation) (Lemanceau et al. 2017). Recently, the core microbiome of potatoes was deciphered and found to be composed of *Sphingobium*, *Bradyrhizobium*, and *Microvirga* (Pfeiffer et al. 2017). Similarly, members of the families Pseudomonadaceae, Micrococcaceae, and Hyphomicrobiaceae are reported to constitute the core microbiome of the grapevine, irrespective of climatic conditions (Zarraonaindia et al. 2015).

Contrary to the core microbiome, certain microbial taxa are found in very low abundance in just a few locations. The geographical distribution and habitat specificity of these microbial satellite taxa were used to characterize them (Edwards et al. 2015; Compant et al. 2019). The microbiota is made up of a variety of prokaryotic and eukaryotic species (Bulgarelli et al. 2013; del Carmen Orozco-Mosqueda et al. 2018; Uroz et al. 2019; Compant et al. 2019). In other terms, a few microbial taxa, which exist in a limited number and are transient in nature, are known as satellite taxa (Hanski 1982; Magurran and Henderson 2003). The population of satellite taxa is highly influenced by geographical location and environmental factors (Jousset et al. 2017). Increasingly, the value of satellite taxa is recognized as a driver of the core ecosystem functions. According to some recent reports, these satellite taxa have a key role in reducing the unnecessary microbial encroachments into the soil biosystem (Mallon et al. 2015; Singh et al. 2020). Few studies also suggested that the loss of these satellite microbes may harm overall plant productivity. These uncommon taxa play an essential role that may not be attributable to their abundance (Hol et al. 2015).

8.6 KEY MICROBIOME RESEARCH STRATEGIES: FROM PETRI PLATE TO DNA SEQUENCERS

As mentioned above in context, multiple factors such as plant genotype, age, environmental factors, soil characteristics, and agricultural practices influence the plant microbiome. The diversity of the microbiome is so much that only 1.0% of the diversity is known, which is mainly due to the lack of appropriate technologies to study the microbial communities (Riesenfield 2004). The microbiome can be studied in two different methods termed culture-dependent and culture-independent methods. In this regard, sterile and careful isolation is the only tool possible to determine the precise diversity, structure, and distribution of microbes in various epiphytic, endophytic, and rhizospheric areas, followed by sequencing. Both next-generation sequencing **(NGS)-based metagenome analysis** (culture-independent studies) and **conventional microbiological methods** (culture-dependent studies) are adopted for deciphering microbiota (Kumar et al. 2021). Both the unculturable and culturable approaches aim toward identifying the precise differences in microbiome groups and composition.

The first phase entails collecting samples from the greenhouse or field conditions in an aseptic condition. After collecting the sample, it is recommended that the surface should be washed with detergents such as PBS (phosphate-buffered saline) and/or PBST (phosphate-buffered saline with Tween 20) to isolate epiphytic microbes from the surfaces of the organs (leaf, stem, roots, panicle). After that, the PBST/PBS wash is collected and split into a 1:99 ratio. The microbes in the collected samples are pelleted down in the 99th section, which is centrifuged at a high speed. The remaining part of the overall wash is used to make dilutions up to 10^{-8} that are cultured on various media using spread or pour-plate methodology.

For the isolation of fungi, archaea, actinomycetes, and bacteria, different types of microbe-specific growth media are used, depending on the assay design and the number of samples. The aim of using a variety of growth media is to separate as many culturable morphotypes of different microbes as possible. For example, nutrient agar, soil-extract agar, King's B agar, Tryptic Soy Agar with heat treatment methods, and Rose Bengal agar are used to separate heterotrophic microbes, soil-specific microbes, Pseudomonads, *Bacillus*-derived genera, and fungi (Compant et al. 2019). With minor changes and amendments to the growth media, various microbial classes such as halophilic, acidophilic, alkaliphilic, psychrophilic, and thermophilic may be extracted. Various growth medium plates are incubated after separate dilutions are poured into a laminar flow hood chamber. After an incubation period of 4–5 days, culturable microbes of various groups are isolated from the plates and pelleted using high-speed centrifugation. Following extraction of microbial cells, community DNA isolation is then performed using either commercially available kits such as Zymo Research DNA MicroPrep or Soil Power kits or optimized lab protocols (Kunadiya et al. 2019).

Genomic tools are most soughtafter today in biology especially in microbiome studies due to their power, speed, and precision. The culture-dependent tools for deciphering the microbiome can help in harnessing the potential of the microbiome for eco-friendly agricultural practices in the future. However, the culture-independent study will help us in understanding the diversity of the microbiome in that niche. Therefore, culture-independent methods are increasingly employed to study microbial diversity. The culture-independent method along with omics tools has contributed extensively to understanding the diversity of microbiome associated with that niche (Kumar et al 2021; Sahu et al 2021). It all started with Denaturation Gradient Gel Electrophoresis (DGGE), then Polymerase Chain Reaction (PCR), and now revolutionary NGS tools have made microbiome studies easier than before. Currently, it is possible to investigate the non-extractable phyllosphere bacteria using marker genes [e.g., 16S ribosomal RNA (rRNA)] in the high-throughput nucleotide sequencing technology termed NGS technologies. Following genomic DNA isolation, primers for amplification of the 16S rRNA (archaea and eubacteria) and 18S rRNA (fungi) genes are used. Genetic markers currently used in the microbial metagenomic investigation are presented in Table 8.1. Purified PCR-amplified products are then

TABLE 8.1
Universal Genetic Markers Used for Microbial Diversity Analysis

Organisms	Markers	Sequence	Databases Used for Mapping and Identification of Reads
Archaea	**16S rRNA** 27f and 1492r	27f 5′- AGAGTTTGATCCTGGCTCAG-3′ 1492r 5′-TACCTTGTTACGACTT-3′ (Lin et al. 2010)	RDP (Maidak et al. 1997), Greengenes (DeSantis et al. 2006), SILVA (Pruesse et al. 2007) -(16S)
Bacteria	**16S rRNA** 27f and 1492r **V3-V4 region** 341f and 785r	341f 5′-CCTACGGGNGGCWGCAG-3′ 785r 5′-GACTACNVGGGTATCTAATCC-3′ (Klindworth et al. 2013)	RDP (Maidak et al. 1997), Greengenes (DeSantis et al. 2006), SILVA (Pruesse et al. 2007)-(16S)
Fungi	**ITS** ITS1 and ITS4	ITS1 5′- TCCGTAGGTGAACCTGCGG- 3′ ITS4 5’-TCCTCCGCTTATTGATATGC-3′ (Glass and Donaldson, 1995)	UNITE (Abarenkov et al. 2010) and GenBank (Benson et al. 2012)-(ITS); SILVA (Pruesse et al. 2007)-(18S)
Oomycetes and Protists	**ITS O** ITS1oo and ITS3oo	ITS1oo5′-GGAAGGATCATTACCACACCGGAAGGATCATTACCAC-3′ ITS3oo 5′-AGTATGYYTGTATCAGTGTC-3′ (Riit et al. 2018)	GenBank (Benson et al. 2012)-ITS; SILVA (Pruesse et al. 2007)-(18S); BOLD (Ratnasingham and Hebert 2007)- (CO1)
Microfauna	**Cox region** COX2F and COX2R FM66 and FM58	COX2F1 5′-GATGTTATTTAAAACAATAGTTC-3′ COX2R1 5′-TAAAGAAGGAATAGCCCAA-3′ (Lee et al. 2017) FM66 5′-TAGGATTTCAAGATCCTGC-3′ FM58 5′-CCACAAATTTCACTACATTGA-3′ (Martin, 2000)	BOLD (Ratnasingham and Hebert 2007) -(CO1); GenBank (Benson et al. 2012)- (18S)

sequenced via outsourcing or installed sequencers. After the sequencing, the phylogenetic profile, group distribution, and novel-gene sequences can be identified. The point of utmost importance is that for the preparation of 16S/18S rRNA gene libraries, Illumina sequencing technology is a well-known choice. Previously, researchers used the Illumina HiSeq platform; but, because of the associated benefits, the MiSeq has been everyone's favorite for gene library preparation (Mehta et al. 2021). Individual microbial taxa and genes have been related to plant colonization, physiology, and fitness through genome-wide association studies (GWASs) and metagenome-wide association studies (Wagner et al. 2016; Bergelson et al. 2019; Roman-Reyna et al. 2019). The outcome could be compositional information about microbial community profiles in a plant niche. The concept of the microbiome describes both the composition of a microbial population within a particular ecological niche (microbiota) and its collective genomic contents (Backhed et al. 2005; Ursell et al. 2012).

Presently, metagenomic NGS (mNGS) tools are frequently used to investigate plant microbiome, which led to an enhanced understanding of microbiome profiles and functions. Nowadays, many researchers have attempted to investigate the microbial diversity of plants using culture-dependent and culture-independent approaches (Bertani et al. 2016; Mano and Morisaki 2008; Sengupta et al. 2017; Sessitsch et al. 2012; Sun et al. 2008). To expand our knowledge of the microbial community, comprehensively combined approaches are required. Beyond the compositional information, i.e., microbial diversity, the functional analysis or functional attributes of these microbial communities are now possible with "omics technologies" such as metagenomics, metatranscriptomics, metaproteomics, and metabolomics (Bäckhed et al. 2005; Ursell et al. 2012).

8.7 SCOPING INTO THE MICROBIOME: WHO IS THERE?

In natural environments, these microorganisms can form complex co-associations with plants and play a significant role in promoting plant productivity and health. At lower phylogenetic resolution, complex plant microbial communities include taxa from a variety of phyla and have deep branching lineages. Overall, bacteria, fungi, actinomycetes, slime-molds, archaea, viruses, phytoplasmas, and other protozoans are among the diverse group that associate with plants (Mueller and Sachs 2015; del Carmen Orozco-Mosqueda et al. 2018; Compant et al. 2019). To understand the association in a better manner, culture-independent high-throughput sequencing has been used in the past 20 years (Lundberg et al. 2012; Vorholt et al. 2012; Peiffer et al. 2013; Ofek-Lalzar et al. 2014; Coleman-Derr et al. 2016; Trivedi et al. 2020). Specific genetic markers have been identified for targeting particular microbial communities (Table 8.1). This has even advanced the knowledge of a range of residing microorganisms in and on plants, as well as in the surrounding world, in recent years (Lundberg et al. 2012; Vorholt et al. 2012; Peiffer et al. 2013; Ofek-Lalzar et al. 2014; Coleman-Derr et al. 2016; Trivedi et al. 2020). Many studies have shown that a subset of microbial lineages known as the "core microbiota" consistently associates with a given host in a range of environments (Edwards et al. 2015; Fitzpatrick et al. 2018; Hamonts et al. 2018; Roman-Reyna et al.2019). Bacteria and, to a lesser degree, fungi, are the most abundant and well-studied members of the plant-associated microbiota, but other species, such as archaea, algae, nematodes, and protists, also play significant roles in plant growth and development (Mendes et al. 2013; Leach et al. 2017). However, the composition, variety, and quantity vary depending on several factors such as genotype, plant's age soil characteristics (pH, structure, precipitation, and organic matter), geographic area, agricultural inputs, environmental conditions (both biotic and abiotic), and pathogen infection are all factors to consider (Fierer 2017; del Carmen Orozco Mosqueda et al. 2018; Yadav et al. 2015; Compant et al. 2019).

8.8 INSIGHTS ON PHYTOBIOME: ABOVE-GROUND VS BELOW-GROUND PHYTOBIOMES

While the below-ground microbiome is termed as rhizosphere microbiome, the above-ground microbial life on the plants is referred to as the phyllosphere microbiome. Even though the phyllosphere is a hostile environment for microbes due to ultraviolet radiation, food and water scarcity, and rapid temperature changes, bacteria, fungi, archaea, and protists colonize it substantially (Muller and Ruppel 2014). However, the vast majority of phyllosphere bacteria are commensal and difficult to cultivate. Thus, culture-independent methods such as 16S or 18S rRNA gene clone libraries and amplicon sequencing in other plants have revealed the compositions of these groups (Vorholt 2012; Rastogi et al. 2013; Copeland et al. 2015; de Souza et al. 2017). This approach has been adopted even in rice microbiome research that attempts to identify the microorganisms that are associated with rice and the ecosystems in which they live. Other molecular methods were used to determine the composition of microbial species in soil and rice tissues before the advent of High Throughput Screening (HTS) platforms. Methods include 16S rRNA gene cloning, polymerase chain reaction–denaturing gel gradient electrophoresis (PCR-DGGE), and terminal restriction fragment duration polymorphism studies.

To the present date, little research on phyllosphere microbiomes has explored the major factors that influence their composition, such as the organisms' origins, their genetic and functional characteristics, and their role in stress adaptation (Rastogi et al. 2013). In rice paddy fields, Proteobacteria, Chloroflexi, Actinobacteria, and Acidobacteria typically dominate the bacterial population (Hussain et al. 2011; Jiang et al. 2016), and Ascomycota, Basidiomycota, and Glomeromycota typically dominate the fungal community (Hussain et al. 2011; Jiang et al. 2016; Yuan et al. 2018). These findings indicate that there could be an environmental commonality that influences the phylum-level composition of paddy soil microbiota. The phyllosphere microbial community composition in rice varies depending on the study. Enterobacteriaceae (Gamma Proteobacteria) occupied the phyllosphere in an amplicon-based profiling sample (Ren et al. 2014). In contrast to the rhizosphere and paddy water, a metaproteomic analysis found that Actinobacteria, Alpha- and Beta-Proteobacteria predominated in the phyllosphere microbiome and showed distinct enriched functions (Knief et al. 2012). A metagenome-based methodology has uncovered a previously unknown aspect of rice's phyllosphere microbiome. Bacterial cultures in the phyllosphere of 3,024 rice accessions were reconstructed using bacterial reads retrieved from available rice genome reads (Roman-Reyna et al. 2019).

Phyllosphere microbiome composition and functional profile were conserved across geographical differences and rice genotypes, according to the researchers.

Additionally, the phyllosphere microbial population of rice plants has also been studied using culture-dependent methods (Harsonowati et al. 2017; Krishanti et al. 2015; Thapa et al. 2018). The majority of research has been focused on establishing the antagonistic nature of certain bacteria and/or fungi against rice pathogens including *M. oryzae, R. solani*, and *X. oryzae* (De Costa et al. 2006; Harsonowati et al. 2017; Krishanti et al. 2015; Thapa et al. 2018). Three bacteria (two *Erwinia* species and one *Bacillus pumilus*) and 18 fungi (*Trichoderma*, *Pestalotia*, and *Alternaria* species) have been found to inhibit the development of *M. oryzae* and *R. solani* (De Costa et al. 2006). Antifungal activity was also found in 18 species of leaf actinomycetes from the genera *Streptomyces*, *Saccharothrinx*, *Gordonia*, and *Lentzea* against the rice blast fungus (Harsonowati et al. 2017).

Furthermore, the bacterial community variation was more closely linked to soil heterogeneity than to geographic distance, while fungal community variation was more closely linked to geographic distance. Another research looked at microbial populations in paddy field bulk soils based on soil depth and cultivation time. Due to the deposition of organic carbon and total nitrogen, microbial abundance grew as the cultivation time lengthened. The structure of bacterial and fungal organisms is also influenced by enzymes such as invertase, urease, and acid phosphatase. Soil pH has been reported to have a major impact on soil microbial communities. In rice paddy fields, methane production and emission are critical processes. Due to the decomposition of root exudates or ashes, 9–19% of all methane emissions come from underwater rice paddy fields (Conrad 2009; Han and Micallef 2016). The main group involved in methane generation in rice paddy fields is methanogenic archaea, such as rice cluster I methanogens (Thauer et al. 2008). Nitrogen metabolisms such as N-fixation, nitrification, denitrification, and bacteria and archaea-mediated ammonium oxidation take place in rice paddy fields, in addition to methane cycling (Ishii et al. 2011). The metabolic behavior of microbial functional groups, as well as plant production, is influenced by nitrogen levels. The behavior of methane-oxidizing bacteria, for example, is affected by nitrogen levels (Bodelier et al. 2000). Methane synthesis affects nitrogen cycling. Methane, for example, has been found to suppress the development and function of nitrifying bacteria (Nitrosospira and Nitrosomonas) (Zheng et al. 2014). Microbial processes can also co-occur with methane and nitrogen metabolism in these situations. Diazotrophic methanotrophs (Type II methanotrophic bacteria) function as nitrogen fixers as well as methane oxidizers under low-nitrogen fertilizer control, contributing to the deposition of both carbon and nitrogen sources in paddy soils (Minamisawa et al. 2016). Several HTS experiments also discovered possible links to other microbial communities (Edwards et al. 2015; Sun et al. 2018). For example, functional groups involved in methane oxidation, iron reduction, and sulfur reduction in paddy soils were tightly correlated with operational taxonomic units (OTUs) classified as methanogens, methanotrophs, syntrophs, and methane-cycling taxa (Edwards et al. 2015; Sun et al. 2018).

In the past few years, the combination of biofertilizers as well as biocontrol properties of phyllosphere inhabiting microbes has been assessed for their intention to boost rice growth and productivity (Venkatachalam et al. 2016). About 34% of 86 isolates of *Achromobacter*, *Stenotrophomonas*, *Pantoea*, *Exiguobacterium*, and *Bacillus* secreted indole-acetic acid (IAA) as a cue to promote rice production. The phyllosphere microbiome also aids plant growth and protection, as well as nutrient cycling and the removal of pesticides and other toxins from the atmosphere that paves for suitability in agriculture as well as the environment (Muller and Ruppel 2014).

8.9 IMPACT OF PHYTOBIOME ON PLANT GROWTH AND DEVELOPMENT

Plant microbiomes are predicted to contribute to plant adaptation and survival. The ecosystem service provided by the microbiome is vast, and the plants are believed to gain short- and long-term benefits (Figure 8.2). The plant microbiome members include beneficial, neutral, or parasitic microorganisms. The development of plant-promoting bacteria (PGPB) can either actively or passively promote plant growth mechanisms. Some PGPBs generate phytohormones such as auxin, cytokinin, and gibberellin, which through modulation affect plant growth in combination with plant endogenous hormone levels (Ghosh et al. 2019). Furthermore, some PGPBs can secrete an enzyme called 1-aminocyclopropane-1-carboxylate (ACC) deaminase, which lowers ethylene levels in the plant. *Pseudomonas* spp., *Arthrobacter* spp., *Bacillus* spp., and other bacteria have been found to increase plant growth by producing ACC deaminase. (Rascovan et al. 2016) discovered a diverse range of bacteria in wheat and soybean roots, including *Pseudomonas* spp., *Paraburkholderia*

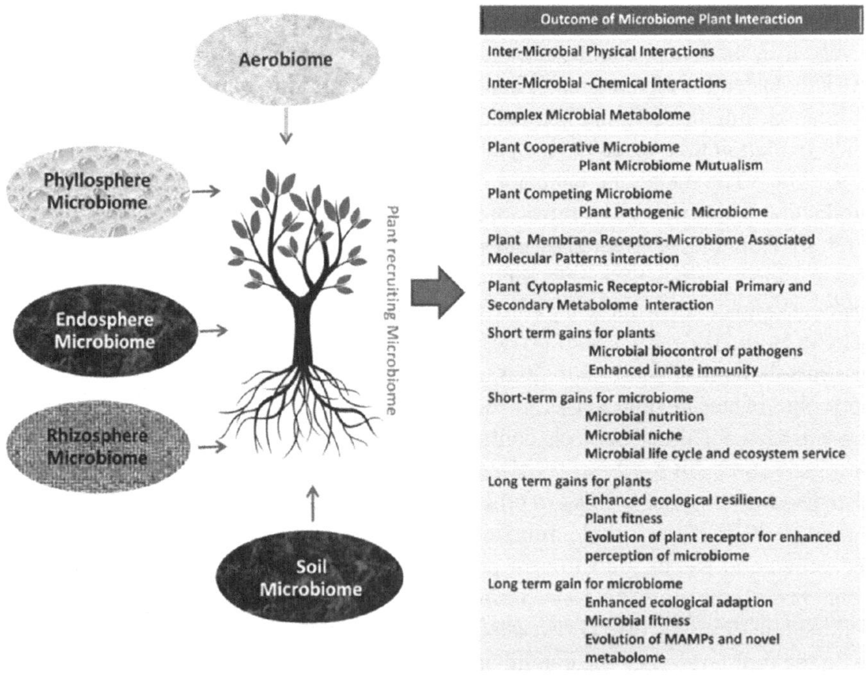

FIGURE 8.2 Plant microbiomes and their predicted functions.

spp., and *Pantoea* spp., that showed important plant growth promotion properties such as phosphate solubilization, N-fixation (Khamwan et al. 2018). Additionally, the microbes provide direct benefits to their host plants, such as the transformation and translocation of essential nutrients in the soil to make them accessible to plants (for example, N-fixation), stabilization of environmental stresses (such as drought), and resistance from plant pathogens through competition, antibiosis, and the development of hydrolytic enzymes (Richardson et al. 2011; Trivedi et al. 2016; Backer et al. 2018; Gouda et al. 2018).

By contrast, the benefits can also be indirect, through the enhancement of a plant's resistance responses (Pieterse et al. 2014). Since the plant-associated microbiome regulates many plant traits, a modern paradigm is developing under which associations between plants and their associated microbiome can be viewed as a way to produce new phenotypes with improved fitness under different environmental conditions, rather than as intrinsically beneficial or deleterious (Ravanbakhsh et al. 2019).

Synthetic communities (SynComs) of microorganisms with broad, persistent, and long-lasting plant-growth-promoting traits have the potential to translate basic scientific findings into applications in greenhouse or field settings (Xue et al. 2015; Santhanam et al. 2015; Zhang et al. 2019). Some of the key roles of the plant microbiome are discussed below:

8.10 PHYTOBIOMES AS PLANT PROBIOTIC PRODUCER

Plant-associated microbiomes play an important role in enhancing plant nutrition. As a result, the chemistry behind the symbiosis of plants with arbuscular mycorrhizal fungi (AMF) and *Rhizobium* bacteria and underlying molecular pathways for nutrient acquisition have been extensively investigated by various researchers (Richardson and Simpson 2011; Pieterse et al. 2014; Trivedi et al. 2016; Backer et al. 2018). Furthermore, non-symbiotic plant-growth-promoting bacteria may either boost the root system architecture of host plants or increase the root's absorbing potential toward water and minerals by increasing the bioavailability of insoluble minerals (Richardson and Simpson 2011; Trivedi et al. 2016). The associated microbiome contributes to functional plant traits related to nutrient acquisition or conservation that distinguish exploitatively (fast-growing) from conservative (slow-growing) plant species very easily (Richardson and Simpson 2011). According to a recent analysis, variations in nitrogen use efficiency of rice varieties are due to the recruitment of a higher proportion of nitrogen-cycle-related bacteria in the root environment of *O. indica* varieties, resulting in more effective nitrogen transformation processes in the root environment than *O. japonica* varieties (Zhang et al. 2019). When organic nitrogen was used as the sole nitrogen source, a 16-member indica-enriched SynCom significantly promoted the growth of an indica variety as compared to a three-member japonica-enriched SynCom. This suggests that the indica-enriched microbiomes' efficient transformation of organic nitrogen into nitrate and ammonium can lead to higher nitrogen use efficiency in indica rice.

As compared to non-microbial controls, multipartite microbial synergies between mycorrhizal fungi and soil microbial communities resulted in a ten-fold increase in nitrogen absorption from organic matter in *Brachypodium distachyon* (Hestrin et al. 2019).

These previously unquantified associations are expected to contribute >70 teragrams of annual plant nitrogen assimilated, with a positive effect on net primary productivity (Hestrin et al. 2019). Differential colonization of plants with two major groups of mycorrhizal fungi is linked to plant traits involved in nutrient use efficiency (Averill et al. 2019).

Plants colonized by ectomycorrhizal fungi use nutrient conservation strategies, while those colonized by AMF use nutrient acquisition strategies (Averill et al. 2019). Converting the precursor tryptophan to the phytohormones such as indole acetic acid (IAA) downregulates genes involved in flowering. Additionally, the rhizospheric microorganisms that prolong nitrogen bioavailability by nitrification delay the flowering period and stimulate plant development (Lu et al. 2018). Additionally, plant-associated microbiota can mobilize readily unavailable nutrients to plants, such as inorganic phosphate and iron via solubilization, mineralization, or excretion through iron-chelating siderophores (Kumar et al. 2018).

Plants rely on microbial cooperation partners, such as AMF and their symbiotic endophytes, to meet their phosphate needs (Bodenhausen et al. 2019). Synonyms of various complexity increase the transcription of plant genes related to phosphate starvation responses under phosphate-limiting conditions that result in increased inorganic phosphate absorption by plants (Castrillo et al. 2017). Plants without mycorrhizal association develop a core inorganic-phosphate-assimilating fungal microbiome that allows them to thrive in phosphate-depleted soils (Hacquard et al. 2016; Almario et al. 2017). Such findings indicate that microbiota-mediated beneficial functions are controlled by fine-tuned interactions between plant nutritional and immune status.

Coumarins are a special class of plant-derived specific secondary metabolites that promote iron mobilization and inhibit the proliferation of a relatively abundant *Pseudomonas* species that compete with plants for the iron and help in increasing the *Arabidopsis thaliana* root bacterial population (Voges et al. 2019). Both rhizobacteria-mediated induced systemic resistance (ISR) and iron acquisition include the root-specific transcription factor MYB72 (Stringlis et al. 2018). Volatile compounds from ISR-inducing rhizobacteria and *Trichoderma* fungi evoke the expression of MYB72, which activates the genetic machinery that allows roots to absorb more iron (Martínez-Medina et al. 2017). Local iron homeostasis changes send systemic signals to plant shoots, triggering a jasmonic-acid-dependent ISR (Martínez-Medina et al. 2017). In contrast to the highly explored role of rhizosphere microbes in nutrient acquisition, the phyllosphere microbiome has long been thought to be relatively inert, playing little role in nutrient provision to the host plant. While surveys of phyllosphere community composition have revealed the existence of nitrogen-fixing bacteria (Ruinen 1965; Sattelmacher et al. 1998; Delmotte et al. 2009; Holland 2011), mechanistic studies of plant-phyllosphere nitrogen dynamics are generally missing.

N-fixation occurs in the internal tissue of the leaf rather than on the leaf surface in temperate environments (Sattelmacher et al. 1998), allowing nitrogenous end products to disperse freely through plant cells. In tropical environments, however, N-fixation occurs in the phyllosphere, potentially because higher moisture availability at the leaf surface allows nitrogen-fixing bacteria to be active (Fürnkranz et al. 2008). While most studies report indirect evidence for phyllosphere

N-fixation, and the mechanisms for microbe–host nitrogen transfer are uncertain (Bentley and Carpenter 1984), the transfer of nitrogenous compounds from nitrogen-fixing microorganisms in the phyllosphere to the host plant has been observed (Bentley and Carpenter 1984). In the canopies of higher plants, tropical epiphytes exhibit N-fixation in their phyllosphere, which possibly provides a significant portion of their nitrogen requirements. The bulk of the microbially fixed nitrogen obtained by these plants is retained, with only a small amount moving to the larger host plant whose canopy they occupy (Hietz et al. 2002). Plants that have nitrogen-fixing microorganisms applied to their leaves can grow faster and have more nitrogen (Giri and Pati 2004). This may, however, be due to the internalization of the bacteria used rather than real phyllosphere N-fixation. As a foliar application of chemical fertilizers increases the nutrient content and yield of plants, such experimental studies on foliar applications demonstrate the potential for leaf uptake of nitrogen (Yildrim et al. 2007; Scagel et al. 2008). Although there are no clear mechanistic demonstrations of the value of N-fixing phyllosphere microorganisms to plant health, there is no doubt that they may be an essential source of nutrients for the plant.

Surprisingly, in areas where N-pollution is a problem, phyllosphere microorganisms can play a protective role by converting ammonia to nitrate through nitrification (Papen et al. 2002; Krupa 2003; Guerrieri et al. 2015). Chemoautotrophic nitrifiers, especially archaea, have been found in the phyllosphere of a variety of plant species (Bowatte et al. 2015; Watanabe et al. 2016), implying that the phyllosphere's ability to mitigate the effects of NH_4^+contamination may be widespread but not always powered by bacteria. Indirect interactions between the phyllosphere microbiota and plant nutrient acquisition might be possible. For example, fertilizer applied to the leaves can promote bacterial growth and the development of cytokinins, which are bioactive within the plant (Holland 2011). These factors may affect overall plant metabolism, as well as nutrient acquisition strategies below ground. In such situations, the phyllosphere microbiome may affect the overall plant's ability to acquire nutrients, rather than just the foliage.

8.11 PHYTOBIOMES AS PLANT IMMUNE-MODULATOR

The entry of pathogens and the subsequent activation of inducible defense responses or the production of disease symptoms is primarily prevented by plant surface barriers such as wax layers, rigid cell walls, cuticular lipids (Reina-Pinto and Yephremov 2009), secondary metabolites (Ahuja et al. 2012; Bednarek 2012) and antimicrobial enzymes (Habib and Fazili 2007). Pathogens that overcome these protective layers then encountered two interconnected resistance mechanisms such as microbial or pathogen-associated molecular patterns (MAMP/PAMPs)-triggered immunity (MTI/PTI) and effector-triggered immunity (ETI).

MTI was earlier known as basal or horizontal resistance that contributes to the recognition of conserved and indispensable elicitors of microbes called MAMPs by a type of plasma-membrane-associated extracellular receptors named pattern recognition receptors (PRRs), and the induction of these PRRs leads to activation of defense responses. On the other hand, these MAMPs are important structures or conserved components of all pathogenic groups (Felix et al. 1993). This contains lipopolysaccharides (LPSs) (Dow et al. 2000), ergosterol (Granado et al. 1995), oligogalacturonides (Galletti et al. 2008), xylanase, bacterial flagellin (Felix et al. 1999), Pep-13, and cold-shock protein (Felix and Boller 2003).

Specific MAMP identification and subsequent elicitation of resistance responses are limited to plant species belonging to a single family (Felix and Boller 2003; Ron and Avni 2004). For example, recognition EF-Tu is found only in the family Brassicaceae. However, other MAMPs, such as chitin, LPS, and flagellin, induce immune responses in different host species, even if a plant family or species has some degree of specificity and perception effectiveness. To break the structural barriers of plant tissues, many plant pathogens release lytic enzymes. The cutin monomers (Kauss et al. 1999), cell wall fragments (Darvill and Albersheim 1984), and peptides (Boller 2005; Huffaker et al. 2006) are generated during this process and work as endogenous elicitors named damage-associated molecular patterns (DAMPs).

These DAMPs typically develop in the apoplast and act as dangerous signals similar to MAMPs to trigger an immune response (De Lorenzo et al. 2018). It is understood that surface receptors named PRRs recognize both MAMPs and effectors. Similarly, receptor-like kinases (RLK) and receptor-like proteins are members of the PRR family (RLP). RLK is located in the plasma membrane and consists of a putative extracellular ligand-binding domain, a separate trans-membrane domain, and an intracellular serine/threonine kinase domain (Shiu and Bleecker 2001). Likewise, RLP comprises an extracellular domain and a membrane-spanning domain. Since they lack an intracellular activation domain, they interact with adaptor molecules for signal transduction.

The earliest event in triggering the MTI signaling downstream is MAMPs-mediated rapid heteromerization of PRRs. Phosphorylation events in PRRs are considered significant regulatory mechanisms in MTI signaling as functional kinases. The kinase domain of FLS2 is rapidly phosphorylated with flg22 peptide stimulation (Gómez-Gómez et al. 2001) in Flg22:FLS2 interaction and the phosphorylated FLS2 rapidly dimerizes with *brassinosteroid intensive* 1 (BRI1)-associated kinase (BAK1) and *Botrytis*-induced kinase 1 (BIK1) (Chinchilla et al. 2007; Heese et al. 2007; Schulze et al. 2010). The rapid interaction between FLS2 and BAK1 (within a span of 2 minutes) indicated the involvement of BAK1 in a preformed membrane complex that gets weakly associated with FLS2. Mutual trans-phosphorylation of the BIK1 and FLS2/BAK1 kinase domains (Schulze et al. 2010) leads to conformation changes, and phosphorylated BIK1 is released to activate the downstream signaling pathway (Belkhadir et al. 2012).

In *Arabidopsis*, two simultaneous downstream MAPK cascades, MKK4/MKK5-MPK3/MPK6 and MEKK1/MKK1/MKK2-MPK4, are triggered by the discharged BIK1 (Asai et al. 2002; Gao et al. 2010), leading to the activation of the WRKY transcription factor family. Being a DNA-binding domain, these WRKYs interact with the W-box (TTGACC/T) motif present in promoters of defense-associated genes (Navarro et al. 2004) and activate the expression of early defense-related genes (Ishihama and Yoshioka 2012). These WRKYs interact with the W-box (TTGACC/T) motif present in the promoter region of the defense-associated gene and activate the expression of early defense-related genes (Ishihama and Yoshioka 2012).

Taking together, these recent studies have shed light on the critical role of the plant-associated microbiome in plant defense function and have opened up new avenues for manipulating plant immunity through microbial interventions.

8.12 PHYTOBIOMES AS ANTIBIOTIC SECRETORS

8.12.1 Secretory Metabolites

Bacteria produce a huge number of secondary metabolites having many several physicochemical and biological properties (Patel et al. 2020). Sometimes, they show stronger biological activities such as toxins or antibiotics due to their high degree of functionalization. For example, non-ribosomal peptides are synthesized by large nonribosomal peptide synthetases (NRPSs). Siderophores and lipopeptides are two important classes of secondary metabolites made by NRPSs (Tyc et al. 2017). Siderophores are low-molecular-weight with high-affinity iron-chelating compounds that are produced by bacteria under iron-limited conditions. These siderophores function in the solubilization, transport, and storage of iron (Chu et al. 2010). It can act in an antagonistic way by scavenging a limited amount of iron from the soil, which in turn reduces the amount of available iron for other organisms in the surroundings. Examples of siderophores are pyoverdines from *Pseudomonas*, bacillibactin from *Bacillus*, desferrioxamine from *Streptomyces*, and ornibactin from *Burkholderia*. Lipopeptides (LPs) are compounds that are made up of a lipid tail with either linear or cyclic oligopeptides. They act as a surfactant, antimicrobial, antipredation, and cytotoxic compound.

Polyketides are another large class of secondary metabolites produced by several bacteria, such as *Actinobacteria* (Stinear et al. 2004) *Pseudomonas* (El-Sayed et al. 2003), *Myxococcus* (Frank et al. 2007), *Bacillus* (Chen et al. 2006), *and Burkholderia* (Partida-Martinez and Hertweck 2005). They are synthesized by bacterial polyketide synthases (PKSs). PKSs are generally classified into three major types (Type I, II, and III). Type I PKSs consist of large, multidomain proteins producing polyketides with one module per successive condensation reaction. The antibiotic erythromycin is one of the best examples of type I PKSs. Type II PKSs are made up of individual proteins that play important role in the formation of aromatic polyketides such as the antibiotics tetracycline. Type III PKSs are involved in the synthesis of polyhydroxy phenols such as 2, 4-diacetylphloroglucinol (2, 4-DAPG) that are produced by several species of *Pseudomonas*. It has been isolated from various Gram-positive and Gram-negative soil bacteria.

8.12.2 Volatile Compounds (VOCs)

Volatile organic compounds (VOCs) are small molecules (<300 Da), which belong to different chemical classes; they can evaporate and diffuse easily through the air and water-filled pores. These physicochemical properties of volatiles make them a suitable medium for inter- and intra-kingdom communication. However, the higher polarity of the soluble secondary metabolites makes them soluble in water. Plant association with both abiotic and biotic stimuli contact is considered to be influenced by VOCs (Farre-Armengol et al. 2016). By creating their own VOCs, the phyllosphere microbes alter the development and emission of plant VOCs (Bringel and Couee 2015; Farre-Armengol et al. 2016). Methane gas, for example, is released when plants dissolve wax, which gets degraded by methanotrophic bacteria (Iguchi et al. 2013). By emitting VOCs, the phyllosphere microbiome can also prevent bacterial and fungal infections from colonizing (Bitas et al. 2013; Kai et al. 2007; Kanchiswamy et al. 2015; Schulz-Bohm et al. 2017). These VOCs are generally secondary metabolites by nature and belong to classes of terpenes.

As per the classification, the terpenes are derived from the two-building unit such as dimethylallyl pyrophosphate (DMAPP) and isopentenyl pyrophosphate. It arises from the mevalonate pathway or the deoxyxylulose phosphate pathway. They are known as plant metabolites, which are produced by the trees as well as by prokaryotes, fungi, and social amoebae. However, bacterial terpenes are known for over a century, but their biological and ecological roles are not studied. For example, geosmin, the well-known terpene released by soil bacteria, an odoriferous *Streptomyces albidoflavus* isolate from maize seeds is reported to produce a novel sesquiterpene called albaflavenone having antibacterial properties (Takamatsu et al. 2011). Many of these terpenes are reported to act as an antagonist for other groups of microbes. Nitrogen compounds are the most widespread compounds produced by many soil bacteria. They are pyrazines, indole, pyrroles, thiazoles, pyridines, and aniline derivatives. Pyrazines are VOCs best known for their antimicrobial activities (Patel et al. 2020; Sheoran et al. 2015; Munjal et al. 2016). They are also produced by plants and a few bacteria such as *Pseudomonas*, *Bacillus*, *Chondromyces*, and *Streptomyces*. Indole is synthesized from its precursor molecule tryptophan by the enzyme tryptophanase by both Gram-positive and Gram-negative bacteria. It can suppress bacterial pathogenesis by inhibiting quorum sensing and virulence factor production. This is also a well-known signaling molecule that modulates spore production, plasmid stability, cell division, antibiotic resistance, and biofilm formation and also controls plant defense signaling and the growth development of host plants.

Several sulfur-containing volatile compounds and alkyl sulfides have a huge structural diversity, which ranges from small compounds such as dimethylsulfide (DMS), dimethyl disulfide (DMDS), and dimethyl trisulfide (DMTS), to large complex volatiles such as 2-methyl tetrahydrothiophene-3-one which originates from homocysteine in bacteria (Nawrath et al. 2010). They play an important role in plant–microbe and interspecific microbe– interactions. For example, DMDS was identified as a quorum-quenching compound reported to enhance the growth of bacteria and inhibit the growth of fungus. Apart from epiphytes, the microbial volatile organic compounds (MVOCs) released by endophytic bacterial species showed excellent antagonistic activity against several plant pathogens such as *Phytophthora capsici*; *Pythium myriotylum*; *Gibberella moniliformis*; *Rhizoctonia solani*; *Athelia rolfsii*; *Colletotrichum gloeosporioides*; *Magnaporthe oryzae*; *Ralstonia solanacearum*; *Xanthomonas axonopodis* pv. *punicae*; *Xanthomonas oryzae* pv. *Oryzae;* and plant-parasitic nematode, *Radopholus similis*. Gas Chromatography/Mass Spectrometry (GC/MS)-based chemical profiling revealed the presence of many aromatic compounds such as dimethyl trisulfide; dimethyl disulfide; and various derivatives of pyrazines such as 2, 5-dimethyl pyrazine; 2-ethyl 3-methyl pyrazine; 2-methyl pyrazine; 2-ethyl pyrazine; 2-ethyl 5-methyl pyrazine; 2-ethyl 3, 6-dimethyl pyrazine (Sheoran et al. 2015; Munjal et al. 2016; Agisha et al. 2019; Sahu et al. 2020; Kumar et al. 2021). The knowledge gained from the volatilome of microbes has contributed to the development of a new method of crop disease management (Archana et al. 2021).

8.13 PLANT MICROBIOME FOR PLANT HEALTH: MICROBIOME TRANSPLANTATION AS A METHOD TO ENHANCE AGRICULTURAL PRODUCTIVITY

The metabolic capability of microbiome components is more diverse than the microbial diversity itself. The complex microbial interactions and the consequent effect they exert on the plant are not understood completely. Nevertheless, several antimicrobial compounds have been identified from the microbiome components that would benefit the plants. Versatile organic compounds have been identified in the microbial cultures that include VOCs, antibiotics, and hormones. Besides, several secondary metabolites belonging to the various classes such as alkaloids, terpenoids, and flavonoids have been identified in microbes.

The success of modern agriculture is largely attributed to the cultivation of elite genotypes of cultivars in a monocropping format in a vast area in a region. Such an industrialized agricultural approach to achieve homogeneity in the "end-product" is fraught with potential threats such as an outbreak of pests and disease epidemics. It is speculated that loss of microbial diversity on genetically homogenous crops grown in a large area could be one of the contributory factors for repeated disease outbreaks. To restore microbial diversity and equilibrium in plant niches, it is essential to reintroduce the microbiome in agricultural fields. Here, the role of microbiome restoration by deliberate transplantation assumes significance. Understanding microbiome structure and composition is the first step in this direction. Microbiome transplantation has been better studied and applied in mammalian systems. Study on the gut microbiome has led to many therapeutic solutions for gut-associated disorders. Fecal microbiome transplantation (FMT) has decreased inflammatory bowel disease and infectious diarrhoea. The microbiome is rich in Firmicutes, which release substances with immunoregulatory properties (Anderson et al. 2012). The selection of microbiome should be depended on the microbiome heritability on specific crops. Co-evolution and co-adaptation are important for plant and microbiome heritability (Mueller and Sachs 2015). The microbiome is now paving way for increased productivity through increased biomass and enhanced tolerance for biotic and abiotic stresses. So far, root-associated microbiome transfer is much studied in various combinations. Mixing disease suppressive soils with disease conducive soils helped in protecting plants from tobacco black rot, *Helminthosporium* foot rot of wheat and potato common scab. A synthetic microbiome can be transferred to plants to protect them from biotic stresses. *Nicotiana attenuata* was protected from sudden wilt disease by transferring a synthetic microbiome consisting of five bacteria, which were associated with roots and were selected (Foo et al. 2017). Consortium-based application of microbiome has increased drought tolerance in grapevine and date palm. Water use efficiency could be improved in *Arabidopsis* by a combination of species of *Pseudomonas*, which resulted in altered post-embryonic root development producing more lateral roots (Gopal and Gupta 2016). Microbiome employment also induces defense signals in plants (Sahu et al. 2020; Sahu et al. 2021; Kumar et al. 2021). By employing *Pseudomonas* isolates, rice blast fungus *Magnaporthe* was successfully controlled by ISR in rice plants (Spence et al. 2014). The phyllosphere core microbiome is reported to suppress blast disease by direct antifungal antagonism and indirect rice defense activation (Sahu et al. 2021). Further, microbiome employment is giving rise to a new plant breeding strategy for increased productivity.

The current tritrophic interaction approaches ignore the microbiome component, which could be attributed to our limited understanding of complex interactions among the plant–pathogen–environment and its impact on disease development. Thus, the study of the interaction behavior of plants and associated microbiome in a holistic manner will help us in drawing a strong inference on the disease dynamics. On the applied side, the microbiome studies would help us in identifying potential microbial agents and their components as **probiotics** for plant growth, **antibiotics** against deleterious microbes as well as an **elicitor** of tolerance against biotic and abiotic stress in plants. Microbiome transplantation-based microbial restoration on plant niches can boost growth, development, innate immunity, and overall productivity of crop plants in future agriculture.

ACKNOWLEDGMENTS

The author is thankful to the Director, IARI, Indian Council of Agricultural Research-Indian Agricultural Research Institute, New Delhi, for support and encouragement. The fund received from SERB for a project on the ***Development and validation of phyllosphere microbiome-based biostimulant for defense activation against blast disease and abiotic stresses in rice*** is thankfully acknowledged.

REFERENCES

Abarenkov, K., Henrik Nilsson, R., Larsson, K.H., Alexander, I.J., Eberhardt, U., Erland, S. and Sen, R., 2010. The UNITE database for molecular identification of fungi–recent updates and future perspectives. *New Phytologist*, 186(2): 281–285.

Agisha, V.N., Kumar, A., Eapen, S.J., Sheoran, N. and Suseelabhai, R., 2019. Broad-spectrum antimicrobial activity of volatile organic compounds from endophytic *Pseudomonas putida* BP25 against diverse plant pathogens. *Biocontrol Science and Technology*, 29(11): 1069–1089.

Ahuja, I., Kissen, R. and Bones, A.M., 2012. Phytoalexins in defense against pathogens. *Trends in Plant Science*, 17(2): 73–90.

Almario, J., Jeena, G., Wunder, J., Langen, G., Zuccaro, A., Coupland, G. and Bucher, M., 2017. Root-associated fungal microbiota of nonmycorrhizal Arabisalpina and its contribution to plant phosphorus nutrition. *Proceedings of the National Academy of Sciences*, 114(44): E9403–E9412.

Anderson, J.L., Edney, R.J. and Whelan, K., 2012. Systematic review: faecal microbiota transplantation in the management of inflammatory bowel disease. *Alimentary Pharmacology & Therapeutics*, 36(6): 503–516.

Archana, T.J., Gogoi, R., Kaur, C., Varghese, E., Sharma, R.R., Srivastav, M., Tomar, M., Kumar, M., and Kumar, A., 2021. Bacterial volatile mediated suppression of postharvest anthracnose and quality enhancement in mango. *Postharvest Biology and Technology*, 177, 111525.

Asai, T., Tena, G., Plotnikova, J., Willmann, M.R., Chiu, W.L., Gomez-Gomez, L., Boller, T., Ausubel, F.M. and Sheen, J., 2002. MAP kinase signaling cascade in *Arabidopsis* innate immunity. *Nature*, 415(6875): 977–983.

Averill, C., Bhatnagar, J.M., Dietze, M.C., Pearse, W.D. and Kivlin, S.N., 2019. Global imprint of mycorrhizal fungi on whole-plant nutrient economics. *Proceedings of the National Academy of Sciences*, 116(46): 23163–23168.

Backer, R., Rokem, J.S., Ilangumaran, G., Lamont, J., Praslickova, D., Ricci, E., Subramanian, S. and Smith, D.L., 2018. Plant growth-promoting rhizobacteria: context, mechanisms of action, and roadmap to commercialization of biostimulants for sustainable agriculture. *Frontiers in Plant Science*, 9: 1473.

Backhed, F., Ley, R.E., Sonnenburg, J.L., Peterson, D.A. and Gordon, J.I., 2005. Host-bacterial mutualism in the human intestine. *Science*, 307(5717): 1915–1920.

Beck, M., Heard, W., Mbengue, M. and Robatzek, S., 2012. The INs and OUTs of pattern recognition receptors at the cell surface. *Current Opinion in Plant Biology*, 15: 367–374.

Bednarek, P., 2012. Chemical warfare or modulators of defence responses–the function of secondary metabolites in plant immunity. *Current Opinion in Plant Biology*, 15(4): 407–414.

Belien, T., Van Campenhout, S., Robben, J. and Volckaert, G., 2006. Microbial endoxylanases: effective weapons to breach the plant cell-wall barrier or rather, triggers of defense systems? *Molecular Plant-Microbe Interactions*, 19:1072–1081.

Belkhadir, Y., Jaillais, Y., Epple, P., Balsemão-Pires, E., Dangl, J.L. and Chory, J., 2012. Brassinosteroids modulate the efficiency of plant immune responses to microbe-associated molecular patterns. *Proceedings of the National Academy of Sciences*, 109(1): 297–302.

Benson, D.A., Cavanaugh, M., Clark, K., Karsch-Mizrachi, I., Lipman, D.J., Ostell, J. and Sayers, E.W., 2012. GenBank. *Nucleic Acids Research*, 41(D1): D36–D42.

Bentley, B.L. and Carpenter, E.J., 1984. Direct transfer of newly-fixed nitrogen from free-living epiphyllous microorganisms to their host plant. *Oecologia*, 63(1): 52–56.

Berg, G., Rybakova, D., Fischer, D., et al., 2020. Microbiome definition re-visited: old concepts and new challenges. *Microbiome*, 8, 103. https://doi.org/10.1186/s40168-020-00875-0

Bergelson, J., Mittelstrass, J. and Horton, M.W., 2019. Characterizing both bacteria and fungi improves understanding of the *Arabidopsis* root microbiome. *Scientific Reports*, 9(1): 1–11.

Bertani, I., Abbruscato, P., Piffanelli, P., Subramoni, S. and Venturi, V., 2016. Rice bacterial endophytes: isolation of a collection, identification of beneficial strains and microbiome analysis. *Environmental Microbiology Reports*, 8(3): 388–398.

Bitas, V., Kim, H.S., Bennett, J.W. and Kang, S., 2013. Sniffing on microbes: diverse roles of microbial volatile organic compounds in plant health. *Molecular Plant-Microbe Interactions*, 26(8): 835–843.

Bodelier, P.L., Roslev, P., Henckel, T. and Frenzel, P., 2000. Stimulation by ammonium-based fertilizers of methane oxidation in soil around rice roots. *Nature*, 403(6768): 421–424.

Bodenhausen, N., Somerville, V., Desiro, A., Walser, J.C., Borghi, L., van der Heijden, M.G. and Schlaeppi, K., 2019. *Petunia*and *Arabidopsis*-specific root microbiota responses to phosphate supplementation. *Phytobiomes Journal*, 3(2): 112–124.

Boller, T., 2005. Peptide signalling in plant development and self/non-self perception. *Current Opinion in Cell Biology*, 17(2): 116–122.

Bowatte, S., Newton, P.C., Brock, S., Theobald, P. and Luo, D., 2015. Bacteria on leaves: a previously unrecognised source of N_2O in grazed pastures. *The ISME Journal*, 9(1): 265–267.

Brader, G., Compant, S., Vescio, K., Mitter, B., Trognitz, F., Ma, L.J. and Sessitsch, A., 2017. Ecology and genomic insights into plant-pathogenic and plant-nonpathogenic endophytes. *Annual Review of Phytopathology*, 55: 61–83.

Bringel, F. and Couée, I., 2015. Pivotal roles of phyllosphere microorganisms at the interface between plant functioning and atmospheric trace gas dynamics. *Frontiers in Microbiology*, 6: 486.

Brunner, F., Rosahl, S., Lee, J., Rudd, J.J., Geiler, C., Kauppinen, S., Rasmussen, G., Scheel, D. and Nürnberger, T., 2002. Pep-13, a plant defense-inducing pathogen-associated pattern from Phytophthora transglutaminases. *The EMBO Journal*, 21(24): 6681–6688.

Bulgarelli, D., Rott, M., Schlaeppi, K. et al., 2012. Revealing structure and assembly cues for *Arabidopsis* root-inhabiting bacterial microbiota. *Nature*, 488: 91–95.

Busby, P.E., Ridout, M. and Newcombe, G., 2016. Fungal endophytes: modifiers of plant disease. *Plant Molecular Biology*, 90, 645–655. https://doi.org/10.1007/s11103-015-0412-0

Castrillo, G., Teixeira, P.J.P.L., Paredes, S.H., Law, T.F., de Lorenzo, L., Feltcher, M.E., Finkel, O.M., Breakfield, N.W., Mieczkowski, P., Jones, C.D. and Paz-Ares, J., 2017. Root microbiota drive direct integration of phosphate stress and immunity. *Nature*, 543(7646): 513–518.

Chen, X.H., Vater, J., Piel, J., Franke, P., Scholz, R., Schneider, K., Koumoutsi, A., Hitzeroth, G., Grammel, N., Strittmatter, A.W. and Gottschalk, G., 2006. Structural and functional characterization of three polyketide synthase gene clusters in *Bacillus amyloliquefaciens* FZB 42. *Journal of Bacteriology*, 188(11): 4024–4036.

Chinchilla, D., Zipfel, C., Robatzek, S., Kemmerling, B., Nürnberger, T., Jones, J.D., Felix, G. and Boller, T., 2007. A flagellin-induced complex of the receptor FLS2 and BAK1 initiates plant defence. *Nature*, 448(7152): 497–500.

Chu, B.C., Garcia-Herrero, A., Johanson, T.H., Krewulak, K.D., Lau, C.K., Peacock, R.S., Slavinskaya, Z. and Vogel, H.J., 2010. Siderophore uptake in bacteria and the battle for iron with the host; a bird's eye view. *Biometals*, 23(4): 601–611.

Coleman-Derr, D., Desgarennes, D., Fonseca-Garcia, C., Gross, S., Clingenpeel, S., Woyke, T., North, G., Visel, A., Partida-Martinez, L.P. and Tringe, S.G., 2016. Plant compartment and biogeography affect microbiome composition in cultivated and native *Agave species*. *New Phytologist*, 209(2): 798–811.

Compant, S., Samad, A., Faist, H. and Sessitsch, A., 2019. A review on the plant microbiome: ecology, functions, and emerging trends in microbial application. *Journal of Advanced Research*, 19: 29–37.

Compant, S., Cambon, M.C., Vacher, C., Mitter, B., Samad, A. and Sessitsch, A., 2021. The plant endosphere world - bacterial life within plants. *Environmental Microbiology*, 23(4): 1812–1829.

Conrad, R., Erkel, C. and Liesack, W., 2006. Rice Cluster I methanogens, an important group of Archaea producing greenhouse gas in soil. *Current Opinion in Biotechnology*, 17(3): 262–267.

Conrad, R., Klose, M. and Noll, M., 2009. Functional and structural response of the methanogenic microbial community in rice field soil to temperature change. *Environmental Microbiology*, 11(7): 1844–1853.

Conrad, R., 2009. The global methane cycle: recent advances in understanding the microbial processes involved. *Environmental Microbiology Reports*, 1(5): 285–292.

Copeland, J.K., Yuan, L., Layeghifard, M., Wang, P.W. and Guttman, D.S., 2015. Seasonal community succession of the phyllosphere microbiome. *Molecular Plant-Microbe Interactions*, 28(3): 274–285.

Darvill, A.G. and Albersheim, P., 1984. Phytoalexins and their elicitors-a defense against microbial infection in plants. *Annual Review of Plant Physiology*, 35(1): 243–275.

De Costa, D.M., Pinto, M.T.C., Geethanjalee, H.D.N. and Dissanayake, N., 2006. Suppression of rice pathogens by phyllosphere-associated microflora of different rice varieties in Sri Lanka. *Tropical Science*, 46: 97–104.

De Lorenzo, G., Ferrari, S., Cervone, F. and Okun, E., 2018. Extracellular DAMPs in plants and mammals: immunity, tissue damage and repair. *Trends in Immunology*, 39(11): 937–950.

De Souza Vandenberghe, L.P., Garcia, L.M.B., Rodrigues, C., Camara, M.C., de Melo Pereira, G.V., de Oliveira, J. and Soccol, C.R., 2017. Potential applications of plant probiotic microorganisms in agriculture and forestry. *AIMS Microbiology*, 3(3): 629.

del Carmen Orozco-Mosqueda, M., del Carmen Rocha-Granados, M., Glick, B.R. and Santoyo, G., 2018. Microbiome engineering to improve biocontrol and plant growth-promoting mechanisms. *Microbiological Research*, 208: 25–31.

Delmotte, N., Knief, C., Chaffron, S., Innerebner, G., Roschitzki, B., Schlapbach, R., von Mering, C. and Vorholt, J.A., 2009. Community proteogenomics reveals insights into the physiology of phyllosphere bacteria. *Proceedings of the National Academy of Sciences*, 106(38): 16428–16433.

DeSantis, T.Z., Hugenholtz, P., Larsen, N., Rojas, M., Brodie, E.L., Keller, K. and Andersen, G.L., 2006. Greengenes, a chimera-checked 16S rRNA gene database and workbench compatible with ARB. *Applied and Environmental Microbiology*, 72(7): 5069–5072.

Dodds, P.N. and Rathjen, J.P., 2010. Plant immunity: towards an integrated view of plant-pathogen interactions. *Nature Reviews Genetics*, 11(8): 539–548.

Dow, M., Newman, M.A. and Von Roepenack, E., 2000. The induction and modulation of plant defense responses by bacterial lipopolysaccharides. *Annual Review of Phytopathology*, 38(1): 241–261.

Edwards, J., Johnson, C., Santos-Medellín, C., Lurie, E., Podishetty, N.K., Bhatnagar, S., Eisen, J.A. and Sundaresan, V., 2015. Structure, variation, and assembly of the root-associated microbiomes of rice. *Proceedings of the National Academy of Sciences*, 112(8): E911–E920.

El-Sayed, A.K., Hothersall, J., Cooper, S.M., Stephens, E., Simpson, T.J. and Thomas, C.M., 2003. Characterization of the mupirocin biosynthesis gene cluster from *Pseudomonas fluorescens* NCIMB 10586. *Chemistry & Biology*, 10(5): 419–430.

Eyre, A.W., Wang, M., Oh, Y. and Dean, R.A., 2019. Identification and characterization of the core rice seed microbiome. *Phytobiomes Journal*, 3(2): 148–157.

Ezenwa, V.O., Gerardo, N.M., Inouye, D.W., Medina, M. and Xavier, J.B., 2012. Animal behavior and the microbiome. *Science*, 338(6104): 198–199.

Farré-Armengol, G., Filella, I., Llusia, J. and Peñuelas, J., 2016. Bidirectional interaction between phyllospheric microbiotas and plant volatile emissions. *Trends in Plant Science*, 21(10): 854–860.

Felix, G. and Boller, T., 2003. Molecular sensing of bacteria in plants: the highly conserved RNA-binding motif RNP-1 of bacterial cold shock proteins is recognized as an elicitor signal in tobacco. *Journal of Biological Chemistry*, 278(8): 6201–6208.

Felix, G., Duran, J.D., Volko, S. and Boller, T., 1999. Plants have a sensitive perception system for the most conserved domain of bacterial flagellin. *The Plant Journal*, 18(3): 265–276.

Felix, G., Regenass, M. and Boller, T., 1993. Specific perception of subnano-molar concentrations of chitin fragments by tomato cells: induction of extracellular alkalinization, changes in protein phosphorylation, and establishment of a refractory state. *The Plant Journal*, 4(2): 307–316.

Fierer, N., 2017. Embracing the unknown: disentangling the complexities of the soil microbiome. *Nature Reviews Microbiology*, 15(10): 579–590.

Fitzpatrick, C.R., Copeland, J., Wang, P.W., Guttman, D.S., Kotanen, P.M. and Johnson, M.T., 2018. Assembly and ecological function of the root microbiome across angiosperm plant species. *Proceedings of the National Academy of Sciences*, 115(6): E1157–E1165.

Fitzpatrick, C.R., Salas-González, I., Conway, J.M., Finkel, O.M., Gilbert, S., Russ, D., Teixeira, P.J.P.L. and Dangl, J.L., 2020. The plant microbiome: from ecology to reductionism and beyond. *Annual Review of Microbiology*, 74: 81–100.

Foo, J.L., Ling, H., Lee, Y.S. and Chang, M.W., 2017. Microbiome engineering: current applications and its future. *Biotechnology Journal*, 12: 1600099.

Frank, B., Knauber, J., Steinmetz, H., Scharfe, M., Blöcker, H., Beyer, S. and Müller, R., 2007. Spiroketal polyketide formation in *Sorangium*: identification and analysis of the biosynthetic gene cluster for the highly cytotoxic spirangienes. *Chemistry and Biology*, 14(2): 221–233.

Fürnkranz, M., Wanek, W., Richter, A., Abell, G., Rasche, F. and Sessitsch, A., 2008. Nitrogen fixation by phyllosphere bacteria associated with higher plants and their colonizing epiphytes of a tropical lowland rainforest of Costa Rica. *The ISME Journal*, 2(5): 561–570.

Galletti, R., Denoux, C., Gambetta, S., Dewdney, J., Ausubel, F.M., De Lorenzo, G. and Ferrari, S., 2008. The AtrbohD-mediated oxidative burst elicited by oligogalacturonides in *Arabidopsis* is dispensable for the activation of defense responses effective against *Botrytis cinerea*. *Plant Physiology*, 148(3): 1695–1706.

Gao, F.K., Dai, C.C. and Liu, X.Z., 2010. Mechanisms of fungal endophytes in plant protection against pathogens. *African Journal of Microbiology Research*, 4(13): 1346–1351.

Ghosh, D., Gupta, A. and Mohapatra, S., 2019. Dynamics of endogenous hormone regulation in plants by phytohormone secreting rhizobacteria under water-stress. *Symbiosis*, 77(3): 265–278.

Giri, S. and Pati, B.R., 2004. A comparative study on phyllosphere nitrogen fixation by newly isolated *Corynebacterium sp.& Flavobacterium sp.* and their potentialities as biofertilizer. *Acta Microbiologica et Immunologica Hungarica*, 51(1–2): 47–56.

Glass, N.L. and Donaldson, G.C., 1995. Development of primer sets designed for use with the PCR to amplify conserved genes from filamentous ascomycetes. *Applied Environmental Microbiology*, 61(4): 1323–1330.

Gómez-Gómez, L., Bauer, Z. and Boller, T., 2001. Both the extracellular leucine-rich repeat domain and the kinase activity of FLS2 are required for flagellin binding and signaling in *Arabidopsis*. *The Plant Cell*, 13(5): 1155–1163.

Gopal, M. and Gupta, A., 2016. Microbiome selection could spur next-generation plant breeding strategies. *Frontiers in Microbiology*, 7: 1971.

Gouda, S., Kerry, R.G., Das, G., Paramithiotis, S., Shin, H.S. and Patra, J.K., 2018. Revitalization of plant growth promoting rhizobacteria for sustainable development in agriculture. *Microbiological Research*, 206: 131–140.

Granado, J., Felix, G. and Boller, T., 1995. Perception of fungal sterols in plants (subnanomolar concentrations of ergosterol elicit extracellular alkalinization in tomato cells). *Plant Physiology*, 107(2): 485–490.

Guerrieri, R., Vanguelova, E.I., Michalski, G., Heaton, T.H. and Mencuccini, M., 2015. Isotopic evidence for the occurrence of biological nitrification and nitrogen deposition processing in forest canopies. *Global Change Biology*, 21(12): 4613–4626.

Habib, H. and Fazili, K.M., 2007. Plant protease inhibitors: a defense strategy in plants. *Biotechnology and Molecular Biology Reviews*, 2(3): 68–85.

Hacquard, S., 2016. Disentangling the factors shaping microbiota composition across the plant holobiont. *New Phytologist*, 209(2): 454–457.

Hacquard, S., Kracher, B., Hiruma, K., Münch, P.C., Garrido-Oter, R., Thon, M.R., Weimann, A., Damm, U., Dallery, J.F., Hainaut, M. and Henrissat, B., 2016. Survival trade-offs in plant roots during colonization by closely related beneficial and pathogenic fungi. *Nature Communications*, 7(1): 1–13.

Hammond-Kosack, K.E. and Jones, J.D., 1996. Resistance gene-dependent plant defense responses. *The Plant Cell*, 8(10): 1773–1791.

Hamonts, K., Trivedi, P., Garg, A., Janitz, C., Grinyer, J., Holford, P., Botha, F.C., Anderson, I.C. and Singh, B.K., 2018. Field study reveals core plant microbiota and relative importance of their drivers. *Environmental Microbiology*, 20(1): 124–140.

Han, S. and Micallef, S.A., 2016. Environmental metabolomics of the tomato plant surface provides insights on *Salmonella enterica* colonization. *Applied and Environmental Microbiology*, 82(10): 3131–3142.

Hanski, I., 1982. Dynamics of regional distribution: the core and satellite species hypothesis. *Oikos*, 38(2): 210–221.
Hardoim, P.R., Hardoim, C.C., Van Overbeek, L.S. and Van Elsas, J.D., 2012. Dynamics of seed-borne rice endophytes on early plant growth stages. *PLoS One*, 7(2): e30438.
Hardoim, P.R., Van Overbeek, L.S., Berg, G., Pirttilä, A.M., Compant, S., Campisano, A., Döring, M. and Sessitsch, A., 2015. The hidden world within plants: ecological and evolutionary considerations for defining functioning of microbial endophytes. *Microbiology and Molecular Biology Reviews*, 79(3): 293–320.
Harsonowati, W., Astuti, R.I. and Wahyudi, A.T., 2017. Leaf blast disease reduction by rice-phyllosphere actinomycetes producing bioactive compounds. *Journal of General Plant Pathology*, 83(2): 98–108.
Heese, A., Hann, D.R., Gimenez-Ibanez, S., Jones, A.M., He, K., Li, J., Schroeder, J.I., Peck, S.C. and Rathjen, J.P., 2007. The receptor-like kinase *SERK3/BAK1* is a central regulator of innate immunity in plants. *Proceedings of the National Academy of Sciences*, 104(29): 12217–12222.
Hestrin, R., Hammer, E.C., Mueller, C.W. and Lehmann, J., 2019. Synergies between mycorrhizal fungi and soil microbial communities increase plant nitrogen acquisition. *Communications Biology*, 2(1): 1–9.
Hietz, P., Wanek, W., Wania, R. and Nadkarni, N.M., 2002. Nitrogen-15 natural abundance in a montane cloud forest canopy as an indicator of nitrogen cycling and epiphyte nutrition. *Oecologia*, 131(3): 350–355.
Hiltner, L., 1904. Uber nevere Erfahrungen und Probleme auf dem Gebiet der Boden Bakteriologie und unter besonderer Beurchsichtigung der Grundungung und Broche. *Arbeiten der Deutschen Landwirtschafts-Gesellschaft. Berlin*, 98: 59–78.
Hol, W.H., De Boer, W., de Hollander, M., Kuramae, E.E., Meisner, A. and van der Putten, W.H., 2015. Context dependency and saturating effects of loss of rare soil microbes on plant productivity. *Frontiers in Plant Science*, 6: 485.
Holland, M.A., 2011. Nitrogen: give and take from phylloplane microbes. In: J.C. Polacco and C.D. Todd (Eds), *Ecological Aspects of Nitrogen Metabolism in Plants*. Wiley-Blackwell, London: 217–230.
Huffaker, A., Pearce, G. and Ryan, C.A., 2006. An endogenous peptide signal in *Arabidopsis* activates components of the innate immune response. *Proceedings of the National Academy of Sciences*, 103(26): 10098–10103.
Hussain, Q., Liu, Y., Zhang, A., Pan, G., Li, L., Zhang, X., Song, X., Cui, L. and Jin, Z., 2011. Variation of bacterial and fungal community structures in the rhizosphere of hybrid and standard rice cultivars and linkage to CO_2 flux. *FEMS Microbiology Ecology*, 78(1): 116–128.
Iguchi, H., Sato, I., Yurimoto, H. and Sakai, Y., 2013. Stress resistance and C1 metabolism involved in plant colonization of a methanotroph *Methylosinus* sp. B4S. *Archives of Microbiology*, 195(10): 717–726.
Ishihama, N. and Yoshioka, H., 2012. Post-translational regulation of *WRKY* transcription factors in plant immunity. *Current Opinion in Plant Biology*, 15(4): 431–437.
Ishii, S., Ohno, H., Tsuboi, M., Otsuka, S. and Senoo, K., 2011. Identification and isolation of active N_2O reducers in rice paddy soil. *The ISME Journal*, 5(12): 1936–1945.
Jiang, Y., Liang, Y., Li, C., Wang, F., Sui, Y., Suvannang, N., Zhou, J. and Sun, B., 2016. Crop rotations alter bacterial and fungal diversity in paddy soils across East Asia. *Soil Biology and Biochemistry*, 95: 250–261.
Jousset, A., Bienhold, C., Chatzinotas, A., Gallien, L., Gobet, A., Kurm, V., Küsel, K., Rillig, M.C., Rivett, D.W., Salles, J.F. and Van Der Heijden, M.G., 2017. Where less may be more: how the rare biosphere pulls ecosystems strings. *The ISME Journal*, 11(4): 853–862.
Kai, M., Effmert, U., Berg, G. and Piechulla, B., 2007. Volatiles of bacterial antagonists inhibit mycelial growth of the plant pathogen *Rhizoctonia solani*. *Archives of Microbiology*, 187(5): 351–360.

Kanchiswamy, C.N., Malnoy, M. and Maffei, M.E., 2015. Chemical diversity of microbial volatiles and their potential for plant growth and productivity. *Frontiers in Plant Science*, 6: 151.

Kauss, H., Fauth, M., Merten, A. and Jeblick, W., 1999. Cucumber hypocotyls respond to cutin monomers via both an inducible and a constitutive H_2O_2-generating system. *Plant Physiology*, 120(4): 1175–1182.

Khamwan, S., Boonlue, S., Riddech, N., Jogloy, S., Mongkolthanaruk, W., 2017. Characterization of endophytic bacteria and their response to plant growth promotion in Helianthus tuberosus L. *Biocatalysis and Agricultural Biotechnology*, 13: 153–159. https://doi.org/10.1016/j.bcab.2017.12.007

Kim, H., Lee, K.K., Jeon, J., Harris, W.A. and Lee, Y.H., 2020. Domestication of *Oryza* species eco-evolutionarily shapes bacterial and fungal communities in rice seed. *Microbiome*, 8(1): 1–17.

Klindworth, A., Pruesse, E., Schweer, T., Peplies, J., Quast, C., Horn, M. and Glöckner, F.O., 2013. Evaluation of general 16S ribosomal RNA gene PCR primers for classical and next-generation sequencing-based diversity studies. *Nucleic Acids Research*, 41(1): e1.

Knief, C., Delmotte, N., Chaffron, S., Stark, M., Innerebner, G., Wassmann, R., Von Mering, C. and Vorholt, J.A., 2012. Metaproteogenomic analysis of microbial communities in the phyllosphere and rhizosphere of rice. *The ISME Journal*, 6(7): 1378–1390.

Krishanti, N.I.P.R.A., Wahyudi, A.T. and Nawangsih, A.A., 2015. Non-pathogenic phyllosphere bacteria producing bioactive compounds as biological control of *Xanthomonas oryzae* pv.*oryzae*. *International Journal of Pharma and Bio Sciences*, 6(1): B-801–B-810.

Krupa, S.V., 2003. Effects of atmospheric ammonia (NH_3) on terrestrial vegetation: a review. *Environmental Pollution*, 124(2): 179–221.

Kunze, G., Zipfel, C., Robatzek, S., Niehaus, K., Boller, T. and Felix, G., 2004. The N terminus of bacterial elongation factor Tu elicits innate immunity in Arabidopsis plants. *The Plant Cell*, 16(12): 3496–3507.

Kumar, M., Kumar, A., Sahu, K.P., Patel, A., Reddy, B., Sheoran, N., Charishma, K., Rajashekara, H., Bhagat, S. and Rathour, R., 2021. Deciphering core-microbiome of rice leaf endosphere: revelation by metagenomic and microbiological analysis of aromatic and non-aromatic genotypes grown in three geographical zones. *Microbiological Research*, 246: 126704.

Kunadiya, M.B., Dunstan, W.D., White, D., Hardy, G.E.S.J., Grigg, A.H. and Burgess, T.I., 2019. A qPCR assay for the detection of *Phytophthora cinnamomi* including an mRNA protocol designed to establish propagule viability in environmental samples. *Plant Disease*, 103(9): 2443–2450.

Laluk, K., Luo, H., Chai, M., Dhawan, R., Lai, Z. and Mengiste, T., 2011. Biochemical and genetic requirements for function of the immune response regulator BOTRYTIS-INDUCED KINASE1 in plant growth, ethylene signaling, and PAMP-triggered immunity in Arabidopsis. *The Plant Cell*, 23(8): 2831–2849.

Leach, J.E., Triplett, L.R., Argueso, C.T. and Trivedi, P., 2017. Communication in the phytobiome. *Cell*, 169(4): 587–596.

Lee, S.J., Jee, B.Y., Son, M.H. and Lee, S.R., 2017. Infection and cox2 sequence of *Pythium chondricola* (Oomycetes) causing red rot disease in *Pyropia yezoensis* (Rhodophyta) in Korea. *Algae*, 32(2): 155–160.

Lemanceau, P., Blouin, M., Muller, D. and Moënne-Loccoz, Y., 2017. Let the core microbiota be functional. *Trends in Plant Science*, 22(7): 583–595.

Lin, Q. and Jianlong, W., 2010. Biodegradation characteristics of quinoline by *Pseudomonas putida. Bioresource Technology*, 101(19): 7683–7686.

Liu, H.T., Meng, J., Bo, W.J., Cheng, D., Li, Y., Guo, L.Y., Li, C.H., Zheng, Y.H., Liu, M.Z., Ning, T.Y., Wu, G.L., Yu, X.F., Feng, S.F., Tana, W.Y., Li, J., Li, L.J., Zeng, Y., Liu, S.V. and Jiang, G.M., 2016. Biodiversity management of organic farming enhances agricultural sustainability. *Scientific Reports*, 6, p.23816

Lu, D., Wu, S., Gao, X., Zhang, Y., Shan, L. and He, P., 2010. A receptor-like cytoplasmic kinase, BIK1, associates with a flagellin receptor complex to initiate plant innate immunity. *Proceedings of the National Academy of Sciences of the United States of America*, 107(1): 496–501.

Lu, T., Ke, M., Lavoie, M., Jin, Y., Fan, X., Zhang, Z., Fu, Z., Sun, L., Gillings, M., Peñuelas, J. and Qian, H., 2018. Rhizosphere microorganisms can influence the timing of plant flowering. *Microbiome*, 6(1): 1–12.

Lundberg, D.S., Lebeis, S.L., Paredes, S.H., Yourstone, S., Gehring, J., Malfatti, S., Tremblay, J., Engelbrektson, A., Kunin, V., Del Rio, T.G. and Edgar, R.C., 2012. Defining the core *Arabidopsis thaliana* root microbiome. *Nature*, 488(7409): 86–90.

Magurran, A.E. and Henderson, P.A., 2003. Explaining the excess of rare species in natural species abundance distributions. *Nature*, 422(6933): 714–716.

Maidak, B.L., Olsen, G.J., Larsen, N., Overbeek, R., McCaughey, M.J. and Woese, C.R., 1997. The RDP (ribosomal database project). *Nucleic Acids Research*, 25(1): 109–110.

Mallon, C.A., Poly, F., Le Roux, X., Marring, I., van Elsas, J.D. and Salles, J.F., 2015. Resource pulses can alleviate the biodiversity–invasion relationship in soil microbial communities. *Ecology*, 96(4): 915–926.

Mano, H. and Morisaki, H., 2008. Endophytic bacteria in the rice plant. *Microbes and Environments*, 23(2): 109–117.

Martin, F., 2000. Phylogenetic relationships among some pythium species inferred from sequence analysis of the mitochondrially encoded cytochrome oxidase II gene. *Mycologia*, 92(4): 711–727.

Martínez-Medina, A., Van Wees, S.C. and Pieterse, C.M., 2017. Airborne signals from *Trichoderma* fungi stimulate iron uptake responses in roots resulting in priming of jasmonic acid-dependent defences in shoots of *Arabidopsis thaliana* and *Solanum lycopersicum*. *Plant Cell & Environment*, 40(11): 2691–2705.

Mehta, S., Singh, B., Patra, A., Tripathi, A., Easwaran, M., Choudhary, J.R., Choudhary, M. and Aggarwal, S.K., 2021. Maize microbiome: current insights for the sustainable agriculture. In: M.K. Solanki et al. (Eds), *Microbiomes and Plant Health*. Academic Press, New York: 267–297.

Mendes, R., Garbeva, P. and Raaijmakers, J.M., 2013. The rhizosphere microbiome: significance of plant beneficial, plant pathogenic, and human pathogenic microorganisms. *FEMS Microbiology Reviews*, 37(5): 634–663.

Minamisawa, K., Imaizumi-Anraku, H., Bao, Z., Shinoda, R., Okubo, T. and Ikeda, S., 2016. Are symbiotic methanotrophs key microbes for N acquisition in paddy rice root? *Microbes and Environments*, 31(1): 4–10.

Mueller, U.G. and Sachs, J.L., 2015. Engineering microbiomes to improve plant and animal health. *Trends in Microbiology*, 23(10): 606–617.

Muller, T. and Ruppel, S., 2014. Progress in cultivation-independent phyllosphere microbiology. *FEMS Microbiology Ecology*, 87(1): 2–17.

Munjal, V., Nadakkakath, A.V., Sheoran, N., Kundu, A., Venugopal, V., Subaharan, K., Rajamma, S., Eapen, S.J. and Kumar, A., 2016. Genotyping and identification of broad spectrum antimicrobial volatiles in black pepper root endophytic biocontrol agent, *Bacillus megaterium* BP17. *Biological Control*, 92: 66–76.

Navarro, L., Zipfel, C., Rowland, O., Keller, I., Robatzek, S., Boller, T. and Jones, J.D., 2004. The transcriptional innate immune response to *flg22*. Interplay and overlap with Avr gene-dependent defense responses and bacterial pathogenesis. *Plant Physiology*, 135(2): 1113–1128.

Nawrath, T., Gerth, K., Müller, R. and Schulz, S., 2010. The biosynthesis of the aroma volatile 2-methyltetrahydrothiophen-3-one in the *Bacterium chitinophaga* Fx7914. *ChemBioChem*, 11(13): 1914–1919.

Ofek-Lalzar, M., Sela, N., Goldman-Voronov, M., Green, S.J., Hadar, Y. and Minz, D., 2014. Niche and host-associated functional signatures of the root surface microbiome. *Nature Communications*, 5(1): 1–9.

Pandey, S.P. and Somssich, I.E., 2009. The role of WRKY transcription factors in plant immunity. *Plant Physiology*, 150(4): 1648–1655.

Papen, H., Gebler, A., Zumbusch, E. and Rennenberg, H., 2002. Chemolithoautotrophic nitrifiers in the phyllosphere of a spruce ecosystem receiving high atmospheric nitrogen input. *Current Microbiology*, 44(1): 56–60.

Partida-Martinez, L.P. and Hertweck, C., 2005. Pathogenic fungus harbours endosymbiotic bacteria for toxin production. *Nature*, 437(7060): 884–888.

Patel, A., Kumar, A., Sheoran, N., Kumar, M., Sahu, K.P., Ganeshan, P., Ashajyothi, M., Gopalakrishnan, S. and Gogoi, R., 2020. Antifungal and defense elicitor activities of pyrazines identified in endophytic *Pseudomonas putida* BP25 against fungal blast incited by *Magnaporthe oryzae* in rice. *Journal of Plant Diseases and Protection*, 128: 1–12.

Pfeiffer, S., Mitter, B., Oswald, A., Schloter-Hai, B., Schloter, M., Declerck, S. and Sessitsch, A., 2017. Rhizosphere microbiomes of potato cultivated in the high Andes show stable and dynamic core microbiomes with different responses to plant development. *FEMS Microbiology Ecology*, 93(2), 242.

Pieterse, C.M., Zamioudis, C., Berendsen, R.L., Weller, D.M., Van Wees, S.C. and Bakker, P.A., 2014. Induced systemic resistance by beneficial microbes. *Annual review of Phytopathology*, 52: 347–375.

Pruesse, E., Quast, C., Knittel, K., Fuchs, B.M., Ludwig, W., Peplies, J. and Glöckner, F.O., 2007. SILVA: a comprehensive online resource for quality checked and aligned ribosomal RNA sequence data compatible with ARB. *Nucleic Acids Research*, 35(21): 7188–7196.

Rascovan, N., Carbonetto, B., Perrig, D., Díaz, M., Canciani, W., Abalo, M., Alloati, J., González-Anta, G. and Vazquez, M.P., 2016. Integrated analysis of root microbiomes of soybean and wheat from agricultural fields. *Scientific Reports*, 6(1): 1–12.

Rastogi, G., Coaker, G.L. and Leveau, J.H., 2013. New insights into the structure and function of phyllosphere microbiota through high-throughput molecular approaches. *FEMS Microbiology Letters*, 348(1): 1–10.

Ratnasingham, S. and Hebert, P.D., 2007. BOLD: The barcode of life data system (https:// www. barcodinglife. org). *Molecular Ecology Notes*, 7(3): 355–364.

Ravanbakhsh, M., Kowalchuk, G.A. and Jousset, A., 2019. Root-associated microorganisms reprogram plant life history along the growth–stress resistance tradeoff. *The ISME Journal*, 13(12): 3093–3101.

Reina-Pinto, J.J. and Yephremov, A., 2009. Surface lipids and plant defenses. *Plant Physiology and Biochemistry*, 47(6): 540–549.

Ren, G., Zhang, H., Lin, X., Zhu, J. and Jia, Z., 2014. Response of phyllosphere bacterial communities to elevated CO_2 during rice growing season. *Applied Microbiology and Biotechnology*, 98(22): 9459–9471.

Richardson, A.E. and Simpson, R.J., 2011. Soil microorganisms mediating phosphorus availability update on microbial phosphorus. *Plant Physiology*, 156(3): 989–996.

Riesenfeld, C.S., Goodman, R.M. and Handelsman, J., 2004. Uncultured soil bacteria are a reservoir of new antibiotic resistance genes. *Environmental Microbiology*, 6(9): 981–989.

Riit, T., Tedersoo, L., Drenkhan, R., Runno-Paurson, E., Kokko, H. and Anslan, S., 2018. Corrigendum for: "Oomycete-specific ITS primers for identification and metabarcoding" published in MycoKeys. *MycoKeys*, 41: 119–120.

Roman-Reyna, V., Pinili, D., Borjaa, F.N., Quibod, I., Groen, S.C., Mulyaningsih, E.S., Rachmat, A., Slamet-Loedin, I.H., Alexandrov, N., Mauleon, R. and Oliva, R., 2019. The rice leaf microbiome has a conserved community structure controlled by complex host-microbe interactions. *Cell-Host-Microbe*, 2019: 615278.

Ruinen, J., 1965. The phyllosphere: III. Nitrogen fixation in the phyllosphere. *Plant and Soil*, 22, 375–394.

Sahu, K.P., Kumar, A., Patel, A., et al., 2020. Rice blast lesions: an unexplored phyllosphere microhabitat for novel antagonistic bacterial species against *Magnaporthe oryzae*. *Microbial Ecology*, 81, 731–745. https://doi.org/10.1007/s00248-020-01617-3

Sahu, K.P., Patel, A., Kumar, M., Sheoran, N., Mehta, S., Reddy, B., Eke, P., Prabhakaran, N. and Kumar, A., 2021. Integrated metabarcoding and culturomic-based microbiome profiling of rice phyllosphere reveal diverse and functional bacterial communities for blast disease suppression. *Frontiers in Microbiology*, 12, 780458. https://doi.org/10.3389/fmicb.2021.780458

Santhanam, R., Weinhold, A., Goldberg, J., Oh, Y. and Baldwin, I.T., 2015. Native root-associated bacteria rescue a plant from a sudden-wilt disease that emerged during continuous cropping. *Proceedings of the National Academy of Sciences*, 112(36): E5013–E5020.

Sattelmacher, B., Mühling, K.H. and Pennewiß, K., 1998. The apoplast- its significance for the nutrition of higher plants. *Zeitschrift für Pflanzenernährung und Bodenkunde*, 161(5): 485–498.

Scagel, C.F., Bi, G., Fuchigami, L.H. and Regan, R.P., 2008. Rate of nitrogen application during the growing season and spraying plants with urea in the autumn alters uptake of other nutrients by deciduous and evergreen container-grown Rhododendron cultivars. *HortScience*, 43(5): 1569–1579.

Schulz-Bohm, K., Martín-Sánchez, L. and Garbeva, P., 2017. Microbial volatiles: small molecules with an important role in intra-and inter-kingdom interactions. *Frontiers in Microbiology*, 8: 2484.

Schulze, B., Mentzel, T., Jehle, A.K., Mueller, K., Beeler, S., Boller, T., Felix, G. and Chinchilla, D., 2010. Rapid heteromerization and phosphorylation of ligand-activated plant transmembrane receptors and their associated kinase *BAK1*. *Journal of Biological Chemistry*, 285(13): 9444–9451.

Sengupta, S., Ganguli, S. and Singh, P.K., 2017. Metagenome analysis of the root endophytic microbial community of Indian rice (*O. sativa* L.). *Genomics Data*, 12: 41–43.

Sessitsch, A., Hardoim, P., Döring, J., Weilharter, A., Krause, A., Woyke, T., Mitter, B., Hauberg-Lotte, L., Friedrich, F., Rahalkar, M. and Hurek, T., 2012. Functional characteristics of an endophyte community colonizing rice roots as revealed by metagenomic analysis. *Molecular Plant-Microbe Interactions*, 25(1): 28–36.

Sheoran, N., Nadakkakath, A.V., Munjal, V., Kundu, A., Subaharan, K., Venugopal, V., Rajamma, S., Eapen, S.J. and Kumar, A., 2015. Genetic analysis of plant endophytic *Pseudomonas putida* BP25 and chemo-profiling of its antimicrobial volatile organic compounds. *Microbiological Research*, 173: 66–78.

Shiu, S.H. and Bleecker, A.B., 2001. Receptor-like kinases from *Arabidopsis* form a monophyletic gene family related to animal receptor kinases. *Proceedings of the National Academy of Sciences*, 98(19): 10763–10768.

Simonin, M., Dasilva, C., Terzi, V., Ngonkeu, E.L., Diouf, D., Kane, A., Béna, G. and Moulin, L., 2020. Influence of plant genotype and soil on the wheat rhizosphere microbiome: evidences for a core microbiome across eight African and European soils. *FEMS Microbiology Ecology*, 96(6), 67.

Singh, S., Kumar, V., Singh, S., Dhanjal, D.S., Datta, S. and Singh, J., 2020. Global scenario of plant–microbiome for sustainable agriculture: current advancements and future challenges. In: A. Yadav, J. Singh, A. Rastegari and N. Yadav (Eds), *Plant Microbiomes for Sustainable Agriculture*. Springer, Cham: 425–443.

Spence, C., Alff, E., Johnson, C., Ramos, C., Donofrio, N., Sundaresan, V. and Bais, H., 2014. Natural rice rhizospheric microbes suppress rice blast infections. *BMC Plant Biology*, 14(1): 130.

Stanley, C.E. and van der Heijden, M.G., 2017. Microbiome-on-a-chip: new frontiers in plant–microbiota research. *Trends in Microbiology*, 25(8): 610–613.

Stinear, T.P., Mve-Obiang, A., Small, P.L., Frigui, W., Pryor, M.J., Brosch, R., Jenkin, G.A., Johnson, P.D., Davies, J.K., Lee, R.E. and Adusumilli, S., 2004. Giant plasmid-encoded polyketide synthases produce the macrolide toxin of *Mycobacterium ulcerans*. *Proceedings of the National Academy of Sciences*, 101(5): 1345–1349.

Stringlis, I.A., Proietti, S., Hickman, R., Van Verk, M.C., Zamioudis, C. and Pieterse, C.M., 2018. Root transcriptional dynamics induced by beneficial rhizobacteria and microbial immune elicitors reveal signatures of adaptation to mutualists. *The Plant Journal*, 93(1): 166–180.

Sun, H., Feng, Y., Ji, Y., Shi, W., Yang, L. and Xing, B., 2018. N_2O and CH_4 emissions from N-fertilized rice paddy soil can be mitigated by wood vinegar application at an appropriate rate. *Atmospheric Environment*, 185: 153–158.

Sun, L., Qiu, F., Zhang, X., Dai, X., Dong, X. and Song, W., 2008. Endophytic bacterial diversity in rice (*Oryza sativa* L.) roots estimated by 16S rDNA sequence analysis. *Microbial Ecology*, 55(3): 415–424.

Takamatsu, S., Lin, X., Nara, A., Komatsu, M., Cane, D.E. and Ikeda, H., 2011. Characterization of a silent sesquiterpenoid biosynthetic pathway in *Streptomyces avermitilis* controlling epi-isozizaene albaflavenone biosynthesis and isolation of a new oxidized epi-isozizaene metabolite. *Microbial Biotechnology*, 4(2): 184–191.

Thapa, S., Ranjan, K., Ramakrishnan, B., Velmourougane, K. and Prasanna, R., 2018. Influence of fertilizers and rice cultivation methods on the abundance and diversity of phyllosphere microbiome. *Journal of Basic Microbiology*, 58(2): 172–186.

Thauer, R.K., Kaster, A.K., Seedorf, H., Buckel, W. and Hedderich, R., 2008. Methanogenic archaea: ecologically relevant differences in energy conservation. *Nature Reviews Microbiology*, 6(8): 579–591.

Toju, H., Peay, K.G., Yamamichi, M., Narisawa, K., Hiruma, K., Naito, K., Fukuda, S., Ushio, M., Nakaoka, S., Onoda, Y. and Yoshida, K., 2018. Core microbiomes for sustainable agroecosystems. *Nature Plants*, 4(5): 247–257.

Trivedi, M.K., Branton, A., Trivedi, D., Nayak, G., Gangwar, M. and Jana, S., 2016. Molecular analysis of biofield treated eggplant and watermelon crops. *Advances in Crop Science and Technology*, 4(1): 1000208.

Trivedi, P., Leach, J.E., Tringe, S.G., Sa, T. and Singh, B.K., 2020. Plant–microbiome interactions: from community assembly to plant health. *Nature Reviews Microbiology*, 18(11): 607–621.

Tyc, O., Song, C., Dickschat, J.S., Vos, M. and Garbeva, P., 2017. The ecological role of volatile and soluble secondary metabolites produced by soil bacteria. *Trends in Microbiology*, 25(4): 280–292.

Uroz, S., Courty, P.E. and Oger, P., 2019. Plant symbionts are engineers of the plant-associated microbiome. *Trends in Plant Science*, 24(10): 905–916.

Ursell, L.K., Metcalf, J.L., Parfrey, L.W. and Knight, R., 2012. Defining the human microbiome. *Nutrition Reviews*, 70(suppl_1): S38–S44.

Venkatachalam, S., Ranjan, K., Prasanna, R., Ramakrishnan, B., Thapa, S. and Kanchan, A., 2016. Diversity and functional traits of culturable microbiome members, including *Cyanobacteria* in the rice phyllosphere. *Plant Biology*, 18(4): 627–637.

Voges, M.J., Bai, Y., Schulze-Lefert, P. and Sattely, E.S., 2019. Plant-derived coumarins shape the composition of an *Arabidopsis* synthetic root microbiome. *Proceedings of the National Academy of Sciences*, 116(25): 12558–12565.

Vorholt, J.A., 2012. Microbial life in the phyllosphere. *Nature Reviews Microbiology*, 10(12): 828–840.

Wagner, J., Coupland, P., Browne, H.P., Lawley, T.D., Francis, S.C. and Parkhill, J., 2016. Evaluation of PacBio sequencing for full-length bacterial 16S rRNA gene classification. *BMC Microbiology*, 16(1): 1–17.

Wallace, J.G., Kremling, K.A. and Buckler, E.S., 2018. Quantitative genetic analysis of the maize leaf microbiome. *BioRxiv*, 2018: 268532.

Watanabe, K., Kohzu, A., Suda, W., Yamamura, S., Takamatsu, T., Takenaka, A., Koshikawa, M.K., Hayashi, S. and Watanabe, M., 2016. Microbial nitrification in throughfall of a Japanese cedar associated with Archaea from the tree canopy. *SpringerPlus*, 5(1): 1–15.

Xue, C., Penton, C.R., Shen, Z., Zhang, R., Huang, Q., Li, R., Ruan, Y. and Shen, Q., 2015. Manipulating the banana rhizosphere microbiome for biological control of Panama disease. *Scientific Reports*, 5(1): 1–11.

Yadav, A.N., Verma, P., Kumar, M., Pal, K.K., Dey, R., Gupta, A., Padaria, J.C., Gujar, G.T., Kumar, S., Suman, A. and Prasanna, R., 2015. Diversity and phylogenetic profiling of niche-specific Bacilli from extreme environments of India. *Annals of Microbiology*, 65: 611–629.

Yadav, A.N., Kumar, V., Dhaliwal, H.S., Prasad, R. and Saxena, A.K., 2018. Microbiome in crops: diversity, distribution, and potential role in crop improvement. In: R. Prasad, S. S. Gill and N. Tuteja (Eds), *Crop Improvement through Microbial Biotechnology*. Elsevier, Amsterdam, The Netherlands: 305–332.

Yildirim, E., Guvenc, I., Turan, M. and Karatas, A., 2007. Effect of foliar urea application on quality, growth, mineral uptake and yield of broccoli (Brassica oleracea L., var. italica). *Plant Soil and Environment*, 53(3): 120.

Yuan, C., Zhang, L., Hu, H., Wang, J., Shen, J. and He, J., 2018. The biogeography of fungal communities in paddy soils is mainly driven by geographic distance. *Journal of Soils and Sediments*, 18(5): 1795–1805.

Zarraonaindia, I., Owens, S.M., Weisenhorn, P., West, K., Hampton-Marcell, J., Lax, S., Bokulich, N.A., Mills, D.A., Martin, G., Taghavi, S. and van der Lelie, D., 2015. The soil microbiome influences grapevine-associated microbiota. *mBio*, 6(2): e00300–e00315.

Zhang, J., Liu, Y.X., Zhang, N., Hu, B., Jin, T., Xu, H., Qin, Y., Yan, P., Zhang, X., Guo, X. and Hui, J., 2019. *NRT*1.1*B* is associated with root microbiota composition and nitrogen use in field-grown rice. *Nature Biotechnology*, 37(6): 676–684.

Zheng, Y., Huang, R., Wang, B.Z., Bodelier, P.L.E. and Jia, Z.J., 2014. Competitive interactions between methane-and ammonia-oxidizing bacteria modulate carbon and nitrogen cycling in paddy soil. *Biogeosciences*, 11(12): 3353–3368.

Zipfel, C., Kunze, G., Chinchilla, D., Caniard, A., Jones, J.D., Boller, T. and Felix, G., 2006. Perception of the bacterial PAMP EF-Tu by the receptor EFR restricts Agrobacterium-mediated transformation. *Cell*, 125(4): 749–760.

Zipfel, C., 2008. Pattern-recognition receptors in plant innate immunity. *Current Opinion in Immunology*, 20(1): 10–16.

9 Recent Developments in Biopesticide Use

Current Status and Future Prospects

Nidhi Rai, Akansha Jain, Sashi Pandey Rai, Birinchi Kumar Sarma, and Harikesh Bahadur Singh

9.1 INTRODUCTION

The world population is increasing at an alarming pace and is likely to touch nearly 9.7 billion by the year 2050 (UNDESA, 2015). This brings a massive concern to the agriculture sector as it obligates an increase in crop provision to match the food requirements. The exhaustive usage of chemical pesticides neglecting the safety guidelines restrains the deleterious effect of pathogens and pests in fields but, on the other hand, has tendered grievous health issues to the consumers and the surrounding environment. As per Food and Agriculture Organization (FAO) of the United Nations (2020), the top four leading pesticide-consuming countries in 2018 were China, the USA, Brazil, and Argentina (FAOSTAT, 2020). Also, the application of pesticides in India has exceptionally expanded from 586,314 metric tons (MT) in 2016–2017 to 62,193 MT in 2020–2021. The Indian state Maharashtra followed by Uttar Pradesh had the highest chemical pesticide consumption in terms of the total quantity (Figure 9.1). Presently, public awareness of food safety and quality has brought in strict safety regulations on imports and the burden of pesticide residues on products.

The traditional pesticide market has endured a huge revolution over the past few years (Jain et al., 2021) with product engagements of biopesticides and bringing in the framework of Integrated Pest Management (IPM). Intensified crop production and crop safety can be achieved by the application of agriculturally important beneficial microbes or biocontrol agents, which are known for enhancing crop yield by decreasing yield loss due to biotic factors (Jain et al., 2012; Kumar et al., 2017). Biopesticides are mainly living microorganisms or plant-based or microbe-based natural products, managing pests and pathogens in the crops along with improving their quality. As green agents, biopesticides have shown promising results and host-specific results in surrogating synthetic pesticides along with comparable crop productivity (Fenibo et al., 2022). Based on the estimates by IMARC Group, the global biopesticides market size was worth at US$ 3.96 Billion in 2020 and is anticipated to exhibit powerful growth during 2021–2026. This chapter provides the current status of biopesticides,

DOI: 10.1201/9781003531944-9

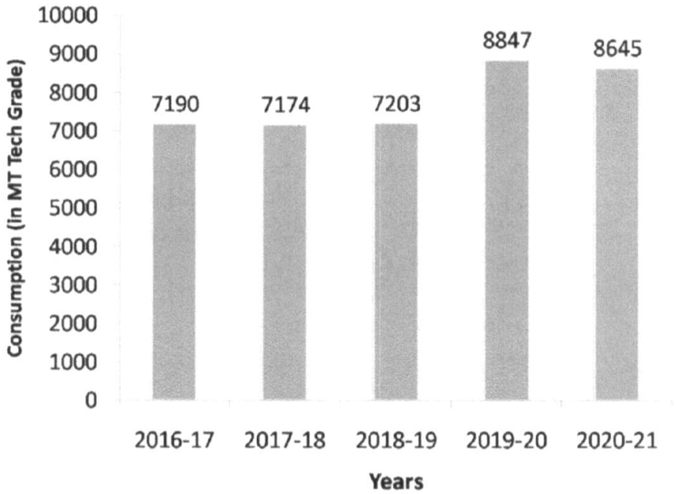

FIGURE 9.1 Consumption of biopesticides formulation in India from 2016–2017 to 2020–2021. (*Source*: Data obtained from Directorate of Plant Protection, Quarantine and Storage, Ministry of Agriculture and Farmers Welfare, Government of India.)

their adequacy in disease management, commercialization hurdles, and regulatory requirements. Furthermore, we have discussed modern technologies applied in the synthesis and improvement of biopesticide products.

9.2 BIO-PRIMING WITH MICROBIAL INOCULANTS FOR CONTROL OF SOIL-BORNE DISEASES

Bio-priming is one of the most approachable methods in the current agricultural farming for sustainable crop production without affecting the environment and improving soil health. Microbial inoculants can be delivered to plants by several methods such as direct application by seed treatment or seedling treatments with liquid or solid microbial formulation for delivery of the microorganisms to the rhizosphere of the plant and their interactions with plant roots to enhance their tolerance to diseases and thereby improving plant health (Figure 9.2) (Philippot et al., 2013; Yadav et al., 2017; Jain et al., 2020 Rakshit and Singh 2018). Wide range of microorganisms having properties of plant growth regulation and antibiotics production are being used as biopesticides individually or in combination against the several seed-borne and soil-borne diseases to achieve food security sustainably (Rakshit et al., 2022). Combination of microbes in the form of consortia can improve soil properties further, and their endurance to survive in various stressed environments also increases. Application techniques, formulation base, and timing are also critical for ensuring settlement, multiplication, and potency of a biopesticides and biostimulants (Rai et al., 2021; Russo et al., 2019; Singh and Vaishnav, 2022a–d).

Several reports have assessed the function of biopesticides on the growth, pathogen tolerance, and crop protection activities on varieties of crops such as cereals, pulses, and vegetables including fruits and oilseeds plant. Numerous biopesticides

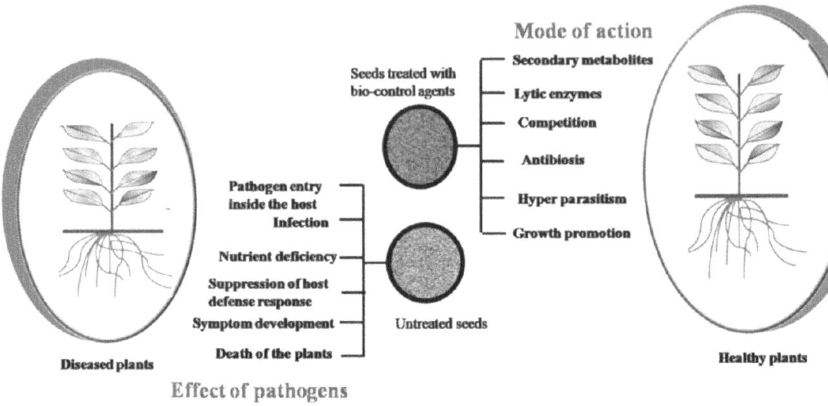

FIGURE 9.2 Effect of biopesticides treatment on plant growth and disease tolerance.

with formulations using bacterial, fungal, viral, and RNAi-based biopesticides are now being used as seed inoculants to mitigate plant diseases and pest menace such as molds, insects, and nematodes (Table 9.1). Some bacteria and fungi such as *Bacillus* spp., *Pseudomonas, Azotobacter, Chryseobacterium, Streptomyces, Trichoderma, Gliocladium, Beauveria, Saccharomyces, and Penicillium are most commonly used as seed inoculants for managing seed-borne pathogens such as Fusarium* spp., *Magnaporthe, Rhizoctonia, Puccinia graminis, Pythium* spp., *Helminthosporium, Sclerotium, and Sclerotinia* (Nagraj et al., 2021; Rojas et al., 2020; Newitt et al., 2019; Amin et al., 2014; Singh et al., 2013; Jain et al., 2015). For example, the application of *T. harzianum, T. gamsii* on garlic suppresses the proliferation of *Fusarium proliferatum, F. oxysporum* and restrains the pathogen (Mondani et al., 2021). Similarly, *Mitsuaria* spp. was observed to be a promising disease control propensity to resist growth of fungal pathogens of plants (Elnahal et al., 2022). Nanoparticle and RNAi-based biopesticides are also being used to control the attack of seed-borne diseases. Nano-primed beneficial microbes have great potential to ease their consistent execution in the field and endure yield across the world (El-Saadony et al., 2021; Yogaiah et al., 2022; Singh et al., 2018). Nano-primed beneficial microbes suggest two possible ways of promoting plant growth by augmenting nutrient use efficiency and coating seeds for a safer environment (Kalia and Kaur, 2019).

9.3 INTEREST OF GOVERNMENT AND PRIVATE SECTOR IN BIOPESTICIDES

Agricultural universities and many ICAR institutes in India are government bodies working on commercialization of biopesticides, but the major share of biopesticide production is still under the private firms. Any new microorganism formulations (biopesticides) are regulated under the Insecticide Act, 1968, and commercial supply needs registration with Central Insecticide Board Registration Committee (CIBRC) (Singh et al., 2016, 2017). Some Indian Institutes/universities such as ICAR-Indian Institute of Horticultural Research, Bangalore, G.B. Pant University of Agriculture

TABLE 9.1
Biocontrol Agents Used for Management of Discrete Soil Borne Pathogens

Crops	Disease/Target Pathogens	Biocontrol Agents	References
Cereals	*Fusarium verticillioides*	*Bacillus subtilis*, *B. mojavensis*, *B. amyloliquefaciens*, *Exophiala spinifera* (ATCC 74269), *Sphingopyxis macrogoltabida* (MTA 144), *Rhinocladiella atrovirens* (ATCC 74270), *Saccharomyces cerevisiae* and *Lacto-bacillus rhamnous*, *Trichoderma harzianum*, *Mycobacterium oleovorans*	Nagraj et al. (2021), Fareid et al. (2011), and Pereira et al. (2011)
	Magnoportheoryzae	*Streptomyces griseus*, *Achromobacter xylosoxidans*	Newitt et al. (2019)
	Xanthomonas oryzae pv. *oryzicola*	*Bacillus amyloliquefaciens*	Zhang et al. (2011)
	Tilletia caries, *Septoria nodorum*, *Fusarium* spp.	*Pseudomonas chlororaphis*	BioAgri AB (2016)
	Puccinia graminis, *Rhizoctonia solani*, *Pythium* spp.	*Streptomyces* spp., *Trichoderma* spp. and Arbuscular mycorrhizal (AM) fungi	Newitt et al. (2019)
	Gaeumannomyces graminis	*Streptomyces Trichoderma*, *Penicillium*, and *Sporidesmium*	Newitt et al. (2019)
	Fusarium spp.	*Streptomyces griseoviridis*, *Streptomyces spororaveus*, *Trichoderma* spp.	Adesina et al. (2017), Modrzewska et al. (2022), and Al Sahli and Abdulkhair (2012)
	Fusarium graminearum	*Sarocladium strictum*, *Anthracocystis floculossa*, *Anthracocystis floculossa and Penicillium olsonii* . *Bacillus*, *Pseudomonas*, *Chryseobacterium*	Rojas et al. (2020), and Besset-Manzoni et al. (2019)
	Urocystis agropyri	*Azobacter* spp., *Gluconacetobacter.*, *Bacillus thuringiensis*	Tao et al. (2014)
	Pyricularia oryzae	*Pseudomonas fluorescens*, *P. syringae*, *T. viride*	Arumugam et al. (2013)
	Helimnthosporium oryzae	*Pseudomonas fluorescens*, *T. viride*	Arumugam et al. (2013)
	Rhizoctonia, *Fusarium spp.*	*Bacillus subtilis.*	Brannen and Kenney (1997) and Bayer Crop Science (2016)

(Continued)

TABLE 9.1 (*Continued*)
Biocontrol Agents Used for Management of Discrete Soil Borne Pathogens

Crops	Disease/Target Pathogens	Biocontrol Agents	References
	Stem borer *Sesamia calamistis*	*Beauveria bassiana*	Cherry et al. (2004)
	Fusarium proliferatum, F. oxysporum	*T. harzianum, T. gamsii*	Mondani et al. (2021)
	Pythium aphanider-matum, Phytophthora capscici, Pythium sylvaticum, Phytophthora sojae, Rhizoctonia solani, Fusarium graminearum, Alternaria solani, and *F. oxysporum*	*Mitsuaria* spp.	Elnahal et al. (2022)
	Fusarium graminearum	*CYP3* target gene specific RNAi	Koch et al. (2016)
	Magnaporthe oryzae	*β2-tubulin X* target gene specific RNAi	Gu et al. (2019)
	Fusarium asiaticum	*Myosin 5* target gene specific RNAi	Song et al. (2018)
	Diabrotica virgifera virgifera	*DvSnf7* genetically modified RNAi	Khajuria et al. (2018)
Vegetables	*Colletrotrichumlindemuthianum*	*Trichoderma harzianum, T. viride, Pseudomonas fluorescens, Gliocladium virens*	Amin et al. (2014)
	Streptomyces scabies	*Pseudomonas, Bacillus*, and *Streptomyces*	Newitt et al. (2019).
	Agrobacterium tumefaciens	*Agrobacterium radiobacter* (K84 strain)	Vicedo et al. (1993)
	Phoma medicaginis	*Bacillus licheniformis*	Slimeneert et al. (2015)
	Ascochyta rabiei	*Chaetomium globosum, T. viride, T. harzianum,*	Bisen et al. (2020)
	Phytophthora capsici and *Sclerotium cepivorum*	*Trichoderma harzianum, Trichoderma pseudokoningii*	Sid Ahmed et al. (1999)
	Meloidogyne species	*Pasteuria penetrance*	Davies et al. (2011)
	Plasmodiophora brassicae	*Bacillus subtilis, Gliocladium catenulatum*	Peng et al. (2011)
	Tobacco mosaic virus	*CP* target gene specific RNAi, Silver nanoparticles from *Pseudomonas fluorescens*	Ahsan (2020); Worrall et al. (2019)
	Powdery mildew pathogens	*Acremonium alternatum, Acrodontium crateriforme, Ampelomyces quisqualis, Cladosporium oxysporum,* and *Gliocladium virens*	Heydari and Pessarakli (2010)

(*Continued*)

TABLE 9.1 (*Continued*)
Biocontrol Agents Used for Management of Discrete Soil Borne Pathogens

Crops	Disease/Target Pathogens	Biocontrol Agents	References
Pulses	*Sclerotinia sclerotiorum*	*Gliocladium virens* and *Coniothyrium minitans*	Whipps et al. (2008)
	Colletotrichum lindemuthianum	*Induratia cofeana* CML 4009	Mota et al. (2021)
	Colletotrichum truncatum	*Pseudomonas* sp., *Trichoderma* sp.	Begum et al. (2010)
	Botrytis spp, *Rhizoctonia solani*, and *Fusarium oxysporum*	*Serratia marcescens*, *Pseudomonas fluorescens*, *T. viride*, *T. harzianum*, *Chaetomium globosum*, *Clonostachys rosea*	Ningaraju (2006) and Sreevidya et al. (2015)
	Rhizoctonia solani, *Fusarium oxysporum*, *Fusarium* graminearum, *Fusarium* spp.	*Bacillus amyloliquefaciens*, *Trichoderma* spp.	Correa et al. (2009) and Kumar et al. (2017)
	Didymella bryoniae	*Pseudomonas chlororaphis*, *Serratia plymuthica*, *Paenibacillus polmyxa*, *Lysobacter gummosus*	Furnkranz et al. (2012)
	Phakopsora pachyrhizi	RNAi based	Koch et al. (2016)
	Plutella xylostella	siRNA as a potential biopesticide	Gong et al. (2011)
	Erysiphe pisi	*Trichoderma asperellum* and *Pseudomonas fluorescens*	Patel et al. (2017) and Sarkar et al. (2014)
Fruits	*Phytophthora palmivora*	*Trichoderma*, *Penicillium*, and *Sporidesmium*	Nega (2014)
	Phytophthora cinnamomi	*Pseudomonas*, *Bacillus*, *Streptomyces*, *photorhabdus* and *Heterorhabditis amazonensis*	Palmieri et al. (2019)
	Monilinia vaccinii-corymbosi	*Bacillus subtilis*	Dedej et al. (2004)
	P. cactorum B. cinerea	*Aureobasidium pullulans*	Iqbal et al. (2021)
	Phytophthora cactorum	*Trichoderma harzianum*, *Trichoderma koningii*, *Flavobacterium*	Alexander and Stewart (2001)
	Pseudocercosporamusae	*Bacillus subtilis B106*	Fu et al. (2010)
	Botrytiscinerea	*DCL1* target gene specific *RNAi*	Wang et al. (2016)
Oil seeds	Sunflower necrosis virus disease	Bacterial consortia—*Bacillus* spp., *Pseudomonas aeruginosa, Streptomyces sp.*	Srinivasan and Mathivanan (2011)
	Fusarium spp., *Rhizoctonia solani, Ralstonia*	*Paenibacillus polymyxa*	Ryu et al. (2006)

(*Continued*)

TABLE 9.1 (*Continued*)
Biocontrol Agents Used for Management of Discrete Soil Borne Pathogens

Crops	Disease/Target Pathogens	Biocontrol Agents	References
	Nematode Rotylenchus reniformis	*Pasteuria* spp.	Schmidt et al. (2010)
	Xanthomonas campestris pv. *Campestris*	*B. subtilis R14, B. pumilus C116, B. megaterium* pv. *cerealis RAB7*, and *B. cereus C210*	Luna et al. (2002)
	B. cinerea, Verticillium dahliae	*Enterococcus* spp.	Fhoula et al. (2013)
	Weissella confuse, Lactococcus lactis, L. plantarum, Lactococcus mesenteroides Enterococcus durans, Enterococcus faecium	*Pseudomonas savastanoi*	Fhoula et al. (2013)

and Technology, Pantnagar, ICAR-Indian Agricultural Research Institute, New Delhi, Assam Agriculture University, Jorhat, Punjab Agricultural University, Ludhiana, Haryana Agriculture University, Hisar, Tamil Nadu Agricultural University, Coimbatore and ICAR-Indian Institute of Oil Seeds Research, Hyderabad, and Banaras Hindu University, Varanasi, are actively involved in research and/or production of biopesticides. The government has also established some Krishi Vigyan Kendras and biocontrol laboratories at the state level for promoting biopesticide production and use. The pivotal companies in the Indian market include Biowrks Inc., Sumitomo Chemical India Pvt. Ltd, Koppert Biological Systems India Pvt. Ltd, Nav Agro Pvt. Ltd, and Kilpest India. The CIBRC has organized the regulation and requirements for biopesticide registration and decided on minimal infrastructural needs for its production (Singh et al., 2016).

9.4 BIOPESTICIDE MARKET FORECAST

Aggravated endeavors by the government and industries to agitate the application of promising, environmentally friendly, and cost-effective agri-bioinputs will expand the market share of biopesticides in the coming years. At present, biopesticides occupy a meager pesticide market share worldwide. The global biopesticides market size was $4.40 billion in 2019 and is expected to reach $10.63 billion by 2027 (Fortune Business Insights, 2020). Interest in the investment by industrial and institutional sectors and the involvement of omics and strain improvement techniques are broadening the horizons of the biopesticide market. Bioinsecticides and biofungicides rule the sales with an approximate share of 43% and 44%, respectively, in 2018. As per the Modor Intelligence report (2021), major biopesticide industries in the U.S. market are Bayer AG, Marrone Bio Innovations, Certis USA L.L.C, Valent BioSciences, and Koppert Biological Systems. In accordant to the Research Institute of Organic Agriculture, the land where organic farming is practiced in India was around 2.7 million ha in 2020 (Modor Intelligence, 2021). The pivotal companies in Indian market include Biowrks Inc., Sumitomo Chemical India Pvt. Ltd, Koppert Biological Systems India Pvt. Ltd, Nav Agro Pvt. Ltd, and Kilpest India. The Indian biopesticides market is predicted to generate $778 Million by 2025, reaching a CAGR of 25.4% (Inkwood Research, 2017). EPA and EU have strict lengthy procedures to be followed before registration, which causes a delay in commercialization and income generation. The biggest reason behind the low success rate in the progress and application of new biopesticides in India is the high cost of registration, regulatory impediment to commercialization, and the hesitation among the farmers to test the new products owing to lack of rational proficiency (Manoharachary et al., 2020).

9.5 NANO-BIOPESTICIDE

Nanotechnology is a fledgling sphere to design upgraded nano-biopesticides and nano gene carriers to restrain disease in the crops. They act in a decisive role in IPM and sustainable disease resistance, after appraising the allied perquisite and risks to persuade their promising impacts on plant health and pathogen control. Moreover, that nanotechnology embellishes the potency of some biopesticides on pathogens and pests and lowers the noxious effects on human beings and the surrounding environment. It also improves efficacy by curtailing the net loss due to physical damage because of the encapsulation of these biopesticides in nano-based systems. In a recent study, eight commercial essential oils formulated in nano-emulsions showed promising action against stored product pest, *Tribolium confusum* (Palermo et al., 2021). Registration, Evaluation, Authorization, and Restriction of Chemicals by EC (REACH Regulation 1907/2006) sets up proceedings for compiling and evaluating knowledge on the properties and threats of substances. The European Commission (EC) manages risk associated with the use of nanomaterials by the best feasible framework by REACH and CLP (Second Regulatory Review on Nanomaterials, 2012; REACH Review, 2013). The EC also adopted Regulation (EU) 2018/1881 to modify REACH Annexes I, III, and VI-XII, introducing nano-specific clarifications (Commission Regulation, 2018). As per US Environmental Protection Agency (USEPA), many nanoscale materials are considered "chemical substances" under the Toxic Substances Control Act (TSCA). In future, nano-biopesticides can be integrated in a prudent manner for sustainable agriculture (Singh et al., 2018; Yogaiah et al., 2022).

The Indian government actively promotes nanotechnology through the Nano Science and Technology Initiative (NSTI) and started a Mission on Nano Science and Technology (Nano Mission) in the year 2007 for laying down a National Regulatory Framework Road-Map for Nanotechnology (NRFR-Nanotech) (PIB, 2014). The Nano Mission is in its Phase-II during the 12th 5-year plan period with a budget of Rs. 650 crores. Up to now, there is no regulatory framework by the government for the use of nanopesticides/nano-biopesticides and strategies are being decided on to consolidate rules within the current rules or design new ones for nanopesticides regulation. To promote nanobiopesticides, the governments have to evolve smooth regulatory system, biosafety, and marketing strategy of commercial products (Yogaiah et al., 2022).

9.6 INVOLVEMENT OF OMICS IN GENERATION OF BIOPESTICIDES

The commencement of the post-genomics era entrusts the improvisation of food safety by aiding in the recognition of various types of biopesticides and building our awareness of functional genes and mode of action associated with restraining disease. Gathering huge data and usage of bioinformatics are key players in the investigation of the emanation received from new-age sequencing platforms (Ngai and McDowell, 2017). Next-generation sequencing and comparison of omics-based data allow the capitalization of species diversity, directing the designing of dynamic biopesticide with the steep precision system under study with negligible effect on the surrounding ecosystem. The availability of draft genomes of pathogens has provided scraping of protein targets for biopesticides. The advent of CRISPR-Cas9 has transformed genome editing and permitted manipulation of genomic sequences in an influential way (Knott and Doudna, 2018) of host/phytopathogens/or biocontrol agents in action. Gene silencing of the *acel* gene in *Trichoderma atroviride* caused upregulation in antibiotics and secondary metabolites synthesis for *Fusarium oxysporum* and *Rhizoctonia solani* management (Fang and Chen, 2018). Future relies on isolating new biocontrol microorganisms and genome engineering for proficient formulation development.

9.7 FUTURE PROSPECTS

The eventual objective of any disease management program is to secure the food quality and quantity without causing any damage to the surrounding ecosystem. Biopesticides have sought worldwide focus as an eco-friendly approach to restrict or decrease the utilization of chemical pesticides. The first commitment in this respect is expected from the government and industrial sectors to expedite the development, mass production, and supply of biopesticide. The application of advanced technology and unique materials in formulation production and distribution would make an increment in the market sales and usage of biopesticides. IPM strategies on assimilating biological agents with other agents into a standard procedure are of utmost importance. Introducing products with microbial consortia can further widen the host and pathogen range and further improve disease management at the farmer's end. The major goal of a researcher and producer should be to decrease the cost of production, and having additional field results with improved product quality and shelf-life. Frameworks issued for biopesticides by the EU and USEPA are less rigid and more pliable than those adopted by the Indian government. Regulatory measures that boost product registration, licensing, catering incentives to improvise biocontrol agent's commercialization, and accessibility are needed with immediate effect. Nanoscience can definitely enhance biopesticides applicability and their application in the field. Future food security shift relies on prompt movement in the direction of biopesticides to process our agricultural lands more sustainably.

ACKNOWLEDGMENTS

AJ is grateful to the Department of Science and Technology, India for grant under the DST Women Scientist under WOSA Scheme (SR/WOS-A/LS-377/2018).

REFERENCES

Adesina, M.F., Lembke, A., Costa, R., Speksnijder, A. and Smalla, K. (2017). Screening of bacterial isolates from various European soils for *in vitro* antagonistic activity towards *Rhizoctonia solani* and *Fusarium oxysporum*: Site-dependent composition and diversity revealed. *Soil Biology and Biochemistry*, 39(11), 2818–2828.

Ahsan, T. (2020). Biofabrication of silver nanoparticles from *Pseudomonas fluorescens* to control tobacco mosaic virus. *Egyptian Journal of Biological Pest Control*, 30(1), 1–4.

Al Sahli, A.A. and Abdulkhair, W.M. (2012). Biocontrol of *Fusarium udum* diseases for some wheat cultivars by *Streptomyces spororaveus*. *African Journal of Microbiology Research*, 6(1), 190–196.

Alexander, B.J.R. and Stewart, A. (2001). Glasshouse screening for biological control agents of *Phytophthora cactorum* on apple (Malusdomestica). *New Zealand Journal of Crop and Horticultural Science*, 29(3), 159–169.

Amin, M., Fitsum, S., Selvaraj, T. and Mulugeta, N. (2014). Field management of anthracnose (*Colletotrichum lindemuthianum*) in common bean through fungicides and bioagents. *Advances in Crop Science and Technology*, 2(2), 2–6.

Arumugam, K., Ramalingam, P. and Appu, M. (2013). Isolation of *Trichoderma viride* and *Pseudomonas fluorescens* organism from soil and their treatment against rice pathogens. *Journal of Microbiology and Biotechnology Research*, 3(6), 77–81.

Bayer Crop Science. (2016). www.cropscience.bayer.us/products/seedgrowth/kodiak. (Accessed on 14th April, 2022).

Begum, M.M., Sariah, M., Puteh, A.B., Abidin, M.Z., Rahman, M.A. and Siddiqui, Y. (2010). Field performance of bio-primed seeds to suppress *Colletotrichum truncatum* causing damping-off and seedling stand of soybean. *Biological Control*, 53(1), 18–23.

Besset-Manzoni, Y., Joly, P., Brutel, A., Gerin, F., Soudière, O., Langin, T. and Prigent-Combaret, C. (2019). Does *in vitro* selection of biocontrol agents guarantee success in planta? A study case of wheat protection against Fusarium seedling blight by soil bacteria. *PLoS One*, 14(12), e0225655.

BioAgri AB. (2016). www.bioagri.se/?p=30933&m=4353&topic=products. (Accessed on 28th April, 2022).

Bisen, K., Singh, V., Keswani, C., Ray, S., Sarma, B.K. and Singh, H.B. (2020). Use of biocontrol agents for the management of seed-borne diseases. In: Kumar, R. and Gupta, A. (eds), *Seed-Borne Diseases of Agricultural Crops: Detection, Diagnosis and Management*. Springer, Singapore, pp. 651–663.

Brannen, P.M. and Kenney, D.S. (1997). Kodiak®—a successful biological-control product for suppression of soil-borne plant pathogens of cotton. *Journal of Industrial Microbiology and Biotechnology*, 19(3), 169–171.

Cherry, A.J., Banito, A., Djegui, D. and Lomer, C. (2004). Suppression of the stem-borer *Sesamia calamistis* (Lepidoptera; Noctuidae) in maize following seed dressing, topical application and stem injection with African isolates of *Beauveria bassiana*. *International Journal of Pest Management*, 50(1), 67–73.

Commission Regulation (EU) 2018/1881 of 3 December 2018 amending Regulation (EC) No 1907/2006 of the European Parliament and of the Council on the Registration, Evaluation, Authorisation and Restriction of Chemicals (REACH) as regards Annexes I, III, VI, VII, VIII, IX, X, XI, and XII to address nanoforms of substances. (Accessed on May 8th, 2022).

Consumption of chemical pesticides in various states/UTs during 2016–17 to 2020–21. Directorate of Plant Protection, Quarantine and Storage. https://ppqs.gov.in/statistical-database. (Accessed on 4th May, 2022).

Correa, O.S., Montecchia, M.S., Berti, M.F., Ferrari, M.C.F., Pucheu, N.L., Kerber, N.L., et al. (2009). *Bacillus amyloliquefaciens* BNM122, a potential microbial biocontrol agent applied on soybean seeds, causes a minor impact on rhizosphere and soil microbial communities. *Applied Soil Ecology*, 41(2), 185–194.

Davies, K.G., Rowe, J., Manzanilla-López, R. and Opperman, C.H. (2011). Re-evaluation of the life-cycle of the nematode-parasitic bacterium *Pasteuria penetrans* in root-knot nematodes. *Meloidogyne* spp. *Nematology*, 13(7), 825–835.

Dedej, S., Delaplane, K.S. and Scherm, H. (2004). Effectiveness of honey bees in delivering the biocontrol agent *Bacillus subtilis* to blueberry flowers to suppress mummy berry disease. *Biological Control*, 31(3), 422–427.

Elnahal, A.S., El-Saadony, M.T., Saad, A.M., Desoky, E.S.M., El-Tahan, A.M., Rady, M.M., et al. (2022). The use of microbial inoculants for biological control, plant growth promotion, and sustainable agriculture: A review. *European Journal of Plant Pathology*, 162, 1–34.

El-Saadony, M.T., Sitohy, M.Z., Ramadan, M.F. and Saad, A.M. (2021). Green nanotechnology for preserving and enriching yogurt with biologically available iron (II). *Innovative Food Science & Emerging Technologies*, 69, 102645.

EU (2012). Communication on the second regulatory review on nanomaterials. (Accessed on May 8th, 2022).

Fang, C. and Chen, X. (2018). Potential biocontrol efficacy of *Trichoderma atroviride* with cellulase expression regulator ace1 gene knock-out. *Biotechnology*, 8, 302.

FAOSTAT. (2020). Pesticides Use and Pesticides Indicators. (Accessed on May 9th, 2022).

Fareid, M.A. (2011). Biocontrol of *Fusarium* moulds and fumonisin B1 production. *World Rural Observations*, 3(3), 58–61.

Fenibo, E.O., Ijoma, G.N. and Matambo, T. (2022). Biopesticides in Sustainable agriculture: Current status and future prospects. In: Mandal, S.D., Ramkumar, G., Karthi, S., and Jin, F. (eds), *New and Future Development in Biopesticide Research: Biotechnological Exploration*. Springer, Singapore, pp. 57-61. https://doi.org/10.1007/978-981-16-3989-0_1.

Fhoula, I., Najjari, A., Turki, Y., Jaballah, S., Boudabous, A. and Ouzari, H. (2013). Diversity and antimicrobial properties of lactic acid bacteria isolated from rhizosphere of olive trees and desert truffles of Tunisia. *BioMed Research International*, 2013, 405708.

Fortune Business Insights. Agricultural Biologicals. Market Research Report 2016–2018. (2020). https://www.fortunebusinessinsights.com/industry-reports/biopesticides-market-100073#:~:text=Biopesticides%20Market%20Size,Forecast%2C%202020%2D2027. (Accessed on 7th May, 2022).

Fu, G., Huang, S., Ye, Y., Wu, Y., Cen, Z. and Lin, S. (2010). Characterization of a bacterial biocontrol strain B106 and its efficacies on controlling banana leaf spot and post-harvest anthracnose diseases. *Biological Control*, 55(1), 1–10.

Furnkranz, M., Adam, E., Müller, H., Grube, M., Huss, H., Winkler, J., et al. (2012). Promotion of growth, health and stress tolerance of Styrian oil pumpkins by bacterial endophytes. *European Journal of Plant Pathology*, 134(3), 509–519.

Gong, L., Yang, X., Zhang, B., Zhong, G. and Hu, M. (2011). Silencing of Rieske iron–sulfur protein using chemically synthesized siRNA as a potential biopesticide against *Plutella xylostella*. *Pest Management Science*, 67(5), 514–520.

Gu, K.X., Song, X.S., Xiao, X.M., Duan, X.X., Wang, J.X., Duan, Y.B., et al. (2019). A β2-tubulin dsRNA derived from *Fusarium asiaticum* confers plant resistance to multiple phytopathogens and reduces fungicide resistance. *Pesticide Biochemistry and Physiology*, 153, 36–46.

Heydari, A. and Pessarakli, M. (2010). A review on biological control of fungal plant pathogens using microbial antagonists. *Journal of Biological Sciences*, 10(4), 273–290.

Imarc. (2021). www.imarcgroup.com/biopesticides-market. (Accessed on 4th May, 2022).

Inkwood Research. (2017). https://inkwoodresearch.com/reports/india-biopesticides-market-forecast-2017-2025/. (Accessed on 4th May, 2022).

Iqbal, M., Jamshaid, M., Zahid, M.A., Andreasson, E., Vetukuri, R.R. and Stenberg, J.A. (2021). Biological control of strawberry crown rot, root rot and grey mould by the beneficial fungus *Aureobasidium pullulans*. *BioControl*, 66(4), 535–545.

Jain, A., Chakrobarty, J. and Das, S. (2020). Underlying mechanism of plant –microbe crosstalk in shaping microbial ecology of the rhizosphere. *Acta Physiologia Plantarum*, 42, 8. https://doi.org/10.1007/s11738-019-3000-0.

Jain, A. and Das, S. (2021). Regulatory requirements for nanopesticides and nanofertilizers. In: Jogaiah, S., Singh, H.B., Fraceto, L.F., and de Lima, R. (eds), *Advances in Nano-Fertilizers and Nano-Pesticides in Agriculture: A Smart Delivery System for Crop Improvement*. Woodhead Publishing Series in Food Science, Technology and Nutrition, Cambridge, pp. 145–152.

Jain, A., Singh, A., Sarma, B.K. and Singh, H.B. (2012). Microbial consortium mediated reprogramming of defense network in pea to enhance tolerance against *Sclerotinia sclerotiorum*. *Journal of Applied Microbiology*, 112(3), 537–550.

Jain, A., Singh, A., Singh, S. and Singh, H.B. (2015). Biological management of *Sclerotinia sclerotiorum* in pea using plant growth promoting microbial consortium. *Journal of Basic Microbiology*, 55, 961–972.

Kalia, A. and Kaur, H. (2019). Nano-biofertilizers: Harnessing dual benefits of nano-nutrient and bio-fertilizers for enhanced nutrient use efficiency and sustainable productivity. In: Pudake, R., Chauhan, N., and Kole, C. (eds), *Nanoscience for Sustainable Agriculture*. Springer, Cham, pp. 51–73.

Khajuria, C., Ivashuta, S., Wiggins, E., Flagel, L., Moar, W., Pleau, M., et al. (2018). Development and characterization of the first dsRNA-resistant insect population from western corn rootworm, *Diabrotica virgifera virgifera* LeConte. *PLoS One*, 13(5), e0197059.

Knott, G.J. and Doudna, J.A. (2018). CRISPR-Cas guides the future of genetic engineering. *Science*, 361, e866–e869.

Koch, A., Biedenkopf, D., Furch, A., Weber, L., Rossbach, O., Abdellatef, E., et al. (2016). An RNAi-based control of *Fusarium graminearum* infections through spraying of long dsRNAs involves a plant passage and is controlled by the fungal silencing machinery. *PLoS Pathogens*, 12, e1005901. https://doi.org/10.1371/journal.ppat.1005901.

Kumar, G., Maharshi, A., Patel, J., Mukherjee, A., Singh, H.B. and Sarma, B.K. (2017). *Trichoderma*: A potential fungal antagonist to control plant diseases. *SATSA Mukhapatra-Annual Technical Issue*, 21, 206–218.

Luna, C.L., Mariano, R.L.R. and Souto-Maior, A.M. (2002). Production of a biocontrol agent for crucifers black rot disease. *Brazilian Journal of Chemical Engineering*, 19(2), 133–140.

Manoharachary, C., Singh, H.B. and Ajit, V. (2020). *Trichoderma: Agricultural Applications and Beyond*. Springer, Singapore, p. 367.

Modor Intelligence. (2021). United States Biopesticides Market - Growth, Trends, COVID-19 Impact, and Forecasts (2022–2027). www.mordorintelligence.com/industry-reports ModorIntelligence. (Accessed on May 7th, 2022).

Modrzewska, M., Bryła, M., Kanabus, J. and Pierzgalski, A. (2022). *Trichoderma* as a biostimulator and biocontrol agent against *Fusarium* in the production of cereal crops: Opportunities and possibilities. *Plant Pathology*, 19(2), 133–140. https://doi.org/0.1111/ppa.13397.

Mondani, L., Chiusa, G. and Battilani, P. (2021). Chemical and biological control of *Fusarium* species involved in garlic dry rot at early crop stages. *European Journal of Plant Pathology*, 160(3), 575–587.

Mota, S.F., Pádua, P.F., Ferreira, A.N., Gomes, L.D.B.W., Dias, M.A., Souza, E.A., et al. (2021). Biological control of common bean diseases using endophytic *Induratia* spp. *Biological Control*, 159, 104629.

Nagraj, D., Achar, P.N. and Sreenivasa, M.Y. (2021). Current perspectives of biocontrol agents for management of *Fusarium verticillioides* and its fumonisin in cereals—A Review. *Journal of Fungi*, 7(9), 776.

Nega, A. (2014). Review on concepts in biological control of plant pathogens. *Journal of Biology, Agriculture and Healthcare*, 4(27), 33–54.

Newitt, J.T., Prudence, S.M., Hutchings, M.I. and Worsley, S.F. (2019). Biocontrol of cereal crop diseases using streptomycetes. *Pathogens*, 8(2), 78.

Ngai, M. and McDowell, M.A. (2017). The search for novel insecticide targets in the post-genomics era, with a specific focus on G-protein coupled receptors. *The Memórias do Instituto Oswaldo Cruz*, 112, 1.

Ningaraju, T.M. (2006). Cloning and Characterization of Chitinase Gene/s from Native Isolates of MSc Thesis Serratiamarcescens. Dharwad, UAS, p. 144.

Palermo, D., Giulia, G., Francesca, L., Vincenzo, P. and Orlando, C. (2021). Essential oil-based nano-biopesticides: Formulation and bioactivity against the confused flour beetle *Tribolium confusum*. *Sustainability*, 13(17), 9746.

Palmieri, D., Portillo, E., Sulbarán, Y., Guerra, M. and San-Blas, E. (2019). Biocontrol of Phytophthora root and stem rot disease in papaya (*Carica papaya*) plants by Photorhabdus, the symbiont bacterium of *Heterorhabditis amazonensis*. *BioControl*, 64(5), 595–604.

Patel, J.S., Kharwar, R.N., Singh, H.B., Upadhyay, R.S. and Sarma, B.K. (2017). *Trichoderma asperellum* (T42) and *Pseudomonas fluorescens* (OKC)-enhances resistance of pea against *Erysiphe pisi* through enhanced ROS generation and lignifications. *Frontiers in Microbiology*, 8, 306.

Peng, G., McGregor, L., Lahlali, R., Gossen, B.D., Hwang, S.F., Adhikari, K.K., et al. (2011). Potential biological control of clubroot on canola and crucifer vegetable crops. *Plant Pathology*, 60(3), 566–574.

Pereira, P., Ibáñez, S.G., Agostini, E. and Etcheverry, M. (2011). Effects of maize inoculation with *Fusarium verticillioides* and with two bacterial biocontrol agents on seedlings growth and antioxidative enzymatic activities. *Applied Soil Ecology*, 51, 52–59.

Philippot, L., Raaijmakers, J.M., Lemanceau, P. and Van Der Putten, W.H. (2013). Going back to the roots: The microbial ecology of the rhizosphere. *Nature Reviews Microbiology*, 11(11), 789–799.

PIB. (2014). www.pib.gov.in. (Accessed on 8th May, 2022).

Rai, N., Rai, S.P. and Sarma, B.K. (2021). Prospects for abiotic stress tolerance in crops utilizing phyto-and bio-stimulants. *Frontiers in Sustainable Food Systems*, 5, 754853.

Rakshit, A., Meena, V.S., Abhilash, P.C., Sarma, B.K., Singh, H.B., Leonardo, F., et al. (2022). *Biopesticides: Advances in Bioinoculants*, Vol. 2. Elsevier, Amsterdam, The Netherlands, p. 430.

Rakshit, A. and Singh, H.B. (2018). *Advances in Seed Priming*. Springer Natutre, Singapore, p. 307.

Rojas, E.C., Jensen, B., Jørgensen, H.J., Latz, M.A., Esteban, P., Ding, Y., et al. (2020). Selection of fungal endophytes with biocontrol potential against Fusarium head blight in wheat. *Biological Control*, 144, 104222.

Russo, M.L., Scorsetti, A.C., Vianna, M.F., Allegrucci, N., Ferreri, N.A., Cabello, M.N., et al. (2019). Effects of endophytic *Beauveria bassiana* (Ascomycota: Hypocreales) on biological, reproductive parameters and food preference of the soybean pest *Helicoverpa gelotopoeon*. *Journal of King Saud University-Science*, 31(4), 1077–1082.

Ryu, C.M., Kim, J., Choi, O., Kim, S.H. and Park, C.S. (2006). Improvement of biological control capacity of *Paenibacillus polymyxa* E681 by seed pelleting on sesame. *Biological Control*, 39(3), 282–289.

Sarkar, A., Patel, J.S., Yadav, S., Sarma, B.K., Srivastava, J.S. and Singh, H.B. (2014). Studies on rhizosphere-bacteria mediated biotic and abiotic stress tolerance in chickpea (*Cicer arietinum* L.). *Vegetos*, 27(1), 158–169.

Schmidt, L.M., Hewlett, T.E., Green, A., Simmons, L.J., Kelley, K., Doroh, M., et al. (2010). Molecular and morphological characterization and biological control capabilities of a *Pasteuria* sp. parasitizing *Rotylenchulus reniformis*, the reniform nematode. *Journal of Nematology*, 42(3), 207.

Sid Ahmed, A., Pérez-Sánchez, C., Egea, C. and Candela, M.E. (1999). Evaluation of *Trichoderma harzianum* for controlling root rot caused by *Phytophthora capsici* in pepper plants. *Plant Pathology*, 48(1), 58–65.

Singh, A., Jain, A., Sarma, B.K., Upadhyay, R.S. and Singh, H.B. (2013). Rhizosphere microbes facilitate redox homeostasis in *Cicer arietinum* against biotic stress. *Annals of Applied Biology*, 163, 33–46.

Singh, H.B., Mishra, S., Leonardo, F. and Renata, D.L. (2018). *Emerging Trends in Agri-Nanotechnology: Fundamentals and Applied Aspects*. CABI, U.K., p. 302.

Singh, H.B., Sarma, B.K. and Keswani, C. (2016). *Agriculturally Important Microorganisms: Commercialization and Regulatory Requirements in Asia*. Springer, Singapore, p. 336.

Singh, H.B., Sarma, B.K. and Keswani, C. (2017). *Advances in PGPR Research*. CABI, U.K., p. 408.

Singh, H.B. and Vaishnav, A. (2022a). *New and Future Developments Microbial Biotechnology and Bioengineering: Sustainable Agriculture as Biostimulants*. Elsevier, Amsterdam, The Netherlands, p. 374.

Singh, H.B. and Vaishnav, A. (2022b). *New and Future Developments in Microbial Biotechnology and Bioengineering: Sustainable Agriculture: Advances in Microbe based Biostimulants*. Elsevier, Amsterdam, The Netherlands, p. 462.

Singh, H.B. and Vaishnav, A. (2022c). *New and Future Developments in Microbial Biotechnology and Bioengineering. Revitalization through Organic*. Elsevier, Amsterdam, The Netherlands, p. 366.

Singh, H.B. and Vaishnav, A. (2022d). *New and Future Developments in Microbial Biotechnology and Bioengineering: Revising Green Chemicals*. Elsevier, Amsterdam, The Netherlands.

Slimene, I.B., Tabbene, O., Gharbi, D., Mnasri, B., Schmitter, J.M., Urdaci, M.C., et al. (2015). Isolation of a chitinolytic *Bacillus licheniformis* S213 strain exerting a biological control against *Phoma medicaginis* infection. *Applied Biochemistry and Biotechnology*, 175(7), 3494–3506.

Song, X.S., Gu, K.X., Duan, X.X., Xiao, X.M., Hou, Y.P., Duan, Y.B., et al. (2018). A myosin5 dsRNA that reduces the fungicide resistance and pathogenicity of *Fusarium asiaticum*. *Pesticide Biochemistry and Physiology*, 150, 1–9.

Sreevidya, M., Gopalakrishnan, S., Melø, T.M., Simic, N., Bruheim, P., Sharma, M., et al. (2015). Biological control of *Botrytis cinerea* and plant growth promotion potential by *Penicillium citrinum* in chickpea (*Cicer arietinum* L.). *Biocontrol Science and Technology*, 25(7), 739–755.

Srinivasan, K. and Mathivanan, N. (2011). Plant growth promoting microbial consortia mediated classical biocontrol of sunflower necrosis virus disease. *Journal of Biopesticides*, 4(1), 65.

Tao, A., Pang, F., Huang, S., Yu, G., Li, B. and Wang, T. (2014). Characterization of endophytic *Bacillus thuringiensis* strains isolated from wheat plants as biocontrol agents against wheat flag smut. *Biocontrol Science and Technology*, 24(8), 901–924.

UNDESA. United Nations Department of Economic and Social Affairs. (2015) World Population Prospects: The 2015 Revision, Key Findings and Advance Tables. United Nations Department of Economic and Social Affairs; New York. Working Paper No ESA/P/WP.

Vicedo, B., Peñalver, R., Asins, M.J. and López, M.M. (1993). Biological control of *Agrobacterium tumefaciens*, colonization, and pAgK84 transfer with *Agrobacterium radiobacter* K84 and the Tra-mutant strain K1026. *Applied and Environmental Microbiology*, 59(1), 309–315.

Wang, M., Weiberg, A., Lin, F.M., Thomma, B.P., Huang, H.D. and Jin, H. (2016). Bidirectional cross-kingdom RNAi and fungal uptake of external RNAs confer plant protection. *Nature Plants*, 2(10), 1–10.

Whipps, J.M., Sreenivasaprasad, S., Muthumeenakshi, S., Rogers, C.W. and Challen, M.P. (2008). Use of *Coniothyrium minitans* as a biocontrol agent and some molecular aspects of sclerotial mycoparasitism. *European Journal of Plant Pathology*, 121(3), 323–330.

Worrall, E.A., Bravo-Cazar, A., Nilon, A.T., Fletcher, S.J., Robinson, K.E., Carr, J.P., et al. (2019). Exogenous application of RNAi-inducing double-stranded RNA inhibits aphid-mediated transmission of a plant virus. *Frontiers in Plant Science*, 10, 265.

Yadav, S.K., Singh, S., Singh, H.B. and Sarma, B.K. (2017). Compatible rhizosphere-competent microbial consortium adds value to the nutritional quality in edible parts of chickpea. *Journal of Agricultural and Food Chemistry*, 65(30), 6122–6130.

Yogaiah, S., Singh, H.B., Leonardo, F. and Renata, D.L. (2022). *Advances in Nano-Fertilizers and Nano-Pesticides in Agriculture: A Smart Delivery System for Crop Improvement*. Elsevier, Amsterdam, The Netherlands, p. 616.

Zhang, R.S., Liu, Y.F. and Chen, Z.Y. (2011). Screening, evaluation and utilization of antagonistic bacteria against *Xanthomonas oryzae* pv. *oryzicola*. *Chinese Journal of Biological Control*, 4, 1–14.

10 Significance of Seed Treatment in Management of Plant Diseases

Sundaram Aravindan and Sanjay Kumar

10.1 INTRODUCTION

Late in the 1960s, the green revolution dramatically improved India's food security situation by increasing domestic food grain production, which shifted the country's status from famine to food and nutritionally secured nation. The country's food grain output increased from 50.82 million tons in 1950–1951 to a new record 308.65 million tons in 2020–2021. However, in order to fulfil the increasing demands of the earth population, which is expected to reach 10.9 billion by 2100, we must increase our food production to twice the level of today while using fewer resources such as land, water, and manpower than at present. Cultural management practices such as summer plowing, sanitation, green manuring, crop rotation, modifying sowing and harvesting date, fallowing, and composting and other environmentally friendly disease management practices used to control plant disease have already lost their importance and are being revalidated as a component to be included in integrated disease management (Sanjeev, 2012; Reddy, 2013). Seed is widely acknowledged as the most cost-effective yet critical input for increasing agricultural productivity. For more than 2,500 years, Indians have been aware of the importance of using good seeds for planting. The importance of seed health in crop husbandry was emphasized strongly by Varahmihira (5th century), Kashyapa (8–9th century), and Surpala (10th century).

Farmers always demand high seed quality, which can result in up to a 30% increase in crop yields (Ellis, 2004; Afzal et al., 2016). Seed- and soil-borne pathogens usually affect most of crop species with a significant impact on germination, seedling establishment, and final output. Seed-borne diseases, such as loose smut of wheat (*Ustilagonuda)* and common bunt of wheat (*Tilletia caries)*, are known from ancient times how significantly it can reduce the crop yield due to its infection in seed before sowing. These examples are giving us importance on seed treatment and how it can be one of the most practical tools for protecting precious seeds from both seed- and soil-borne pathogens (Sanjeev, 2012). Seed treatment refers to the exposure of seeds to certain physical, chemical, or biological agents that are not only used to make the seeds disease-free but are also used to bring forth the

DOI: 10.1201/9781003531944-10

potential of disease control. With judicious use, it can protect and enhance establishment, growth, and potential productivity of plants, thereby playing a key role in the history of mankind, in preventing hunger and starvation (Forsberg et al., 2003). Seed treatments are also most efficient when they are goal-oriented and crop-specific in order to ensure optimal stand establishment and yield improvement under changing climatic conditions.

There are various advantages to seed treatment above other disease management or crop management practices, such as seed protection during its preservation in storage as well as after planting of seeds in soil, initial inoculum was reduced, toxic effect on non-target organisms was reduced, pesticide drifts during spray are eliminated, the exposure of lands and water surface toward agrochemical active compounds was also reduced with highest effectiveness in controlling of all seed, soil, and foliar diseases. Even, uniform, and combination treatments can be applied with greater precision through seed treatment (DPPQS, 2007). Many plant breeders focused a significant portion of their efforts in the early 20th century on breeding for resistance to many diseases such as common bunt. But when pathogens are able to manage through seed treatment, now their focus is shifted toward breeding for those traits that can't be controlled through seed treatments or other important traits such as grain quality. (Mathre et al., 2001). Seed treatment will thus play a critical role in protecting seeds and seedlings from both seed- and soil-borne illnesses that wreak havoc on crop establishment and growth. Commercial seed treatment to deliver agrochemicals has been widely employed for a number of crops, and it will almost certainly continue in future also. But the physical seed treatments such as aerated heat, hot water treatment, and radiation along with biological agents might hold great future as an alternative to agrochemical seed treatment methods in crop management. Alternative crop protection systems are currently being researched all around the world, with more emphasis on food safety along with conservation of natural resources and protection of global ecosystem (Nicholas and Groot, 2013). Biological seed treatments are mostly made from naturally available beneficial microorganisms and its active components that give protection against both biotic, abiotic stresses, and also in plant growth (Schwinn, 1994).

Present scenario: A crucial success element for the seed treatment business was the invention of a full protection solution against diverse stresses in a single product that is farmer- and crop-friendly and environmentally sustainable (Schwinn, 1994). The fastest-growing agricultural chemicals sector is seed treatment, with a global market worth of $1.8 billion and an anticipated growth rate of $3.0 billion by 2025, it is becoming an important agriculture input sector. Chemical seed treatments are the most common active component, accounting for around $3–$5 billion in 2020 and at least two-thirds of the entire seed treatment market. Biofertilizers, biopesticides, and biostimulants are among the biologicals used in seed treatment. In 2020, the biological seed treatment market was expected to be worth $1–$1.5 billion (Taylor and Trimmer, 2020). In India, 70% of the seed requirement is currently met by the farmers by their own seeds stored from previous season, which are used for planting without any seed treatment. Except for high-cost hybrid seeds, most of the seeds that are mostly sourced from private or public sector agencies are also mostly untreated (Upadhyaya, 2013). According to estimates, 80% of the crop seeds planted in the country are not treated with any seed treatment methods, as against the 100% in developed nations (DPPQS, 2007). Given the importance of seed treatment in achieving better crop protection and crop management of major crops, many farmers in India across the states are mostly ignorant and not following this important practice in the crop protection practices. With proper extension activities and strategies if farmers made aware of importance seed treatment in crop protection practices, farmers can maximize production while also achieving food security. Furthermore, the idea behind this chapter is to define and understand different seed treatment methods, technologies with their advancement in their field, and how we can assist in the development of improved and more homogeneous crop production in the near future.

10.2 HOW SEED TREATMENTS ARE APPLIED?

The term "seed treatment" refers to both processes and products. Specific products and techniques can be used to improve the cultivation environment for seeds, seedlings, and young plants. The complexity of seed treatment ranges from basic dressing to coating and pelleting (Figure 10.1).

10.2.1 Seed Dressing

Among the different methods pursued for seed treatment, this is most widely used method. In this method, the seed is either dry-treated or wet-treated with a slurry or liquid formulation (Upadhyaya, 2013). Industries and farms both use dressings normally for seed treatment. Normally, low-price earthen containers are used to mix agrochemicals/biological agents with seed; otherwise, the seeds are placed over a plastic/linen/piece of paper, and then required quantities of seed treatment agents were dusted and thoroughly mixed by farmers. Use of a motorized or manually operated seed treatment drum is the preferred way to get evenly treated seeds with seed treatment agents, which can't be done through physical mixing by using shovel (leads to uneven mixing), which is normally used by many farmers in Indian villages. Usually, seeds must be weighed before the seed treatment in order to apply the correct quantity (Thippenswamy and Lokesh, 1997).

10.2.2 Seed Coating

To improve adherence of seed treatment agents with formulation, a special type of binder is used to the seeds. Among the different methods, seed coating requires modern advanced technology determined by the different types of industry such as in big-scale commercial treatment, automatic treaters such as Gustafson Treaters, which can automatically calibrate the right quantity of seed and agents required for the treatment (Upadhyaya, 2013).

Usually, for better seed coating, dextrans, vegetable oil, carboxy methyl cellulose, paraffin, gum Arabic are used as adhesives (Mathre et al., 2001).

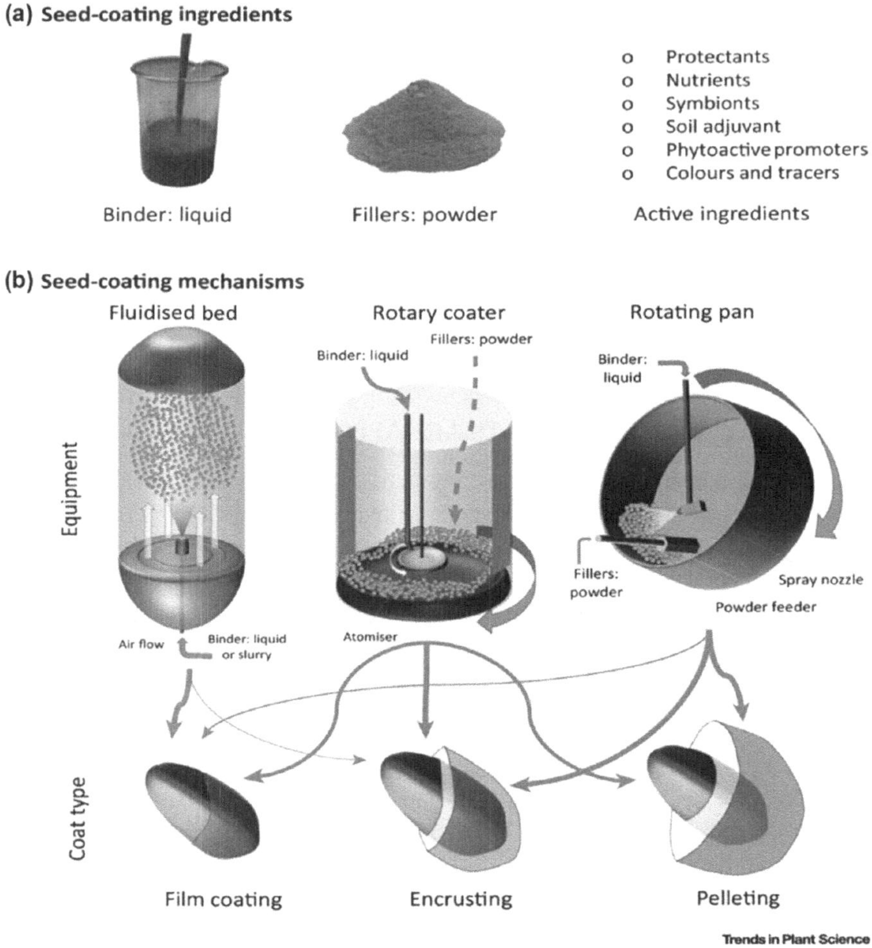

FIGURE 10.1 Examples of seed coating equipment and seed coating type. (Images adapted from Pedrini et al. (2017).)

10.2.3 Seed Pelleting

Among the seed treatment methodologies, the most advanced one is seed pelleting as it results in treated seeds with better physical shape, thereby improving seed handling and palatability. The advanced technology results in need of necessitating more specialized machinery and techniques that make it more expensive among different methods (DPPQS, 2007). Usually, in seed coating, seeds and seed treatment agents such as chemicals come in direct contact, so those chemicals that have phytotoxic effects can't be used in seed coating; here seed pelleting is plays a very good substitute to overcome this disadvantage. Seed pelleting also plays a great role on those seeds that are very small and can't be accurately metered by mechanical planting equipment; here by using greater material loading around the seeds, size and weight of seeds are manipulated in such way that it can be easily used with modern advanced planting machines for sowing or planting (Halmer, 1988; Upadhyaya, 2013).

10.2.4 Permeation

Despite being a very advanced technique in seed pelleting, we need to dry the seeds after pelleting, which is labor- and time-intensive, thereby not making seed pelleting an economically viable option; here permeation is playing a very good substitute. In permeation, the seed treatment agents are dissolved in organic solvents, which helps them carry the agents beneath the different layers of seeds. In permeating, we don't need to dry the seeds separately like in seed pelleting as after treatments, the organic solvents evaporated themselves, which makes permeating a very good alternative over seed pelleting (Meyer and Meyer, 1971; Shao et al., 2007).

10.2.5 Fluid Drilling

In fluid drilling method, the seeds were kept in aerated water until its radicle emerged from seed. Normally, the seed is kept soaked in viscous gel with suitable drill before sowing. The seeds are then soaked in a viscous gel before being sown with a suitable drill, where the gel plays a very important role of protecting the radicle (by keeping seed moisture in stable) during the seed germination (Gray, 1981), delivering the agrochemicals and biological agents, thereby protecting seed from its seed- and soil-borne disease. Damping off disease caused by *Pythium* spp. was effectively managed when both chemicals captan and metalaxyl were mixed into a magnesium silicate gel compared with normal seed treatment methods such as coating and pelleting (Giammichele and Pill, 1984).

10.2.6 Seed Priming

Seed priming is a broad term for use of different moistening techniques in improving seed performance by regulating the seed moisture and temperature. It also refers to the different biological changes that occur during seed priming treatments (Hacisalihoglu et al., 1999). Seed priming is most commonly used in cultivation of different vegetable crops such as onion, tomato, carrot, pepper, flowering and ornamental plants. Its use is very much followed especially in commercial protected cultivation. In conjunction with solid matrix priming, agrochemicals and biological agent can be used for better management of pests and diseases. They are mostly added to seeds initially through slurry treatment; later they solid particulate, and priming will be carried out for better performance of seed germination and seedling establishment (Figure 10.2) (Taylor and Harman, 1990).

10.3 METHODS OF SEED TREATMENT

Seed treatment refers to the application of different agents such as agrochemicals (fungicide, nematicides, etc.), physical agents such as solar radiation, hot water treatment into the seeds to disinfect and protect them from seed- and soil-borne pathogens. It involvement

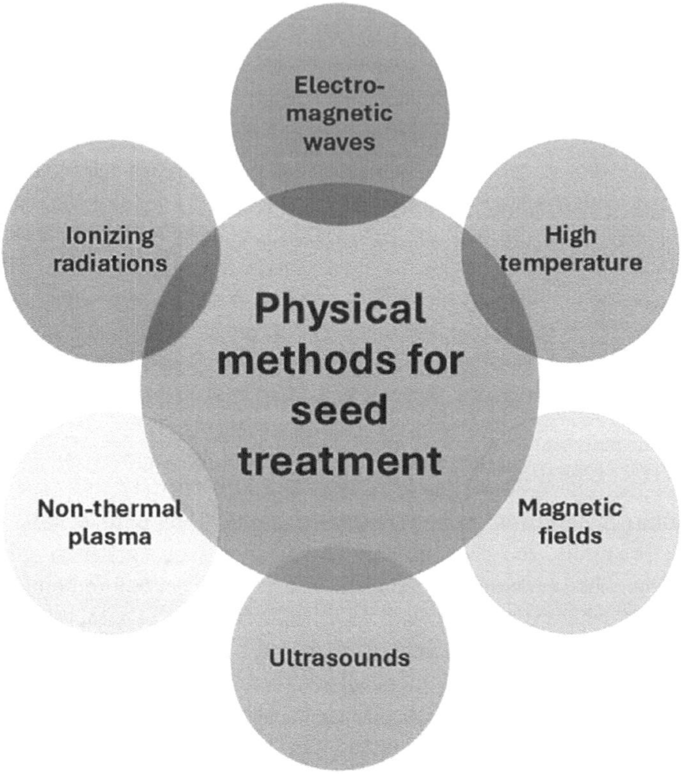

FIGURE 10.2 Seed priming methods for better enhancement of seed germination and plant development.

includes from a basic seed coating, seed pelleting, and seed dressing (ASF, 2010; Krishna, 2011). Although seed treatment is like baby care that begins with the mother (Heydecker and Coolbear, 1977), the treatment with which we are mostly concerned in this chapter is mostly applied to the seed before it is planted into the field to manage the seed- as well as soil-borne pathogens. There are various seed treatment methods available to manage plant diseases, which are discussed in the following chapter.

10.3.1 Physical Seed Treatment

Taking into the consideration of the aftereffects of chemicals on environment and living organisms, some stand-in methods were developed and currently used for seed treatment (Jindal et al., 1991; Elwakil, 2003; Aladjadjiyan, 2007). As we know, the pesticides and fungicides, which are widely used for treating seeds, have many disadvantages such as damaging the ecosystem, land, and water. (Chapman and Harris, 1981; Vasilevski, 2003). It is important to look into the use of alternate methods that are economically viable as well as sustainable in nature; such as physical methods with new interventions into the management of plant diseases are the need of this 21st century (Figure 10.3) (Amein et al., 2011).

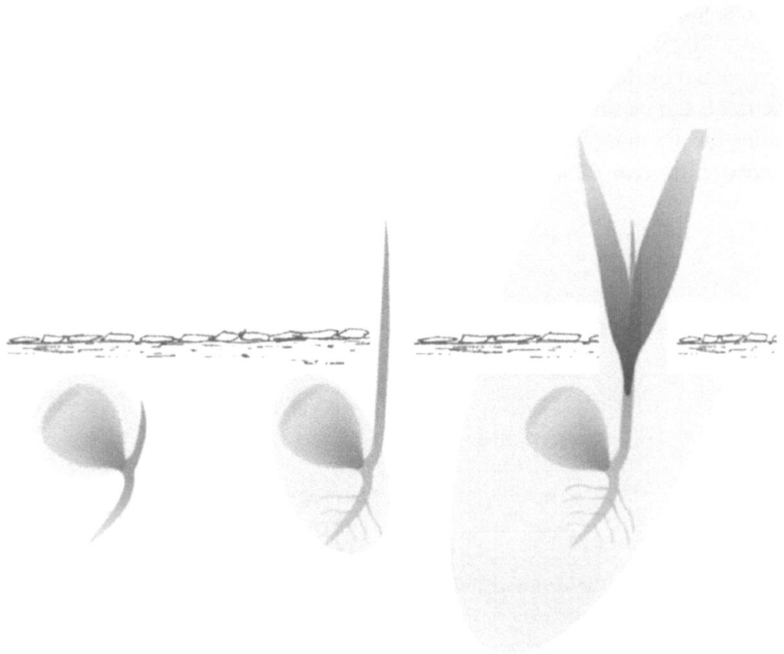

FIGURE 10.3 Schematic representation of various physical methods used for seed treatment.

10.3.2 Hot Water Treatment

Hot water treatment is a well-established practice to control many seed-borne pathogens, here the hot temperature with right amount and time period is used just enough to disinfect the seed from seed-borne pathogens without affecting the viability of seed. This method is still a very effective method in managing seed-borne pathogens (Floyd, 2005; Muniz, 2001). Here the pathogen elimination is more sustainable and environment-friendly in comparison with other treatments. But enough care to be taken during this practice; otherwise, the seed will lose its viability (Erdey et al., 1997; Meah, 2004). Treatment diseases such as blackleg and the black rot of crucifers through hot water treatment are a classic example (Napoles et al., 1991). Before offering the treatment, the seeds should be prewarmed in porous cloth bag for nearly 10 minutes at 20°C in water; it is a very importance step. Later, the prewarmed seeds of right quantity will be kept in water bath at exact temperature and period as per the procedure. Even a slight change in the procedure will lead to significant loss of seed viability. After treatment, the seeds are cooled down by immersing in cold water to avoid heating action. Once seeds have cooled down, the seed are fanned out in thin polythene sheet or paper cloth for drying (Jindal et al., 1991). Once the treatment is over (seed dried), the seeds should be sown immediately, and storing of seed for next time is not advised (Floyd, 2005). This method is not uniformly applicable to different crop seeds; every crop needs specific procedure; otherwise, it will do the harm. For some of the crops such as cucumbers, sweet corn, lettuce, peas, beets, and beans (Nega et al., 2003; Miller and Lewis Ivey, 2005; Floyd, 2005), this method is not feasible. Age-old seeds may be very badly damaged by this treatment so before going for bulk treatment, it is best to always test the small samples especially that are over 1 year old for its viability, thereby we can determine the amount of injury caused by this treatment (Table 10.1). Seeds treated by hot water are listed in Table 10.2 (Floyd, 2005).

TABLE 10.1
Hot Water Treatment for Some Seed-Borne Pathogens

Disease	Causal Organisms	Treatment	References
Tomato canker	*Clavibacter michiganensis subsp. Michiganensis*	60 minutes at 53°C	Bryan (1930)
Tomato canker	*C. m. subsp. michiganensis*	Soaking infected seeds 30 minutes in hot water at 56°C	Shoemaker and Echandi (1976)
Tomato	*Pseudomonas syringae* pv *tomato*	48°C for 60 minutes	Devash et al. (1979)
Pea bacterial blight	*Pseudomonas syringae* pv *pisi*	Dry heat at 65°C for 72 hour and soaking in water at 55°C for 15 minutes	Grondeau et al. (1992)
Moongbean black rot	*Pseudomonas syringae* pv. *Phaseolicola*	50°C for 45–60 minutes	Tamietti (1982)
Cabbage, cauliflower bacterial black rot	*Xanthomonas campestris* pv. *Campestris*	50°C for 30 minutes	Shekhawat et al. (1982)
Carrot bacterial blight	*Xanthomonas campestris* pv *carotae*	52°C for 10 minutes	Ark and Gardner (1944)
Pumpkin bacterial blight	*Xanthomonas campestris* pv *cucurbitae*	56°C for 30 minutes	Moffett and Wood (1979)
Rice brown leaf spot	*Helminthosporium oryzae*	51°C for 7 minutes	Flyod (2005)
Rice blast	*Magnaporthe grisea*	6–12 hours in cool water + 1–2 minutes at 50°C	Flyod (2005)
Cereal loose smut	*Ustilagosegatum var. tritici*	5 hour at 21°C presoak + 1 minutes at 49°C + 11 minutes at 52°C	Flyod (2005)
Brassica leaf spot	*Alternaria brassicicola*	18 minutes at 50°C	Flyod (2005)
Brassica leaf spot	*Alternaria brassicae*	20 minutes at 50°C	Flyod (2005)
Brassica canker	*Leptosphaeria maculans*	30 minutes at 50°C	Flyod (2005)
Millet downy mildew	*Sclerospora graminicola*	10 minutes at 55°C	Flyod (2005)

Source: Adopted from Flyod (2005).

TABLE 10.2
Chemical Compounds Currently Used as Cereal Seed Treatments (Mathre et al., 2001)

Common Name	Trade Name	Chemical Name
Captan	Agrosol, Agrox, Granox, Orthocide	N-trichloromethylthio-4-cyclohexene-1, 2-dicarboximide
Carboxin	Vitavax	5,6-dihydro-2-methyl-N-phenyl-1, 4-oxathiin-3-carboxamide
Difenoconazole	Dividend	cis,trans-3-chloro-4-[4-methyl-2-(1H-1,2,4-triazol-1-ulmethyl)-1,3-dioxolan-2-yl]phenyl 4-chlorophenyl ether
Imazalil	FloPro IMZ, Double R, Deccozil, Nu-Zone, Fungaflor	(+)-allyl 1-(2,4-dichlorophenyl)-2-imidazol-1-ylethyl ether
Mancozeb	Dithane M-45, Mankocide, Mansul, Penncozeb	Zinc Manganese ethylenebisdithiocarbamate
Maneb	DB Green, Granol NM, Trinox, Pro-Tex	Manganese ethylenebisdithiocarbamate
Metalaxyl	Apron, Allegiance	methyl N-(2-methoxyacetyl)-N-(2,6-xylyl)-DLalaninate
Tebuconazole	Raxil, Preventol, Tebuject	1-(4-chlorophenyl)4,4-dimethyl-3-(1H-1,2,4-triazol-1-ulmethyl) pentan-3-ol
Thiabendazole	Mertect, Metasol	2-(4-Thiazolyl)-benzimidazole
Thiram	Arasan, Vertagard, Thiramad	Tetramethylthiuramdisulfide
Triadimenol	Baytan	1-(r-chlorophenoxy)-3,3-dimethyl-1-(1H-1,2,4-triazol-1-yl) butan-2-ol
Triticonazole	Charter	(+)-(E)-5-(4-chlorobenzylidene)-2-dimethyl-1-(1H-1,2,4-triazol-1-ylmethyl) cyclopentanol

10.3.3 Dry Heat Treatment

Dry heat seed treatment has been reasonably applied in different ways. One of the simple ways is solarization of seeds; here the seeds are kept in irradiation of Sun (Luthra, 1953; Luthra and Sattar, 1934), which is mostly used in tropical countries, but its low precision and difficulties in large-scale application lead to little interest in modern industrial agriculture. Dry heat treatment is very effective in inactivation of seed-borne viruses and pathogens, so it plays a key role in vegetable and floriculture crops that are mainly cultivated from seeds especially in developed countries such as Korea and Japan (Seung-Hee Lee et al., 2004). The Thomas and Adcock (2004) study in lupine seeds revealed that dry heat treatment at 65°C for a period of 4–7 days eradicated the anthracnose pathogen from seeds.

10.3.4 Aerated Heat Treatment

Seed companies in the early 20th century mostly used hot water method for managing seed-borne pathogens (Johnsson, 1990; Neergaard, 1977); however with the disadvantages such as increased treatment cost and low precision, reduced germination leads hot water treatment was nearly totally neglected for most of the crops when cheap and efficient agrochemicals had become available in 1960s. A new method using aerated heat has been put forward as a way to remove the difficulties that are inherent in hot water treatment (Tapke, 1926; Baker, 1969; Kobayashi, 1990). The new devices were developed in such a way to precisely control all the parameters such as air flow, humidity, temperature, cooling duration, which helps to carry out this seed treatment method in more commercial and practical way. Weimer (1952) reported that in lupin seeds with treatment of 7 hour at 70°C, the anthracnose pathogens were controlled nearly 80% successfully. Forsberg et al. achieved successful control of combined infections of *T. caries*, *S. nodorum*, and *Fusarium* spp., on wheat and *B. sorokiniana* on barley through aerated steam treatment. In sugarcane, aerated steam methods are used commercially to control ratoon stunting and other seed-borne pathogens (Edison and Ramakrishnan, 1972; Damann, 1983; Viswanathan, 2001).

10.3.5 Radiation Treatment

The agrochemicals used in seed treatment lead to different undesirable effects such as non-target organism, lethal chemical components persistence in the ecosystem. Radioactive irradiation method played a successful role in few cases (Cuero et al., 1986; Bagegni et al., 1990), but the issue is that irradiation exposures also limit the seed viability. The alternated seed treatment methods used in management of seed-borne pathogens include different types of electromagnetic radiation such as high-energy electrons (Sitton et al., 1995), ultrasonic radiation (Nagy et al., 1995), microwave (Aladjadjiyan, 2010) and gamma ray (Harwalker et al., 1995), and UV radiation (Bhaskara Reddy et al.,1998;). Rao et al. (1994) and Sinha et al. (1994) reported that gamma radiation helps in inactivation of pathogens and pests in many crops. The Singh and Singh (2005) study revealed that there is reduced pathogens infestation when rice varieties were treated with non-injurious levels of gamma irradiation. The highest reduction in pathogens infestation was observed when seed was irradiated with 0.10 kiloGray (kGy). Bel'skii and Mazulenko (1984) reported that laser treatment was found to be very successful in controlling seed-borne pathogens, but its commercial applications were found to be very limited as laser beams are very narrow and seeds' surfaces need to be covered very evenly for better result, which is not possible commercially. In Maize, Hernández Aguilar Claudia et al. (2011) found that low-intensity laser beam was very effective in controlling seed-borne pathogens.

10.3.6 Chemical Seed Treatment

Nowadays, chemical seed treatment is very familiar and worldwide used due to its broad spectrum ability in managing plant pathogens with short time and high level of effectiveness, which makes it cost-effective and economically viable (Nameth, 1998).

Seeds are treated with different types of agrochemicals such as insecticides, fungicides, and nematicides, but among these, majority of seed treatment chemicals are fungicides. Fungicides are an important component of integrated disease management practice for crops (Lamichhane et al., 2020), Fungicidal seed treatment mostly eradicates the seed-borne pathogens and also forms protective zone around the seed by which it also protects seeds from soil-borne pathogens also, which results in vigorous seedlings (Figure 10.4), (Nene and Thapliyal, 1979). In history, seed treatment methods started nearly around 17th century, when wheat seeds infected with smut (*Ustillagospp*) were successfully disinfected using brine solution by Tillet (1755). Prevost discovery in 1807, using copper sulfate for management of smuts, was major breakthrough in managing smut for another century. Kittleson (1952) discovered Captan (first-contact fungicide) for seed treatment to protect seeds from different fungal diseases. Among all the available systemic fungicides, the most widely used for seed treatment belong to methyl benzimidazolecarbamate (MBC) group of fungicides (Campos et al., 2015). The MBCs, such as carbendazim, bind to β-tubulin, which results in cell division suppression. As per estimate, India's yearly usage of MBC group fungicide is nearly higher than 2,000 metric tons, and its use is registered for more than 18 crops such as apple, cassava, wheat, rice, mango, cucurbits, cotton, grape, brinjal, groundnut, pea, rose, walnut, and tapioca (Bhushan et al., 2013) (Table 10.3).

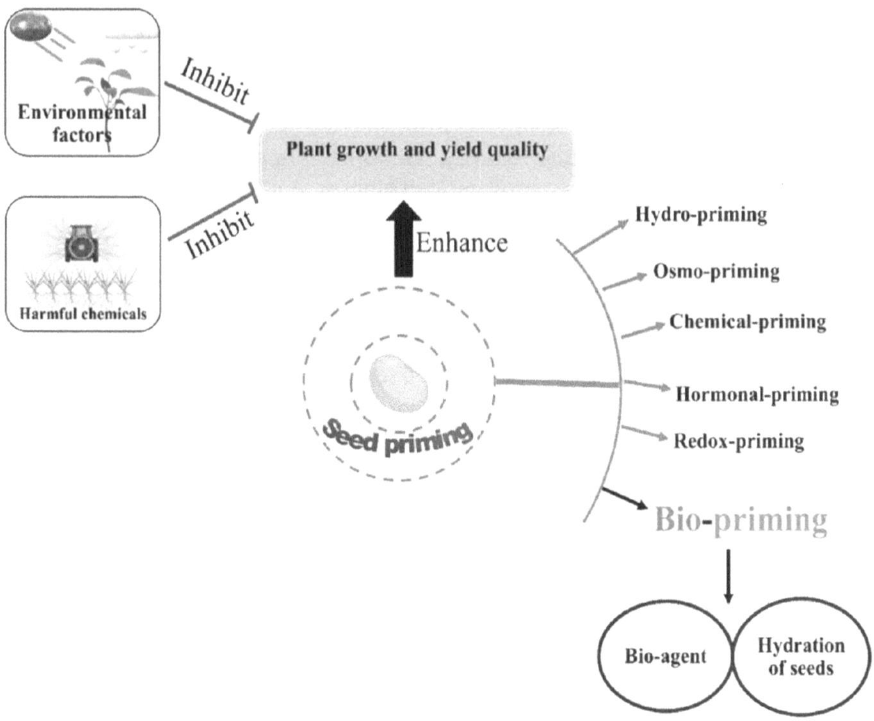

FIGURE 10.4 Mobility of fungicides used for seed treatment. The light blue area shows the zone of protection covered by contact (left), locally systemic (middle), and fully systemic fungicides (right), respectively (Lamichhane et al., 2020).

TABLE 10.3
Mode of Action of Seed Treatment Fungicides (Mathre et al., 2001)

Fungicides	Mode of Action
Difenoconazole	Sterol Biosynthesis Inhibitor
Imazalil	Sterol Biosynthesis Inhibitor
Tebuconazole	Sterol Biosynthesis Inhibitor
Triadimenol	Sterol Biosynthesis Inhibitor
Triticonazole	Sterol Biosynthesis Inhibitor
Carboxin	Inhibits microchondrial function by disruption complex II of succinic dehydrogenase
Metalaxyl	Inhibits RNA synthesis by disrupting incorporation of uridine into RNA
Fludioxonil	Impairs membrane bound transport processes by inhibiting the accumulation and incorporation of glucose and mannose into hyphal wall glucans
Thiabendazole	Arrests nuclear division by interfering with microtubule assembly by binding to the protein tubulin
Captan	Nonspecific
Mancozeb	Nonspecific
Maneb	Nonspecific
PCNB	Nonspecific
Thiram	Nonspecific

Randhawa et al. (1985) reported that, when stored wheat seeds were treated with fungicides Captan, Thiram, and Vitavax, Thiram was found highly successful in ensuring the seed germination and viability. Gupta et al. (1990) studied four different fungicides against seed-borne pathogens associated with wheat seed. Among the fungicides, Dithane M-45 and Captan lessened the incidence of *Bipolaris sativum* and *Alternaria alternata*, and at the same time, they raised germination rates at 87% and 88%, respectively, than Benlate. El-Tayed et al. (1987) found that seed dressing fungicides Benlate and Dithane M-45 were very successful in managing *Fusarium roseum* and *Alternaria alternata*. Izhar et al. (1987) reported that fungicides such as Vitavax have very good effect on both plant characteristic and grain yield in wheat.

10.3.7 Biological Seed Treatment

Farmers were using agrochemicals for seed treatment widely in the past decades. Because of their toxic effect on human and environmental health, they promote biological control of both seed- and soil-borne diseases. Currently, the use of beneficial biological agents for seed treatment is a most popular method, which is becoming increasingly paramount in managing plant diseases (Howell et al., 1997). Biological seed treatment includes beneficial microorganisms such as different species of *Trichoderma* and *Pseudomonas, Bacillus*. They improve seed germination percentage, enhance seedling vigor, and manage soil- and seed-borne pathogens (Figure 10.5) (Müller and Berg, 2008). Seed bio-priming with *Trichoderma harzianum* reported to reduce the incidence of root rot disease in cowpea (El-Mohamedy et al., 2006).

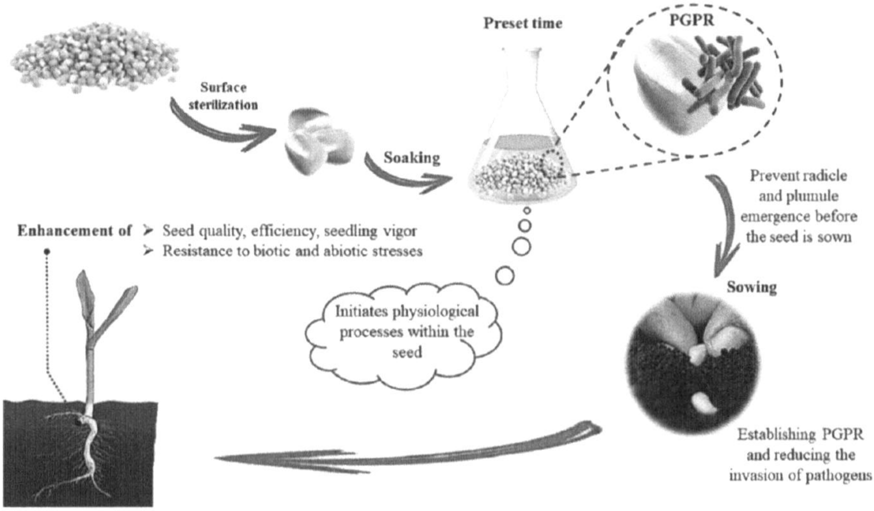

FIGURE 10.5 Method of seed treatment in bio-priming with PGPR (DebasisMitra et al., 2021).

Biological seed treatment is a promising method that enhances stress tolerance of seed before germination. Bio-priming of seeds with *Pseudomonas flourescens* @ 1 × 107 cfu/seed provided good protection against *Pythium ultimum* (damping off) pathogen in maize plant (Callan et al., 1990). In pearl millet, bio-priming with isolates of *Pseudomonas fluorescens* showed improved germination, vigor and elicited disease resistance against downy mildew disease (Raj et al., 2004). Treatment of seed with *Trichoderma* spp. gave better management of biotic and abiotic stresses through secretion of antimicrobial compound (Ahmad et al., 2015; Mastouri et al., 2012). Bio-priming of seed with *Trichoderma* spp. activates reactive oxygen species detoxification system and induces pathogenesis-related proteins accumulation at early stage of root colonization (De Palma et al., 2019; Yedidia et al., 2000). Many studies showed that treatment of seed with microbes augments the plant resistance level through lignification and creation of ROS (Meshram et al., 2019; Patel et al., 2017).

10.4 FUTURE BIOPROTECTANTS FOR SEED TREATMENT

The effective use of biological agents to manage plant pathogens through seed treatment necessitates the evolution of systems that are functional, dependable, and cost-effective (Harman, 1991). In recent times, some biological agents are in the level of large-scale commercial production after their successful field-level testing and registration (Taylor and Harman, 1990), whereas most of the other biological agents are in the beginning of their commercial exploitation for the usage as seed treatment (Harman and Nelson, 1994). Despite the wide use of biological agents as seed treatment pesticides (Table 10.4) commercially for a variety of crops (TNAU Agritech Portal, 2013), the agrochemicals use for seed treatment will continue for long time as an integral part of future disease management.

TABLE 10.4
Recommendation of Seed Treatment for Major Crops

S. No.	Name of Crop	Pest/Disease	Seed Treatment	Remarks
1	Sugarcane	Root rot, wilt	Carbendazim (0.1%) *Trichoderma* spp. 4–6 gm/kg. seed	For seed dressing metal seed dresser/earthen pots or polythene bags are used.
2	Maize	Soil and seed-borne disease	*Trichoderma viride*, *T. harzianum* 4 g/kg seed	-do-
3	Groundnut	Stem rot, Seed rot, Seedling rot	Soil application of caster cake @ 1,000 kg/ha or Neem cake. Seed treatment *Trichoderma viride* @ 4 gm/kg seed	-do-
		White grubs	Chlorpyriphos/Quinalphos @ 2.5–12 mL/kg seed	
4	Cotton	Soil- and Seed-borne diseases,	Acid delinting should be followed before sowing @ 1 L commercial H_2So_4 for 10 kg seed. *Trichoderma* spp 4 gm/kg seed. Captan 3 g/kg seed Carbendazim 2 g/kg seed	-do-
		Black arm	Streptocycline 0.01% dipping of seeds	-do-
5	Rice	Root rot disease and other insects/pests	*Trichoderma* 5–10 gm/kg. seed (before transplanting) Chloropyriphos 30 gm/10 kg seed	-do-
6	Chilli	Anthracnose spp. Pseudomonas spp.	Seed treatment with *Trichoderma viride* 4 g/kg, Carbendazim @ 1 g/100 gm seed.	-do-
		Soil-borne infection of fungal disease	*Trichoderma viride Trichoderma harzianum @ 2 gm/kg seed.*	-do-
7	Pigeon pea	Wilt, Blight	*Trichoderma* spp. @ 4 gm/kg. seed	For seed dressing metal seed dresser/earthern pots or polythene bags are used.
8	Pearl Millet (Bazara)	Soil-borne disease	*Trichoderma harzianum T. viride* @ 4 gm/kg seed.	For seed dressing metal seed dresser/earthern pots or polythene bags are used.
9	Maize	Soil-borne diseases	*Trichoderma harzianum T. viride* @ 4 gm/kg. seed.	-do-

(Continued)

TABLE 10.4 (*Continued*)
Recommendation of Seed Treatment for Major Crops

S. No.	Name of Crop	Pest/Disease	Seed Treatment	Remarks
10	Sesame	Root rot disease, Seedling blight, *Cercospora* leaf spot, Dry root rot, *Alternaria* leaf spot, Bacterial blight, Bacterial leaf spot	1. *T. viride* @ 4 gm/kg. Seed. 2. Seed treatment with Thiram 2–2.25 g/kg, Agrimycin – 100 (250 ppm) or Streptocycline suspension 0.05%	-do-
11	Sorghum	Soft-borne	*Trichoderma harzianum* *T. viride* @ 4 gm/kg. seed.	-do-
12	Pea	Root rot, White rot	Seed treatment with *Bacillus subtilis* *Pseudomonas fluorescens* Soil application @ 2.5–5 kg in 100 kg FYM or Carbendazim or Captan 2 gm/kg. seed *Trichoderma harzianum* 4 gm/kg seed Thiram + Carbendazim 2 gm/kg seed Carbendazim or Captan 2 gm/kg seed	-do-
13	Onion	Smut	*T. viride* @ 2 gm/100 gm. seed. Benlate or Vitavax @ 0.01%	-do-
14	Okra	Root knot nematode Collar Rot	Carbosulfan (25 ST) @ 30% (w/w) *Trichoderma viride* @ 2 g/100 g seed	-do-
15	Tomato	Soil-borne infection of fungal disease Early blight Damping off	*Trichoderma harzianum* and *T. viride* @ 2 gm/100 gm seed.	For seed dressing metal seed dresser/earthen pots or polythene bags are used.
16	Coriander	Wilt	*Trichoderma viride Trichoderma harzianum* @ 4 gm/kg seed.	-do-
17	Brinjal	Soil-borne infection of fungal disease	*Trichoderma viride Trichoderma harzianum* @ 2 gm/100 gm. seed.	-do-
18	Cucurbits	Soil-borne disease	*Trichoderma viride* @ 2 gm/100 gm.seed.	-do-
19	Leguminous Vegetables	Soil-borne infection Nematode	*Trichoderma viride* @ 2 gm/100 gm.seed. Carbofuran/ Carbosulfan 3% (w/w)	-do-
20	Sunflower	Diseases	*Trichoderma viride*, *Trichoderma harzianum* @ 4 gm/kg seed.	-do-

(*Continued*)

TABLE 10.4 (*Continued*)
Recommendation of Seed Treatment for Major Crops

S. No.	Name of Crop	Pest/Disease	Seed Treatment	Remarks
21	Soybean	Seedling disease	*Rhizobium* spp. and Phosphate Solubilizing Bacteria (PSB) @ 5 + 5 gm/kg seed	For seed dressing metal seed dresser/earthen pots or polythene bags are used
22	Wheat	Termite	Teat the seed before sowing with any one of the following insecticides. i) Chlorpyriphos @ 4 ml/kg seed	-do-
23	Sorghum	Soil/Seed-borne diseases	Seed treatment with *Trichoderma viride* , *T. harzianum* @ 4 g/kg seeds	-do-
24	Cruciferous vegetables	Soil/Seed-borne diseases	Seed treatment with *Trichoderma viride* @ 2 g/100 g seeds	-do-
25	Gram	Root knot nematode, Lesion Nematode, Wilt	Seed treatment with *Trichoderma viride* , *T. harzianum* @ 4 g/kg seeds Combination of Carbendazim with Carbosulfan@ 0.2% Carbendazim with Thiram with carbosulfan @ 0.2%	
26	Potato	Soil- and Tuber-borne diseases	Seed treatment with MEMC 3% WS @ 0.25% or boric acid 3% for 20 minutes before storage.	

10.5 FUTURE PROSPECTS

Among the different parts of the globe, most of agrochemical companies are heavily investing their money into development of new seed treatment technologies; this is one of the booming sectors in plant disease management. Especially when around the globe, there is severe pressure over crop protection companies to make crop production as safe and ecologically sustainable. To achieve this, we need to develop new technologies that result in use of very little amount of toxic agrochemical for disease management. So, we need to develop technologies where the different agents can be mixed with seed itself for controlling pathogens. In recent years, many studies were carried out to know how seed treatment helps at great level in improving seed germination, establishing uniform plant growth, along with controlling seed- and soil-borne pathogens in many cereal and vegetable crops. Scientists have developed many methods to reduce agrochemicals use, often influenced by the ecological concerns (Basra et al., 2003). The agrochemicals use in modern agriculture can be limited only by limiting of treatment based on the need; such seed treatment should be done only on those seed lots that are already infected; the amount of fungicides used for management of plant disease can be limited to great extent. We need to conduct more basic and advanced research in the area of seed infestation, relationship between seed infestation and disease development under normal conditions. In order to reduce the risk of seed- and soil-borne pathogens, we must develop an economic threshold level based on the environmental conditions in which seeds are sown.

10.6 CONCLUSION

This chapter covers all the areas regarding seed treatment, its different methods, and its advances in research around different parts of the globe in brief. It begins with basic seed dressing, seed coating with the most basic unrefined methods, and progresses to technological evolution over the period. Furthermore, the latest seed treatment technologies such as film coating, pelleting, and priming come up to rectify and master the limitations or drawbacks of earlier technologies. Seed treatment due to its precision and effectiveness such as reducing the rate of pesticide application to land area, has become an upfront technology in precision agriculture today. We can conclude from our knowledge of seed treatment that it will become a practical, economical, and simple method of delivery through the latest seed treatment technologies, largely for small and medium farmers in emerging countries like India. In coming times, development and commercialization of seed treatment technologies mainly depend how much the new technologies were practically useful in cultivation of the particular crops along with how safe they are toward environment, socially and economically. As a result, in the upcoming days, seed treatment research may focus on state-of-the-art methods such as bio-priming, laser, gamma ray, and ozone treatment as a substitute to chemical seed treatment. While the advanced developments in seed treatment technology refine existing strategies, future research should focus mainly on biological seed treatments as well as agrochemicals treatment using microorganisms as disease-suppressing and seed-enhancing materials though either alone or in combination with other methods. By this way, seed treatment can be transformed into a system not just parts that are added into seeds before sowing. Finally, keeping the view of multiple advantages of seed treatment in management of plant disease, we cannot overlook its role in sustainable crop cultivation, so it should be a first step in cultivation of any crops.

REFERENCES

Afzal, I.; Rehman, H.U.; Naveed, M.; Basra, S.M.A. Recent advances in seed enhancements. In: Susana, A.; Balestrazzi, A. (Eds), *New Challenges in Seed Biology-Basic and Translational Research Driving Seed Technology*; InTechOpen: London, UK, 2016; pp. 47–74.

Ahmad, P.; Hashem, A.; Abd-Allah, E.F.; Alqarawi, A.; et al. Role of *Trichoderma harzianum* in mitigating NaCl stress in Indian mustard (*Brassica juncea* L) through antioxidative-defense system. *Frontiers in Plant Science*, 2015, 6, 868.

Aladjadjiyan, A. The use of physical methods for plant growing stimulation in Bulgaria. *Journal of Central European Agriculture*, 2007, 8(3), 369–380.

Aladjadjiyan, A. Effect of microwave irradiation on seeds of lentils (lens culinaris, med.). *Romanian Journal Biophysics*, 2010, 20(3), 213–221.

Amein, T.; Wright, S.A.I.; Wikström, M.; et al. Evaluation of non-chemical seed treatment methods for control of *Alternaria Brassicicola* on cabbage seeds. *Journal of Plant Diseases and Protection*, 2011, 118(6), 214–221.

Ark, P.A.; Gardner, M.W. Carrot bacterial blight as it affects the roots. *Phytopathology*, 1944, 34, 416–420.

ASF. National code of practice for the use of seed treatments. *Australian Seed Federation Limited*, 2010, pp. 1–7, Retreived: September 4, 2014 from https:// www.asf.asn.au/userfiles.

Bagegni, A.M.; Sleper, D.A.; Kerr, H.D.; et al. Viability of *Acremonium Coenophialum* in tall fescue seed after ionising radiation treatments. *Crop Science*, 1990, 30, 1272–1275.

Baker, K.F. Aerated-steam treatment of seed for disease control. *Horticulture Research*, 1969, 9, 59–73.

Basra, S.M.A.; Zia, M.N.; Mehmood, T.; et al. Comparison of different invigoration techniques in wheat *Triticumaestivum* L. seeds. *Pakistan Journal of Arid Agriculture*, 2003, 5, 11–17.

Bel'skii, A.I.; Mazulenko, N.N. Effects of presowing treatment of barley seeds on the incidence of fungal diseases on the plants. *Mikologiya Fitopatologiya*, 1984, 18, 312–316.

Bhaskara Reddy, M.V.; Raghavan, G.S.V.; Kushalappa, A.C.; et al. Effect of microwave treatment on quality of wheat seeds infected with *Fusarium Graminearum*. *Journal of Agricultural Engineering Research*, 1998, 71(2), 113–117.

Bhushan, C.; Bhardwaj, A.; Misra, S.S. State of pesticide regulations in India. Report of Centre for Science and Environment, New Delhi. 2013, Retreived: January 02, 2021 from www.cseindia.org.

Bryan, M.L. Studies on bacterial canker of tomato. *Journal of Agricultural Research*, 1930, 41(12), 825–851.

Callan, N.W.; Mathre, D.; Miller, J.B. Bio-priming seed treatment for biological control of *Pythiumultimum* pre emergence damping-off in sh-2 sweet corn. *Plant Disease*, 1990, 74, 368–372.

Campos, E.V.R.; De Oliveira, J.L.; Da Silva, C.; et al. Polymeric and solid lipid nanoparticles for sustained release of carbendazim and tebuconazole in agricultural applications. *Scientific Reports*, 2015, 5, 1–14. https://doi.org/10.1038/srep13809.

Chapman, R.A.; Harris, C.R. Persistence of four pyrethroid insecticides in a mineral and an organic soil. *Journal of Environmental Science and Health*, 1981, 16, 605–615.

Claudia, H.A.; Liliana, R.P.C.; Arturo, D.P.F.; et al. Laser light on the mycoflora content in maize seeds. *African Journal of Biotechnology*, 2011, 10(46), 9280–9288.

Cuero, R.G.; Smith, J.E.; Lacey, J. The influence of gamma irradiation and sodium hypochlorite sterilisation on maize seed microflora and germination. *Food Microbiology*, 1986, 3, 107–113.

Damann, K.E. Evaluation of commercial heat-treatment methods for control of ratoon stunting disease of sugarcane. *Plant Disease*, 1983, 67, 966–967.

De Palma, M.; Salzano, M.; Villano, C.; et al. Transcriptome reprogramming, epigenetic modifications and alternative splicing orchestrate the tomato root response to the beneficial fungus *Trichoderma harzianum*. *Horticulture Research*, 2019, 6, 5. https://doi.org/10.1038/s41438-018-0079-1.

Devash, Y.; Bashan, Y.; Okon, Y.; et al. Survival of *Pseudomonas* tomato in soil and seeds. *Hassadeh*, 1979, 60(3), 597–601.

DPPQS (Department of Plant Protection Quarantine and Storage). Campaign for Seed Treatment: Campaign for 100% Seed Treatment during Rabi, 2007. Retrieved: August 31, 2014 from https://ppqs.gov.in/ Seedtreatment.htm.

Edison, S.; Ramakrishnan, K. Aerated steam therapy for the control of grassy shoot disease (GSD) of sugar cane. *Mysore Journal of Agricultural Research*, 1972, 6, 492–494.

Ellis, R.H. Seed and seedling vigor in relation to crop growth and yield. *Journal of Plant Growth Regulation*, 2004, 11, 249–255.

El-Mohamedy, R.S.R.; AbdAlla, M.A.; Badiaa, R.I. Soil amendment and seed biopriming treatments as alternative fungicides for controlling root rot diseases on cowpea plants in Nobaria Province. *Research Journal of Agriculture and Biological Science*, 2006, 2(6), 391–398.

El-Tayed, L.M.; Musa, A.; Makki, Y.M. Effect of seed treatments on growth and yield of two wheat varieties. *Review of Plant Pathology*, 1987, 10, 445–445.

Elwakil, M.A. Use of antioxidant hydroquinone in the control of seed–borne fungi of peanut with special reference to the production of good quality seed. *Plant Pathology Journal*, 2003, 2, 75–79.

Erdey, D.P.; Mycock, D.J.; Bolkan, H.A. The elimination of *Fusarium Moniliforme* (Sheldon) infection in maize caryopses by hot water treatments. *Seed Science and Technology*, 1997, 25, 485–501.

Floyd, R. Vegetable Seed Treatments, Farm note 90/1990. Department of Agriculture and Food, Western Australia. 2005, Retrieved: September 5, 2014 from https://archive.agric.wa.gov.au/PC_92733.html.

Forsberg, G.; Kristensen, L.; Eibel, P.; et al. Sensitivity of cereal seeds to short duration treatment with hot, humid air. *Journal of Plant Disease and Protection*, 2003, 110(1), 1–16.

Giammichele, L.A.; Pill, W.G. Protection of fluid - drilled tomato seedlings against damping - off by fungicide incorporation in a gel carrier. *Horticultural Science*, 1984, 19(6), 877–879.

Gray, D. Fluid drilling of vegetable seeds. *Horticultural Reviews*, 1981, 3, 1–27.

Grondeau, C.; Ladonne, F.; Fourmond, A.; et al. Attempt to eradicate *Pseudomonas Syringae* pv. *Pisi* from pea seeds by heat treatments. *Seed Science and Technology*, 1992, 20, 515–525.

Gupta, R.B.L.; Majumdar, V.L.; Bhatnagar, G.C. Influence of seed dressing fungicides on mycoflora and viability of wheat seed under storage. *Seed Research*, 1990, 18, 157–159.

Hacisalihoglu, G.; Taylor, A.G.; Paine, D.H.; et al. Embryo elongation and germination rates as sensitive indicators of lettuce seed quality: priming and aging studies. *Horticultural Science*, 1999, 34, 1240–1243.

Halmer, P. *Technical and Commercial Aspects of Seed Pelleting and Film-Coating, in Application to Seeds and Soil, Mono.39*. UK: BCPC, Thornton Health; 1988, pp. 191–204.

Harman, G.E. Seed treatments for biological control of plant disease. *Crop Protection*, 1991, 10, 166–171.

Harman, G.E.; Nelson, E.B. *Mechanisms of Protection of Seed and Seedlings by Biological Seed Treatments: Implications for Practical Disease Control, in Seed Treatment: Progress and Prospects, Mono. 57*. UK: BCPC, Thornton Health; 1994, pp. 283–292.

Harwalker, M.R.; Donger, T.K.; Padwal-Desai, S.R. Radiation disinfestations of spice and spice products. I. Radiation sensitivity of developmental stages of *Lasioderma Serricorne* and *Stegobium Panicium*. *Journal of Food Science & Technology*, 1995, 32, 249–251.

Heydecker, W.; Coolbear, P. Seed treatments for improved performance – survey and attempted prognosis. *Seed Science and Technology*, 1977, 5, 353–425.

Howell, C.R.; DeVay, J.E.; Garber, R.H.; et al. Field control of cotton seedling disease with *Trichoderma virens*in combination with fungicide seed treatment. *The Journal of Cotton Science*, 1997, 1, 15–20.

Izhar, A.; Bashir, M.; Bakhtmand, K.; et al. Effect of seed dressing fungicides on yield and yield components of wheat. *Pakistan Journal of Agriculture Research*, 1987, 8, 361–365.

Jindal, K.K.; Thind, B.S.; Soni, P.S. Physical and chemical agents for the control of *Xanthomonas Campestrispv. vignicola* from cowpea seeds. *Science and Technology*, 1991, 17(2), 371–382.

Johnsson, L. Washing and treating with warm water to clean wheat inoculated with common bunt (*Tilletia caries* DC Tul.). *Växtskyddsnotiser*, 1990, 54(1), 26–28.

Kittleson, A.R. A new class of organic fungicides. *Science*, 1952, 115, 84–86. https://doi.org/10.1126/science.115.2978.84-a.

Kobayashi, T. Impact of seedborne pathogen on quarantine in Japan. *Seed Science and Technology*, 1990, 18(2), 427–433.

Krishna, D. Insecticide and Fungicide Seed Treatment. 2011, Retrieved: 1 September, 2014 from https://agropedia.iitk.ac.in/content/insecticide-and-fungicideseed-treatment.

Lamichhane, J.R.; You, M.P.; Laudinot, V.; et al. Revisiting sustainability of fungicide seed treatments for field crops. *Plant Disease*, 2020, 104(3), 610–623.

Luthra, J.C. Solar energy treatment of wheat loose smut, *Ustilagotritici* (Pers.) Rostr. *Indian Phytopathology*, 1953, 6, 49–56.

Luthra, J.C.; Sattar, A. Some experiments on the control of loose smut, *Ustilagotritici* (Pers.) Jens. of wheat. *Indian Journal of Agricultural Science*, 1934, 4, 177–199.

Mastouri, F.; Björkman, T.; Harman, G.E. *Trichoderma harzianum* enhances antioxidant defense of tomato seedlings and resistance to water deficit. *Molecular Plant-Microbe Interactions*, 2012, 25(9), 1264–1271.

Mathre, D.E.; Johnston, R.H.; Grey, W.E. Small grain cereal seed treatment. American Phytopathological Society, The Plant Health Instructor. 2001, Retrieved: 31 September, 2014 from https://www.apsnet.org/education/AdvancedPlantPath/Topics/SeedTreatment/. https://doi.org/10.1094/PHI-I-2001-1008-01.

Meah, M.B. Vegetable Seed Treating Plant. USAID, Bangladesh Collaborative Research Project and IPM Lab. Department of Plant Pathology, BAU, Mymensingh, 2004.

Meshram, S.; Patel, J.S.; Yadav, S.K.; et al. *Trichoderma* mediate early and enhanced lignification's in chickpea during *Fusarium Oxysporum* f. sp. *ciceris* infection. *Journal of Basic Microbiology*, 2019, 59(1), 74–86.

Meyer, H.; Mayer, A.M. Permeation of dry seeds with chemicals: use of dichloromethane. *Science*, 1971, 171(3971), 583–584.

Miller, S.A.; Lewis Ivey, M.L. *Hot Water Treatment of Vegetable Seeds to Eradicate Bacterial Plant Pathogens in Organic Production Systems*. Columbus, OH: The Ohio State University Extension; 2005, p. 3.

Moffett, M.L.; Wood, B.A. Seed treatment for bacterial spot of pumpkin. *Plant Disease Reporter*, 1979, 63(7), 537–539.

Müller, H.; Berg, G. Impact of formulation procedures on the effect of the biocontrol agent *Serratiaplymuthica* HRO-C48 on *Verticillium* wilt in oilseed rape. *BioControl*, 2008, 53(6), 905–916.

Muniz, M.F.B. Control of microorganisms associated with tomato seeds using thermotherapy. *Revista Brasileira de Sementes*, 2001, 23(1), 176–280.

Nagy, J.; Rathos, J.; Toth, O. Effect of ultrasonic irradiation and vacuum infiltration combined with fungicides on fungi-infected sunflower seeds. *Acta Agronomica Hungarica*, 1995, 43, 93–102.

Nameth, S.T. Priorities in seed pathology research. *Scientia Agricola*, 1998, 55, 94–97.

Napoles, P.; Amat, Z.; Ramirez, P. The use of different treatments of control *Xanthomonas campestris* pv. *campestrisin* Cabbage seeds. *Protection De-Plants*, 1991, 1(3): 33–41.

Neergaard, P. *Seed Pathology. Vols. 1 and 2*. London, UK: Macmillan; 1977.

Nega, E.; Ulrich, R.; Werner, S.; et al. Hot water treatment of vegetable seed – an alternative seed treatment method to control seed-borne pathogens in organic farming. *Journal of Plant Diseases and Protection*, 2003, 110, 220–234.

Nene, Y.L.; Thapliyal, P.N. *Fungicides in Plant Disease Control*. 2nd Edition. New Delhi, India: Oxford and IBH Pub. Co.; 1979.

Nicholas, K.R.; Groot, S.P.C. *Assessment of Non-Chemical Seed Treatment Methods in Organic Seeds: Assessment of Physiological Sensitivity of Seeds to Physical Treatment*. Germany: LAP Lambert Academic Publishing; 2013, pp. 88.

Patel, J.S.; Kharwar, R.N.; Singh, H.B.; et al. *Trichoderma asperellum* (T42) and *Pseudomonas fluorescens* (OKC)-enhances resistance of pea against *Erysiphepisi* through enhanced ROS generation and lignifications. *Frontiers in Microbiology*, 2017, 8, 306.

Pedrini, S.; Merritt David, J.; Stevens, J.; Dixon, K. Seed coating: science or marketing spin? *Trends in Plant Science*, 2017, 22, 106–116.

Raj, S.N.; Shetty, N.P.; Shetty, H.S. Seed bio-priming with *Pseudomonas fluorescens* isolates enhances growth of pearl millet plants and induces resistance against downy mildew. *International Journal of Pest Management*, 2004, 50(1), 41–48.

Randhawa, H.S.; Sharma, H.L.; Kaur, J.; et al. Effect of fungicides on germination and seed mycoflora on wheat under different storage conditions. *Pesticides*, 1985, 19, 36–38.

Rao, Y.S.; Srinangarajan, A.K.; Kamal, A.S.; et al. Studies on extension of shelf life of raw a by gamma irradiation. *Journal of Food Science and Technology*, 1994, 31, 311–315.

Reddy, P.P. *Recent Advances in Crop Protection*. India: Springer; 2013, 259 p. https://doi.org/10.1007/978–81–322–0723–8.

Research and Markets. (2013). Seed treatment market by type (chemical non-chemical), by application (fungicide, insecticide, bio-control and others) and by crop (cereals, oilseeds and others): global trends, forecasts and technical insights up to 2018.

Sanjeev, K. Cultural approaches for plant disease management. *Research & Reviews: Journal of Agricultural Science and Technology*, 2012, 1(2), 12–21.

Schwinn, F. Seed treatment – a panacea for plant protection? Seed Treatment: Progress and Prospects. UK: BCPC Publications. 1994, Monograph 57, 3. Retrieved: September 9, 2014 from https://www.amazon.com/gp/ search?

Seung-Hee, L.; Jin-Seok, K.; Jung-Myung, L. Moisture contents in bottle gourd seeds during dry heat treatment and subsequent germination. *Horticultural Science*, 2004, 39(4), 758–759.

Shao, S.; Meyer, C.J.; Ma, F.; et al. The outermost cuticle of soybean seeds: chemical composition and function during imbibitions. *Journal of Experimental Botany*, 2007, 58(5), 1071–1082.

Shekhawat, P.S.; Jain, M.L.; Chkravarti, B.P. Detection and seed transmission of *Xanthomonas campestris* pv. *campestris* causing black rot of cabbage and cauliflower and its control by seed treatment. *Indian Phytopathology*, 1982, 35(3), 442–447.

Shoemaker, P.B.; Echandi, E. Seed and plant bed treatments for bacterial canker of tomato. *Plant Disease Report*, 1976, 60(2), 163–166.

Singh, M.T.; Singh, M.S. Effect of gamma irradiation on seed micoflora seed germination and seedling growth of rice. *Oryza*, 2005, 42(2), 129–132.

Sinha, P.; Sharma, R.P.; Roy, M.K. Management of storage rot in onion through gamma irradiation and chemicals. *Journal of Food Science and Technology*, 1994, 31, 341–343.

Sitton, J.W.; Borsa, J.; Schultz, T.R.; et al. Electron beam irradiation effects on wheat quality, seed vigor and viability and pathogenetity of teliospores of *Tilletiacontroversa* and *T. tritici*. *Plant Disease*, 1995, 79(6), 586–589.

Tamietti, G. Seed treatment trials to control halo blight of bean. *Informatore Fitopatologico*, 1982, 32(6), 47–50.

Tapke, V.F. Single-bath hot-water and steam treatments of seed wheat for the control of loose smut. *USDA Bulletin*, 1926, 1383, 1–29.

Taylor, A.G.; Harman, G.E. Concepts and technologies of selected seed treatments. *Phytopathology*, 1990, 28, 312–339.

Taylor, A.G.; Trimmer, M.; DunhamTrimmer International Bio Intelligence, Lakewood Ranch, Florida. Global chemical and biological seed treatments market. Personal communication, 2020.

Thippenswamy, T.; Lokesh, S. Efficacy of different seed dressing chemicals on seed mycoflora, seed germination and seedling vigor of sunflower (*Helianthus annus* L.). *Seed Research*, 1997, 25(1), 64–67.

Thomas, G.J.; Adcock, K.G. Exposure to dry heat reduces anthracnose infection of lupin seed. *Australian Plant Pathology*, 2004, 33, 537–540.

Tillet, M. *Dissertation on the Cause of the Corruption and Smutting of the Kernels of Wheat in the Head, and on the Means of Preventing These Untoward Circumstances (no. 5).* United States: American Phytopathological Society; 1755.

TNAU Agritech Portal, Seed Technology. Seed Treatment: Insecticides & Fungicides. 2013, Retrieved: November 19, 2013 from https://agritech.tnau.ac.in/seed_certification/seed_treatment_Insecticides%20&% 20Fungicides.html.

Upadhyaya, P. Seed Treatment for Controlling Seed Borne Diseases. 2013, Retreived: September 4, 2014 from https://ekduniya.net/sites/lifelines/seed-treatmentcontrolling-seed-borne-diseases/#.VAoB6X-zfIU/.

Vasilevski, G. Perspectives of the application of biophysical methods in sustainable agriculture. *Bulgarian Journal of Plant Physiology*, 2003, 2003, 179–186.

Viswanathan, R. Different aerated steam therapy (AST) regimes on the development of grassy shoot disease symptoms in sugarcane. *Sugar Technology*, 2001, 3(3), 3–91.

Weimer, J. *Lupine Anthracnose*. United States Department of Agriculture Circular No. 904, 1952.

Yedidia, I.; Benhamou, N.; Kapulnik, Y.; et al. Induction and accumulation of PR proteins activity during early stages of root colonization by the mycoparasite*Trichoderma harzianum* strain T-203. *Plant Physiology and Biochemistry*, 2000, 38(11), 863–873.

11 Disease-Suppressive Soils and Plant Disease Management

An Overview

Somasundaram Jayaraman,
Thangavelu Muthukumar, Anandkumar Naorem,
and Davis Joseph Bagyaraj

11.1 INTRODUCTION

Soil is a complex ecosystem providing food for all living things, namely microorganisms, plants, animals, and humans (Chandrashekara et al., 2012). In order to fulfill food and nutritional requirement as well as increasing food productivity, it is important to maintain soil biodiversity and sustainability of the soil environment (Lal, 2020). The tripartite interaction between soil–microbe–plant plays a vital role in maintaining productivity of agricultural crops (Devi and Meetei, 2018) (Figure 11.1). Soil harbors millions of organisms such as bacteria, fungi, protozoa, algae, and nematodes, the vast majority of which are still undescribed/understood. They are

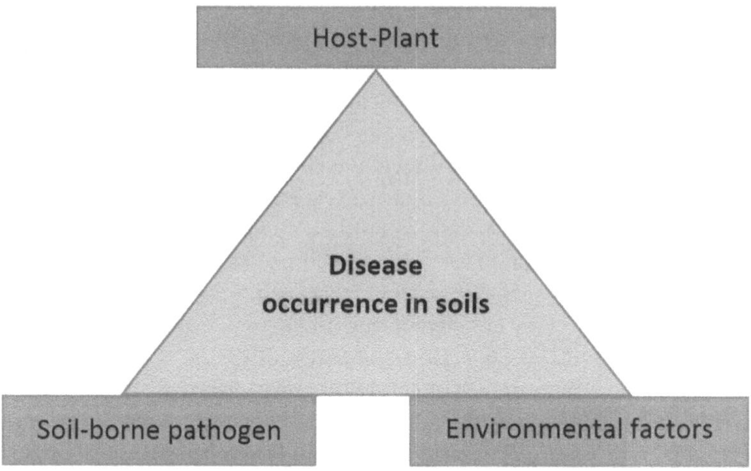

FIGURE 11.1 Tripartite interaction between- Soil- Plant-Host microbe (Source: Authors developed)

DOI: 10.1201/9781003531944-11

important drivers of essential ecological functions and services. Plants mainly provide carbon and energy sources to the microbial communities in soil. However, the soil properties such as texture, structure, pH, and electrical conductivity also determine the type and activity of microorganisms in soil. Further, agricultural management practices such as tillage, organic amendments/manures, crop rotation, irrigation, fertilizer, and pesticide application also influence the type and population of soil microorganisms. "Soil microbiome" refers to microbial communities in the soil. Microorganisms in the microbiome can interact among themselves and also with their hosts to form synergistic relationships. As a dynamic living resource, soil is the basis of sustainable agriculture, as well as the physical support for most other human activities. Soil microorganisms play an essential role in decomposing organic matter, nutrients cycling, thereby enriching/fertilizing the soil. Without the cycling of elements, continuation of life on earth would be impossible. Soil is essential for maintenance of biodiversity not only below-ground but also above- ground. Biodiversity in below-ground is vast and unappreciated. Because of the impact of microbiome on ecosystems, currently there is considerable interest in studying soil microbiomes and their role in altering soil ecosystem. By engineering soil microbiome, more diverse and balanced microbial communities can be established for enhancing soil health and productivity (Jayaraman et al., 2021). Some of the microorganisms in soil cause disease of plants but many promote plant growth, nutrition, and tolerance to plant diseases and abiotic stresses (Rangaswami and Bagyaraj, 2017).

Soil-borne plant pathogens are very difficult to control compared to above-ground pathogens. Plant diseases caused by soil-borne pathogens result in substantial losses to agricultural production to the extent of 50%–70% yield loss in many crops worldwide (Panth et al., 2020). Soil-borne plant pathogens causing root and crown rots, wilts, and damping-off are major yield-limiting factors in agriculture, horticulture, and forestry system. Most soil-borne pathogens are difficult to control by conventional strategies such as the use of resistant host cultivars and synthetic fungicides. The lack of reliable chemical controls, the occurrence of fungicide resistance in pathogens, and the breakdown of host resistance by soil-borne pathogen populations are among the key factors underlying efforts to develop other control measures (Cook and Baker, 1983). Soil fumigation is the most effective chemical treatment but is too expensive for many crops, and fumigants such as methyl bromide are being phased out for environmental reasons. Soil solarization is another effective method recommended to protect plants from soil-borne plant pathogens. This is also an expensive treatment and recommended only for higher-value horticultural crops. Further, it can be used only in areas with a hot climate. On the other hand, every natural soil possesses some ability to suppress the activity of plant pathogens due to the presence and activity of soil microorganisms. Some soils that are very effective in suppressing plant pathogens are referred as "disease-suppressive soils" (DSSs) (Devi and Meeti, 2018; Liu et al., 2019a, 2019b).

DSSs provide the best examples of natural microbe–based plant defense, whereby via rhizodeposition, plant roots stimulate, enrich, and support soil microorganisms as the first line of defense mechanism against soil-borne pathogens (Schlatter et al., 2017). In this context, much is to be learned from DSSs, where susceptible plants are protected from soil-borne pathogens by the indigenous "microbiota," which is antagonistic to the pathogen.

11.2 BACKGROUND ON DISEASE-SUPPRESSIVE SOILS

The existence of DSS has been recognized by farmers and scientists since early in the 20th century (Cook, 2014). DSSs are defined by Cook and Baker (1983) as soils in which because of their microbial community and activity, the pathogen does not establish or persist, the pathogen establishes but causes no damage, or the pathogen causes some disease damage, but the disease becomes progressively less severe even though the pathogen persists in soil. The mechanism of suppression of soil includes antibiosis, competition, parasitism, and predation (Figure 11.2).

Mechanisms of antibiosis postulated to be involved in pathogen-suppressive soils are inhibition by production of antibiotic (Thomashow and Weller, 1995) and iron starvation by production of an iron-binding siderophore (Schroth and Hancock, 1982). For antibiosis to play a role in DSS, the bioactive substance would need to be produced in the niche(s) occupied by the pathogen, and the pathogen would need to be sensitive to the antibiotic or siderophore. Further, the antibiotics produced modulate other processes essential for survival and function of the producing bacteria (Raaijmakers and Mazzola, 2012).

Strains and species within the *Pseudomonas fluorescens* group have received the greatest attention because of their ability to produce a wide array of antibiotics and siderophores. At least four antibiotics produced by the *P. fluorescens* group, namely

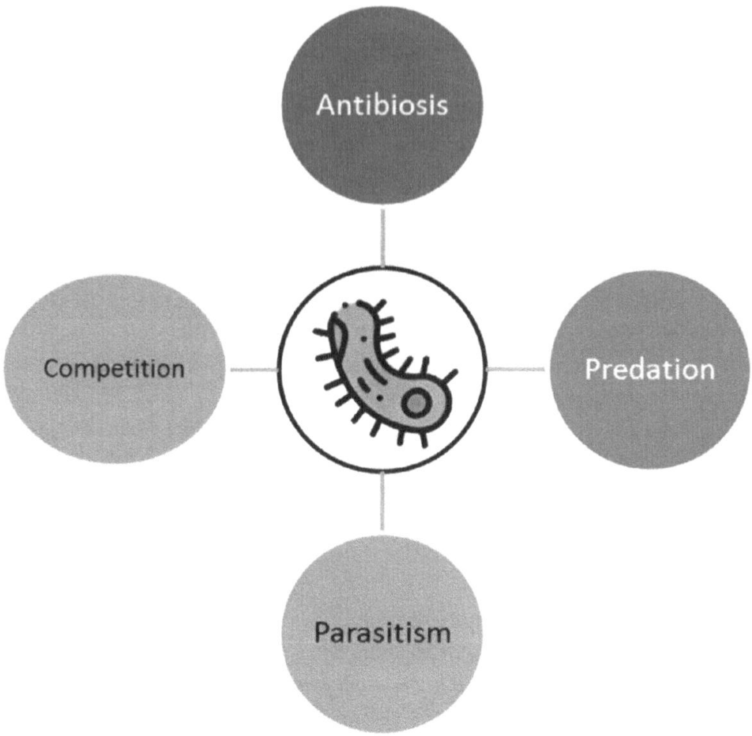

FIGURE 11.2 Mechanism in disease suppressive soils (Source: Authors developed).

pyoluteorin, pyrolnitrin, 2,4-diacetylphloroglucinol (DAPG), and phenazine-1-carboxylate (PhZ), were early targets of research on their roles in disease suppression (Weller et al., 2002). The role of siderophores produced by *P. fluorescens* in DSS received a great deal of attention by researchers in the 1980s. Reflecting on the lack of genes for host plant resistance to soil-borne plant pathogens, it was hypothesized that the plants may have evolved a different strategy for defense against root pathogens, namely interactions with antibiotic and siderophore producing rhizosphere microorganisms (Cook, 2014). Though some argue for limiting use of the term "disease suppressive" to situations involving a clear biological component (Bruehl, 1987), there is plentiful evidence for the role of both biotic and abiotic elements of the soil having a role in disease suppression. Similarly, chemical and physical attributes of soil, including pH, organic matter, and clay content, can also operate in the suppression of plant diseases directly or indirectly through their impact on soil microbial activity. Although these abiotic characteristics of soil can contribute to disease suppression, soil suppressiveness is often times directly or indirectly a function of the activity of soil microorganisms or microbial metabolites.

The natural characteristic of suppressive soil is accepted worldwide as a management strategy to create sustainable food production ensuring high agricultural productivity levels (Guerra et al., 2020). Soil disease suppression can be regarded as good indicator of soil biological property conferred by the microbial community (microbiome) because sterilization of DSS destroys the disease controlling capacity of the suppressive soils, and in addition, the transfer of disease suppressiveness could be possible when conducive soil receives an amount of suppressive soil (Stutz et al., 1986).

Therefore, DSS could potentially work in reducing the loss of productivity due to the soil-borne pathogens (Persson and Olsson, 2000). Besides, the development of DSS could be one of the effective examples in which the indigenous soil microbiota effectively protects the host plant against the infections from soil-borne pathogens (Weller et al., 2002). The so-called DSSs have the ability to suppress the various plant pathogens by improving the plant health, inducing natural plant defense, producing antibiotics, competing against pathogens, modulating plant immune system, or hyper-parasitizing the pathogen (Liu et al., 2021). Biocontrol-based microbiota includes bacteria (*Bacillus*, *Paenibacillus* and *Pseudomonas*), actinomycetes (*Streptomyces*), and filamentous fungi (*Trichoderma*, *Fusarium*, *Gliocladium*, *Aspergillus*, and *Penicillium*) that can trigger all mechanisms associated to disease suppression. These organisms not only suppress the disease-causing pathogens by different mechanisms associated with disease suppression but also promote plant growth through improved nutrient uptake and production of phytohormones. The DSSs have been observed for several soil-borne pathogens such as *Streptomyces scabies*, *Pythium splendens*, *Pythium ultimum*, *Thielaviopsis basicola*, *Heterodera avenae*, *Phytophthora cinnamomi*, *Criconemella xenoplax*, *Phytophthora infestans*, *Fusarium oxysporum*, *Rhizoctonia solani*, *Meloidogyne* spp., *Gaeumannomyces graminis* var. *tritici*, *Heterodera schachtii*, *Ralstonia solanacearum*, *Aphanomyces euteiches*, and *Plasmodiophora brassicae* (Jayaraman et al., 2021).

The understanding of disease-suppressive mechanisms is a crucial step to enhance the suppressive effect by manipulation of the soil microbiota. The mechanisms underlying the suppressive effect are primarily associated with the biological activity of soil microbiota, which interacts with the soil organic matter as well as the host plant. The most important factors are represented by the increased microbial activity and the release of diffusible antibiotic-like compounds and the activation of systemic disease resistance in the host by them and also soil physicochemical properties, cultivation practices, and environmental conditions (Cook, 2014; Jayaraman et al., 2021). The recent next generation sequencing (NGS) and high-throughput sequencing (HTS) techniques offering wider opportunity of soil microbiome will help in understanding and to engineer DSSs better in the years to come. With this background, different aspects of DSSs are covered in this chapter.

11.3 TYPES OF DISEASE-SUPPRESSIVE SOILS

The suppressiveness of soils to soil-borne pathogens has been classified into two types, namely general suppression and specific suppression (Weller et al., 2002; Chandrashekara, 2012). General suppression relates usually to the total microbiological activity of the soil in contrast to specific suppression, which relates to the antagonistic activity of a specific group of microorganisms or possibly a single microbial species suppressive to a specific soil-borne pathogen, operating in addition to and against a background of general suppression. Another key distinguishing feature is that general suppression is a natural and preexisting characteristic of soil and is not transferrable from field to field, whereas specific suppression develops in response to the presence of the pathogen or disease and is transferable from one soil to another (Schattler, 2017). Of the two types of DSSs, general and specific suppression, only specific suppression can be managed with any degree of certainty and then largely or entirely by choice of the cropping system.

Amending soil with large quantities of compost or other biomass is about the only management option available to reliably increase the level of general suppression. This may be part of the long-term success of direct seeding where crop residue is left to decompose on the soil surface or is mixed superficially into the top few centimeters of soil with the planting operation in many developing countries. Categorizing suppressive soils into these two types is useful for discussion; they should more appropriately be viewed not just as a combination but also as a continuum from general to specific suppression and as the likelihood of general suppression giving rise to specific suppression in response to specific cropping systems or management practices. Although general suppression is a component of DSS, the understanding and potential exploitation of the phenomenon termed specific suppression are more commonly the subject of interest (Cook and Baker, 1983; Weller et al., 2002).

11.3.1 General Suppression

General suppression is often effective against a broad spectrum of soil-borne diseases and cannot be transferred by adding suppressive soil to a conducive soil. It is reduced by steaming and eliminated by sterilizing the soil and can be enhanced by management practices that increase the population size, diversity, and/or activity of the soil microbiome (Weller et al., 2002; Cook, 2014). A general suppression mechanism (GSM) effect is attributed to microbial biomass affecting more than one pathogen simultaneously where diversified mechanisms are activated offering a basal protection against a broad range of pathogens. The GSM refers to the widespread but limited potential of the soil in suppressing the disease-causing activity of soil-borne pathogens (Cook and Rovira, 1976; Cook and Baker, 1983). It is also referred to as nonspecific antagonism or biological buffering (Jayaraman et al., 2021). This category of mechanism refers to the disease suppression through the competition between the resident soil microbes and pathogens for a common resource such as nutrients and/or space and the release of antibiotic compounds and toxins from an active soil microbial community that hampers the growth and development of the pathogen (Hadar and Papadopoulou, 2012).

It is important to remember that most soil-borne pathogens exist primarily as dormant structures, waiting for root or seed exudates or added nutrients to stimulate and enhance germination, growth, and root infection. General suppression is easily visualized and quantified when inoculum of a soil-borne pathogen is added to an unamended field soil and to the same soil that has been sterilized. Consequently, less disease develops on the host plant in the unamended compared to the sterilized soil (Weller et al., 2002), and the amount of suppressiveness can be quantified by calculating the difference in disease incidence or severity between the two soils.

An excellent example was the control of Phytophthora root rot in an avocado grove established in the early 1940s in Queensland, Australia, which remained healthy after more than 40 years despite growing in soil infested with *Phytophthora cinnamomi* in an environment highly favorable for disease development. In contrast, Phytophthora root rot was severe in neighboring groves (Schlatter et al., 2017). Suppressiveness was associated with soil organic matter, maintained at about 12% by adding large amounts of biomass such as chicken manure, corn stalks, and other plant materials, which stoked the microbial activity. Essentially, the orchard was managed organically, and the soil was described as a "really live soil" based on the diversity and size of the microbial biomass/turnover of nonliving microbial biomass (i.e., necromass) (Cook, 2014). No particular organism or group of organisms could be shown to account for the suppressive effect. However, this general suppressiveness was not transferable, and it was overcome by adding large quantities of pathogen inoculum (Cook, 2014).

11.3.2 Specific Suppression

The specific suppression refers to the effects of an individual or specific group of soil microorganisms on disease suppression (Cook, 2007). Specific suppression mechanism (SSM) operates against only certain types of pathogens. The SSM has been described for pathogens such as *Gaeumannomyces graminis* var. *tritici*, *Rhizoctonia solani*, *Thielaviopsis basicola*, *Fusarium oxysporum*, *Pythium* spp., and few others. In all cases, a particular pathogen causes significantly less disease in suppressive soils than in other soils (conducive soils); the effect is lost when soil is treated with biocides, indicating the involvement of microorganisms. The distinctive characteristic of specific suppression is its transferability factor (Andrade et al., 1994); i.e., it can be transferred from suppressive soil to a conducive soil. The mechanism of specific soil suppression includes antibiosis, competition, parasitism, and predation (Chandrashekara et al., 2012; Jayaraman et al., 2021).

Specific suppression is highly effective; it results from individual species or select groups of microorganisms and is transferable by adding pure cultures or very small amounts of 1%–10% suppressive soil (Cook, 2007, 2014). Suppressiveness is eliminated by treating the soil by pasteurization with moist heat (60°C for 30 minutes) or fumigation with methyl bromide or chloropicrin (Raaijmakers and Weller, 1998; Weller et al., 2002) and reduced or eliminated by growing a non-host crop (Bagyaraj, 1981; Schattler, 2017), but interestingly, the field regains suppressiveness once the susceptible host is grown again. One of the early studied diseases for specific suppression is take-all decline (TAD) caused by *Gaeumannomyces graminis var. tritici*, which is taken as an example and briefly discussed here. TAD, one of the classical examples of induced-specific suppression, is defined as the spontaneous reduction in the incidence and severity of take-all and increase in yield occurring with continuous monoculture of wheat following a severe attack of the disease in the first 3–4 crops, followed by the disappearance with continuous monocrop of wheat (Cook, 2014). However, the first clear documentation of a microbiological basis for TAD was by Gerlagh (1968) working with polder soils in the Netherlands.

The development of TAD follows a consistent pattern everywhere, but cropping history, environmental conditions, and soil factors will impact the intensity (robustness) of the suppressiveness as well as the length of time before its onset, which typically averages 4–6 years (Kwak and Weller, 2013). TAD suppressiveness can be transferred to conducive soils in both the field and the greenhouse by adding 1%–10% suppressive soil to a conducive soil and generally is eliminated by pasteurization at 55°C–60°C for 23 minutes (with some exceptions; Cha et al., 2016) or by fumigating (using methyl bromide) the soil (Shipton, 1975; Weller et al., 2002). Transferability by adding a small amount of soil or inoculum of the responsible microbial species is the key characteristic that separates "specific suppression" from "general suppression." Because specific suppression is due to a population rather than a community, it does not take much time for that population of a specific organism to become established in its niche associated with the soil, and suppressiveness persists in the absence of plants (Weller et al., 2002; Schattler, 2017). Changes in the soil or rhizosphere microbiome resulting in inhibition of *G. graminis* var. *tritici* have been reported as mechanisms of TAD. Similarly, Cook and Rovira (1976) demonstrated that suppression of the take-all pathogen occurred in the rhizosphere of wheat and not in the bulk soil and provided the first evidence that root-associated fluorescent pseudomonads played a significant role in disease suppression.

There is clear microbiological and biochemical evidence for the key role played by fluorescent *Pseudomonas* spp. producing the antibiotics 2,4-diacetylphloroglucinol (DAPG) and phenazine-1-carboxylate (PhZ) (Raaijmakers and Weller, 1998; Raaijmakers and Mazzola 2012) in TAD. Mutant strains no longer able to produce the antibiotics lost the ability to protect wheat against take-all, but regained the ability to protect against take-all when the respective mutant genes were complemented with a wild-type gene (Weller et al., 2002). DAPG is highly active against *G. graminis* var. *tritici*, and the antibiotic acts on multiple basic cellular processes of the pathogen (Kwak et al., 2011). Adding 10% TAD soil to conducive soils established populations of DAPG producers above the threshold needed for take-all control and transferred suppressiveness to the conducive soils (Schattler, 2017). The occurrence of high precipitation or irrigation is ideal for TAD (Mavrodi et al., 2012), and infection of wheat roots by *G. graminis* var. *tritici*, which also initiates an enrichment of DAPG producing pseudomonads to a density above 10^5 CFU/g root (CFU = colony forming unit), the threshold required for take-all control (Cook, 2014). Whole genome sequencing has brought more clarity to diversity within the DAPG-producing pseudomonads (Loper et al., 2012), placing them into at least three species: *Pseudomonas fluorescens*, *Pseudomonas protegens*, and *Pseudomonas brassicacearum* provoking a question as to how one genotype of the *Pseudomonas fluorescens* "complex" can be responsible for such a powerful natural suppression of take-all. The answer may lie in the mutual affinity of this bacterium and wheat roots.

11.4 FACTORS AFFECTING DSS

Soil abiotic and biotic factors influence the occurrence of pathogens and subsequent disease development in crops. Many of these factors also affect the capacity of soils to suppress plant diseases (Legrand et al., 2019). The effect of soil factors on disease suppressiveness may be direct but often governed by the interaction among and between the biotic and abiotic factors. Moreover, there is an urgent need to focus on factors influencing soil suppressiveness in order to understand the role of different factors and to explore avenues for modifying soil conditions that resist disease development in crops leading to increased productivity. The role of biotic and abiotic factors on soil suppressiveness to plant diseases is discussed here under, with consideration on explicating the mechanisms involved.

11.4.1 Abiotic Factors

11.4.1.1 Soil Temperature

Soil temperature in association with moisture is an important abiotic factor that determines the progress of disease in plants (Manici et al., 2012). The effect of temperature on disease suppressiveness indicates the involvement of microbes that are either temperature-resistant and/or sensitive in disease suppression. Available evidences do indicate a reduction in soil suppressiveness at 50°C and its complete elimination at 80°C (Mendes et al., 2011). Döring et al. (2020) found that soil suppressiveness to *Pythium ultimum* differed with soil type, and an increasing temperature (40°C) along with moisture stress (–50%) reduced the disease suppressiveness. This reduction in soil suppressiveness with increasing temperature suggests the involvement of heterogeneous microbial community with varied temperature sensitivity (Yin et al., 2013). For instance, a group of Gram-negative bacteria could dominate suppressive soils till the soil temperature reaches 50°C, which is then replaced by less temperature-sensitive spore-forming group (Manici et al., 2012). Soil temperature was also found to play a significant role in the control of *Macrophomina phaseolina*, the causative agent for charcoal root rot in crops (Kanaan et al., 2016). The colony-forming units of *M. phaseolina* remained unchanged at soil temperature 50°C; however, an increase in temperature to 52°C eliminated *M. phaseolina* (Kanaan et al., 2016).

11.4.1.2 Soil Reaction (pH)

The soil pH is one of the important factors that influence nutrient release, plant growth, and yield under the pH range of 4–9 in most cultivated soils (Husson, 2013). However, many agronomic practices including fertilization or organic amendment can influence soil pH (Bai et al., 2018). Soil pH can indirectly affect the incidence of plant diseases by affecting the abundance, diversity, and activity of disease-suppressive microbes prevailing in the soil (Gordon et al., 2019) and the progression of diseases in plants (Avilés and Borrero, 2017). Therefore, it has been proposed that soil pH also influences the development and severity of diseases directly by affecting the soil-borne plant pathogens (Ghorbani et al., 2008). For instance, soil pH was significantly and negatively correlated to occurrence of rhizome rot and wilt disease complex in ginger (Sharma et al., 2010). The Fusarium wilt in banana was greately reduced in high soil pH compared to acidic range (4.6–5.6) (Deltour et al., 2017). A significant linear correlation between soil pH and soil suppressive indices was reported by Bongiorno et al. (2019) from *Pythium* suppressive European soils. Variations in soil pH levels also induce changes in soil microbial diversity and their activities (Zhou et al. 2020). Therefore, soil pH can directly influence disease suppression by affecting the pathogens or indirectly by modulating the nutrient availability to phytopathogens and/or diversity and abundance of microorganisms in the soil.

11.4.1.3 Electrical Conductivity

The electrical conductivity (EC) indicating the salt concentration in the soil is a crucial factor for soil health and crop productivity. Soil EC affects crop growth and yield through its influence on the nutrient availability to plants and activities of microorganisms in the soil. It is considered as a consistent predictor of suppressiveness to Fusarium wilt in banana as soils suppressive to Fusarium wilt are often characterized by high EC values than those that are conducive for the disease development (Heck et al., 2019). Similarly, high EC values were also associated with the soils suppressive to *Pythium irregulare* causing root rot in baby leaf lettuce (Ros et al., 2020). The role of EC in suppressive soils could be due to its direct influence on the pathogen life cycle in the soil or through the modulation of microbial community composition and their activities thereby affecting disease incidence (Wang et al., 2020).

11.4.1.4 Cation Exchange Capacity

The capacity of soils to hold cations through electrostatic attraction and thereby preventing soil acidification is known as cation exchange capacity (CEC). Generally, soils with higher clay content or organic matter have a higher CEC. Similar to EC, soil suppressiveness to *Pythium ultimum* and *Ralstonia solanacearum* is often correlated to higher CEC (Bongiorno et al., 2019). The CEC also depends on the type of clay, as soils containing smectite clay often have high CEC values, which are suppressive to Fusarium wilt of banana than soils lacking smectite clay, and are conducive (Orr and Nelson, 2018). Moreover, base CEC values must be atleast 70% to control Fusarium wilt in banana (Dita et al., 2018). Contrarily, Gatch and du Toit (2015) did not find any correlation between the CEC of the soil to the severity of Fusarium wilt disease or population of *Verticillium dahliae* populations in spinach. The suppression of Fusarium wilt through iron acquisition by the siderophore-producing *Pseudomonas* or the synthetic chelating agent ethylenediamine-N,N′-bis (2-hydroxyphenylacetic acid) (EDDHA) is more successful in soils containing smectite clays characterized by high CEC (Ferret et al., 2014). In addition, high CEC values also decrease the micronutrients availability to the pathogens, thereby reducing their uptake (Orr and Nelson, 2018).

11.4.1.5 Redox Potential

The redox potential is a measure of the oxidation reduction potential (Eh) of the soil. Aerobic soils have an oxidizing environment while anaerobic soils have a reducing environment (Husson et al., 2016). Soils conducive for disease incidence are oxidative compared to suppressive soils, which are reductive in nature. For instance, prevalence of Fusarium wilt in watermelon was correlated positively to soil Eh (Cao et al., 2016), while anaerobic soil disinfection increases suppressiveness against various pathogens (Rosskopf et al., 2015). Subjecting soils to reductive conditions increases suppressiveness against different pathogens such as *Colletotrichum*, *Cylindrocarpon*, *Fusarium*, *Ralstonia*, *Verticillium*, *Rhizoctonia*, *Pythium*, and *Sclerotinia* (Rosskopf et al., 2015). Changes in the composition of microbial communities in response to a drop in Eh under reductive conditions appear to be the main mechanism behind Eh-mediated soil suppressiveness. For instance, anaerobic conditions during the reductive phase in anaerobic soil disinfestation process favor the dominance of bacterial species in phylum Firmicutes such as the *Bacillus* and the obligately anaerobic *Clostridium* spp. that produce organic acids such as the acetic acid and butyric acid that are known to effectively suppress pathogens such as *Fusarium oxysporum* f. sp. *lycopersici* (Momma, 2015). Moreover, *Clostridium* bacteria also produce chitosanase and β-1,3-glucanase that can act against the fungal pathogens (Ueki et al., 2019).

11.4.1.6 Soil Texture and Clay Content

Soil texture plays a pivotal role in disease development. For instance, *Rhizoctonia solani* induced root rot of wheat in glasshouse conditions was more prevalent in sandy soil than in sandy clay loam or loamy sand soils (Gill et al., 2000). Similarly, the prevalence of Fusarium wilt in tomato, watermelon, and marigold was more in sandy soils when compared to silt clay soils. As soil suppressive indices were significantly and linearly correlated to the contents of clay and silt in *Pythium*-infested soils of Europe, the suppressive indices were significantly and negatively correlated to sand content in these sites (Bongiorno et al., 2019). Decreased disease suppressiveness to the Panama wilt of banana was evident with increasing silt and sand content in the soil as evidenced by the presence of an inverse Spearman's rank correlation of these soil textural variables to disease suppressiveness (Deltour et al., 2017). In contrast to these observations, seven of the 19 sandy arable soils examined from the Pernambuco state of Brazil were found to be suppressive to *Fusarium solani* causing root rot disease in cassava (Barros et al., 2014). Similarly, the incidence of *Fusarium* wilt in fields that were under continuous commercial strawberry monoculture was lower in soils containing a higher percentage of sand (>70%) and a lower percentage of silt (<10%) compared to those containing a comparatively low percentage of sand (<60%) and a high percentage of silt (>10%) (Cha et al., 2016). The effect of soil texture on soil-borne pathogens is ascribed to its effect on the water-holding potentiality and low nutrient content of these soils (Barros et al., 2014).

Increasing the clay content of the soil improves disease suppressiveness. The positive influence of soil clay content on disease suppressiveness is an indirect effect as fine-textured clay soils have more water and nutrient retaining capacities than sandy soils and are thus more conducive for plant and microbial growth. In a study, the relative suppressiveness of six important Australian soil types on the pathogen *Fusarium oxysporum* f.sp. *cubense* causing wilt disease in banana was examined. Bowen et al. (2019) found that the severity of the disease was inversely related to the soil clay content. Nevertheless, this suppressive effect of soil clay content can vary with the phytopathogen involved. For example, the severity of root rot caused by *Pythium myriotylum* in *Xanthosoma sagittifolium was correlated positively to the clay content of the* ferralsol *soils* (Adiobo et al., 2007). Also, an increased incidence of Pythium damping-off in sugar beet was reported with an increasing clay content of the soil (Knudsen et al., 2002). *This positive effect of clay content on the increased Pythium* disease severity was attributed to the increased moisture-holding ability of the clay soils, as diseases caused by *Pythium* are very common in wet soils (Bickel and Koehler, 2021). The incidence of root rot diseases and Fusarium wilt in chickpea during various phases of plant growth in Ethiopia was related to the clay content of the vertisol soils (Bekele et al., 2021). Similarly, the type of clay can influence disease suppression through its effect on microbial activity. For instance, the smectite type of clay increases soil disease suppressiveness through the increased bacterial activity as evidenced through increased respiration (Mueller, 2015). Suppression of Aphanomyces root rot was significantly correlated to the ratio of illite–kaolinite and vermiculite–smectite ratio in the soil (Persson and Olsson, 2000). The role of clay on suppression of soil-borne fungal pathogens involves interactions with other soil factors such as pH, redox potential, and bioavailability of the micronutrients (Orr and Nelson, 2018).

11.4.1.7 Organic Matter

Several studies have found increased suppressiveness in soils that were under organic management than those under conventional farming (Bongiorno et al., 2019). For instance, soils conducive to Fusarium wilt of banana were distinguished by high organic carbon content of the soils (Deltour et al., 2017). Different fractions of soil organic carbon (SOC) in addition to total organic cabon can play an important role in soil suppressiveness. In a bioassay study, growth of *Fusarium culmorum,* the causative agent for several cereal and non-cereal crop diseases, was inversely correlated to SOC and its fractions such as particulate organic matter (POM) C and microbial biomass C (Palojärvi et al., 2020). This negative influence of SOC fractions on *F. culmorum* translated into a positive fungistatic activity with the strongest and weakest influence being conferred by POM-C fraction and SOC, respectively (Palojärvi et al., 2020). Among the different organic C fractions in the soil, the labile fractions and their characteristics were specifically related to soil suppressiveness (Dignam et al., 2018). The labile C, a fraction of the total organic C, serves as an energy source to microorganisms and is often correlated to the abundance and activity of microorganisms in the soil (Haynes, 2005). The labile C fraction therfore has attracted increasing attention in recent years as an unique indicator of soil quality and often linked to suppressiveness to plant diseases (Bongiorno et al., 2019). Nevertheless, the actual mechanisms untangling the relationship of labile organic

C fractions to microbial activity and their biomass and soil suppressiveness are yet to be workedout. But, based on avialble information, it could be suggested that the positive influence of labile carbon on soil suppressiveness could be ascribed to the activity of soil microbial biota on the activity and incidence of pathogen antagonist. Some aspects of organic-C-mediated soil suppressiveness that needs consideration in organic amended soils are the great variation in the activity based on the quality of the organic amendments, frequency of application, and the disappearance of the suppression with time (Bonanomi et al., 2018).

11.4.1.8 Soil Nutrient Content

Studies have shown that both macro- and micronutrient contents of soils could influence soil suppressiveness. For instance, soils that were free from banana Fusarium wilt exhibited higher concentrations of P, Ca, and Mg than soils that were conducive for the disease (Furtado et al., 2009). Fe, a crucial nutrient for fungal growth and the aggressiveness of pathogenic fungi such as *Fusarium graminearum*, is associated with effective Fe uptake through siderophore production (Comensoli et al., 2017). Therefore, competition for the acquisition of Fe through siderophore production is a strategy adopted by antagonistic microorganisms in suppressing the activities of *Fusarium* pathogens (Legrand et al., 2019). This is evident in the case of *Pseudomonas* spp., with high siderophore-producing ability (Arya et al., 2018). Similarly, Mn content in soils of France affected the growth of *F. graminearum* (Legrand et al., 2019). Soil Mn content is a crucial factor determining the composition of microbial communities in soils under different crop species (Marschner et al., 2003). Although the fungistatic effect of Mn is not clear, it is presumed that the reduced availability of this element might affect the production and activity of Mn peroxidase enzymes that are involved in the hydrolysis of lignin substrates (Li et al., 2008). A negative correlation was recorded between the severity of root rot disease caused by *Pythium myriotylum* in cocoyam (*Xanthosoma sagittifolium*) and Ca, K, Mg, and N content of Cameroon soil (Adiobo et al., 2007). Contrarily, high soil N content decreases the soil suppressiveness to Fusarium wilt disease in banana (Deltour et al., 2017). The incidence of Fusarium wilt in strawberry was lower in soils containing more Cu and Zn than in soils containing more amounts of C and N (Cha et al., 2016). Among the major plant nutrients, application of K was shown to suppress the development of *Fusarium oxysporum* root rot in several crops (Cai et al., 2020). Increased K status in plant tissue was correlated with increased phenolic contents that stimulated plant resistance and suppressed sporulation and conidial germination of some plant pathogens (Elmer and Datnoff, 2014). In addition to these, the interaction between nutrients and their ratios influence the disease suppressiveness in the soils (Krištůfek et al., 2015).

11.4.2 Biotic Factors

11.4.2.1 Microbial Interactions

The abundance, quality, diversity, and community composition of soil microbiome are important soil quality indicators that play significant roles in nutrient cycling, plant growth, and soil health. The composition of bacterial and fungal communities in suppressive soils is significantly different from those of conducive soils as revealed by culture-dependent and culture-independent approaches. Suppression of plant diseases is frequently related to the biomass and activity of the entire soil microbial biome (Bongiorno et al., 2019) or to specific microbial groups. For example, the diversity of bacterial communities in the soil was negatively correlated to both abundance of *Ralstonia solanacearum* and occurrence of wilt disease incidence in tomato (Liu et al., 2015). Contrarily, suppression to *Pythium*, the causative agent for root rot and damping-off, was related to the higher abundance of Mucoromycotean (*Mortierella*) or Ascomycotean (*Acremonium*, *Chaetomium*, *Humicola*) fungi (Ros et al., 2020). Nevertheless, populations and diversity of soil microbiome are highly dynamic as it changes with space and time. Taxa in bacterial families Clostridiaceae, Lachnospiraceae, and Ruminococcaceae associated with soil suppressiveness in reductive soils produce toxic compounds that antagonize phytopathogens (Huang et al., 2019). Similarly, the relative abundances of bacteria Sphingobacteriales and fungi Sordariales are also higher in reductive soils when compared to conducive soils and are inversely related to disease prevalence (Liu et al., 2019a). Moreover, the abundances of ammonia monooxygenase and nitrogenase genes involved in nitrification tend to be higher in DSSs (Poret-Peterson et al., 2020).

11.4.2.2 Endophytic Fungi and Bacteria

Besides soil microbiome, root endophytic microbiome has a pivotal function in controlling populations of pathogens in the soil. For example, the arbuscular mycorrhizal (AM) fungi associating with a wide range of crop species are shown to reduce the incidence and intensity of diseases in plants. The AM fungi reduce disease severity owing to its direct effect on plant's induced systemic resistance or indirectly through changes in soil rhizosphere microbiome (Chen et al., 2018). Like AM fungi, another group of widely root associating dark septate endophytic fungi associating with a wide range of AM and non-AM plants are also shown to involve in plant disease control (Harsonowati et al., 2020). Endophytic fungi modulate soil microbiome through their influence in the quantity and quality of plant root exudates. Like endophytic fungi, bacteria residing in endorhizosphere are also known to defend against the development of soil-borne diseases in plants. Several researchers have demonstrated the control of wilt pathogens (*Fusarium* and *Verticillium*) by endophytic bacteria in few crop plants (Thangavelu and Gopi, 2015; Ben Abdallah et al., 2016). In addition, endophytic bacteria belonging to *Pseudomonas*, *Burkholderia*, *Bacillus*, *Brevibacillus* are also known to suppress damping-off and root rot, scab, Crown gall, and nematode infestation in a number of crop species (Latha et al., 2019).

11.4.2.3 Induced Systemic Resistance

The induced systemic resistance (ISR) is the enhanced state of the resistance mechanism (chemical or physical) in plants in response to an infection or presence of an elicitor (Bakker et al., 2013). It is well known that several biotic and abiotic agents can protect plants against soil-borne pathogens. Biocontrol bacteria (*Bacillus*, *Pseudomonas*, *Serratia*) *and* fungi (*Trichoderma*, *Piriformospora*, *Penicillium*, *Phoma*), non-pathogenic strains of pathogens (*Fusarium* sp.), and the symbiotic AM fungi elicit ISR response in plants. Likewise, microbial determinants involved in

ISR include antibiotics, biosurfactants, lipopolysaccharides, Fe-regulated metabolites, siderophores, and volatile organic compounds (VOCs) (Tamm et al., 2011). In a recent study, *Bacillus* sp. 2P2, an endophytic bacterium, elicited ISR in tomato plants when challenged with the collar rot pathogen *Sclerotium rolfsii* (Sahu et al., 2019). *The ISR in tomato plants resulted in an increased level of* ascorbate oxidase, ammonia lyase, phenylalanine, peroxidase, and polyphenol oxidase activities. Moreover, cell wall strengthening occurred *via* lipid peroxidation, lignification, and suberization in addition to the upregulation of glucanases and chitinases genes *that are* involved in pathogen suppression (Sahu et al., 2019).

11.5 MECHANISMS INVOLVED IN DSS

11.5.1 Antibiosis

Antibiosis is one of the most investigated mechanisms in DSS. It involves the production of specific or non-specific metabolites such as the antibiotics, VOCs, or other toxic substances by an organism to inhibit the growth and/or activity of another organism (Thomashow and Pierson, 1991). Among the different types of antibiotics produced by soil microorganisms, the role of phenazines (PHZ) and 2,4-diacetylphloroglucinol (DAPG) produced by several strains of fluorescent *Pseudomonas* has been investigated in depth (Raaijmakers and Mazzola, 2012). Antibiotics produced by fluorescent *Pseudomonas* are involved with the suppression in the incidence of Fusarium wilt and TADs in different types of soils (Biessy et al., 2021). Moreover, other antibiotics such as pyrrolnitrin or pyoluteorin along with DAPG or PHZ were associated with the suppression of *Rhizoctonia solani* (Latz et al., 2012) and *Thielaviopsis basicola* (Haas and Defago, 2005).

Soil microorganisms, apart from producing antibiotics and lantipeptides, produce VOCs belonging to various chemical groups such as alcohols, alkenes, benzenoids, ketones, terpenes, sulfides, and pyrazines (Schulz-Bohm et al., 2017). Many of these VOCs possess antimicrobial properties and are involved in the DSS. Initial studies suggested the role of hydrogen cyanide and ammonia in DSS (Voisard et al., 1989), and since then, there is mounting evidence that an array of VOCs can inhibit the development and establishment of phytopathogens. Species of *Streptomyces* inhibit the growth of *Rhizoctonia* hyphae thorough the production of VOCs, specifically 1,3,5-trichloro-2-methoxybenzene and methyl 2-methylpentanoate (Cordovez et al., 2015). *Paraburkholderia graminis* (Burkholderiaceae) uses sulfurous volatile compounds such as dimethyl di- and trisulfide encoded by *dsr*-gene (dimethyl sulfoxide reductase gene) and *cd*-gene (cysteine desulfurase gene) to inhibit the growth of *Rhizoctonia solani* (Carrión et al., 2018). Nevertheless, the sensitivity to VOCs tends to vary with the phytopathogens involved. For example, Van Agtmaal et al. (2018) suggested that *R. solani* and *Pythium ultimum* are mostly susceptible to volatile mediated suppression when compared to *F. oxysporum*. Likewise, VOCs diffused by *Bacillus amyloliquefaciens* sub sp. *plantarum* strain showed a strong and significant growth inhibition against *Alternaria brassicicola*, *Botrytis cinerea*, *Alternaria brassicae*, and *Sclerotinia sclerotiorum, but not against Verticillum longisporum* (Asari et al., 2016). A comparative study on the modifications in gene expression during *Trichoderma gamsii–R. solani/Pythium irregulare* interaction suggested the involvement cysteine-rich proteins encoded by genes associated with antibiosis (Zhang et al., 2015).

11.5.2 Competition

Competition among microorganisms could be either indirect through the exploitation of resources or direct through causing damage to one another (interference) (Ghoul and Mitri, 2016). A key mechanism operating in DSS includes resource competition by non-pathogenic strains of the pathogens. For instance, competition for Fe through siderophore production by Pseudomonas bacteria and utilization of carbon by non-pathogenic Fusarium isolates constitute the vital mechanisms in soils suppressive to Fusarium wilt (Sajeena et al., 2020).

11.5.3 Predation

In soils, microorganisms face several biotic interactions that constrain/affect their fitness. Free-living soil protists are one among the main modulators of rhizosphere microbial activities and can enhance the abundance and activity of phytopathogen-suppressive microbes through selective predation (Gao et al., 2019). Investigations carried out so far revealed that soil protists can improve soil function and plant health (Amacker et al., 2020). The protists' presence induces the biosynthesis of the antimicrobial lipopeptidic surfactants such as viscosin and massetolide that act against plant pathogens (Mazzola et al., 2009). Further, the extracellular protease secreted by *Pseudomonas fluorescens* to prevent predation also inhibits plant-parasitic nematodes (Siddiqui et al., 2005). Similarly, introduction of Amoebozoa protists such as *Naegleria* spp., and *Cercomonas* spp. into soil has been shown to support bacteria with high biocontrol activity against fungi through exoprotease production (Amacker et al., 2020). The capacity of bacterial strains to hinder fungal pathogens and *Pythium ultimum* and *Fusarium* firmly overlaps with their capability to inhibit the predatory protist isolates (Bahroun et al., 2021). For instance, the prevalence of protists induced antifungal activity of *Pseudomonas yamanorum* B12 against *Fusarium* despite its reduced abundance (Bahroun et al., 2021).

11.5.4 Parasitism

The involvement of parasitism has been investigated in soils suppressive to pathogens such as *Sclerotinia*, *Rhizoctonia*, *Fusarium*, *Cochliobolus,* and *Pythium* (Gómez Expósito et al., 2017). Most of the microbes occurring in the parasitic suppression of phytopathogens are either oomycetes such as *Pythium oligandrum* or fungi such as *Trichoderma*, *Coniothyrium*, and *Verticillium*. These parasitic microorganisms produce a diverse range of mycolytic enzymes that deter the growth of pathogenic fungi. For example, the mycolytic enzymes chitinase and β-1,3, glucanase produced by *Trichoderma asperellum* isolated from the *Fusarium* suppressive soils of banana were involved in the antifungal activity against different species of phytopathogenic *Fusarium* (Win et al., 2021). Similarly, β-1, 3-glucanase is shown to be involved in the penetration of *Sclerotinia sclerotiorum sclerotial cell wall by Coniothyrium minitans* (Giczey et al., 2001).

In spite of widespread occurrence of soil bacteria with parasitic potentials, such as production of enzymes capable of degrading fungal cell walls, there are no specific studies to demonstrate their involvement in suppressive soils. For instance, proteases produced by isolates of *Stenotrophomonas maltophilia* and *Pseudomonas fluorescens* were shown to be involved in the suppression of the oomycete's pathogen (*Pythium ultimum*) and plant pathogenic nematodes (*Bursaphelenchus xylophilus*, *Meloidogyne incognita*) (Siddiqui et al., 2005; Huang et al., 2009). Moreover, the bacterial genus *Collimonas* can produce chitinases that could parasitize fungi. In a recent study, the *Bacillus* sp. NS-22 and *Pseudomonas* sp. NS-1 that are capable of producing the chitinases induced damages in the hyphal cells of the pigeon pea wilt disease pathogen *Fusarium udum* (Dukare et al., 2020). However, the role of most mycoparasitic rhizobacteria in disease suppressive soils is yet to be fully ascertained.

11.6 DIFFERENT APPROACHES IN EVALUATING DISEASE SUPPRESSIVE SOIL

11.6.1 Classical Approaches

There are well-established research studies into understanding the organisms and the genes that are fundamental for suppressing disease occurrence in soil. In many cases, the soil suppressiveness to plant diseases could be traced back to the antagonistic activities of the soil microbiota against pathogens. There are several approaches to unravel the mechanisms associated with DSSs (Toyota and Shirai, 2018). A large number of studies have demonstrated that the microbial basis for DSSs by using biocides, soil pasteurization, and/or soil transplantation after this phenomenon was first addressed by Henry in 1931 (Henry, 1931a,b; Gómez Expósito et al., 2017). Following this, studies have targeted large-scale isolation of microorganisms from the roots and root zone soils of plants growing in DSS and screened them for their activities against selected target pathogens in plate assays or using soils that were conducive for disease development. However, several microorganisms that significantly controlled the pathogens under controlled environment often failed to do so at the field level (Gómez Expósito et al., 2017). This variation in pathogen controlling activity of microorganisms under field conditions was ascribed to their inability to survive and establish in the introduced soil or to manifest their activity at the right place and right time (Alabouvette et al., 2009).

11.6.2 Modern Approaches

In recent times, culture-independent approaches are used to understand the diversity and activities of microorganisms functioning in DSSs. The technologies used are, for example, metataxoneme, metagenome, metatranscriptome, metaproteome, and metabolome analyses (Heyer et al., 2019; Shakya et al., 2019). Culture-independent methods to study the soil microbial biota on disease suppressiveness involve extraction of DNA directly from the soil or indirectly from the bacterial cells (Kowalchuk et al., 2005). Therefore, these methods are termed as direct and indirect methods of DNA extraction. The direct extraction of DNA from soil yields high-molecular-weight undamaged clonable DNA that is useful for metagenomics approaches (Van Elsas et al., 2008a). Efficient screening follows cloning to select and further study interested clones. Suitably, the target traits are identified though functional screening in the metagenomics host. Nevertheless, such functional screening can be unsuccessful if the metagenomics host fails to recognize the expression signals on the cloned fragments (Courtois et al., 2003). To overcome this shortfall, rapid genetic screening of the metagenomics library for the occurrence of specific gene regions against the probe comparator regions based on homology is used. However, the success of soil metagenomics approach depends on the efficacy of all the steps starting from the extraction of the soil DNA up to the cloning and screening process (Van Elsas et al., 2008b). Most of the earlier soil metagenomics approaches had limited success rates pertaining to the inconsistent actual hit rates compared to expected target gene numbers.

One of the approaches to understand DSS is to identify pathways that produce substances with antimicrobial properties. Van Elsas et al. (2008b) proposed a metagenomics protocol for unrevealing the antibiotic biosynthetic genes that were implicated in the soil suppressiveness to *Rhizoctonia solani*. This approach revealed many gene inserts (~40-kb size) that were involved in potential antibiotic production in controlling the pathogens. Similarly, Van Elsas et al. (2008a) reviewed the experiences from METACONTROL project on DSS and highlighted the challenges faced by the metagenomics approach in understanding the mechanism involved in it. However, advancement in throughput sequencing technologies over the years has resulted in quick evaluation of the target genes throwing more light on soil genetic diversity and their potential applications.

Peptides generated by the modular enzymes (NRPS: Non-ribosomal peptide synthetases) are one of the crucial group of natural substances that are associated in disease suppression. The functional amplicon sequencing of NRPS-related adenylation domains is used to soils that were either suppressive or conducive to the wheat root pathogen *Fusarium culmorum* (Tracanna et al., 2021). The results of the study indicated that NRPS repertoires were similar among suppressive soils, especially the several gene clusters involved in siderophore biosynthesis. Finally, 10× metagenomic sequencing approach was performed to validate functional metagenomics unbiasedly resulting in the recognition of several gene clusters related to disease suppressiveness (Tracanna et al., 2021). Similarly, bacterial taxa with NRPS genes such as those in Actinobacteria (*Streptomyces*) and Proteobacteria (*Pseudomonas*) were correlated with Fusarium wilt disease suppression in banana (Zhao et al., 2018; Liu et al., 2019b).

In a metagenomics study, different types of functional genes associated with the synthesis of antibiotics, antifungal compounds, and siderophores in addition to chemotaxis were examined by Enebe and Babalola (2021) under organic and inorganic fertilization in the rhizosphere of maize. The findings of the study revealed that the expressions of all the examined functional genes were upregulated under organic amendments and were downregulated in inorganic fertilization, suggesting the importance of organic manuring on the abundance and expression of disease-suppressive genes in the soil (Enebe and Babalola, 2021). Cordovez et al. (2015)

used indirect approach to characterize and delineate VOC producing *Streptomyces* isolates involved in the *Rhizoctonia*-suppressive soil by sequencing 16S rRNA and the house-keeping genes *recA* and *atpD*.

An examination into natural biological suppression of *Rhizoctonia solani* AG 8 by 454 pyrosequencing targeting the fungal 28S LSU rRNA gene in Australia revealed considerable differences between the fungal group composition in suppressive and unsuppressive soils (Penton et al., 2014). Soils conducive to *Rhizoctonia solani* were dominated by fungi belonging to *Penicillum*, *Alternaria*, and *Gibberella*, while those suppressive to the pathogen were characterized by fungal genera that either mycoparasites (*Bionecteria*, *Didymostilble*, *Hypocera*, *Peethambara*, *Xylaria*), biocontrol agents (*Ceratobasidium*, *Davidiella*, *Eupenicillium*, *Lobulomyces*, *Penicillium*), producing antifungal compounds (*Cercospora*, *Chaetomium*, *Podospora*), or fungicidal activity (*Kurtzmanomyces*, *Subramaniomyces*, *Chaetomium*, *Corynascus*) (Penton et al., 2014).

11.7 FARM MANAGEMENT PRACTICES ON DISEASE SUPPRESSION IN SOIL

Although the underlying mechanism of farm management practices in disease suppression has not been fully understood, there are several studies that indicate there are some relationships between them. The possibility of improving the disease suppressive ability of soil through farm management practices would offer an affordable and eco-friendly approach to resilient crop production (Figure 11.3). Application of organic amendments not only improves the soil physical and chemical properties but also manipulates the biological factors that create several pathways for improved

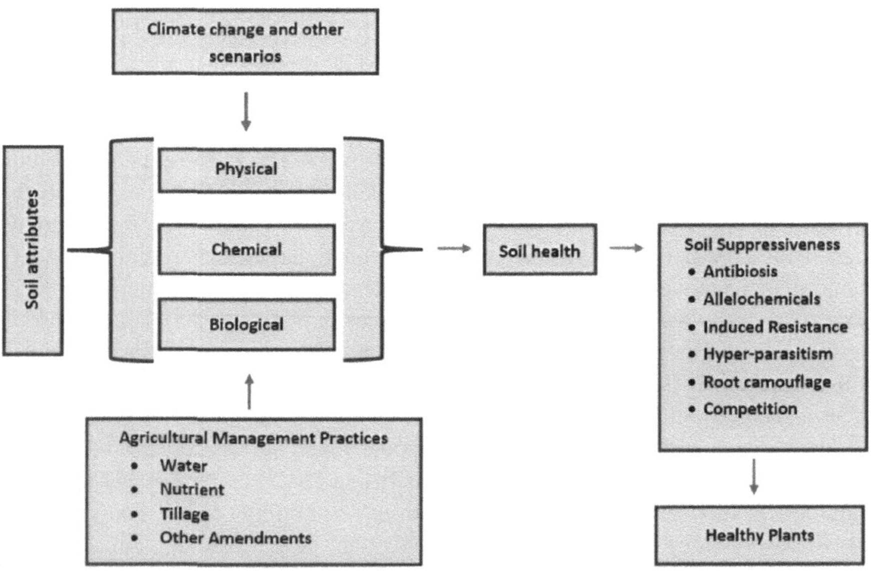

FIGURE 11.3 Biotic and abiotic factors on disease suppressiveness in soil (Source: Authors developed).

disease suppression (Jayaraman et al., 2021). Organic amendments such as compost, vermicompost, and farmyard manure improve the plant's defense system (Yogev et al., 2010). A well-prepared and good quality manure can deliver multifaceted plant growth promoting microorganisms to the plants, thus improving both disease resistance and plant growth. Therefore, it is often recommended to apply manure rich in plant growth promoting bacteria (PGPB) than applying a single inoculant (Barnett et al., 2006). Organic manure harbors specific biocontrol agents such as *Trichoderma hamatum* (Hadar and Papadopoulou, 2012) and has been proven by comparing the effect of sterilized compost and unsterilized compost on disease development on crops (Tilston et al., 2002). El-Masry et al. (2002) reported the disease-suppressive quality of water extracts from composts, although there were no significant levels of antibiotics and siderophores in the water extract. Such findings indicate the presence and contribution of beneficial microorganisms in organic amendments.

Owing to the sole crop sequence, intensive tillage, and no crop rotation, there has been increased incidence of plant diseases, which are mostly caused by necrotrophic soil-borne pathogens (Cha et al., 2016). Most of the crops lack genetic resistance to these necrotrophic pathogens (Cha et al., 2016). The shift from conventional tillage to conservation tillage has also gained the attention of researchers to evaluate the benefits of conservation tillage on disease suppression, which ultimately affects production costs. Reduction in severity of plant diseases allows lesser applications of pesticides and reduces production costs. There are several studies that have addressed the significant effects of conservation tillage on disease suppression (Johnson et al., 2001). Monfort et al. (2004) conducted a field experiment to test the effects of different tillage practices on disease suppression of early leaf spot (*Cercospora arachidicola*) in peanut. Reduced tillage could show higher suppression of leaf spot as compared to conventional tillage. Culbreath et al. (1999) and Johnson et al. (2001) had shown reduced incidence of spot wilt of peanuts in reduced tillage plots. However, southern stem rot in peanuts, caused by *Sclerotium rolfsii*, cannot be reduced with conservation tillage (Ferguson and Shew, 2001). Tillage tends to affect the distribution and diversity of soil microbial communities. This is due to the changes in carbon allocation (substrate availability), soil moisture, and soil temperature conditions with the change in tillage intensities. Sharma-Poudyal et al. (2017) suggested that tillage practices can have profound impact on some soil fungal communities. Conservation tillage promotes the development of beneficial fungi and offers an advantage in competing with pathogens in colonizing the roots of the crops. On the other hand, intensive tillage disrupts the fungal population and the hyphal growth is negatively impacted. Conservation tillage with crop residue retention and crop rotation increase soil organic matter content and are shown to increase soil microbial biomass quantity and improve disease suppressive ability of soil (Janvier et al., 2007). Long-term conservation tillage had shown to improve disease suppressiveness of soil as compared to intensive tillage (Palojärvi et al., 2020). The reasons for improved disease suppression in conservation tillage plots are improved labile carbon content and increased soil microbial biomass. The suppression of disease in conservation tillage plots can be depth-dependent. Sipilä et al. (2012) highlighted that due to more soil microbial biomass in surface layers, strong disease suppression can be expected on surface layers in conservation tillage plots. Most of studies on disease suppression in conservation tillage had indicated the strong relation between soil disease suppressiveness and labile carbon (Bongiorno et al., 2019). This is attributed to the fact that an active soil microbiome requires labile carbon and is essential for the function of disease suppression.

The difference in disease resistance ability of one crop cultivar to another influences the occurrence and diversity of rhizospheric microorganisms (Kobayashi et al., 2015). This is one of the reasons why crop rotation is a general practice followed to improve disease suppressive ability of the soils. The presence of non-host crops in crop rotation breaks the chain of disease development and leads to better disease suppression in crop rotation. The composition of soil microbial community might differ during the different stages of plant growth due to release of different types of rhizodeposits during different plant growth stages. This is related to the incorporation of intercrops in farm management practices involved in disease suppression.

11.8 CHALLENGES AND COMPLEXITIES IN DISEASE SUPPRESSION

The natural disease suppression in the soil involves a diverse community of soil microorganisms that are highly complex (Ramette et al., 2006). Although there are several reports of disease suppression, owing to its complexities, there are several unanswered questions/issues in DSS. The complexity starts from the interaction between the host plant, pathogen, and the environmental conditions. This disease triangle is yet fully not explored to understand the disease suppression because there are several soil parameters/properties, climatic conditions, complex behavior of the plants, and the pathogens. The difficulty in identifying the actual causal agent of disease suppression is one of the old issues every researcher has been addressing. The soil system is one of the largest reservoirs of biological diversity, and manipulating a certain section of microbial community to improve disease suppression is a hard task. Presence of certain beneficial microbes in the rhizosphere does not necessarily mean the development of disease suppression. Moreover, the nature of the soil-related disease suppressiveness is highly unpredictable because of the complex interactions of different pathogenic and beneficial microbes and the heterogeneities of soil and weather at the field, landscape, and regional levels. Moreover, application of organic amendments to improve disease suppression must be studied further. It is so because the organic amendments prepared from the same base material may differ in disease-suppressive potential.

11.9 PERSPECTIVES OF DISEASE-SUPPRESSIVE SOIL MICROBIOMES

Since 1970s, the soil microbiome has been explored to fight the crop diseases caused by several soil-borne pathogens. Soil disease suppression involves both biotic and abiotic factors, although there are more studies on microbial-driven disease suppression. However, environmental conditions can be manipulated to improve disease-suppressive ability. Therefore, soil physical and chemical parameters are studied in predicting the disease-suppressive ability of the soil (Löbmann et al., 2016). The environmental conditions alter the strength and the direction of plant–soil feedbacks and influence the disease suppression (Mavrodi et al., 2012).

Similarly, soil properties have profound influence on disease suppression. In fact, changes in clay content, minerals, CEC, soil pH, soil drainage have greater influence on disease suppression in soils (Martin and Hancock, 1986; Tamm et al., 2010). Manipulating soil pH can help in restricting the development of soil pathogens. For example, acidic soils restrict motility, sporangium formation, and zoospore release of *Phythopthora* and therefore reduce the severity of root rot diseases (Jambhulkar et al., 2015). On the other hand, higher soil pH restricts *Fusarium* wilt. In addition, the soil nutrient status is one of the main contributing factors in development of DSS. High soil nutrient status has been associated with higher disease suppressiveness (Tamm et al., 2010). Thus, altering the content of soil nutrient to offer nutrient stress to microbial community in soil and to suppress the spore germination and growth of the pathogen is called as "microbiostasis" or "fungistasis" (Jambhulkar et al., 2015). Further, low nutrient status in the soil leads to improper nutrition of the crops and thus increases the susceptibility of the plant to several soil-borne diseases. Disease suppression is generally low in arable lands compared to natural ecosystem. Therefore, land use system, crop cultivars, organic manure application, and management practices have strong influence on disease suppressiveness in soils. Moreover, farming practices "mimicking" natural ecosystem always have a greater potential in disease suppressiveness (e.g., grasslands have been recognized as "preserver of microbial diversity").

One of the future perspectives in DSS studies is to develop crop cultivars that allow selective microbial communities in the rhizosphere, so that the plant could increase its resistance toward disease incidence. The answer to how some crop monoculture enhances soil disease suppressiveness needs further investigation. The complexities involved in soil disease suppressiveness must be addressed, taking soil, plant, and environment as contributing factors with advanced molecular assessment tools to reduce biasness and error. Besides, it is imperative to focus on factors influencing disease suppressiveness in order to understand the role of different factors and to explore possibilities of modifying soil conditions that resist disease development in crops leading to increased productivity.

11.10 SUMMARY AND CONCLUSIONS

DSS is one of the interesting subjects and attracts greater attention of researchers worldwide. It can be regarded as good indicator of soil biological property conferred by the microbial community (microbiome); however, the complexities of soil microbial community interaction in DSS are always fascinating and challenging. Some of the salient observations and conclusions are listed hereunder:

- Different mechanisms, namely general and specific suppressions, are helpful in understanding the disease suppression in soil. Besides, both biotic and abiotic factors also greatly govern the DSS.
- DSSs have the ability to suppress the various plant pathogens by improving the plant health, inducing natural plant defense, producing antibiotics, competing against pathogens, modulating plant immune system, or hyper-parasitizing the pathogen.

- These microorganisms not only suppress the disease-causing pathogens by different mechanisms associated with disease suppression but also promote plant growth through improved nutrient uptake and production of phytohormones.
- VOCs produced by microorganisms possessing antimicrobial properties need in-depth understanding in the DSS.
- The understanding of disease-suppressive mechanisms is a crucial step to enhance the suppressive effect by manipulation of the soil microbiota.
- Although soil disease suppressiveness is closely linked with the microbiota/microbial community, a few questions remain unanswered with respect to interaction and compatibility of the soil microbes on disease suppression. How monoculture crop can augment the population of 2, 4–DAPG producers remains unclear? Future biological studies on soil suppressiveness need to generate new insights into the microbial soil ecology, which can serve as a foundation for the development of cost-effective and eco-friendly management strategies for disease suppression.
- Besides, the complex interactions between two main drivers of DSS (i.e., abiotic and biotic factors) and their variations need in-depth study and understanding. Yet, it is a challenge to further develop molecular assessment tools (including specialized DNA microarrays and metagenomics) to unravel the antagonistic behavior of soil toward soil-borne pathogens.
- However, recent advancement in NGS/HTS technologies over the years has resulted in quick evaluation of the target genes throwing more light on soil genetic diversity and their potential applications in DSS. Indeed, DSS is a potential option requiring in-depth knowledge of the complexity of the biochemical and ecological interactions between the host plant and the pathogens.
- Therefore, more detailed studies on molecular and culture-independent approaches/techniques are essential to decode the basic mechanisms of DSS for harnessing the benefits of "soil microbiome."

REFERENCES

Adiobo, A.; Oumar, O.; Perneel, M. et al. Variation of Pythium-induced cocoyam root rot severity in response to soil type. *Soil Biology & Biochemistry*, **2007**, *39*(11), 2915–2925.

Alabouvette, C.; Olivain, C.; Migheli, Q. et al. Microbiological control of soil-borne phytopathogenic fungi with special emphasis on wilt inducing *Fusarium oxysporum*. *New Phytologist*, **2009**, *184*(3), 529–544.

Amacker, N.; Gao, Z.; Agaras, B.C. et al. Biocontrol traits correlate with resistance to predation by protists in soil pseudomonads. *Frontiers in Microbiology*, **2020**, *15*(11), 3164.

Andrade, O.; Mathre, D.E.; Sands, D.C. Suppression of *Gaeumannomyces graminis* var. *tritici* in montana soils and its transferability between soils. *Soil Biology & Biochemistry*, **1994,** *26*, 397–402.

Arya, N.; Rana, A.; Rajwar, A. et al. Biocontrol efficacy of siderophore producing indigenous *Pseudomonas* strains against *Fusarium* wilt in tomato. *National Academy Science Letters*, **2018**, *41*(3), 133–136.

Asari, S.; Matzén, S.; Petersen, M.A. et al. Multiple effects of *Bacillus amyloliquefaciens* volatile compounds: plant growth promotion and growth inhibition of phytopathogens. *FEMS Microbiology Ecology*, **2016**, *92*(6), fiw070.

Avilés, M.; Borrero, C. Identifying characteristics of Verticillium wilt suppressiveness in olive mill composts. *Plant Diseases*, **2017**, *101*(9), 1568–1577.

Bagyaraj, D.J. Studies on the biological control of take all of wheat. In: Veeresh, G.K. (Ed), *Progress in Soil Biology and Ecology in India*. UAS Press: Bangalore, India, **1981**; pp. 20–24.

Bahroun, A.; Jousset, A.; Mrabet, M. et al. Protists modulate Fusarium root rot suppression by beneficial bacteria. *Applied Soil Ecology*, **2021**, *168*, 104158.

Bai, Z.; Caspari, T.; Gonzalez, M.R. et al. Effects of agricultural management practices on soil quality: a review of long-term experiments for Europe and China. *Agriculture, Ecosystems & Environment*, **2018**, *265*, 1–7.

Bakker, P.A.; Doornbos, R.F.; Zamioudis, C. et al. Induced systemic resistance and the rhizosphere microbiome. *Plant Pathology Journal*, **2013**, *29*(2), 136–143.

Barnett, S.J.; Roget, D.K.; Ryder, M.H. Suppression of Rhizoctonia solani AG-8 induced disease on wheat by the interaction between Pantoea, Exiguobacterium, and Microbacteria. *Australian Journal of Soil Research*, **2006,** *44*, 331–342.

Barros, J.A.; Medeiros, E.V.; Notaro, K.A. et al. Different cover promote sandy soil suppressiveness to root rot disease of cassava caused by *Fusarium solani*. *African Journal of Microbiology Research*, **2014**, *8*(10), 967–973.

Bekele, D.; Tesfaye, K.; Fikre, A. et al. The extent and association of chickpea *Fusarium* wilt and root rot disease pressure with major biophysical factors in Ethiopia. *Journal of Plant Pathology*, **2021**, *103*(2), 409–419.

Ben Abdallah, R.A.; Jabnoun-Khiareddine, H.; Nefzi, A. et al. Endophytic bacteria from *Datura stramonium* for *Fusarium* wilt suppression and tomato growth promotion. *Journal of Microbial and Biochemical Technology*, **2016**, *8*(1), 30–41.

Bickel, J.T.; Koehler, A.M. Review of *Pythium* species causing damping-off in corn. *Plant Health Progress*, **2021**, *22*(3), 02.

Biessy, A.; Novinscak, A.; St-Onge, R. et al. Inhibition of three potato pathogens by phenazine-producing *Pseudomonas* spp. is associated with multiple biocontrol-related traits. *Msphere*, **2021**, *6*(3), e00427–21.

Bongiorno, G.; Postma, J.; Bünemann, E.K. et al. Soil suppressiveness to *Pythium ultimum* in ten European long-term field experiments and its relation with soil parameters. *Soil Biology & Biochemistry*, **2019**, *133*, 174–187.

Bonanomi, G.; Lorito, M.; Vinale, F.; Woo, S.L. Organic amendments, beneficial microbes, and soil microbiota: toward a unified framework for disease suppression. *Annual Review of Phytopathology*, **2018**, *56*, 1–20.

Bowen, A.; Orr, R.; McBeath, A.V. et al. Suppressiveness or conduciveness to *Fusarium* wilt of bananas differs between key Australian soils. *Soil Research*, **2019**, *57*(2), 158–165.

Bruehl, G.W. *Soilborne Plant Pathogens*. Macmillan Publishin Co., New York, **1987**.

Cai, H.; Tao, N.; Guo, C. Systematic investigation of the effects of macro-elements and iron on soybean plant response to *Fusarium oxysporum* infection. *Plant Pathology Journal*, **2020**, *36*(5), 398–405.

Cao, Y.; Wang, J.; Wu, H. et al. Soil chemical and microbial responses to biogas slurry amendment and its effect on *Fusarium* wilt suppression. *Applied Soil Ecology*, **2016**, *107*, 116–123.

Carrión, V.J.; Cordovez, V.; Tyc, O. et al. Involvement of Burkholderiaceae and sulfurous volatiles in disease-suppressive soils. *ISME Journal*, **2018**, *12*(9), 2307–2321.

Cha, J.Y.; Han, S.; Hong, H.J. et al. Microbial and biochemical basis of a *Fusarium* wilt-suppressive soil. *ISME Journal*, **2016**, *10*, 119–129. https://doi.org/10.1038/ismej.2015.95.

Chandrashekara, C.; Bhatt, J.C.; Kumar, R.; Chandrashekara, K.N. Suppressive soils in plant disease management. In: Singh, A. (Ed), *Eco-Friendly Innovative Approaches in Plant Disease Management*. International Book Distributors: New Delhi, India, **2012**; pp. 241–256.

Chen, M.; Arato, M.; Borghi, L. et al. Beneficial services of arbuscular mycorrhizal fungi–from ecology to application. *Frontiers in Plant Science*, **2018**, *9*, 1270.

Comensoli, L.; Bindschedler, S.; Junier, P. et al. Iron and fungal physiology: a review of biotechnological opportunities. *Advances in Applied Microbiology*, **2017**, *98*, 31–60.

Cook, R.J. Take-all decline: A model system in biological control and clue to the success of intensive cropping. In: Vincent, C.; Goettel, M.; Lazarovits, G. (Eds.), *Biological Control a Global Perspective: Case Studies from Around the World*. CABI Publishing: UK, **2007**; pp. 399–414.

Cook, R.J. Plant health management: pathogen suppressive soils. In: Alfen, V., Neal, K., (Eds), *Encyclopedia of Agriculture and Food Systems*. Academic Press: London, **2014**; pp. 441–455.

Cook, R.J.; Baker, K.F. *The Nature and Practice of Biological Control of Plant Pathogens*. The American Phytopathological Society: St Paul, MN, **1983**.

Cook, R.J.; Rovira, A.D. The role of bacteria in the biological control of *Gaeumannomyces graminis* by suppressive soils. *Soil Biology & Biochemistry*, **1976**, *8*(4), 269–273.

Cordovez, V.; Carrion, V.J.; Etalo, D.W. et al. Diversity and functions of volatile organic compounds produced by Streptomyces from a disease-suppressive soil. *Frontiers in Microbiology*, **2015**, *6*, 1081. https://doi.org/10.3389/fmicb.2015.01081.

Courtois, S.; Cappellano, S.; Ball, C.M. et al. Recombinant environmental libraries provide access to microbial diversity for drug discovery from natural products. *Applied and Environmental Microbiology*, **2003**, *69*, 49–55.

Culbreath, A.K.; Todd, J.W.; Brown, S.L.; Baldwin, J.A.; Pappu, H. A genetic and cultural "package" for management of to- mato spotted wilt virus in peanut. *Biological and Cultural Tests for Control of Plant Diseases*, **1999**, *14*, 1–8.

Deltour, P.; França, S.C.; Pereira, O.L. et al. Disease suppressiveness to *Fusarium* wilt of banana in an agroforestry system: influence of soil characteristics and plant community. *Agriculture, Ecosystems & Environment*, **2017**, *239*, 173–181.

Devi, Y.B.; Meetei, T.T. Biosciences and plant biology. *International Journal of Current Research in Biosciences and Plant Biology*, **2018**, *5*, 67–75.

Dignam, B.E.; O'Callaghan, M.; Condron, L.M. et al. Effect of land use and soil organic matter quality on the structure and function of microbial communities in pastoral soils: implications for disease suppression. *PLoS One*, **2018**, *13*(5), e0196581. https://doi.org/10.1371/journal.pone.0196581.

Dita, M.; Barquero, M.; Heck, D. et al. *Fusarium* wilt of banana: current knowledge on epidemiology and research needs toward sustainable disease management. *Frontiers in Plant Science*, **2018**, *9*, 1468.

Döring, T.F.; Rosslenbroich, D.; Giese, C. et al. Disease suppressive soils vary in resilience to stress. *Applied Soil Ecology*, **2020**, *149*, 103482. https://doi.org/10.1016/j.apsoil.2019.103482.

Dukare, A.; Paul, S.; Arambam, A. Isolation and efficacy of native chitinolytic rhizobacteria for biocontrol activities against *Fusarium* wilt and plant growth promotion in pigeon pea (*Cajanus cajan* L.). *Egyptian Journal of Biological Pest Control*, **2020**, 30, 1–2.

El-Masry, M.H.; Khalil, A.I.; Hassouna, M.S.; Ibrahim, H.A.H. In-situ and in vivo suppressive effect of agricultural composts and their water extracts on some phytopathogenic fungi. *World Journal of Microbiology and Biotechnology*, **2002**, *18*, 551–558.

Elmer, W.H.; Datnoff, L.E. Mineral nutrition and suppression of plant disease. In: Van Alfen, N. (ed), *Encyclopedia of Agriculture and Food Systems*; vol. 4. Elsevier: San Diego, CA, **2014**; pp. 231–244.

Enebe, M.C.; Babalola, O.O. Metagenomics assessment of soil fertilization on the chemotaxis and disease suppressive genes abundance in the maize rhizosphere. *Genes*, **2021**, *12*(4), 535.

Ferguson, L.M.; Shew, B.B. Wheat straw mulch and its impacts on three soilborne pathogens of peanut in microplots. *Plant Diseases*, **2001,** *85*, 661–667.

Ferret, C.; Sterckeman, T.; Cornu, J.Y. et al. Siderophore-promoted dissolution of smectite by fluorescent Pseudomonas. *Environmental Microbiology Reports*, **2014**, *6*(5), 459–467.

Furtado, E.L.; Bueno, C.J.; Luiz De Oliveira, A. et al. Relações entre ocorrência do Mal-de-Panamá em bananeira da cv. Nanicão e nutrientes no solo e nas folhas. *Tropical Plant Pathology*, **2009**, *34*(4), 211–215.

Gao, Z.; Karlsson, I.; Geisen, S. et al. Protists: puppet masters of the rhizosphere microbiome. *Trends in Plant Science*, **2019**, *24*(2), 165–176.

Gatch, E.W.; du Toit, L.J. A soil bioassay for predicting the risk of spinach *Fusarium* wilt. *Plant Diseases*, **2015**, *99*(4), 512–526.

Gerlagh, M. Introduction of *Ophiobolus graminis* into new polders and its decline. *Netherlands Journal of Plant Pathology*, **1968**, *74*, 1–97.

Ghorbani, R.; Wilcockson, S.; Koocheki, A. Soil management for sustainable crop disease control: a review. *Environmental Chemistry Letters*, **2008**, *6*, 149–162.

Ghoul, M.; Mitri, S. The ecology and evolution of microbial competition. *Trends in Microbiology*, **2016**, *24*(10), 833–845.

Giczey, G.; Kerényi, Z.; Fülöp, L. et al. Expression of cmg1, an exo-β-1, 3-glucanase gene from *Coniothyrium minitans*, increases during sclerotial parasitism. *Applied and Environmental Microbiology*, **2001**, *67*(2), 865–871.

Gill, J.S.; Sivasithamparam, K.; Smettem, K.R. Soil types with different texture affects development of Rhizoctonia root rot of wheat seedlings. *Plant Soil*, **2000**, *221*(2), 113–120.

Gómez Expósito, R.; De Bruijn, I.; Postma, J. et al. Current insights into the role of rhizosphere bacteria in disease suppressive soils. *Frontiers in Microbiology*, **2017**, *8*, 2529. https://doi.org/10.3389/fmicb.2017.02529.

Gordon, T.R.; Stueven, M.; Pastrana, A.M. et al. The effect of pH on spore germination, growth, and infection of strawberry roots by *Fusarium oxysporum* f. sp. *fragariae*, cause of *Fusarium* wilt of strawberry. *Plant Diseases*, **2019**, *103*(4), 697–704.

Guerra, C.A.; Heintz-Buschart, A.; Sikorski, J.; Chatzinotas, A.; Guerrero-Ramírez, N.; Cesarz, S.; Beaumelle, L.; Rillig, M.C.; Maestre, F.T.; Delgado-Baquerizo, M.; Buscot, F. Blind spots in global soil biodiversity and ecosystem function research. *Nature Communications*, **2020**, *11*(1), 1–13.

Haas, D.; Défago, G. Biological control of soil-borne pathogens by fluorescent pseudomonads. *Nature Reviews Microbiology*, **2005**, *3*(4), 307–319.

Hadar, Y.; Papadopoulou, K.K. Suppressive composts: microbial ecology links between abiotic environments and healthy plants. *Annual Review of Phytopathology*, **2012**, *50*, 133–153.

Harsonowati, W.; Marian, M.; Narisawa, K. The effectiveness of a dark septate endophytic fungus, *Cladophialophora chaetospira* SK51, to mitigate strawberry *Fusarium* wilt disease and with growth promotion activities. *Frontiers in Microbiology*, **2020**, *15*(11), 585.

Haynes, R.J. Labile organic matter fractions as central components of the quality of agricultural soils: an overview. *Advances in Agronomy*, **2005**, *23*(52), 21–68.

Heck, D.W.; Ghini, R.; Bettiol, W. Deciphering the suppressiveness of banana *Fusarium* wilt with organic residues. *Applied Soil Ecology*, **2019**, *138*, 47–60.

Henry, A. Occurrence and sporulation of *Helminthosporium sativum* PKB in the soil. *Canadian Journal of Research*, **1931a**, *5*, 407–413. https://doi.org/10.1139/cjr31-078.

Henry, A. The natural microflora of the soil in relation to the foot-rot problem of wheat. *Canadian Journal of Research*, **1931b**, *5*(4), 69–77.

Heyer, R.; Schallert, K.; Büdel, A. et al. A robust and universal metaproteomics workflow for research studies and routine diagnostics within 24 h using phenol extraction, FASP digest, and the Meta Proteome Analyzer. *Frontiers in Microbiology*, **2019**, *10*, 18–83.

Huang, X.; Liu, J.; Ding, J. et al. The investigation of nematocidal activity in *Stenotrophomonas maltophilia* G2 and characterization of a novel virulence serine protease. *Canadian Journal of Microbiology*, **2009**, *55*(8), 934–942.

Huang, X.Q.; Liu, L.L.; Zhao, J. et al. The families Ruminococcaceae, Lachnospiraceae, and Clostridiaceae are the dominant bacterial groups during reductive soil disinfestation with incorporated plant residues. *Applied Soil Ecology*, **2019**, *135*, 65–72.

Husson, O. Redox potential (Eh) and pH as drivers of soil/plant/microorganism systems: a transdisciplinary overview pointing to integrative opportunities for agronomy. *Plant Soil*, **2013**, *362*(1), 389–417.

Husson, O.; Husson, B.; Brunet, A. et al. Practical improvements in soil redox potential (Eh) measurement for characte risation of soil properties. Application for comparison of conventional and conservation agriculture cropping systems. *Analytica Chimica Acta*, **2016**, *906*, 981–989.

Jambhulkar, P.P.; Sharma, M.; Lakshman, D.; Sharma, P. Natural mechanisms of soil suppressiveness against diseases caused by Fusarium, Rhizoctonia, Pythium, and Phytophthora. In: Meghvansi, M.K., Varma, A. (Eds), *Organic Amendments and Soil Suppressiveness in Plant Disease Management, Soil Biology*. Springer International Publishing: Switzerland, **2015**; pp. 95–123.

Janvier, C.; Villeneuve, F.; Alabouvette, C.; Edel-Hermann, V.; Mateille, T.; Steinberg, C. Soil health through soil disease suppression: which strategy from descriptors to indicators? *Soil Biology & Biochemistry*, **2007,** *39*, 1–23. https://doi.org/10.1016/j.soilbio.2006.07.001.

Jayaraman, S.; Naorem, A.K.; Lal, R.; Dalal, R.C.; Sinha, N.K.; Patra, A.K.; Chaudhari, S.K. Disease-suppressive soils - beyond food production: a critical review. *Journal of Plant Nutrition and Soil Science*, **2021**, *21*, 1437–1465.

Johnson, W.C., III.; Brenneman, T.B.; Baker, S.H.; Johnson, A.W.; Sumner, D.R.; Mullinex, B.G., Jr. Tillage and pest management considerations in a peanut-cotton rotation in the southeastern coastal plain. *Agronomy Journal*, **2001,** *93*, 570–576.

Kanaan, H.; Medina, S.; Minz, D. et al. Soil suppressiveness against *Macrophomina Phaseolina* as affected by cyclic temperature regimes. *Phytoparasitica*, **2016**, *44*(3), 305–311.

Knudsen, I.M.; Larsen, K.M.; Jensen, D.F. et al. Potential suppressiveness of different field soils to Pythium damping-off of sugar beet. *Applied Soil Ecology*, **2002**, *21*(2), 119–129.

Kobayashi, S.D.; Malachowa, N.; DeLeo, F.R. Pathogenesis of Staphylococcus aureus abscesses. *The American Journal of Pathology*, **2015**, *185*(6), 1518–1527.

Kowalchuk, G.A.; De Bruijn, F.J.; Akkermans, A.D.L. et al. *Molecular Microbial Ecology Manual II*. Springer Verlag, Heidelberg, Germany, **2005**.

Krištůfek, V.; Diviš, J.; Omelka, M. et al. Site, year and cultivar effects on relationships between periderm nutrient contents and common scab severity. *American Journal of Potato Research*, **2015**, *92*, 473–482.

Kwak, Y.S.; Han, S.; Thomashow, L.S.; Rice, J.T.; Paulitz, T.C.; Kim, D.; Weller, D.M. Saccharomyces cerevisiae genome-wide mutant screen for sensitivity to 2, 4-diacetylphloroglucinol, an antibiotic produced by Pseudomonas fluorescens. *Applied and Environmental Microbiology*, **2011**, *77*(5), 1770–1776.

Kwak, Y.S.; Weller, D.M. Take-all of wheat and natural disease suppression: a review. *Plant Pathology Journal*, **2013**, *29*, 125–135.

Latha, P.; Karthikeyan, M.; Rajeswari, E. Endophytic bacteria: prospects and applications for the plant disease management. In: Ansari, R.; Mahmood, I. (Eds), *Plant Health under Biotic Stress*. Springer: Singapore, **2019**; pp. 1–50.

Latz, E.; Eisenhauer, N.; Rall, B.C. et al. Plant diversity improves protection against soil-borne pathogens by fostering antagonistic bacterial communities. *Journal of Ecology*, **2012**, *100*(3), 597–604.

Legrand, F.; Chen, W.; Cobo-Díaz, J.F. et al. Co-occurrence analysis reveal that biotic and abiotic factors influence soil fungistasis against *Fusarium graminearum*. *FEMS Microbiology Ecology*, **2019**, *95*(5), fiz056.

Li, L.; Li, X.Z.; Tang, W.Z. et al. Screening of a fungus capable of powerful and selective delignification on wheat straw. *Letters in Applied Microbiology*, **2008**, *47*(5), 415–420.

Liu, H.; Li, J.; Cavalhais, L.C.; Percy, C.; Verma, J.P.; Schenk, P.M.; Singh, B. Evidence for the plant recruitment of beneficial microbes to suppress soil-borne pathogen. *New Phytologist*, **2021**. *229*(5), 2873–2885.

Liu, L.; Sun, C.; Liu, S. et al. Bioorganic fertilizer enhances soil suppressive capacity against bacterial wilt of tomato. *PLoS One*, **2015**, *10*(4), e0121304. https://doi.org/10.1371/journal.pone.0121304.

Liu, L.L.; Huang, X.Q.; Zhao, J. et al. Characterizing the key agents in a disease-suppressed soil managed by reductive soil disinfestation. *Applied and Environmental Microbiology*, **2019a**, *85*, e02992–18.

Liu, Y.; Zhu, A.; Tan, H. et al. Engineering banana endosphere microbiome to improve *Fusarium* wilt resistance in banana. *Microbiome*, **2019b**, *7*(1), 1–5.

Löbmann, M.T.; Vetukuri, R.R.; de Zinger, L.; Alsanius, B.W.; Grenville-Briggs, L.J.; Walter, A.J. The occurrence of pathogen suppressive soils in Sweden in relation to soil biota, soil properties, and farming practices. *Applied Soil Ecology*, **2016**, *107*, 57–65.

Loper, J.E.; Hassan, K.A.; Mavrodi, D.V. et al. Comparative genomics of plant-associated *Pseudomonas* spp.: insights into diversity and inheritance of traits involved in multitrophic interactions. *PLoS Genetics*, **2012**, *8*, e1002784.

Manici, L.M.; Donatelli, M.; Fumagalli, D. et al. Potential response of soil-borne fungal pathogens affecting crops to a scenario of climate change in Europe. In: Seppelt, R.; Voinov, A.A.; Lange, S.; Bankamp, D. (Eds.), *Proceedings of the Sixth Biannial Meeting of the International Environmental Modelling and Software Society, Managing Resources of a Limited Planet*, Leipzig, Germany, 1–5, **2012**; pp. 649–657.

Marschner, P.; Fu, Q.; Rengel, Z. Manganese availability and microbial populations in the rhizosphere of wheat genotypes differing in tolerance to Mn deficiency. *Journal of Plant Nutrition and Soil Science*, **2003**, *166*(6), 7127–7118.

Martin, F.N.; Hancock, J.G. Association of chemical and biological factors in soils suppressive to *Pythium ultimum*. *Phytopathology*, **1986**, *76*, 1221–1231.

Mavrodi, O.V.; Walter, N.; Elateek, S.; Taylor, C.G.; Okubara, P.A. **2012**. Suppression of Rhizoctonia and Pythium root rot of wheat by new strains of Pseudomonas. *Biological Control*, **2012**, *62*, 93–102.

Mazzola, M.; de Bruijn, I.; Cohen, M.F. et al. Protozoaninduced regulation of cyclic lipopeptide biosynthesis is an effective predation defense mechanism for *Pseudomonas Fluorescens*. *Applied and Environmental Microbiology*, **2009**, *75*, 6804–6811.

Mendes, R.; Kruijt, M.; De Bruijn, I. et al. Deciphering the rhizosphere microbiome for disease-suppressive bacteria. *Science*, **2011**, *332*(6033), 1097–1100.

Momma, N. Studies on mechanisms of anaerobicity-mediated biological soil disinfestation and its practical application. *Journal of General Plant Pathology*, **2015**, *81*(6), 480–482.

Monfort, W.S.; Culbreath, A.K.; Stevenson, K.L.; Brenneman, T.B.; Gorbet, D.W.; Phatak, S.C. Effects of reduced tillage, resistant cultivars, and reduced fungicide inputs on progress of early leaf spot of peanut (Arachis hypogaea). *Plant Diseases*, **2004**, *88*, 858–864.

Mueller, B. Experimental interactions between clay minerals and bacteria: a review. *Pedosphere*, **2015**, *25*(6), 799–810.

Orr, R.; Nelson, P.N. Impacts of soil abiotic attributes on *Fusarium* wilt, focusing on bananas. *Applied Soil Ecology*, **2018**, *132*, 203–233.

Palojärvi, A.; Kellock, M.; Parikka, P. et al. Tillage system and crop sequence affect soil disease suppressiveness and carbon status in boreal climate. *Frontiers in Microbiology*, **2020**, *12*, 693341.

Panth, M., Hassler, S.C.; Baysal-Gurel, F. Methods for management of soilborne diseases in crop production. *Agriculture*, **2020**, *10*, 16.

Penton, C.R.; Gupta, V.V.; Tiedje, J.M. et al. Fungal community structure in disease suppressive soils assessed by 28S LSU gene sequencing. *PLoS One*, **2014**, *9*(4), e93893.

Persson, L.; Olsson, S. Abiotic characteristics of soils suppressive to Aphanomyces root rot. *Soil Biology & Biochemistry*, **2000**, *32*(8–9), 1141–1150.

Poret-Peterson, A.T.; Sayed, N.; Glyzewski, N.; Forbes, H.; Gonzalez-Orta, E.T.; Kluepfel, D.A. Temporal responses of microbial communities to anaerobic soil disinfestation. *Microbial Ecology*, **2020**, *80*, 191–201.

Raaijmakers; J.M.; Mazzola, M. Diversity and natural functions of antibiotics produced by beneficial and plant pathogenic bacteria. *Annual Review of Phytopathology*, **2012**, *50*, 403–424.

Raaijmakers, J.M.; Weller, D.M. Natural protection by 2,4- diacetylphloroglucinol-producing *Pseudomonas* spp. in take-all decline soils. *Molecular Plant-Microbe Interactions*, **1998**, *11*, 144–152.

Ramette, A.; Moe¨nne-Loccoz, Y.; Dé fago, G. Genetic diversity and biocontrol potential of fluorescent pseudomonads producing phloroglucinols and hydro- gen cyanide from Swiss soils naturally suppressive or conducive to *Thielaviopsis basicola* mediated black root rot of tobacco. *FEMS Microbiology Ecology*, **2006,** *55*, 369–381.

Rangaswami, G.; Bagyaraj, D.J. *Agricultural Microbiology*. PHI Learning Pvt. Ltd: New Delhi, **2017**.

Ros, M.; Almagro, M.; Fernández, J.A. et al. Approaches for the discrimination of suppressive soils for *Pythium irregulare* disease. *Applied Soil Ecology*, **2020**, *147*, 103439.

Rosskopf, E.N.; Serrano-Pérez, P.; Hong, J. et al. Anaerobic soil disinfestation and soilborne pest management. In: Meghvansi, M.; Varma, A. (Eds), *Organic Amendments and Soil Suppressiveness in Plant Disease Management*, *Soil Biology*, vol 46. Springer: Cham, **2015**; pp. 277–305.

Sahu, P.K.; Singh, S.; Gupta, A. Antagonistic potential of bacterial endophytes and induction of systemic resistance against collar rot pathogen *Sclerotium rolfsii* in tomato. *Biological Control*, **2019,** *137*, 104014.

Sajeena, A.; Nair, D.S.; Sreepavan, K. Non-pathogenic *Fusarium oxysporum* as a biocontrol agent. *Indian Phytopathol*. **2020**, *73*(2), 177–183.

Schlatter, D.; Kinkel, L.; Thomashow, L.; Weller, D.; Paulitz, T. Disease suppressive soils: new insights from the soil microbiome. *Phytopathology*, **2017**, *107*, 1284–1297.

Schroth, M.N.; Hancock, J.G. Disease-suppressive soil and root-colonizing bacteria. *Science*, **1982**, *216*, 1376–1381.

Schulz-Bohm, K.; Martín-Sánchez, L.; Garbeva, P. Microbial volatiles: small molecules with an important role in intra-and inter-kingdom interactions. *Frontiers in Microbiology*, **2017**, *8*, 2484.

Shakya, M.; Lo, C.C.; Chain, P.S. Advances and challenges in metatranscriptomic analysis. *Frontiers in Genetics*, **2019**, *10*, 904.

Sharma, B.R.; Dutta, S.; Roy, S.; Debnath, A.; Roy, M.D. The effect of soil physico-chemical properties on rhizome rot and wilt disease complex incidence of ginger under hill agro-climatic region of West Bengal. *Plant Pathology Journal*, **2010**, *26*(2), 198–202.

Sharma-Poudyal, D.; Schlatter, D.; Yin, C.; Hulbert, S.; Paulitz, T. Long- term no-till: a major driver of fungal communities in dryland wheat cropping systems. *PLoS One*, **2017**, *12*, e0184611.

Shipton, P.J. Take-all decline during cereal monoculture. In: Bruehl, G.W. (Ed), *Biology and Control of Soilborne Pathogens*. American Phytopathological Society: USA, **1975**; pp. 137–144.

Siddiqui, I.A.; Haas, D.; Heeb, S. Extracellular protease of *Pseudomonas fluorescens* CHA0, a biocontrol factor with activity against the root-knot nematode *Meloidogyne incognita. Applied and Environmental Microbiology*, **2005**, *71*, 5646–5649.

Sipilä, T.P.; Yrjälä, K.; Alakukku, L.; Palojärvi, A. Cross-site soil microbial communities under tillage regimes: fungistasis and microbial biomarkers. *Applied and Environmental Microbiology*, **2012**, *78*, 8191–8201. https://doi.org/10.1128/AEM.02005-12.

Stutz, E.W.; Defago, G.; Kern, H. Naturally occurring fluorescent pseudomonads involved in suppression of black root rot of tobacco. *Phytopathology*, **1986**, *76*, 181–185.

Tamm, L.; Thürig, B.; Bruns, C.; Fuchs, J.G.; Köpke, U.; Laustela, M.; Leifert, C.; Mahlberg, N.; Nietlispach, B.; Schmidt, C.; Weber, F.; Fließbach, A. Soil type, management history, and soil amendments influence the development of soil-borne (*Rhizoctonia solani, Pythium ultimum*) and air-borne (*Phytophthora infestans, Hyaloperonospora parasitica*) diseases. *European Journal of Plant Pathology*, **2010**, *127*, 465–481.

Tamm, L.; Thürig, B.; Fliessbach, A. et al. Elicitors and soil management to induce resistance against fungal plant diseases. *NJAS - Wageningen Journal of Life Sciences*, **2011,** *58*(3–4), 131–137.

Thangavelu, R.; Gopi, M. Field suppression of *Fusarium* wilt disease in banana by the combined application of native endophytic and rhizospheric bacterial isolates possessing multiple functions. *Phytopathologia Mediterranea*, **2015**, *54*(2), 241–252.

Thomashow, L.S.; Weller, D.M. Current concepts in the use of introduced bacteria for biological disease control: mechanisms and antifungal metabolites. In: Stacey, G., Keen, N. (Eds), *Plant–Microbe Interactions*. Chapman & Hall: New York, **1995**; pp. 187–235.

Thomashow, L.S.; Pierson, L.S. Genetic aspects of phenazine antibiotic production by fluorescent pseudomonads that suppress take-all disease of wheat. *Molecular Plant-Microbe Interactions*, **1991**, *1*, 443–449.

Tilston, E.L.; Pitt, D.; Groenhof, A.C. Composted recycled organic matter suppresses soil-borne diseases of field crops. *New Phytologist*, **2002**, *154*, 731–740.

Toyota, K.; Shirai, S. Growing interest in microbiome research unraveling disease suppressive soils against plant pathogens. *Microbes and Environments*, **2018**, *33*, 345–347.

Tracanna, V.; Ossowicki, A.; Petrus, M.L.C. et al. Dissecting disease-suppressive rhizosphere microbiomes by functional amplicon sequencing and 10× metagenomics. *mSystems*, **2021**, *6*(3), e0111620.

Ueki, A.; Takehara, T.; Ishioka, G. et al. Production of beta-1,3-glucanase and chitosanase from clostridial strains isolated from the soil subjected to biological disinfestation. *AMB Express*, **2019**, *9*, 1–14.

Van Agtmaal, M.; Straathof, A.L.; Termorshuizen, A. et al. Volatile-mediated suppression of plant pathogens is related to soil properties and microbial community composition. *Soil Biology & Biochemistry*, **2018**, *117*, 164–174.

Van Elsas, J.D.; Costa, R.; Jansson, J. et al. The metagenomics of disease-suppressive soils–experiences from the metacontrol project. *Trends in Biotechnology*, **2008a**, *26*(11), 591–601.

Van Elsas, J.D.; Speksnijder, A.J.; Van Overbeek, L.S. A procedure for the metagenomics exploration of disease-suppressive soils. *Journal of Microbiological Methods*, **2008b**, *75*(3), 515–522.

Voisard, C.; Keel, C.; Haas, D. et al. Cyanide production by *Pseudomonas fluorescens* helps suppress black root rot of tobacco under gnotobiotic conditions. *EMBO Journal*, **1989**, *8*(2), 351–358.

Wang, S.; Sun, L.; Ling, N. et al. Exploring soil factors determining composition and structure of the bacterial communities in saline-alkali soils of Songnen Plain. *Frontiers in Microbiology*, **2020**, *10*, 2902.

Weller, D. Take-All Decline and Beneficial Pseudomonads. In: Lugtenberg, B. (Ed.), *Principles of Plant-Microbe Interactions*; Springer: Cham, **2015**; https://doi.org/10.1007/978-3-319-08575-3_38.

Weller, D.M. Take-all decline and beneficial pseudomonads. In: Lugtenberg, B. (Ed), *Principles of Plant-Microbe Interactions -Microbes for Sustainable Agriculture*. Springer: Switzerland, **2015**; pp. 363–370.

Weller, D.M.; Raaijmakers, J.M.; Gardener, B.B.M.; Thomashow, L.S. Microbial populations responsible for specific soil suppressiveness. *Annual Review of Phytopathology*, **2002**, *40*, 309–348.

Win, T.T.; Bo, B.; Malec, P. et al. The effect of a consortium of *Penicillium* sp. and *Bacillus* spp. in suppressing banana fungal diseases caused by *Fusarium* sp. and *Alternaria* sp. *Journal of Applied Microbiology*, **2021**, *131*(4), 1890–1908.

Yin, C.; Hulbert, S.H.; Schroeder, K.L. et al. Role of bacterial communities in the natural suppression of *Rhizoctonia solani* bare patch disease of wheat (*Triticum aestivum* L.). *Applied and Environmental Microbiology*, **2013,** *79*, 7428–7438.

Yogev, A.; Raviv, M.; Hadar, Y.; Cohen, R.; Wolf, S.; Gil, L.; Katan, J. Induced resistance as a putative component of compost suppressiveness. *Biological Control*, **2010**, *54*(1), 46–51.

Zhalnina, K.; Dias, R.; de Quadros, P.D. et al. Soil pH determines microbial diversity and composition in the park grass experiment. *Microbial Ecology*, **2015**, *69*(2), 395–406.

Zhang, X.; Harvey, P.R.; Stummer, B.E. et al. Antibiosis functions during interactions of *Trichoderma afroharzianum* and *Trichoderma gamsii* with plant pathogenic *Rhizoctonia* and *Pythium*. *Functional & Integrative Genomics*, **2015**, *15*(5), 599–610.

Zhao, M.; Yuan, J.; Zhang, R. et al. Microflora that harbor the NRPS gene are responsible for *Fusarium* wilt disease-suppressive soil. *Applied Soil Ecology*, **2018**, *132*, 83–290.

Zhou, Z.; Wang, C.; Luo, Y. Meta-analysis of the impacts of global change factors on soil microbial diversity and functionality. *Nature Communications*, **2020**, *11*, 3072.

12 Advances in Genetic Improvement of Legumes to Biotic Stresses

Mamta Sharma, Avijit Tarfdar, Raju Ghosh, Mediga Kasi Rao, K. Srinivas, and Rajeev K. Varshney

12.1 INTRODUCTION

Legumes belong to the family Fabaceae or Leguminosae, the third-largest angiosperm family in the plant kingdom. The seeds of legumes (pulses) contain 16%–50% dietary protein depending on the species, important minerals, dietary fiber (10%–23%), and vitamins (Vance et al. 2000; Maphosa and Jideani 2017) contributing one-third of human dietary protein. Grain legumes cover an area of 81 million hectares, with a global production of more than 92 million tonnes (FAOSTAT 2018). India produces approximately a 1/4th of the world's grain legumes and is also the top consumer. China, Myanmar, Canada, Australia, Brazil, Argentina, the United States, and Russia are the world's leading grain legume producers (FAOSTAT 2018). Holistically, the legume crops could be categorized into pulses (chickpea, pigeon pea, cowpea, mung bean, black gram, common bean, and lentil), vegetables (pea and common bean), oilseed crops (soybean), and forage legumes (clover and alfalfa). Pulses are among the most widely used food grains throughout the world. With the rising importance of everyone's nutritional security and bearing in mind the nutritional values of pulses, the pulses are considered to be "poor man's meat." Hence, pulses in combination with cereals provide one of the best solutions for eliminating malnutrition, especially in developing countries. Hence, there is a necessity to upscale the production and productivity of pulses through more intensive interventions. Besides alleviating nutrition deficiencies, legume plants can improve soil health by fixing nitrogen and reducing the use of fertilizer. In India, over a dozen pulse crops are grown, e.g., chickpea, pigeonpea, mungbean, urdbean, lentil, cowpea, and fieldpea. Although over 12%–15% of the Earth's arable land is deployed for legume cultivation, their yield is limited primarily due to challenges in environmental adaptability and secondarily damages caused by different biotic stresses (insects and pathogens) (Graham and Vance 2003). Among legume diseases, 50% of the diseases are caused by fungi, 38% by viruses, and little by bacteria. Some of the major diseases in legumes caused by the fungus are wilts root rot, mildews, rusts, anthracnose, and blights. Similarly, bacterial diseases are mainly wilts, leaf blights, leaf spots, etc. Mosaic, stripe, mottle, and stunt are among the major viral diseases in legumes. Global distribution of the pathogens and the occurrence

DOI: 10.1201/9781003531944-12

of the diseases in grain legumes are well-documented. But under changing climate scenarios, the nature of pathogens and their distribution are shifting very rapidly. Sometimes, minor and sporadic diseases are also becoming a major threat to achieving maximum yields from crops because the pathogens are becoming more aggressive. The varieties that are resistant to diseases could be the most efficient, safe, and economical method for the management of pulse diseases. In the past few decades, the production of pulses increased considerably, and the disease-resistant varieties had the major contribution to that. In this chapter, we summarize the information about the major diseases in grain legumes and their distribution. We also describe here the molecular basis of breeding for disease resistance in legumes and the genes behind the genetics of disease resistance. Some emerging challenges have also been addressed including the adaptation of advanced technologies in pulse farming systems for narrowing the yield gaps in pulses.

12.2 IMPACT OF PLANT DISEASES ON MAJOR GRAIN LEGUMES

Legumes are adapted to a wide range of environmental conditions, from cool to warm seasons. Various grain legumes are grown in India, including chickpea (*Cicer arietinum* L.), pigeonpea (*Cajanus cajan* L.), field pea (*Pisum sativum* L.), cowpea (*Vigna unguiculata* L.), mungbean (*Vigna radiata* L.), urdbean (*Vigna mungo* L. Hepper), and faba bean (*Vicia faba* L.). These grain legumes are harmed by a variety of plant pathogens, including viruses, bacteria, fungi, and pathogenic weed species, which affect all parts of the plant at all stages of growth (Pande et al. 2009). In some cases, total yield losses have been reported. The magnitude of the biotic stress effect on plant growth and development is determined by the plants' ability to resist diseases in a given environment. More importantly, as a result of the effects of climate change, plants will be more susceptible to disease. In such cases, gaining a better understanding of plant physiology could lead to long-term control of plant diseases. Grain legume scientists have identified several biotic agents that significantly limit crop productivity and cause instability over the last few decades while working with various species of grain legumes. Some of these agents also pose a serious threat to crop survival. Farmers in many countries have abandoned the cultivation of some grain legumes due to yield instability and high-risk cultivation caused by various biotic stresses, posing a serious threat to the sustainability of the entire crop production system. Taking all of this into account, several authors have reported the effects of numerous major and minor biotic factors, disease symptoms, and losses caused by major diseases in grain legumes, which are presented and summarized in Figure 12.1 and Table 12.1. *Mungbean yellow mosaic virus* (*MYMV*), for example, is a prevalent issue in tropical legumes such as mungbean and urdbean. Ascochyta blight (AB) causes significant damage to chickpea and faba bean crops. Chickpea, pigeonpea, and field pea are all susceptible to fusarium wilt (FW). In subtropical pigeonpea farming, sterility mosaic disease (SMD) caused by a virus is an endemic problem. The creation of resistant cultivars in the above pulse crops is an immediate requirement to save pulse production from these plant infections. Recent genome sequencing efforts in major pulse crops have resulted in massive amounts of marker data and a

FIGURE 12.1 Picture showing the major diseases in different grain legumes crops.

TABLE 12.1
List of Major Diseases in Different Grain Legumes Crops

Crop	Disease	Causal Organism	Symptoms	Distribution	Yield Loss (%)	References
Pigeonpea	Fusarium wilt (FW)	*Fusarium oxysporum f. sp. udum*	Patches of dead plants in the field when the crop is flowering or podding are the first indications of wilt. The most characteristic symptom is a purple band extending upwards from the base of the main stem. Vascular tissues exhibit brown discolouration in the region of the purple band.	Asia, Africa, Tanzania, Uganda, Germany, Italy, Vietnam, Kenya, Thailand, Indonesia and Trinidad	0–70	Ravikumara et al. (2022)
	Phytophthora Blight (PB)	*Phytophthora drechsleri f. sp. cajani*	Water-soaked lesions on their leaves and brown to black, slightly sunken lesions on their stems and petioles. The disease resembles damping-off and sudden death occurs.	Asia, Southern and Eastern Africa, Australia and North America	50–100	Sharma and Ghosh (2018)
	Sterility Mosaic disease (SMD)	*Pigeonpea sterility mosaic virus*	Diseased plants appear as bushy, pale green plants without flowers or pods. In severe cases, leaves become smaller and cluster near tip because of shortened internodes and stimulation of auxillary buds.	Bangladesh, India, Myanmar, Nepal, Sri Lanka and Thialand	25–100	Patil and kumar (2015)
	Dry root rot (DRR)	*Rhizoctonia bataticola*	Complete drying of the plants, rotting of the root system, presence of sclerotial bodies on the root system.	South Asia, Africa, Iran, Mexico and the United States	10–100	Smitha et al. (2015)
	Alternaria blight	*Alternaria alternata*	Small brown-dark brown lesions on the leaves, typical concentric rings (target board symptom), defoliation and drying of infected leaves, branches and flowers.	Asia and Africa	20–80	Sharma et al. (2013)

(*Continued*)

TABLE 12.1 (*Continued*)
List of Major Diseases in Different Grain Legumes Crops

Crop	Disease	Causal Organism	Symptoms	Distribution	Yield Loss (%)	References
Chickpea	Dry root rot (DRR)	*Rhizoctonia bataticola*	Leaf yellowing and premature drying, patches of straw coloured, dried plants in the field.	South Asia, Africa, Iran, Mexico, the United States, and Spain	60–80	Rai et al. (2022)
	Fusarium wilt (FW)	*Fusarium oxysporum f.sp. ciceri*	Drooping of petioles, rachis and leaflets. Infected leaves turn yellow-light brown, Vascular discolouration. death of seedlings or adult plants in patches	Indian subcontinent, Spain, Ethiopia, Mexico, Tunisia, Turkey, and the United States	10–100	Patil et al. (2014) and Sharma et al. (2019)
	Rust	*Uromyces ciceris arietini*	A ring of small oval pustules can be seen around larger pustules which occur on both leaf surfaces, petioles, stems and pods.	India, Mediterranean, southeast Europe, south Asia, east Africa, Italy and Mexico	0–50	Kumbar et al. (2021)
	Ascochyta blight (AB)	*Ascochyta rabiei*	Small water-soaked necrotic spots on leaves, petioles, flowers, pods, branches and stems.	India, Bangladesh, Algeria, Israel, Italy, Morocco, Nepal, Spain, Pakistan, Syria, Iran, USA, Australia and Canada	0–100	Sharma and Ghosh (2016)
	Botrytis grey mould (BGM)	*Botrytis cinerea*	Grey to brown lesions on leaflets, branches and pods covered with masses of spores. Drooping of infected branches is common. Patches of dead plants.	India, Nepal, Bangladesh, Pakistan, North Africa, Australia, America	10–80	Johansen et al. (2008)
Mungbean	Powdery mildew (PM)	*Erysiphe polygoni*	White powdery patches of fungus on the leaves. In severe cases defoliation of leaves occurs.	Philippines, Australia, South Asia and South-East Asia	9–50	Pande et al. (2009)

(*Continued*)

TABLE 12.1 (*Continued*)
List of Major Diseases in Different Grain Legumes Crops

Crop	Disease	Causal Organism	Symptoms	Distribution	Yield Loss (%)	References
	Leaf spots	*Alternaria alternata* *Cercospora canescens* *Corynespora cassicola* *Protomycopsis phaseoli*	Severe leaf spotting and defoliation at the time of flowering and pod formation.	South Asia	10–50	Trivedi et al. (2014)
	Dry root rot (DRR)	*Rhizoctonia bataticola*	Yellowing and drooping of the leaves. affected plants can be easily pulled out leaving dried, rotten root portions in the ground which contain a large number of black minute sclerotia.	India, China and Pakistan	> 60	Sunil Kulkarni et al. (2019)
	Viral diseases	*Mungbean yellow mosaic virus (MYMV)* *Leaf curl virus (LCV)* *Urdbean Leaf crinkle virus (ULCV)*	Stunting, yellowing, leaf puckering and curling. In severe cases, no flower and pod formation.	South Asia, Africa and North America	10–100	
Field pea	Downy mildew (DM)	*Peronospora viciae*	Development of oospore and greyish mycelium in the intercellular spaces of stems and leaves. In severe infection, defoliation of leaves occurs.	UK and Australia, Sweden, New Zealand, Canada and India	30–45	Chang et al. (2014)
	Powdery mildew (PM)	*Erysiphe pisi*	Formation of white, floury patches on leaves, tendrils, pods and stems.	India and Nepal	0–25	Teshome and Tegegn (2017)
	Rust	*Uromyces viciae-fabae*	Orange-brown pustules with a light green halo on the leaves and stems.	Europe, North and South America, India, China, Australia, and New Zealand	25–50	Nongmaithem et al. (2017)

(*Continued*)

TABLE 12.1 (*Continued*)
List of Major Diseases in Different Grain Legumes Crops

Crop	Disease	Causal Organism	Symptoms	Distribution	Yield Loss (%)	References
	Fusarium wilt (FW)	*Fusarium sp.*	Downward curling of the leaves, which become pale green and flaccid, then turn yellow starting at the bottom and progressing upward until the entire plant is yellow and eventually becomes dry and crisp.	South Asia, Canada, the USA, and Lithuania	25–50	Nongmaithem et al. (2017)
Faba bean	Chocolate spot	*Botrytis fabae*	Discrete dark-brown spots surrounded by an orange-brown ring on leaves, flowers and stems. In severe cases, defoliation and killing of the plants.	Mediterranean countries	10–75	
	Ascochyta blight (AB)	*Ascochyta fabae*	Round lesions on the leaves, and elongated lesions on the stems, both usually grey in colour and presenting distinctive rings of black pycnidia.	Mediterranean countries	5–30	Ahmed et al. (2016)
	Rust	*Uromyces viciae fabae*	Red coloured blister-like spots on the leaves and stems. Presence of chlorotic halo around the spots. Premature defoliation of leaves occurs.	Mediterranean countries	20–75	Torres et al. (2006)

molecular breeding or genomics platform. The use of these has aided in the development of fewer enhanced variants and has a bright future in the development of disease-resistant pulse crops.

12.3 THE ROLE OF CLIMATE CHANGE IN THE EMERGENCE AND RE-EMERGENCE OF LEGUME DISEASES

A disease is said to be emerging if it is new, it occurs in a novel host, there is an unpredicted outbreak, its economic importance upsurges, if it fascinates public opinion and the scientific community irrespective of economic importance, or if it appears in an area for the first time (It is referred to as geographic emergence). Climate inconsistency also adds a novel dimension to the plant disease scenario by fluctuating the equilibrium of host–pathogen interactions. This eventually results in either a higher frequency of epidemic outbreaks or several novel pathogens surfacing as threats or less recognized minor pathogens causing severe yield losses. Climate change affects various factors involved in the population dynamics of the pathogen such as survival, vigour, sporulation, rate of multiplication, distance and direction of dispersal of inoculum, rate of spore germination, and penetration of pathogens. The environmental change, particularly when combined with the pathogen and host introductions, may result in unprecedented effects (Figure 12.2).

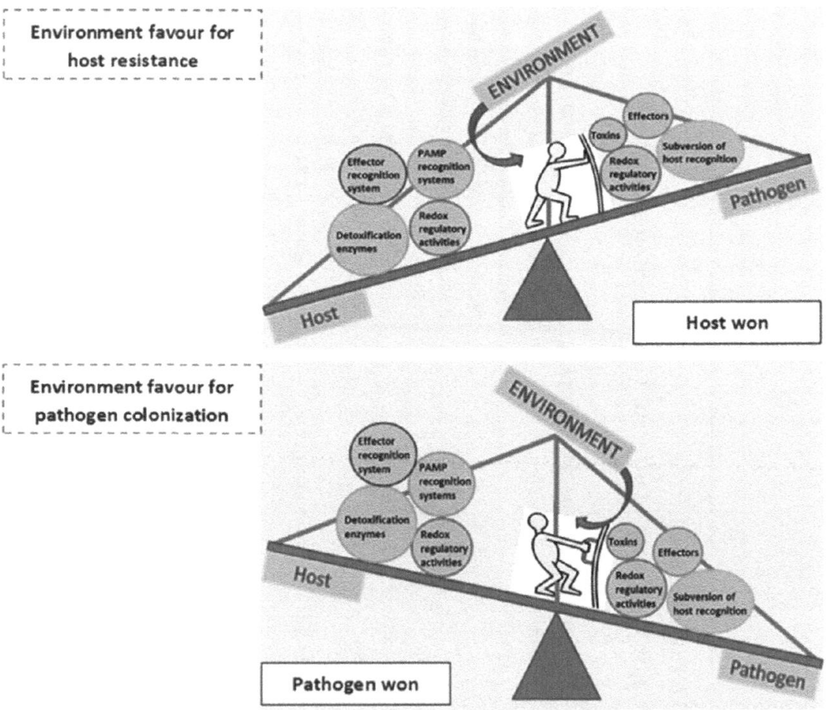

FIGURE 12.2 Picture showing the disease triangle of host–pathogen–environment interactions.

Host and pathogens are influenced by the interactive effects of several climatic factors. Climate change is just one of the several ways in which the environment can change in the long term from disease-suppressive to disease-conducive or vice versa (Chakraborty and Datta 2003; Chakraborthy 2013). The ongoing fluctuations in climatic conditions such as upsurge in CO_2 emissions, unpredicted rainfall patterns, temperature increase, decrease or increase in relative humidity, and low soil moisture stress (Zhao et al. 2010) are likely to impact the plant diseases establishment, its dispersal and epidemiology (Graham and Vance 2003). These variations will influence pathogen emergence in new localities. Climate modifies the susceptibility of the host and influences the host cultivars planted. It also drives the spreading of hosts, both cultivated and wild, alters trade patterns, and determines the ranges of competitor and biocontrol species (Chakraborty 2005). The climate also disturbs the virulence of many pathogens (Coakely et al. 1999). In response to these changes in climate and their direct and indirect effects on crops and their pathogens, new crop varieties may be developed. This global climate changes by various factors (Collins et al. 2013; Cook et al. 2016), and change or increased CO_2 levels can influence both the host and the pathogen in various ways. The fourth assessment report of the Intergovernmental Panel on Climate Change (IPCC) has been predicted that due to 21st century emissions, the concentration of atmospheric CO_2 will rise from the current 368–600 ppm, and the temperature would increase between 0.9°C and 3.5°C leading to the world in the direction of global climate change (Meehl et al. 2007). This rise in atmospheric CO_2 has direct biological impacts on various plants and is the key driver for warming our planet at an enhanced rate, changing many biological functions of both plants and animals (Bergot et al. 2004).

Consequently, the evidence suggests a major shift in the soil-borne diseases such as DRR (*Rhizoctonia bataticola*), FW (*Fusarium oxysporum f. sp. ciceris*), collar rot (*Sclerotium rolfsii*), wet root rot (*Rhizoctonia solani*), and black root rot (*Fusarium solani*) in many chickpea growing regions. As we know, plants in fields are continuously exposed to multiple biotic and abiotic stresses where plants exhibit certain unique and convergent physiological and molecular responses that interact and impact each other to withstand the combined effect of these stresses. These combined stress interactions in plants may either have a direct or indirect effect on pathogens through other community interactions (Sharma and Ghosh 2017) leading to either positive or negative effects on plant responses (Table 12.2 and Figure 12.3). Till now, it is not clear in the literature, whether the disease severity is enhanced or diminished by a higher CO_2 and its impact on plant–disease interaction. According to Thompson and Drake (1994), elevated CO_2 reduced the PM (*Erysiphe graminis*) on wheat and the severity of rust (*Puccinia sparganioides*) on C3 sedge (*Scirpus olneyi*). On the other side, higher temperature and augmented CO_2 concentration are also posing a higher threat by increasing fungal disease perception, namely late blight of potato, rice blast, and sheath blight. The report suggests that the elevated CO_2 condition increases the reproduction rate of polycyclic diseases that could produce highly virulent pathotypes, which can break down the resistance of pulse crops.

Preliminary reports on the effect of temperatures on ascochyta-chickpea and ascochyta lentil pathosystems exhibited that both host resistance and pathogen virulence had varied with the variable levels of temperature. Low soil moisture gives a reduced

TABLE 12.2
Effect of Change in Climatic Factors on Pathogens Incidence

Climatic Factor	Positive Effect (Increase)
Elevated CO_2	PB in pigeonpea
Soil moisture (low)	DRR disease (*Rhizoctonia bataticola*) in chickpea (Tarafdar et al. 2018)
Moisture stress and higher temperatures	DRR of chickpea
High humidity and moderate temperatures	*Sclerotium rolfsii* chickpea
Erratic and heavy rainfall	PB of pigeonpea (Sharma et al. 2016)
Unusual rainfall	Macrophomina stem blight of pigeonpea
Changing climatic conditions	*Uromyces viciae-fabae* lentil
Increase in temperature of 3°C	FW resistant lines susceptible, and races of *Fusarium oxysporum* f. sp. *ciceris* (Foc) showed different differential reactions or crossover interactions with the temperature due to change in their virulence on different cultivars (Landa et al. 2006)

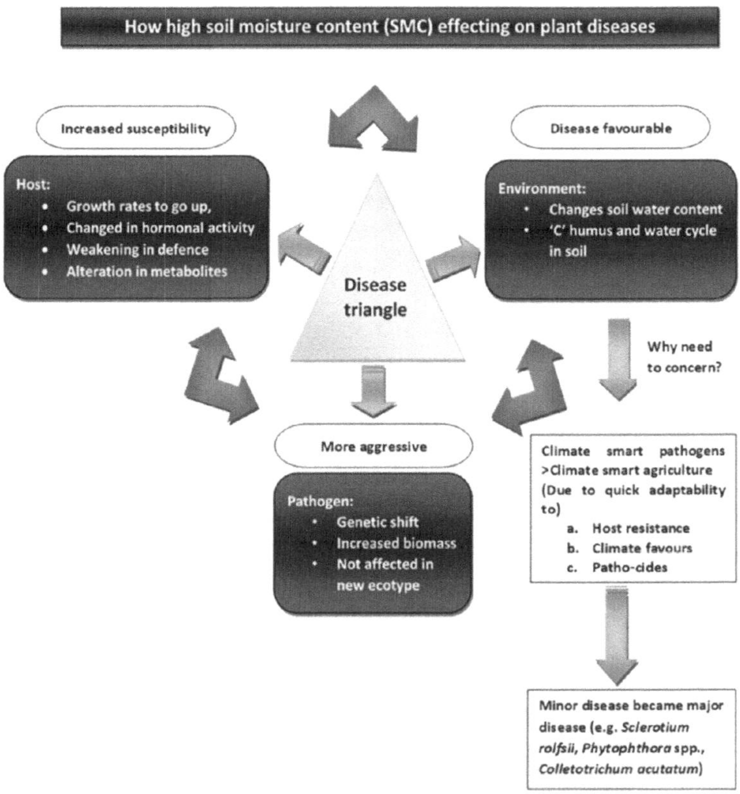

FIGURE 12.3 The spatial and temporal effects of climate change on the establishment of a new or altered ecosystem.

effect (negative) on collar rot disease (*Sclerotium rolfsii*). DRR in mungbean and urdbean caused by *Macrophomina phaseolina* is also reported as an emerging disease in South Asia and South-East Asia (Iqbal et al. 2003), causing 10%–44% disease incidence in both legumes in India and Pakistan (Bashir and Malik 1988; Raguchander et al. 1993; Indira and Gayatri 2003). In China and Australia (Nobel et al. 2019), halo blight is an emerging disease, and surveys revealed 30%–50% crop failure in mungbean. Leaf crinkle, whitefly (*Bemisia tabaci*) and aphid (*Aphis craccivora*) transmitted disease, is caused by urdbean leaf crinkle virus (ULCV) (Unidentified). The disease is nowadays an emerging viral disease of mungbean and urdbean (Bashir et al. 2005; Karthikeyan et al. 2014; Gautam et al. 2016) in South Asia and is reported to cause 85% yield loss in urdbean production in Pakistan. In addition, endemic and devastative foliar and soil-borne diseases in faba bean, new diseases are being reported in some countries. The further most destructive disease affecting faba bean production is chocolate spot. The first report of this disease was done in a few areas in 2012, and nowadays, it extends its coverage in most growing regions in Ethiopia.

12.4 IMPLICATION OF WHOLE-GENOME SEQUENCES FOR STRATEGIC MANAGEMENT OF LEGUME DISEASES

12.4.1 Host Perspective

Fungi are among the most challenging plant pathogens to tackle owing to their genetic elasticity and plasticity, which permit them to adjust quickly to their varying environments. Substantial effort has been made to understand the plant–fungal interaction mechanisms in both model and legumes crops. Large-scale genomic studies have allowed an understanding of the various plant disease-resistance mechanisms against hemibiotrophic, biotrophic, and necrotrophic fungal pathogens.

Numerous races of *F. oxysporum* f. sp. *Ciceri* (Foc) have been identified across the chickpea-cultivated regions of the world. While most *F. oxysporum* strains are measured as necrotrophic or hemibiotrophic pathogens, Foc race 1 (Foc1) is reported to be an obligate biotrophic pathogen of chickpea (Gupta et al. 2009). Analyses based on cDNA-amplified restriction fragment length polymorphism (cDNA-AFLP), cDNA-based microarrays, and cDNA-random amplified polymorphic DNA (cDNA-RAPD) technique have been deployed to study chickpea interactions with Foc1 (Gurjar et al. 2012). Although host responses during biotrophic infections are facilitated by salicylic acid (SA)-dependent pathways, gene expression analyses in the above-referenced studies indicate non-traditional responses. The genes encoding sucrose synthases, invertases, and β-amylase were induced in resistant interactions and identified by the cDNA-AFLP technique. The 14-3-3 gene expression was overrepresented in the incompatible interaction, signifying potential nutrient starvation, and the resistant interaction potentially copes with this sugar starvation by over-inducing sugar-metabolism genes. This study was highly indicative that sugar also acts as a signaling molecule in response to pathogen perception (Gupta et al. 2010). A comparative study of compatible and incompatible interactions with Foc races 1, 2, and 7 was conducted in chickpea using the cDNA-RAPD method. This study identified a role of plant glucosyltransferase genes in resistance response. Further, race-dependent defence responses were detected (Gurjar et al. 2012). A similar study with compatible and incompatible interactions with Foc race 1, 2, and 4 was conducted recently with the Long SAGE method. Clustering analysis and interaction networks of differentially expressed genes (DEGs) identified that the resistant interaction is categorized by reactive oxygen species (ROS) production, SA production, lignification, R-gene expression, and hormone homeostasis. The incompatible interaction was enhanced for actin depolymerization genes, aquaporin genes, and tetrapyrrole synthesis genes (Upasani et al. 2017). Another study describing the transcriptional alterations during compatible and incompatible chickpea interactions with Foc1 was done using RNAseq analysis. Plant pathogen interaction networks created with this transcriptional data identified numerous nodal hub genes that modulate defense responses and could be further characterized for resistance (Gupta et al. 2017). A microarray-based study of compatible and incompatible chickpea interaction transcriptomes with Foc1 was used to create regulatory gene networks. A total number of 76 disease and immunity-related genes were identified by this study. During wilt interactions, the gene regulatory networks identified transcriptional plasticity in immune pathways and disease pathways. This work also emphasized that the primary metabolic components are shared between defense and disease (Ashraf et al. 2018).

In some fungal genera, including Fusarium, genes encoding the production of some molecules have evolved on chromosomal regions that are not required for saprophytic growth and are thus known as "conditionally dispensable chromosomes" (CDCs, also known as supernumerary, accessory, lineage-specific, B-chromosomes or mini-chromosomes) in contrast to "core" chromosomes whose gene content is vital and preserved across generation. Dispensable genomic regions encoding genes that play a role in pathogenicity and host specificity, including effector genes, have been identified in Fusarium isolates infecting a range of host plants. CDCs have also been identified in other fungal species including several Ascomycete phytopathogens and have been found to play vital roles in pathogenicity and host-range delineation. The first Fusarium CDC identified was from *F. solani* and had a cluster of six pea-pathogenicity (PEP) genes involved in detoxifying the plant defense compound pisatin produced by *Pisum sativum* (garden pea). For some Fusarium species together with the tomato-infecting *F. oxysporum* f. sp *lycopersici* (Fol), the genes that exist on CDCs define its host range, and these chromosomes when transferred to non-pathogenic species can confer pathogenicity on a new host. Dispensable regions of the genome are presumed to be sustained while they convey an evolutionary benefit, to allow for adaptation of novel genes separately from regions containing core conserved genes for the transfer of genetic material, e.g., conferring pathogenicity on a new host. The clustering of genes important for pathogenicity on a small CDC chromosome that could be transferred would provide a highly selective advantage for a "one step" horizontal transfer event that could allow an isolate to become pathogenic on a new host. Probably, genes that do not confer pathogenicity on the new host would be more susceptible to shuffling and subsequent loss, as has been detected for Fol and *F. solani* CDCs relative to core chromosomes. Genomic mapping and sequencing of Fusarium species have discovered chromosome numbers to be highly plastic, ranging from

4 to 17. The common ancestor species has been projected to have only 11 chromosomes, with the increase in chromosome number due in part to the presence of CDCs, which are supposed to have originated in *F. oxysporum* via horizontal gene transmission from other Fusarium species. Gene content in CDCs is often relatively sparse but enhanced in transposable elements. For example, less than 1/8 of Fol pathogenicity CDC 14 is predicted to encode protein-coding genes, and these are predominately proteins of unknown function. In contrast to core chromosomes, CDCs are enriched for pathogenicity-related proteins, secreted proteins, and proteins involved in secondary metabolism. Some Fol CDC genes vital for pathogenicity encode the secreted in xylem (SIX) effector proteins and are often associated with distinct repeat types. SIX proteins, first identified in the xylem sap of Fol-infected tomato plants, are small, secreted, and often cysteine-rich. So far, 14 families of SIX proteins have been identified, with low similarity with each other or other known fungal proteins. Several have characterized roles in virulence and/or avirulence with their cognate host R genes identified, although for the most part, their biological function within the host remains unknown.

In faba bean (*Vicia faba*) where whole-genome sequence information is not available, Super SAGE and Deep Super SAGE coupled with next-generation sequencing have been used effectively to study transcriptomes during Ascochyta infections. During resistant interactions in the faba bean-*Ascochyta fabae* infection development, genes associated with jasmonic acid (JA) signaling and pectin esterase-encoding genes were identified with the Super SAGE technique (Madrid et al. 2013). In another study, de novo transcriptome assembly was used to identify transcripts in susceptible and resistant interactions of faba bean and *A. fabae*. Genes encoding LRR proteins, Rho2 GTPase-activating protein 2 (RGA2), numerous plant growth regulators, heat shock proteins, chitin elicitor-binding protein, and those genes that produce chlorogenic acid, scopoletin, and flavonoids were among the important genes involved in fava bean defenses. Functional characterization of these candidate genes will be the next crucial step toward including them in breeding programs.

Aphanomyces root rot caused by *Aphanomyces euteiches* is a major soil-borne oomycete pathogen that infects several legume species, including pea. In an attempt to know the strategies employed during *A. euteuches* interactions with *Medicago truncatula* in a compatible interaction, a cDNA-AFLP approach was first deployed to know the optimal infection time for estimation, followed by cDNA enrichment with suppression subtractive hybridization (SSH). This study identified classical PR- and defense genes. The molecular analysis showed abscisic-acid-mediated signaling that could induce PR-10 protein. The PR-4 protein-containing hevein the domain, which could bind chitin, was also identified. A recent study was shown to compare the transcriptional responses in compatible interactions of pea plants with both the oomycete pathogens deliberated here—*Phytophthora pisi* and *A. euteiches* using *M. truncatula* microarray. The study discovered different recognition and signaling components in the host against the two pathogens. Pathogen-associated molecular pattern (PAMP)-triggered immunity (PTI) and effector-triggered immunity (ETI) responses were detected in the initial stages of infection with both pathogens. JA- and ET-hormone signaling was associated in both interactions, while the auxin-induced small auxin upregulated RNA (SAUR) family proteins were specific to *A. euteiches*.

The interactions of downy mildew pathogen, *Peranospora viciae* f. sp. p*isi*, with pea leaves, were studied with SSH cDNA libraries. The study identified downy mildew resistance genes RPP6/6/27 involved in this interaction.

Phakospora pachyrhizi causal agent of Asian soybean rust (ASR) is a devastating disease that is recorded among the top five biotic threats to agriculture. Although six resistance genes, resistance to *P. pachyrhizi* (Rppl–6), have been identified in soybean that confers resistance to ASR in a race-specific manner, no single soybean genotype can confer resistance to entire races of the rust fungus. Numerous studies have been done to identify key players in resistance to ASR. Primary studies to identify key players in the R-gene-mediated response of ASR deployed the SSH cDNA method. These studies showed a time-dependent coordinated gene expression pattern in Rpp-mediated resistant and susceptible interactions and identified the role of peroxidases and lipoxygenases in resistance. Other work using whole-genome microarrays confirmed these two results and provided further details (Soria-Guerra et al. 2010a,b; Morales et al. 2013). Numerous of these studies have reported an overrepresentation of transcription factors (TFs) and the roles of flavonoids and cell-wall lignification in the active resistance mechanisms. Metabolite analysis of the ASR interactions has established some of these findings. Numerous genomic studies involving TFs identified roles for WRKY, the Basic Leucine Zipper (bZIP) domain, and predicted TF families in resistant interactions (Aoyagi et al. 2014). The Super SAGE technique was used to identify numerous antimicrobial peptides such as defensins, thionin, and lipid transfer protein (LTP) family genes in cowpea and soybean infected with the rust pathogen *P. pachyrhizi* (Kido et al. 2010).

12.4.2 Pathogen Perspective

Members of the kingdom of fungi, bacteria, or viruses comprise numerous plant pathogens causing diseases in several crops. Among them, few are having a broad host range and few cause diseases in a specific host. For developing a disease, the interaction between host and pathogen is the most important part during the course of infection. Pathogenomics provide detailed information on genetic factors that are regulating different pathways involved in disease epidemiology and pathogen biology. Analyzed data generated from several sequencing projects provide deep insight into the genomes of pathogens infecting different hosts including legumes and that have a diverse way of the life cycle. The pathogen genome studies revealed that the pathogen possesses several hundred genes, which encoded candidate secreted effector proteins that can be highly divergent even among related species. Also, the pathogenomic studies (i) explore the putative genes, effectors, and regulators that are associated with pathogenicity, virulence, and secretion of primary and secondary metabolism; (ii) help in comparison of such genes among related pathogens; (iii) identify the potential genetic targets for chemical control; and (iv) characterize the complex dynamics of host × pathogen interactions during establishment of disease in a better way. In addition, the studies help to understand what are the necessity of lineage-specific genes on mobile and partly dispensable chromosomes and how the pathogenicity and host range determinants are constituted and maintained intraspecifically and possibly interspecifically through transferable lineage-specific genes.

Also, it gives detailed information about pathogens' convergent lifestyle-specific adaptations and how shaping the genomes maximizes pathogenic success according to the different infection strategies employed. The comparative analysis of the pathogen genomes provides new insights into how the common and divergent virulence strategies are evolved within the pathogens for invading related plant species. The genomic information of some economically important pathogens infecting legumes is discussed below.

12.4.2.1 Fusarium Oxysporum f. sp. Ciceris

FW and root rot caused by members of the *Fusarium oxysporum* species complex are major constraints to the production of different economically important crops worldwide including legumes. *F. oxysporum* spp. are the most important globally ubiquitous soil-borne fungal pathogens in plants. Among them, *Fusarium oxysporum* f. sp. *ciceris* is commonly known for its pathogenicity to chickpea. The genome of *Fusarium oxysporum* (Fo) consists of a set of 11 "core" chromosomes, shared by most strains and responsible for housekeeping, and one or several accessory chromosomes. The whole genome of a Foc race 1 isolate (Foc-38-1) was sequenced and assembled into ~54.8 Mb encoding for 16,124 proteins. Further analysis to aid identification of proteins common to the legume-infecting ff. spp. through ortholog comparisons across 44 diverse fungal species both closely and distantly related but sharing similar hosts identified 1,090 paralog groups that are unique to Foc 38-1. Most of the candidate effector genes are arranged into a common chromosome. Among the effector genes, SIX genes encoded SIX effector proteins are very important for the pathogenicity of *Fusarium oxysporum* group of fungi. A total of reported 14 SIX genes in *Fusarium oxysporum* spp., five SIX genes, namely SIX 5, SIX 8, SIX 11, SIX 13, and SIX 14, were identified in the *Foc*. Among them, SIX 8 is the most important gene for virulence of *Foc*.

12.4.2.2 Macrophomina Phaseolina

M. phaseolina is a soil-borne plant pathogenic fungus that infects more than 500 plant hosts including major food, fiber, and oil crops. The pathogen genome laid the foundation to elucidate its specialized mechanism to infect diverse hosts. For that reason, it is necessary to understand the genetic basis underlying pathogenicity, which will ultimately help in controlling the pathogen and making resistance strategies. Whole-genome analysis revealed that *M. phaseolina* is distinct from other known phytopathogenic fungi. The assembled sequence of the complete genome was ~49 Mb in size with predicted 14,249 open reading frames (ORFs). The genes encode secreted oxidases, peroxidases, and hydrolytic enzymes involved in degrading cell wall polysaccharides and lignocelluloses to penetrate into the host tissue were also identified. A significant number of genes encode MFS-type membrane transporters, P450s, transposases, secondary and metabolites glycosidases were also found. The genomic data indicate that *M. phaseolina* has a large set of genes that helps to adapt a wide range of osmotic and pH environments. In addition, *M. phaseolina* possesses a large number of genes for broad host range pathogen, pathogen–host interaction genes including those for adhesion, cell wall breakdown, signal transduction, purine biosynthesis, and potent mycotoxin patulin.

12.4.2.3 Scelorotium rolfsii

The soil-borne pathogen *S. rolfsii* has a wide host range and can infect more than 500 plant species due to its ability of high growth rate along with the production of cell degrading enzymes. It mostly causes rot on the stem or collar region of the plants. The sexual state of the fungus is *Athelia rolfsii*. But it is rarely seen in nature. Along with pulses, the *S. rolfsii* is an economically important pathogen for groundnut also. In 2017, Equebal et al. drafted a *S. rolfsii* genome sequence isolated from groundnut for revealing virulence and pathogenicity-causing genes for better understanding of the host × pathogen interaction. The draft genome assembly was ~73 Mb in size. The sequence annotation identified 16,830 genes, which are involved in virulence and pathogenicity, fungicide resistance, along with putative effector and lethal genes. Secretome analysis revealed several CAZY genes including 1,085 enzymatic genes, e.g., carbohydrate esterases, glycoside hydrolases, auxiliary activities, carbohydrate-binding modules, polysaccharide lyases, and glycosyl transferases. Comparative analysis with other existing ascomycotina genome predicted conserved domain family of WD40, CYP450, Pkinase, and ABC transporter revealing insight into the evolution of pathogenicity and virulence. The secretome analysis identified 349 secretory proteins involved in various activities such as signaling, host–pathogen interactions. Of them, 306 were classical and 43 were non-classical secretory proteins. The enzyme annotation revealed four classes of enzymes, namely transferases, oxidoreductases, isomerases, and hydrolases particularly involved in the metabolism of the host cell wall, which is an important phenomenon required for fungal invasion into host tissues. All such enzymes along with virulence factors or toxins have been reported to degrade the plant cell wall in necrotrophic phase of fungal attack.

12.4.2.4 Ascochyta rabiei

A. rabiei (Pass.) Labr. [teleomorph: *Didymella rabiei* (Kovatsch.) Arx], the necrotrophic pathogen, is commonly known for infecting cultivated annual chickpea, causing chickpea AB. But the pathogen can infect other wild perennial chickpea species also, such as *Cicer monbretti, C. ervoides, C. judaicum*, and *C. pinnatifidum*. Other than chickpea, *A. rabiei* infects several leguminous plants such as lentil (*Lens culinaris*), field pea (*Pisum sativum*), cowpea (*Vigna unguiculata*), faba bean (*V. faba*), alfalfa (*Medicago sativa*), and hairy tare (*Vicia hirsuta*). In 2015, for the first time, the draft genome of *A. rabiei* was assembled at the National Institute of Plant Genome Research (NIPGR), New Delhi, India. The genome size was 34.6 Mb. The assembly covered almost 99% of the total genome. The annotated sequence predicted that a total of 10,596 protein-coding genes encode a large and diverse inventory of secretory proteins, several transporters, and enzymes for different primary and secondary metabolites that help *A. rabiei* for living their necrotrophic lifestyle. A wide range of carbohydrate-active enzymes responsible for complex polysaccharides degradation was also predicted through sequence analysis. The fungus produces several phytotoxins, and the genes responsible for the production of solanapyrones A, B, C, cytochalasin D and a proteinaceous toxin were also recognized. The complete analysis of the *A. rabiei* sequence identified a set of 758 secretory proteins including both classical and non-classical secreted proteins. Of them, several are secretory proteins that showed high cysteine content with numerous tandem repeats.

12.4.2.5 Botrytis cinerea

The fungal pathogen Botrytis cinerea causes serious losses in more than 200 crop species worldwide. The development of strategies for its control is difficult, due to the broad genetic plasticity of the fungus. *B. cinerea* genome sequences have proven to be essential in dissecting the genetic basis of pathogenicity. The first genome assemblies of pathogen were 41.2 Mb in size with overall GC content being 42.75%. Sequence annotation predicted that 13,353 genes encoded several proteins including different pathogenicity-causing CAZyme. But no effectors were identified.

12.4.2.6 Fusarium udum

F. udum is a very specific pathogen to pigeonpea. The draft genome of the pathogen was sequenced first time at the National Bureau of Agriculturally Important Microbes (NBAIM), Mau Nath Bhanjan, Uttar Pradesh, India, with the Illumina NextSeq sequencer, using a HiSeq 2000 platform (Srivastava et al. 2018). The assembled genome was 56.75 Mb in size with 42.44% GC content. Analysis of the overall genome predicted 11,829 protein-coding genes, 296 tRNAs, and 53 rRNAs. Among the annotated genes, a total of 8,928 genes were categorized into functional groups using the KOG database. Furthermore, the genome encompassed 15,649 transmembrane helices, 3,682 transporter genes, 2,858 carbohydrate-active enzymes, 1,439 signal peptides, and 1,060 putative genes related to its virulence and pathogenicity.

12.4.2.7 Fusarium solani

F. solani, a soil-borne fungal pathogen, causes Fusarium root rot in various legumes crops. But the fungus sometimes lives as an endophytic fungus and shows significant antifungal activity. The genome of an endophytic *F. solani* strain JS-169, isolated from a mulberry twig, was sequenced. The draft genome of the strain JS-169 was 45.81 Mb in size with a GC content of 49.91%. The genome was estimated to have 15,465 protein-coding genes. Of them, 1,449 genes were identified for encoding secretory proteins. Besides, 780 genes were for encoding different transcription factors and 133 genes were for cytochrome P450. A total number of 86 genes were also identified, which encode different enzymes responsible for plant cell wall degradation. Other than those genes, 27 genes were identified for encoding peroxidases, and nine genes for encoding laccases. A total of 14 polyketide synthases and 26 nonribosomal peptide synthases were also identified within the draft genome sequence of *F. solani*.

12.4.2.8 Rhizoctonia solani

Infection and incidence of *R. solani* are generally low in legumes except for common bean (*Phaseolus vulgaris* L.). It causes hypocotyl rot, especially in common bean, and sometime in chickpea. But if conditions are favorable, it can cause huge yield loss in legumes. The pathogen also causes sheath blight in rice. The genome of AG1-1A strain of *R. solani* isolated from a popular Malaysian rice variety was sequenced. The assembled draft genome of AG1-1A strain was 28.93 Mb in size with 46.99% GC content. After annotation of sequence data, a total of 10,037 genes were predicted in the genome assembly, of which 7,072 genes were mapped to the fungal nonredundant databases (NRDB). Furthermore, two rRNA- (one from 18S and one from 28S rRNA) and 117 tRNA-coding genes were predicted in the assembly.

12.4.2.9 Alternaria alternata

A. alternata is an opportunistic pathogen on numerous hosts and is reported to cause diseases in more than 380 plant species. The pathogen causes leaf spots, rots, and blights on many plant parts. It causes economical losses in legumes mainly in mungbean, urdbean, field pea, and pigeonpea. An isolate (strain Z7) of *A. alternata* from citrus causing Alternaria brown spot was sequenced through the combination of Oxford Nanopore sequencing and Illumina sequencing technologies. The pathogen consists of 11 essential and two CDCs. The chromosome-level genome assembly of strain Z7 comprised 165 contigs with a GC content of 51.01% and contig N50 of 3.08 Mb. The total size of the genome was 34.28 Mb. The genome encompassed 12,067 protein-coding genes, 34 ribosomal RNAs, and 107 transfer RNAs. The genome sequencing revealed that the production of host-specific ACT toxin is regulated by the ACT toxin gene cluster (ACTT1, ACTT2, ACTT3, ACTT5, ACTT6, ACTTS2, and ACTTS3) located in the CDC (Hatta et al. 2002; Wang et al. 2019b). Genetic inactivation of the ACTT gene can block the biosynthesis of ACT toxin, leading to complete loss of virulence (Ajiro et al. 2010; Miyamoto et al. 2010).

12.4.2.10 Cercospora canescens

C. canescens is a hemibiotrophic pathogen causing leaf spot disease on mungbean and urdbean. In 2015, a group of researchers from Benaras Hindu University, Uttar Pradesh, India, drafted genome sequences of a *C. canescens* isolate. The assembled genome sequence of the pathogen was ~33.97 Mb with 10,627 protein-coding genes. Sequence annotation identified a total of 2,842 proteins involved in biological processes and molecular functions. The identified proteins were mainly involved in infection process used to compromise nutrients or destroy host tissues. Besides, the genes encoded gycosidases, transposases, and cytochrome P450s, which are functionally involved with signal transduction, cell wall breakdown, transporters, host stomata perception, adhesion, polyketide synthase, and cercosporin. A total of 528 simple sequence repeats were also identified from the genome sequence assembly of *C. canescens*. The gene annotation study provides insights into the pathogenic mechanism and a better understanding of the virulence differentiation of *C. canescens*.

12.5 NOVEL GENES INVOLVED IN DISEASE SUSCEPTIBILITY/ RESISTANCE IN LEGUMES CROPS

Looking at the losses caused by various diseases, the development of disease-resistant varieties is the best way to solve these problems, but the use of conventional breeding methods is limited due to a lack of efficient and reliable screening methods. The traditional approach to plant breeding focuses on selecting desirable plants based on their phenotypic characteristics. The majority of agronomically important features are quantitative traits that are influenced heavily by climatic circumstances. As a result, phenotypic selection is less precise and takes longer. To make the breeding program more robust, modern biotechnological technologies such as next-generation sequencing, whole-genome sequencing, whole-transcriptome analysis (Sharma et al. 2016), genetic engineering, genome editing, and marker-assisted selection (MAS) should be used. Different molecular markers in grain legumes were identified using these methods to ensure disease resistance in grain legumes (Table 12.3). The availability of

TABLE 12.3
List of Mapped-Based Genes Conferring Disease Resistance in Major Grain Legumes Using Various Molecular Markers

Crop	Disease	QTLs/Resistant Genes	Type of Marker(s)
Pigeonpea	Fusarium wilt	Fw Gene	RAPD
		C.cajan_03203	SNP
		qFW11.1, qFW11.2 and qFW11.3	SNP
		NBS-AD and LRR	TRAP
		Mapped as a QTL	RAPD
	Sterility mosaic disease	qSMD3 qSMD4 qSMD5 qSMD6	SSR
		C.cajan_01839	SNP
		CcLG11	SNP
		SMD	AFLP
Chickpea	Ascochyta blight	QTL-1, QTL-2	RAPD and ISSR
		QTLs	RAPD
		Ar19	RAPD
		QTLar2b	SSR
		Ar19	STMS
		QTLar1, QTLar2	SSR
		QTL	SSR
		QTLAR3	SSR
		QTL2	SSR
		QTL	SSR
		QTL	STMS
		AB-Q-SR-4-1	SSR
		QTLs	SNP
		CaETR	SCAR
	Fusarium wilt	Foc3	SSR
		TR59	STMS
		QTLfoc02, QTLfoc5	SSR
		Foc1, Foc2, Foc3	SSR
		FW-Q-APR – 6-2	SSR
		Foc-1	STMS
		QTL (GSSR 18-TC14801)	SSR
		QTLCaLG02	SSR and SNP
	Botrytis grey mould	LG3, LG 6A	SSR
Mungbean	Powdery mildew	qPMR-1 and qPMR-2	SSR
		QTLs	RFLP
		QTLs	RAPD, CAP, AFLP
		V4718	ISSR-RGA
	Mungbean mosaic virus	MYMV resistance gene	SCAR

(Continued)

TABLE 12.3 (*Continued*)
List of Mapped-Based Genes Conferring Disease Resistance in Major Grain Legumes Using Various Molecular Markers

Crop	Disease	QTLs/Resistant Genes	Type of Marker(s)
		qYMIV1, qYMIV2, qYMIV3, qYMIV4, and qYMIV5	SSR
		OPB07-SCAR_583 (MYMVR-583)	SCAR
	Cercospora leaf spot	qCLS	SSR
Urdbean	Yellow mosaic virus	Monogenic	STS-RGA
	Mung bean yellow mosaic India virus	QTL	SSR and RGH markers
	Mungbean yellow mosaic virus	MYMV resistance gene	RAPD
Field pea	Powdery mildew	er	SCAR
		er	RFLP
		er (Sc-OPO-181200)	RAPD/SCAR
		er (OPD-10650)	RAPD
		Er (PSMPSAD60, PSMPSAA374e, PSMPA5, PSMPSAA369, PSMPSAD51	SSR
		er1-6	SNP
		er-1	STMS
	Ascochyta blight	QTLs	SSR, RAPD and CAPS
		QTLs	STS
		Asc2.1, Asc4.2, Asc4.3 and Asc7.1	Candidate defense-related sequences
		QTL abIII-1 and abI-IV-2	SNP
		abI-IV-2.1 and abI-IV-2.2	SNP
		QTLs	SSR and SNP
		MpII.1, MpIII.5, MpV.2 and MpV.3	SNP
	Fusarium wilt	Fw	RAPD
		Fw	RAPD, AFLP
		Fnp	SSR and RAPD
		w_Trap_480, Fw_ Trap_340, and Fw_Trap_220	SCAR
	Pea rust	Up1	RAPD
	Pea seed-borne mosaic virus	Sbm-1	STS
	Pea common mosaic virus	mo	RFLP
Faba bean	Bean golden mosaic virus	Bgm-1	RAPD
	Angular leaf spot	Mapped as a QTL	SNP
	Bacterial blight	Mapped as a QTL	RAPD

good sources of resistance materials and the identification of markers associated with disease resistance major genes or QTLs that can be used in an MAS breeding program to accelerate resistant screening for large populations are all important factors in developing biotic-stress-resistant varieties. The molecular markers have opened new avenues for improving and speeding up resistance breeding programs, particularly for disease resistance. They have a great potential to meet the demand for host–plant resistance exploitation (Viashalakshi et al. 2017). As a result, while performing robust classical genetic analysis, genetic markers such as RAPD, RFLP, AFLP, SNP, SSR, SCAR, STS, CAPS, STMS, ISSR, and TRAP are an important component of molecular breeding programs including grain legumes (Table 12.3). These markers could be used to find resistant genes and their locations in the genome (Sharma et al. 2018). In this section, we summarize the available data on the identification of molecular markers associated with disease resistance genes/QTLs and their potential application to improving disease resistance traits in legume crops.

12.5.1 Pigeonpea

FW is the most important and destructive disease in Indian Subcontinent (Saxena 2008). Due to five different variants of *F. udum* (Tiwari and Dhar 2011), precise phenotyping at the field level becomes difficult for genetic studies. Hence, different reports were found on the genetics of FW resistance. Resistance to FW is controlled by different gene actions in various genetic backgrounds, ranging from single to multiple genes with complementary to duplicate gene actions (Saxena 2008). Several resistance theories against FW have been established. Among the theories, the single dominant gene theory (Kotresh et al. 2006), parallel control of two complementary genes and major genes theory (Bhanu et al. 2018), duplicate genes, two genes, and multiple factors theory (Okiror 2002), and lastly, a single recessive gene theory (Bhanu et al. 2018) were most relevant and indicated the genetic base of FW resistance in pigeonpea (Singh et al. 2016). Recently, an association-based mapping approach has detected a significant association of the SSR marker HASSR18 (accounting for 5%–6% phenotypic variation due to wilt resistance across the years) with the genetic resistance against FW variant 2 in pigeonpea (Patil et al. 2017). Furthermore, the development of varieties resistant to SMD caused by *Pigeonpea sterility mosaic virus (PPSMV)* becomes very difficult through conventional breeding because pigeonpea is a long-duration and often cross-pollinated crop, and phenotyping is tedious due to the existence of three different strains (Patancheru, Bangalore, Coimbatore strains) of *PPSMV* in India (Kulkarni et al. 2003). *PPSMV* resistance was reported to be controlled by a single gene (Ganapathy et al. 2009), oligo-genes (Gnanesh et al. 2011), and two genes with inhibitory gene action (Daspute et al. 2014). While four QTLs for Patancheru *PPSMV* isolate and two QTLs for Bangalore *PPSMV* isolate were also identified (Gnanesh et al. 2011).

Furthermore, the resistant trait was governed by two independent nonallelic genes, designated SV1 and SV2, with inhibitory gene interaction, during the inheritance study of pigeonpea SMD in the segregating population of two crosses, Gullyal white (susceptible) × BSMR 736 (resistant) and BSMR 736 (resistant) × ICP 8863 (susceptible) (Bhairappanavar et al. 2014). In addition, two F2 populations, ICP 8863 × ICPL

20097 (segregating for Patancheru SMD isolate) and TTB 7 × ICP 7035 (segregating for both Patancheru and Bangalore SMD isolates), were established, and F2:3 families were phenotyped for resistance to respective isolates of SMD. In addition, each mapping population's parental genotypes were screened for over 3,000 SSR markers, intra-specific genetic maps with 11 linkage groups were generated, and 120 and 78 SSR loci were developed for the ICP 8863 ICPL 20097 and TTB 7 ICP 7035 populations, respectively. The creation of intra-specific genetic maps and the identification of QTLs for SMD resistance in pigeonpea were the first investigation of its kind.

In a recent mini-core collection of pigeonpea germplasm, novel sources of resistance to Fusarium wilt and SMD were discovered (Sharma et al. 2012). Because of the limited success in creating SMD-resistant cultivars, major breeding efforts are still required in this regard. Another disease, Phytophthora blight (PB), has been discovered to be more severe in short-duration cultivars than in long/medium-duration genotypes (Ratnaparkhe and Gupta 2007). The few reports on the genetics of PB resistance in pigeonpea imply that resistance to PB in the host may be controlled by a few key genes. The unavailability of a genetic map is the main constraint to make molecular markers or combine molecular and conventional genetic techniques in pigeonpea improvement programs. Recent advancements in pigeonpea genomics pave the potential for disease-resistant breeding using markers to identify resistance genes for several illnesses. With the release of the pigeonpea draught genome sequence (genome size = 833.07 Mb), this crop has become a gold mine of genetic materials (Varshney et al. 2012). Thus, 3,072 SSRs from 88,860 BESs (Bohra et al. 2011), 3,583 SSR markers from ESTs (Raju et al. 2010), and 309,052 SSRs from scanning the draught genome sequence of pigeonpea are now accessible. Furthermore, the pigeonpea research community has 10,000 SNPs available (Varshney et al. 2013). These markers will be extremely helpful in saturating genetic maps with a large number of molecular markers and labeling QTL/genes for critical traits such as disease resistance. For these diseases, several markers associated with resistance genes have been identified, which can be used for multiple disease resistance marker-assisted selection and gene pyramiding. Two RAPD markers (OPM03704 and OPAC11500) were identified to be linked with FW resistance (Kotresh et al. 2006), six SSRs (ASSR-1, ASSR-23, ASSR-148, ASSR-229, ASSR-363, and ASSR-366) (Singh et al. 2016), and five SSR markers (PFW 26, PFW 31, PFW 38. SNP mapping revealed two genes, *C. cajan* 01839 for SMD resistance and *C. cajan* 03203 for FW resistance (Singh et al. 2016). Mapping for SMD was attempted using AFLP markers, and four markers, E-CAA/M-GTG150, E-CAA/M-GTG60, E-CAG/M-GCC120, and E-CAG/M-GCC150, were discovered to be linked with the SMD resistance gene at distances of 5.7, 4.8, 5.2, and 20.7 cM from the resistance loci, respectively (Ganapathy et al. 2009). SMD reaction was found to co-segregate with a single coupling phase short decamer random DNA marker (IABTPPN7414) and a repulsion phase marker (IABTPPN7983) (Daspute and Fakrudin 2015). SSR markers were used to identify six QTLs (qSMD1, qSMD2, qSMD3, qSMD4, qSMD5, and qSMD6) linked to SMD in two different populations (Gnanesh et al. 2011). Another ten QTLs, including three major QTLs linked to SMD resistance, were discovered in three different populations (Saxena et al. 2017). So far, reported linked markers have not been successfully used to transfer resistance genes toward the development of disease-resistant pigeonpea varieties. Validation of these markers across different genetic backgrounds is equally important for their use in breeding programs. The tightly linked markers to disease resistance genes should be developed using multi-parent mapping populations (MAGIC) with very precise phenotyping so that they can be applied across pigeonpea populations. With the availability of high-throughput approaches and tightly linked markers in the near future, marker-assisted back cross (MABC) will be a viable option for breeding resistance to serious diseases such as FW and SMD in pigeonpea.

12.5.2 Chickpea

MAS has been used to isolate near-isogenic lines in chickpea using sequence-tagged microsatellite markers. To aid in the creation of NILs in chickpea, SSR markers firmly related to foc5 (TA59) and foc01 (TR59, TS35) were utilized to aid in the selection of resistant and susceptible genotypes (Jendoubi et al. 2016). ICRISAT generated MABC lines resistant to FW (focl and foc3) and AB in the genetic background of the C 214 cultivar (Varshney et al. 2014). At various agricultural universities in India, MABC lines resistant to foc4 were established in the genetic backgrounds of JG 74, Phule G12, and Annigeri 1. In the background of Pusa 256, another group of MABC lines resistant to foc2 has been established at the Indian Institute of Pulses Research in Kanpur, India (Varshney et al. 2013; Saxena et al. 2016). Globally, eight Foc races have been identified. Races 1A, 2, 3, and 4 have been recorded from India, whereas races 0, 1A, 1B/C, 5, and 6 have been reported from the United States and Spain, respectively. India was the first to report an Foc race-specific resistant cultivar. Major resistance genes in chickpea have been found to control wilt resistance. Races 3 and 5 have monogenic resistance, while races 0, 1A, 2, and 4 have digenic or trigenic resistance. Chickpea wilt resistance was first studied genetically in race 1, where it was discovered to be inherited via a recessive gene. However, it has recently been discovered that resistance to race 0 in chickpea cultivar JG 62 is given by one gene and another gene in CA 2139. Resistance to race 4 is conferred by a recessive, monogenic component in WR 315, but resistance to race 4 is conferred by recessive, digenic factors in cultivar Surutato 77. The resistance of WR 315 to race 2 was discovered to be reliant on the digenic allele in a study (a and b). In WR 315, when both genes are in recessive form (aabb), the cultivar is resistant; however, in cultivar C 104, AABB causes susceptibility. It was also discovered that the third gene C has a role in race 2 resistance. The genetics of resistance to race 5 in C 104 and WR315 is still unknown. Race 1A causes late wilting in several vulnerable chickpea genotypes. Race 1A late wilting was found to be regulated by three genes: two recessive genes, h1 and h2, and one dominant gene, H3, with any two of these three genes conferring resistance. The presence of any two independent genes from the three genes a, b, and C is discovered to govern resistance in race 2, but the presence of any one of these three genes leads to late wilt.

Several researchers have investigated the genetic basis of resistance to AB in chickpea, and several sources of resistance have been identified. Because multiple genes have been identified as conferring resistance to AB in chickpea, knowledge of their genomic locations and linkage map is required to facilitate molecular breeding for developing resistant cultivars. Blight resistance in chickpea is conferred by either a single dominant gene (in genotypes ILC-72, ILC-183, ILC-200, and ILC-4935) or a single recessive gene (in genotypes ILC-191) or by two dominant genes (in genotypes EC-26446, PG-82-1, P-919, P-1252-1, and NEC-2451) or by one recessive gene (in genotype BRG-8). In various chickpea lines (in genotypes GLG84038 and GL84099), two dominant complementary genes provided AB resistance, while one dominant and one recessive independent gene were also discovered (in genotype ICC 1468). Kusmenoglu (1990) indicated that AB was regulated by at least two recessive genes, but Dey and Singh (1993) discovered two complementary recessive alleles in chickpea that provided resistance to AB. At least two key QTLs involved in AB resistance in chickpea now have molecular markers (Millan et al. 2006), which could be utilized in marker-assisted breeding programs or for pyramiding AB resistance genes. Botrytis gray mold (BGM) resistance has been attributed to a single dominant gene or complementary dominant genes (Rahul et al. 1995). To detect resistance to dry root rot (DRR) in chickpeas, several researchers used molecular markers such as SSR and RAPD in resistance breeding programs (Talekar et al. 2017). Because phenotypic selection of DRR-resistant lines is always inconvenient due to its significant association with soil moisture depletion, molecular markers are particularly beneficial in screening *R. bataticola*-resistant lines. The identification of DRR resistance gene genetic loci will aid in the molecular understanding of the chickpea *R. bataticola* interaction. For map-based gene cloning and marker-assisted resistance breeding, gene tagging for DRR resistance is required.

12.5.3 Mungbean and Urdbean

Various theories have been proposed to explain resistance to *MYMV* in mungbean and urdbean. According to these theories, *MYMV* resistance in both crops is primarily determined by the inheritance of either a single recessive gene (Dhole and Kandali 2012) or a single dominant gene (Prasad et al. 2015) or two recessive genes (Ammavasai et al. 2004). In bacterial leaf blight, cercospora leaf spot, and powdery mildew of mungbean, researchers discovered that the inheritance of resistance is controlled by either a single dominant gene (Humphry et al. 2003), a single recessive gene (Mishra et al. 1988), or quantitative genes (Chankaew et al. 2011). In urdbean, a single dominant gene controls anthracnose resistance, while a single recessive gene controls powdery mildew resistance (Bindra et al. 2016). The study of genotypic diversity and the discovery of linked markers for the R gene and thc building of QTL maps using molecular markers has enhanced the efficiency of *MYMV* resistance breeding programs (Sudha et al. 2013). With qMYMIV2 on linkage group 2 and qMYMIV7 on linkage group 7, two significant QTLs influencing gene resistance to *MYMV* have been identified (Alam et al. 2014). New *MYMV* resistance donors have also been discovered from interspecific sources. In mungbean variety V4718, Chankaew et al. (2013) discovered a large QTL related with powdery mildew resistance on LG 9 and two minor QTLs on LG4. The first QTL mapping for cercospora leaf spot resistance in mungbean was done in Thailand, where they discovered one main QTL (qCLS) on LG3 between the markers CEDG 117 and VR 393, which explained 66%–81% of the phenotypic variation (Chankaew et al. 2011). As a result, they can be utilized in mungbean breeding programs to detect QTLs connected to *MYMV*, cercospora leaf spot, and powdery mildew resistance as a selection tool in early generations and for subsequent use in creating segregating material.

In urdbean, a single recessive gene (Souframanien and Gopalakrishna 2006) and two recessive genes are found to control yellow mosaic disease (YMD) resistance (Verma and Singh 1986). In mungbean, RAPD markers associated with the YMD resistance gene were discovered and converted to SCAR markers (MYMVR-583) for greater repeatability in MAS (Dhole and Reddy 2013). ISSR marker associated with YMD resistance was converted into SCAR marker in urdbean and validated in several resistant urdbean genotypes (Souframanien and Gopalakrishna 2006). Resistance to YMD has been linked to the resistant gene analogue (RGA) markers YR4 and CYR1 in urdbean (Maiti et al. 2011). Cowpea, azuki bean, and common bean markers were found to be beneficial in both mungbean and urdbean before the availability of SSR markers in these crops (Gupta and Gopalakrishna 2010). In urdbean, the SSR marker CEDG180 from cow peas was discovered to be linked to YMD resistance (Gupta et al. 2013). Through composite interval mapping of mung bean, three QTLs for MYMIV resistance were found in India: qYMIV1, qYMIV2, and qYMIV3; and two QTLs in Pakistan: qYMIV4 and qYMIV5 (Kitsanachandee et al. 2013). Four main QTLs for MYMIV resistance were identified using AFLP and SSR markers (Chen et al. 2013). ISSR 8111357, YMV1-FR, and CEDG180 were discovered to distinguish between YMV resistant and susceptible urdbean genotypes and can be exploited for MAS (Gupta et al. 2015).

At the Bhabha Atomic Research Centre (BARC) in Mumbai, India, three separate dominant genes (Pm1, Pm2, and Pm3) regulating resistance reaction to powdery mildew illness were discovered in mungbean (Reddy 2009). In mungbean, the RFLP markers were the first markers used to identify the linkage between a major powdery mildew resistance locus and the marker (Humphry et al. 2003), whereas two QTLs for powdery mildew resistance, qPMR-1 and qPMR-2, were reported in mungbean (Kasettranan et al. 2010). In the F2 population of urdbean, the SSR markers DMBSSR 130 and VM 27 were shown to be related to powdery mildew-resistant plants (Savithramma and Ramakrishnan 2016). Cercospora leaf spot resistance is conferred by a single dominant gene (Chankaew et al. 2011). There have been few investigations on cercospora-resistant gene tagging in mungbean and urdbean. CEDC031, CEDG044, CEDG084, CEDG117, CEDG305, VR108, and VR393 were shown to be related to cercospora leaf spot resistance in the F2 and BC1F1 mungbean populations (Chankaew et al. 2011).

12.5.4 Field Pea

In pea germplasm, two recessive genes (er1 and er2) and one dominant gene (Er3) have been found for powdery mildew resistance (Fondevilla et al. 2007). Gene er1 is found in all resistant pea genotypes, but er2 is found in only a few resistant individuals, according to genetic investigations of resistance to *E. pisi*. For powdery mildew resistance, the newly found dominant gene Er3 is now being described and introduced into a cultivated pea. Several studies have been conducted to identify the genomic areas related with the er1 locus using RFLP, RAPD, SCAR, and SSR markers to aid

MAS in pea breeding efforts (Srivastava et al. 2012). Nisar and Ghafoor (2011) evaluated the efficiency of MAS for powdery mildew in the F2 population of the hybrid Fallon (erl)/11760-3(ER1) with RAPD marker OPB18430, which is connected to the erl gene at a distance of 11.2 cM. Sun et al. (2016) recently found erl–6, a new erl allele that confers powdery mildew resistance in Chinese pea. Through transcript analysis, they discovered that the resistance impact of the erl–6 allele was consistent with that of the erl–2 allele.

AFLP, RAPD, SSR, STS, and EST-SSR markers were used to create many linkage maps in pea to identify genomic areas linked with AB resistance (Fondevilla et al. 2008). Scientists are working on AB resistance in pea and have discovered more than 30 QTLs linked to AB resistance across all seven linkage groups (LGs) (Timmerman-Vaughan et al. 2004). In two pea mapping populations, Timmerman-Vaughan et al. (2004) found 19 QTLs for AB resistance on LGs I, II, III, IV, V, and VII, as well as Group A, and VI. Fondevilla et al. (2008) reported six QTLs on LGs II, III, IV, and V in *P. sativum ssp. syriacum*, whereas Fondevilla et al. (2011) reported three additional QTLs on LGs III and VI. Carrillo et al. (2014) discovered four novel QTLs on LGs II, III, and V that influence cellular processes in *P. sativum ssp. syriacum* resistant to AB. According to Fondevilla et al. (2011), the QTLs MpIII.1, MpIII.3, and MpIII.2 found in *P. sativum ssp. syriacum*. QTLs mpIII-1, mpIII-3, and mpIII-5 found in *P. sativum*. In peas, co-localization of QTLs for disease resistance with candidate genes implicated in defense responses to *P. pinodes*, including RGAs, PsDof1 (a putative transcription factor), and DRR230-b (a pea defensin), was reported (Timmerman-Vaughan et al. 2016). Furthermore, Jha et al. (2015) found a significant link between AB scores and SNPs found in candidate genes PsDof1 (PsDof1p308) and RGA-G3A (RGA-G3Ap103). In an interspecific pea population (PR-19) produced from a hybrid between Alfetta (*P. sativum*) and wild pea accession P651 (*P. fulvum*), nine QTLs for AB resistance were recently found (Jha et al. 2016). Jha et al. (2017) fine-mapped QTLs abI-IV-2 and abIII-1 in RIL-based HIF populations using SNP-based GBS. Within the abI-IV-2 QTL for AB resistance, they discovered two additional QTLs, abI-IV-2.1 and abI-IV-2.2, which accounted 5.5%–14% of the total phenotypic variation.

According to McClendon et al. (2002), most Fop races have a single gene that controls resistance. A single gene, Fw, was found to provide resistance to FW race 1 on linkage group III. Resistance to Fop race 2 was thought to be qualitative, and a single gene (Fnw) named Fusarium near wilt was attributed to it. The main locus Fnw has now been localized to LG IV of pea and called Fnw4.1 (McPhee et al. 2012). McPhee et al. (2012) discovered two further major minor QTLs on LG III for Fop race 2, namely Fnw 3.1 and Fnw 3.2. The gene Fwf, which confers resistance to Fop race 5, has been assigned to LG II (McClendon et al. 2002). The genetics of Fop race 6 resistance are unknown; however, few scientists believe it is driven by a single dominant gene. Using the TRAP marker technique in conjunction with BSA, Kwon et al. (2013) successfully developed SCAR markers tightly related to Fw in pea. They described the development of three relevant SCAR markers in pea that are associated to Fop race 1 resistance. Later, the accuracy for MAS was increased by combining two SCARs, Fw Trap 480 and Fw Trap 220, in a multiplex PCR (Kwon et al. 2013). Rakshit et al. (2001) used a RAPD marker OPD 10650 connected to the powdery mildew resistance locus at 3.6 cM to perform marker-assisted breeding for powdery mildew resistance in pea. However, OPD10650 was not found to be effective for MAS in progeny resulting from a cross of the resistant cultivar Highlight (er1) with the susceptible cultivar Radley by Tiwari et al. (1998). Because er1 is a recessive gene, obtaining homozygous resistant BCnF2 parents for the next backcross cycle necessitates a generation of selfing after each backcross generation. MAS is an excellent method for introducing the er1 gene into superior cultivars that are susceptible to powdery mildew (Ghafoor and McPhee 2012).

12.5.5 Faba Bean

Traditional breeding procedures have been successfully used to generate disease-resistant faba bean cultivars for a long time. However, recent advances in faba bean genetics and genomics, as well as the possible use of molecular breeding methodologies to generate disease-resistant cultivars, have proven to be the way forward in generating disease-resistant cultivars. During the embryo growth stage of the faba bean variety "Windsor," the first transcriptome of faba bean emitted roughly 5,000 expressed sequence tags (EST) (Ray and Georges 2010). The functional classification and relative expression of the transcripts were determined using transcriptional profiling of the faba bean embryo during the early middle stage of plant development (Braich et al. 2017). Following that, investigations began to include numerous, distinct inbred genotypes, as well as multiple organs (Webb et al. 2016). Illumina sequencing of a library of mixed tissues enriched with embryo transfer cells produced the most comprehensive transcriptome coverage. The generation of more molecular markers has been enabled by the growing amount of transcriptome and genome sequencing datasets, resulting in an increase in map coverage and marker density (O'Sullivan and Angra 2016). In addition, a vast number of mapping populations are currently being used to mine molecular diversity in target regions of interest, such as SNPs and SSRs. A total of 551 SNPs and 71 SSRs were discovered in the faba bean genome using SNP mining and mapping techniques on a genome-wide scale in a study (Webb et al. 2016). SNPs were recently linked to Orobanche and Ascochyta resistance, according to Ocana-Moral et al. (2017). This new breakthrough will make early selection for these two biotic stresses easier, which will help to raise selection intensity and improve faba bean breeding efficiency.

12.6 MOLECULAR BREEDING APPROACHES FOR DISEASE RESISTANCE IN LEGUMES

12.6.1 Going Back to the Nature: Using Wild Relatives for Legumes Diseases Resistance

The legume family is the third-largest flowering family with over 18,000 species worldwide that are a rich source of proteins, oils, and nutrients. However, the production potential of legume-derived food cannot meet increasing global demand (Zhang et al. 2019). Wild legumes represent a large assemblage of wild species adaptive to various habitats and harbor amusing genetic diversity for the enhancement of the agronomic, nutritional, and medicinal values of the domesticated legumes. Former domestication processes instigated a high degree of relatedness among modern varieties of crops, leading to a narrower genetic base of cultivated germplasm predisposed to pests and diseases.

Crop wild relatives (CWRs) harbor genetic diversity experienced by natural selection in a range of environments. To wholly understand and exploit local adaptation in CWRs, various studies in geographical centeres of origin combining ecology, physiology, and genetics are desired. With the arrival of modern genomics and computation, united with systematic phenotyping, it is possible to revisit wild accessions and landraces and prioritize their usage for breeding, providing sources of disease resistances; tolerances of drought, heat, frost, and salinity abiotic stresses; nutrient densities across major and minor elements; and food quality traits. The establishment of hybrid populations with CWRs provides breeders with a substantial benefit of a pre-breeding tool for identifying and harnessing wild alleles and provides enormously valuable long-term resources. There is a necessity for further collection and both ex-situ and in situ conservation of CWRs diversity of these taxa in the face of habitat loss and degradation and climate change. The usage of genetic resistance is considered to be the furthermost effective and sustainable strategy to control plant pathogens in agricultural practice. Domestication of wild relatives led to crop distribution going away from their original centeres (Vavilov 1926), and their pathogens shadowed this distribution (Turcotte et al. 2017). Wild legumes signify a large group of wild species adaptive to diverse habitats and harbor rich genetic diversity for the improvement of the agronomic, nutritional, and medicinal values of the domesticated legumes. Accumulating evidence advocates that the genetic variation retained in these under-exploited leguminous wild relatives can be used to expand crop yield, nutrient status, and resistance to tolerance to biotic and abiotic stresses through the integration of omics, genetics, and genome-editing modern technologies.

Wild legumes also represent a genetic reservoir by providing exotic genes/alleles conferring broad-spectrum resistance to many different diseases, such as pathogens and fungi. Many wild relatives (Figure 12.4) are resistant to various diseases. For example, *L. orientalis* is resistant to stemphylium blight (*Stemphylium botryosum*) (Kant et al. 2017), and *L. ervoides* and *L. nigricans* are resistant to AB (*A. lentis*) (Tullu et al. 2010; Dadu et al. 2017, 2018). Wild lentils were also found to be resistant to anthracnose (*Colletotrichum truncatum*), FW (*F. oxysporum*), PM (*E. polygoni*), and rust (*U. fabae*) (Singh et al. 2018). Further assessment of these resistance sources is needed to evaluate the breadth of the resistance spectrum and to discover the resistance loci for lentil crop improvement.

Domesticating plants to raise as crops can turn out to be a double-edged scythe. On the one hand, selecting exact desirable traits, such as high yields, can raise crop productivity. But other vital traits, such as resistance to pests, can be vanished. To alleviate this, researchers frequently turn to the wild relatives of crops. These wild relatives continue to evolve in nature, often under adverse situations. They own several useful genes for desirable traits. In a novel study, scientists report substantial strides in transferring disease- and stress-resistance traits from wild relatives of numerous legumes to their domesticated varieties. This research was conducted at the International Crops Research Institute for the Semi-Arid Tropics (ICRISAT) in Patancheru, India. There are numerous wild relatives of these crops that are resistant to pests and diseases.

FIGURE 12.4 Picture showing a list of wild relatives of *Cajanus* spp.

Combining those traits would improve the resilience of domestic legume varieties, but variances in their genetic makeup make it difficult to interbreed wild and domestic strains. Sharma et al. (2015) and her contemporaries used their skills and contemporary technology to breed wild groundnut varieties, then recognize which of these were also resistant to fungal infections. These were then crossed with cultivated groundnut varieties to develop novel breeding lines with sound resistance and yields. CWRs are the reservoir of numerous useful genes and traits. Using this concealed treasure, disease-resistant cultivar could be developed in future.

Three single resistance genes, named *er*1, *er*2, and *er*3, have been reported so far in pea and its wild relatives (Fondevilla and Rubiales 2012) to afford incomplete resistance to powdery mildew caused by *E. pisi*. Of these genes, *er*1 was recognized as an MLO gene (Rispail and Rubiales 2016). A combination of knowledge of pea germplasm diversity with that of the *eIF4E* gene for virus resistance (Ashby et al. 2011) and screening of nearly 3,000 accessions with acknowledged geographical origin including of wild *Pisum sp.* led to the identification of new alleles of resistance (Konecna et al. 2014). List of wild relatives and their detected resistances toward biotic stresses in grain legumes is summarized and presented in Table 12.4.

TABLE 12.4
List of Wild Relatives and Detected Resistances

Crop	Wild Relative	Trait of Interest	Present Status	Proposed Research Strategy	References
Chickpea	Primary genepool *C. reticulatum*	Biotic stresses Fungal resistances	Ascochyta, Fusarium wilt, Botrytis gray mold, and Phytopthora resistance screening (see refs) & currently underway in GRDC-funded projects at Curtin Uni, SARDI & NSW DPI	Phenotype characterization QTL and GWAS mapping	Amalraj et al. (2019)
	Secondary genepool *C. echinospermum*	Viral resistances	–	RNAseq analysis	Sharma et al. (2005)
Field Pea	Secondary genepool *P. abyssinicum*	Fungal resistances	Powdery mildew resistance identified in P. fulvum, Fusarium wilt resistance identified	To continue in CWR germplasm testing and screening	Barilli et al. (2018)
	P. fulvum	Viral resistances	Source of resistance to PSbMV virus identified	To identify underlying loci to be used in marker assisted breeding	Konecna et al. (2014)
Lentil	Primary genepool *L. orientalis*	Fungal pathogens	Identified resistance to Ascochyta blight, powdery mildew; rust, Fusarium wilt, Stemphylium blight	Phenotype characterization	Kumari et al. (2018)
	Tertiary genepool *L. ervoides*	Fungal pathogens	Identified resistance to anthracnose, Stemphylium blight, powdery mildew; rust, Fusarium wilt, Ascochyta blight	–	Bhadauria et al. (2017)
	Quaternary genepool *L. nigricans*	Fungal pathogens	Identified resistance to Fusarium wilt, powdery mildew, rust, Ascochyta blight, Stemphylium blight, Anthracnose	–	Podder et al. (2013)

12.6.2 Characterization of Germplasms and Breeding Lines for Resistance Donor

Since the dawn of human civilization till the present, the conventional breeding approach has been utilized in plant breeding. Through the application of this technique, numerous crop varieties have been developed. Resistance breeding is greatly aided by the use of hybridization with locally suited varieties/genotypes and the use of resistance selection in different varieties against various diseases. Although the method was initially created for self-pollinated crops, pure-line selection is the greatest way to maintain the consistency in the characteristics of the existing cultivars that are utilized to create hybrids in both self- or cross-pollinated crops. The pedigree approach is better suited for features such as seed size, shape, disease, pest, height, and maturity that, if controlled by a single gene, are highly heritable in nature and are frequently employed to create disease-resistant cultivars. The backcross approach is frequently employed for inter-varietal character transfer, inter-specific disease resistance transfer, cytoplasm transfer, generation of transgressive sergeants, and isogenic line production. The majority of the time, this technique is employed to enhance locally prevalent cultivars that are deficient in one to two features. The technique involves combining numerous related lines chosen from several pure-line selections and lines with various disease-resistance genes. It is known as multiple disease-resistance selection. Several pure lines with various resistance genes are developed through backcross programs to create multiline varieties. According to Borlaug (1958), a total of eight to ten lines are necessary to exploit horizontal resistance, and all eight to ten lines must be developed through backcrossing of several multiline varieties with varying degrees of disease tolerance. Further, the molecular marker-assisted selection breeding (MASB) is particularly helpful in revealing the genetic diversity of germplasm and labeling genes with economic significant. Figure 12.5 shows different screening techniques followed for characterization of resistant donors.

Nearly, 7.4 million germplasm accessions of different crop species have been collected and/or assembled and conserved in over 1,750 ex-situ genebanks worldwide (FAO 2010) and the grain legumes found to be second largest collection (15%)

a - Pigeonpea screening against wilt and SMD

b - chickpea screening in wilt sick plot

FIGURE 12.5 Picture showing screening of germplasm and breeding lines under artificial and natural epiphytotic conditions. (a) Pigeonpea screening against wilt and sterility mosaic disease in a sick plot and (b) Chickpea scrrening in a sick soil against wilt.

next to cereals (45%). Globally, approximately 1.1 million grain legume germplasm accessions are preserved in various genebanks; of these, ICRISAT genebank holds approximately 50,000 accessions, including domesticated and wild relatives of chickpea (20,267 accessions), pigeonpea (13,771 accessions), and groundnut (15,445 accessions) from 133 countries. Other significant genebanks that house grain legume germplasm in addition to ICRISAT include the National Bureau of Plant Genetic Resources (NBPGR), New Delhi, India (16,881 Cicer, 12,900 Cajanus, and 14,593 Arachis accessions), the International Centre for Agricultural Research in Dry Areas (ICARDA), Aleppo, Syria (13,818 Cicer accessions), the United States Department of Agriculture (USDA), USA (9,964 Arachis accessions), and the Directorate of Groundnut Research (DGR), Junagadh, India (8,934 Arachis accessions).

In 1982, the first genotypes of pigeonpea resistant to Fusarium wilt were identified. Thirty-three pigeonpea genotypes were found to be resistant to *F. udum* after Nene and Kannaiyan (1982) screened more than 11,000 different kinds. With only a 5% disease incidence, the five pigeonpea genotypes ICP 7182, ICP 7336, ICP 8863, ICP 8869, and BDN 1 were discovered to be the most effective resistant lines (Nene and Kannaiyan 1982). Following that, ICRISAT screened a broad range of germplasm to find accessions that were tolerant to *F. udum*. ICP 4769 (IC 306520), Maruti (ICP 8863), IC 306497, ICP 7118, ICP 9168, ICP 7182, ICP 11299, ICP 10958, ICP 9145, Arhar, NP (WR) 15, Birsa, Pragati (IC 306505), and Asha (ICPL 87119) are some examples of promising accessions for resistance that were found. Six germplasm accessions (ICP 6739, 8860, 11015, 13304, 14638, and 14819) were found to be combined resistance to FW and SMD (Sharma et al. 2012). Four genotypes (ICPL 20094, ICPL 20106, ICPL 20098 and ICPL 20115) have been found to be the most stable and resistant to SMD in multi-environment field testing (Sharma et al. 2015). The performance of three genotypes (ICPL 20096, ICPL 20107, and ICPL 20110) against SMD was generally steady. All of these lines have a medium maturation period and may be excellent sources of pigeonpea breeding programs resistance to FW and SMD. Several varieties were released via the pedigree breeding method, and some farmers continue to use these cultivars today. A few of these iconic cultivars are ICP 7035, C 11, T 7, BDN 1, Bahar, NA 1, ICP 8863 (Maruti), and C 11 (Bohra et al. 2017). Also found in the germplasm were some significant sources of disease resistance, including IPA 8F, IPA 9F, IPA 16F, and ICP 7035 for both fusarium wilt and sterility mosaic disease, 15-3-3 for FW, ICPL 366 for SMD, ICP 7105 for Alternaria blight, PB 9 for PB, and many more. Additionally, combined resistance was identified in ICPL 87119 (Asha) 1993 C 11 × ICPL 6 Resistant to FW and SMD, Malviya Chamatkar (MAL 13) (MA 2 × MA 166) × Bahar)) tolerant to FW and SMD, LRG 41 Selection from Chilakaluripet in Guntur (Andhra Pradesh) resistant to FW and SMD, Jawahar Tur (JKM 189) (ICPL 87119 × Plant 142) Moderately resistant to FW, SMD and PB, Bahar selection from Motihari (Bihar) Compact, resistant to SMD, and BSMR 736 (CTP 7217 × No 148) resistant to FW and SMD.

Chickpea FW resistance lines have been screened using efficient field, greenhouse, and laboratory techniques (Nene et al. 1981). In several nations, a number of sources with high wilt resistance have been discovered. Adequate characterization and evaluation for agromorphological traits are necessary to facilitate utilization of germplasm by breeders. In chickpea, elite genetic resources were identified for several biotic stresses, namely FW (Halila and Strange 1997), AB (Singh and Reddy 1993), and collar rot (Sugha et al. 1991). Over 13,500 chickpea germplasm accessions from 40 different countries were tested at ICRISAT for resistance to Foc race 1, and 160 resistant accessions, including 150 desi and 10 kabuli, were identified (Haware et al. 1992). FW resistant cultivars that are long-lasting and stable have been identified. Chefe in Ethiopia, BG 212, BG 362, and BG 372, Pusa 212, Avrodhi, JG 315, JG 74, Vihar and KAK 2 in India, Senora, UC 15 and UC 27 in Mexico, Punjub 2,000 in Pakistan, ICCV 2, UC 15 and FLIP 85-20C in Sudan are some of the resistant cultivars that are widely grown in many nations. Due to the lack of true resistance in germplasm and the existence of a different variable Ascochyta strain, the development of cultivars with high level and long-lasting resistance to AB is particularly challenging. However, ICRISAT has discovered sources of AB resistance in kabuli and desi chickpea. Pande et al. (2010) developed effective controlled-environment and field screening techniques to identify chickpea genotypes for resistance to AB. Using these resistance screening techniques, 150 elite chickpea breeding lines were evaluated and 29 lines with high and stable resistance to AB were identified. In addition to CM 72, C 727, and C 44 in Pakistan, GNG 469, PbG 1, Gaurav, BG 261, ICCV 98818, and ICCV 05503 in India, CDC Frontier and CDC Anna in Canada, Sanford, Dwelley, Sierra, and Myles in the United States, ILC 3279 in Cyprus, and Yorker, Genesis 836, and Flipper in Australia are some of the well-known resistant cultivars that have been released. Recently, Gayacharan et al. (2020) evaluate the existing chickpea germplasm diversity conserved in Indian National Genebank against the AB disease under artificial epiphytotic conditions. Screening has resulted in identification of some promising chickpea accessions such as IC275447, IC117744, EC267301, IC248147, and EC220109, which have shown the disease resistance (AB) in multiple seasons and locations. Lack of highly resistant chickpea lines has slowed down the progress of BGM resistance breeding programs. At ICRISAT, more than 12,000 breeding and germplasm accession lines have been evaluated; none were discovered to be extremely resistant (Pande et al. 2002a, 2002b). However, chickpea line ICC 14344 has shown some mild resistance. In India, the line was made available as the cultivar Avarodhi. Aside from this, it was discovered that BG 276, GL 90159, GL 91040, and GL 91071 were moderately resistant BGM. Such a host–plant resistance can offer beneficial disease protection. Barichola 5 and ICCL 87322, two cultivars that are moderately resistant, were released in Bangladesh. Although certain wild Cicer species, such as the closely related *C. echinospermum*, have higher levels of resistance, these have not yet been integrated into the cultivated species (Pande et al. 2002a, 2002b).

A few South Asian researchers have also found MYMV-resistant lines of mungbean, including those from Bangladesh (Mondol et al. 2013), Pakistan (Munawwar et al. 2014), Thailand (Akhtar et al. 2009), and India (Nair et al. 2017). According to research by Bashir et al. (2005) from Pakistan, the mungbean lines NM-2, 98-CMH-016, BRM-195, VC-3960 (A-89), and VC-3960 (A-88) were highly resistant to ULCV. There have also been reports of MYMV and leaf crinkle disease-resistant lines of urdbean in Pakistan and India (Gupta 2003; Kumar and Bal 2012). Recently, by Pavishna et al. (2019), in order to identify the resistant variety of MYMV, germplasms, wild relatives, and breeding lines have been screened under natural field conditions during summer 2018 and 2019 and revealed 28 genotypes, namely KKB

14034, KKB 14003, KKB 14004, KKB 14045, KKB 14051, KKB 14014, KKB 14009, KKB 14020, KKB 14038, KKB 14015, KKB 14001, KKB 14043, KKB 14042, KKB 14041, KKB 14053, KKB 14048, VBG 10010, VBG 10053, RU 1511, RU 1513, VBN 6, and VBN 4 that exhibited resistance in both the years. There are various lines of mungbean and urdbean from India that are resistant to serious fungal diseases including powdery mildew and Cercospora leaf spot (Marappa 2008; Ramakrishnan and Savithramma 2014). In mungbean (Choudhary et al. 2011; Kaur et al. 2011) and urdbean (Iqbal et al. 2003; Sharma et al. 2014) in Thailand and Pakistan, few research studies have been conducted to discover resistant lines against anthracnose and DRR. But researchers need to work on finding reliable sources of resistance. In Pakistan, the mungbean lines NCM 15-11, AZRI-1, NCM 11-8, and 14,063 have demonstrated possible resistance to bacterial leaf spot (Munawwar et al. 2011). Akhtar et al. (2014) conducted field screening during kharif, 2008 and 2009, and revealed that out of 31 genotypes of greengram, only 1 genotype ML 1299 and out of 14 genotypes of blackgram, only 3 genotypes, namely BS 2-3, IPU 02–43, and B 3-8-8 showed resistant or highly resistant response against multiple diseases including cercospora leaf spot, web blight, and powdery mildew diseases.

The most popular technique at ICARDA and other breeding programs is conventional breeding, which aims to create germplasm that is resistant to biotic stress. In order to expose the materials to high disease pressure, it entails analyzing a large number of breeding lines and germplasm accessions in both greenhouse and disease hotspot sites. The breeding program at ICARDA uses the discovered donors for resistance to important diseases in faba bean (chocolate spot, rust, Ascochyta blight, and Orobanche) with high yield potential, and shares these donors with other breeding programs through platforms operated by international nurseries (Hanounik and Robertson 1989; Maalouf et al. 2016). As a result, several varieties are released in Australia, Canada, China, Egypt, Ethiopia, and Spain (Sillero et al. 2008; Villegas-Fernandez et al. 2009; Sillero et al. 2010; Rubiales et al. 2012) from these international public goods. Recently, germplasms with multiple disease resistance have been discovered through multi-location testing for Ascochyta blight, chocolate spot, and rust. The majority of breeding programs are focused on creating resistant genotypes for a particular disease of economic value (Maalouf et al. 2016). Some national programs, such as those in Ethiopian Institute of Agricultural Research (EIAR), were able to cross their landraces with resistant sources from ICARDA chocolate spot resistant lines with big seed sizes to create high-yielding cultivars with high levels of chocolate spot resistance (Beyene et al. 2016). Further, EIAR researchers released several high-yielding faba bean varieties through direct selection from germplasm supplied by ICARDA or by transferring good levels of resistance from ICARDA germplasm into locally adapted varieties. Among the faba bean varieties released (Temesgen et al. 2015) with partial resistance to chocolate spot are "Moti" (ILB 4432 × Kuse-2-27-33), "Gebelcho" (ILB 4726 × "Tesfa"), Obsie (ILB 4427 × CS20DK) and "Walki" (ILB 4615 × Bulga 70). Additionally, another variety named "Gora" (ILB2717- 1 × R878-1) has been released in Ethiopia with higher degree of resistance to chocolate spot and larger seed size than traditional cultivars (Temesgen et al. 2015).

In cultivated lentil, there are genetic variations for resistance to the Ascochyta blight. Blight resistance has also been identified in wild lentil species. Thirty lentil accessions with high blight resistance were found by Erskine and Bayaa in 1991. They discovered 12 more resistance lines in 1993. Further testing revealed that 36 accessions of *Lens ervoides*, 24 accessions of *L. orientalis*, 12 accessions of *L. odemensis*, and 3 accessions of *L. nigricans* were resistant to Ascochyta blight. Some popular resistant lines cultivated in Canada are ILL 358, ILL 5588, and Laird; in Chile are ILL 358 and ILL 4605; in Ethiopia are ILL 358 and ILL 857; in India are HPL 5, L 442, LG 169, LG 170, LG 171, LG 172, and Pant L406; in Morocco are ILL 5698, ILL 5700, and ILL 5883; in New Zealand are ILL 5684, ILL 5588, and ILL 5714; in Pakistan are ILL 358, ILL 4605, ILL 5588, and ILL 5684; and in Syria are ILL 857, ILL 2439, and ILL 4605. Germplasms, wild relatives, and breeding lines have all been tested in greenhouses and on the ground in an effort to find the resistant variety of lentil Fusarium wilt. Through systematic use of resistant sources of lentil for wilt from farmed accession, such as ILL 5588, ILL 5883, ILL 590, and ILL 4400, ICARDA has generated a broad range of Fusarium-wilt-resistant cultivars. Ebla 1, Idleb 2, Idleb 3, and Idleb 4 in Syria; Hala, Rachayya, and Talya 2 in Lebanon; Syran 96 and Firat 87 in Turkey; Teshale, Alemtena, Adaa, and Assano in Ethiopia; and Priya, Seri, Pant L4, and L 4174 in India are examples of well-known wilt-resistant varieties grown in their respective countries. One of the primary sources of lentil wilt resistance is wild species. To find donor-resistant sources of F*usarium oxysporum f. sp. lentis*, 70 accessions from four wild species or subspecies were screened in India (Fol). Several wild lentil accessions were discovered to be wilt-resistant. Few are from *L. culinaris* ssp. *orientalis* (IL WL76), *L. nigricans* (e.g., IL WL 22, IL 26, IL 430), *L. ervoides* (e.g., IL WL 40, IL WL 41, IL 133, IL WL204, IL WL 299), and *L. odemensis* (e.g. IL WL 35, IL WL 39, IL WL 300).

12.6.3 Genetics of Legumes for Disease Resistance

The constant interactions between plants and pathogens in the environment and the resulting outcomes are of significant importance for agriculture and agricultural scientists. Disease-resistance genes in plant cultivars can break down in the field due to the evolution of pathogens under high selection pressure. Thus, the protection of crop plants against pathogens is a continuous arms race. Like any other type of crop plant, legumes are susceptible to many pathogens (Kankanala et al. 2019). Plants have evolved robust defense mechanisms against pathogen attack that are triggered by initial recognition of the pathogen. These mechanisms involve a cascade of signaling responses known as Pathogen-Associated Molecular Pattern (PAMP)-Triggered Immune (PTI) response, which eventually leads to changes in the gene expression of the host. Depending on the type of interaction, this can result in either disease susceptibility or disease resistance. The pathogens, on the other end of the spectrum, have evolved several mechanisms involving effector delivery to evade the host defenses. The host defense response to effectors is called Effector-Triggered Immunity (ETI) (Young et al. 2005). The continuous arms race between the host and pathogen eventually determines the outcome of the interaction (Jones and Dangl 2006). The host responses also vary based on pathogen infection strategies. The current understanding is that successful defense responses against biotrophic pathogens are predominantly mediated by the SA-dependent pathway and that those against hemibiotrophs and necrotrophs involve ethylene and JA signaling (Glazebrook 2005).

Leguminosae includes a diverse variety of plants. Medicago truncatula and Lotus japonicus have been chosen as model species to advance the study of legumes (Zhu et al. 2005). Several genetic and genomic resources have been developed in these two model legumes to assist breeding programs for enhanced tolerance/resistance to abiotic and biotic stresses in legume crop species. These include genome sequences (Sato et al. 2008; Young et al. 2011), expressed sequence tags (ESTs) (Gamas et al. 2006), physical and genetic maps (Shah et al. 2016), and insertional mutagenesis lines (Tadege et al. 2008; Urbanski et al. 2013), among others. In addition to the model plants, the genome sequences of crop plants such as Glycine max (cultivated soybean), Glycine soja (wild soybean), Cajanus Cajun (pigeon pea), Cicer arietinum (chickpea), Vigna radiata (mung bean), V. angularis (adzuki bean), V. unguiculata (cowpea), Arachis hypogaea (cultivated peanut), A. duranensis (wild peanut A genome), A. ipaensis (wild peanut B genome), Medicago sativa (alfalfa), Phaseolus vulgaris (common bean), Trifolium pretense (red clover), Lupinus angustifolius (lupin), and Lens culinaris (lentil) are currently available at https://legumeinfo.org/genomes. The macrosynteny and microsynteny studies among some of these genomes have been useful for translating the knowledge from model to crop plants (Zhu et al. 2005). The availability of genome sequences coupled with recent advancements in affordable NGS techniques and bioinformatics tools has enabled extensive study of genome-wide expression changes during plant–pathogen interactions to identify the pathways involved in plant defense. Macroarrays, microarrays, RNAseq, suppressive subtractive hybridization (Mehrtens et al. 2005), cDNA-amplified fragment length polymorphism (AFLP) techniques, and gene-expression atlases have been used extensively to identify candidate genes for disease resistance. In this review, we focus on the interactions of legumes with plant pathogens such as fungi, oomycete, bacteria, nematodes, and viruses at the genomic level and the use of genomic technologies in breeding for resistance.

12.7 ADVANCEMENT IN BIOTECHNOLOGY FOR ENHANCING RESISTANCE IN LEGUMES

Legumes are an excellent source of different health-promoting phytochemicals and proteins as well. Keeping in mind their nutritional importance and sustainable agriculture production, considerable efforts are being made in this recent period to accelerate their genetic gains related to nutritional quality, production, and stress resilience. Till now, the total production of legumes worldwide is up to the mark. Biotic stress is one of the major concerns that significantly affected the yields of legumes. The recent escalations in genomic resources for multiple legume crops have laid a concrete ground for the use of transformative breeding technologies such as genomic selection, transgenic and genome editing for crop protection as well as improvement. The strategies deployed for genetic engineering have emerged nowadays as a core of plant biology and remarkably facilitate crop improvement programs. Several researchers and scientists worldwide are working on developing genomic tools for genetic advancement through the genomic-assisted selection of different traits of legumes including biotic stresses. But the genetic variation within the landraces of the legumes to key diseases and the wild relatives to be exploited for novel resistance genes as well are very limited. Hence, genetic alteration through the development of transgenic or gene-edited plants should be the possible solution in the upcoming days. For that, the identification of standardized protocol is foremost important to the introgression of resistant genes against biotic constraints through transformation.

12.7.1 Transgenic Development

The genetically modified (GM) plants developed from the genetic transformation will allow breeders to release the varieties resistant to several diseases (Figure 12.6 and Table 12.5). However, making transgenic through regeneration is very stringent in legumes. Over the past decade, considerable progress toward standardizing an efficient transformation protocol for legumes has been made, but in comparison to other crops, the legumes are still underutilized. The recalcitrance to in vitro

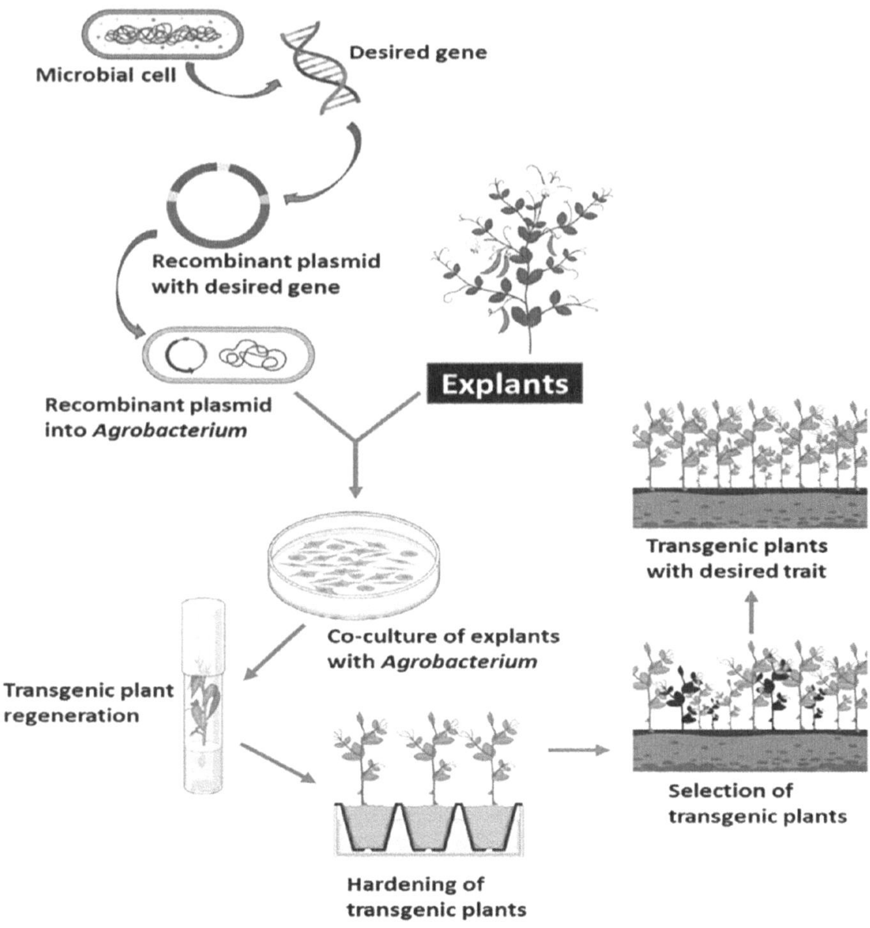

FIGURE 12.6 Schematic diagram of developing disease-resistant transgenic plant.

TABLE 12.5
The Details of Transgenic Lines Developed for Disease Resistance in Legume Crops

Crop	Cultivar	Explant Used	Trait for Transgenic	Gene Introduce	Gene Activity	References
Soybean	Pusa 16	Immature cotyledon	Fungal resistance	Tobacco osmotin	Over-expression	Subramanyam et al. (2012)
	Kwangan	Half-seed	*Soybean mosaic virus* (SMV) resistance	SMV coat protein	RNAi	Kim et al. (2013)
	Tianlong 1	Half-seed	SMV resistance	Soybean xanthine dehydrogenase	Over-expression	Zhou et al. (2015)
	Shennong 9 and Williams 82	Cotyledonary node	SMV resistance	SMV P3 cistron	RNAi	Yang et al. (2018)
Cowpea	Nova Era	Embryonic axis	Resistance to *Cowpea severe mosaic virus* (CPSMV)	Proteinase cofactor gene	RNAi	Cruz and Aragao (2014)
	Nova Era	Embryonic axis	Resistance to *Cowpea aphid-borne mosaic virus* (CABMV)	Coat protein gene	RNAi	Cruz and Aragao (2014)
	Pusa Komal	Cotyledonary node	*Mungbean yellow mosaic India virus* (MYMIV)	Transcriptional activator protein	RNAi	Kumar et al. (2017)
Urdbean	PS-1	Cotyledonary node	Resistance to *Corynespora cassiicola*	Barley chitinase and ribosome-inactivating protein	Over-expression	Chopra and Saini (2014)
Mungbean	ML-267	Cotyledonary node	Resistance to *Rhizoctonia solani*	BjNPR1 gene	Over-expression	Vijayan and Kirti (2012)

regeneration along with high genotype-dependent responses is a the major constraint for the successful transformation and generation of transgenic legume crops. Stable transformation, the competency to pass on the integrated gene to the offspring of the next generations to give similar to itself, meets the necessities of functional genomics and transgenic breeding. The first report on transgenic pulse crop was published in the mid-80s of the 19th century by Kohler et al. in *V. aconitifolia*. However, the progress in transgenic development is still a struggle. Over the past decade, some of the commonly cultivated legume species, e.g., chickpea, pigeonpea, lentil, and field pea, have been whole-genome sequenced, which further helps the trait improvement by identification and incorporation of desirable genes using the transformation methods. To date, a total of 49 GM events have been approved in legumes, among them, 42 events belong to soybean (*Glycine max*) (https://www.isaaa.org). Over the past few years, some of the varieties developed through genetic modification with desired traits were approved for cultivation. However, success rates in chickpea (*C. arietinum*), pigeonpea (*C. cajan*), mungbean (*V. radiata*), urdbean (*V. mungo*), cowpea (*V. unguiculata*), field pea (*P. sativum*), and some other forage legumes (alfalfa and clover) are still not adequate. Apart from that, some attempts have been made to standardize a suitable protocol for regeneration and transgenic development of the FBYN virus as well but no outcome so far.

12.7.2 Gene Editing

The gene-editing method is based on the use of engineered nucleases and cellular DNA repair pathways to make precise, targeted changes to the genome of an organism. The method began nearly three decades ago with the key discovery that specific double-stranded breaks can be introduced in chromosomes using a meganuclease I (Table 12.6). However, for gene editing, the application of meganucleases was restricted due to the low availability of target sites in most genes. However, a revolution occurred within the technology when the successful adaptation of RNA-guided Cas9 nuclease harnessed from the type II prokaryotic clustered regularly interspaced short palindromic repeats (CRISPR) adaptive immune system for genome editing in eukaryotic cells. The recent plant-specific advances in CRISPR/Cas9-based gene-editing technology have been successfully established for targeted improvement of traits in legume crops, e.g., soybean, cowpea, chickpea, and model legumes such as *M. truncatula* and *Lotus japonicus*. However, the successful implication of gene-editing technology in other legumes is a big challenge due to their recalcitrance nature to *in vitro* gene transfer and regeneration. Though, various modifications to *in vitro* culture methods in terms of media composition, explant choice, DNA delivery, and gene-editing detection methods, etc., can potentially improve the rate of regeneration of the whole plant and gene transformation in legume crops. To global food security, public acceptance and a favorable regulatory framework are important factors in realizing CRISPR's potential benefits. Although gene-editing technology can bring enormous benefits to legume breeding, regulatory hurdles are a cause for serious concern. Among a large number of legume crops, soybean is the only species for which transgenic cultivars have been commercialized after passing through the different regulatory processes. Soybean is the first and only legume crop for which

TABLE 12.6
The Traits for Which Gene-Editing Technology Is Used in Legume Crops

Legume	Scientific Name	Genes Mutated	Trait for Alteration	References
Chickpea	*Cicer arietinum*	4-coumarate ligase (4CL) and Reveille 7 (RVE7) genes	Drought tolerance	Badhan et al. (2021)
Cowpea	*Vigna unguiculata*	Symbiosis receptor-like kinase and VuSPO11-1	Meiosis gene to develop asexual plants suitable for hybrid development	Che et al. (2021)
Soybean	*Glycin max*	Fatty acid desaturase 2 (FAD2-2) microsomal omega-6 desaturase	Modulation in oleic acid content	Al Amin et al. (2019)
		Squamosa promoter binding protein-like	Alters plant architecture	Bao et al. (2019)
		Lipoxygenase genes	Lipoxygenase-free seeds to improve the quality of oil and protein	Wang et al. (2020)
Wild legume	*Lotus japonicus*	Leghemoglobin 1/2/3, symbiosis receptor-like kinase (LjSYMRK),	Symbiotic nitrogen fixation	Wang et al. (2019a)
		Lotus histidine kinase 1 (LHK1)-interacting protein (LjCZF1-2)	Symbiotic nitrogen fixation	Cai et al. (2018)
Medicago	*Medicago truncatula*	Candidate genes for nodulation	Genom-wide association study	Curtin et al. (2017)
		Nodule-specific PLAT domain (NPD1–5) and nitrate peptide family (NPD) genes	Symbiotic nitrogen fixation	Trujillo et al. (2019)

CRISPR technology has been widely reported, and a gene-edited product specifically for changing oil profile that is high in oleic acid and reduced saturated fat has reached the market. Microspore is the potential explant in legumes that could be used for gRNA delivery and CRISPR/Cas9-mediated gene editing. However, due to the recalcitrant nature of the legume species, the production of double haploid (DH) plants is very limited in legumes. Using anther culture and androgenesis regeneration, the DH plants have only been successful in chickpea.

12.8 CONCLUSION

The growing global population is expected to reach 9 billion in 2050. To meet the demand for everyone's nutrition security in the upcoming days, the role of legumes is not avoidable. But in recent days, the production of legumes is at risk owing to increasing disease pressure due to climate change. Several diseases usually soil-borne and foliar are known to attack the legume crops in almost all the growth stages. The distribution of the pathogens and patterns of disease occurrence are different in each location. So, it is very important to develop cultivars with multiple diseases resistance. Conventional breeding with the support of other advanced technology will help in achieving the target of developing resistant varieties.

REFERENCES

Ahmed S, Abang MM, Maalouf F (2016) Integrated management of Ascochyta blight (*Didymella fabae*) on faba bean under Mediterranean conditions. *Crop Prot* 81:65–69.

Ajiro N, Miyamoto Y, Masunaka A, Tsuge T, Yamamoto M, Ohtani K et al (2010) Role of the host-selective ACT-toxin synthesis gene ACTTS2 encoding an enoyl-reductase in pathogenicity of the tangerine pathotype of *Alternaria alternata*. *Phytopathol* 100:120–126.

Akhtar J, Lal HC, Kumar Y, Singh PK, Ghosh J, Khan Z et al (2014) Multiple disease resistance in greengram and blackgram germplasm and management through chemicals under rain-fed conditions. *Legume Res* 37(1):101–109.

Akhtar KP, Kitsanachandee R, Srinives P, Abbas G, Asghar MJ, Shah TM et al (2009) Field evaluation of mungbean recombinant inbred lines against mungbean yellow mosaic disease using new disease scale. *Thailand Plant Pathol J* 25:422–428.

Al Amin N, Ahmad N, Wu N, Pu X, Ma T, Du Y et al (2019) CRISPR-Cas9 mediated targeted disruption of FAD2-2 microsomal omega-6 desaturase in soybean (Glycine max L). *BMC Biotech* 19:9.

Alam AKMM, Somta P, Srinives P (2014) Identification and confirmation of quantitative trait loci controlling resistance to mungbean yellow mosaic disease in mungbean (*Vigna radiata* (L.) Wilczek). *Mol Breed* 34:1497–1506.

Amalraj A, Taylor J, Bithell S, Li Y, Moore K, Hobson K et al (2019) Mapping resistance to Phytophthora root rot identifies independent loci from cultivated (*Cicer arietinum* L.) and wild (*Cicer echinospermum* PH Davis) chickpea. *Theor Appl Genet* 132(4):1017–1033.

Ammavasai S, Phogat DS, Solanki IS (2004) Inheritance of resistance to *Mungbean yellow mosaic virus (MYMV)* in green gram (*Vigna radiata* L Wilczek). *Indian J Genet* 64:146.

Ashby JA, Stevenson CE, Jarvis GE, Lawson DM, Maule AJ (2011) Structure-based mutational analysis of eIF4E in relation to sbm1 resistance to pea seed-borne mosaic virus in pea. *PLoS One* 6(1):e15873.

Badhan S, Ball AS, Mantri N (2021) First report of CRISPR/Cas9 mediated DNA-free editing of 4CL and RVE7 genes in chickpea protoplasts. *Int J Mol Sci* 22:396.

Bao A, Chen H, Chen L, Chen S, Hao Q, Guo W et al (2019) CRISPR/Cas9-mediated targeted mutagenesis of GmSPL9 genes alters plant architecture in soybean. *BMC Plant Biol* 19:131.

Barilli E, Cobos MJ, Carrillo E, Kilian A, Carling J, Rubiales D (2018) A high-density integrated DArTseq SNP-based genetic map of *Pisum fulvum* and identification of QTLs controlling rust resistance. *Front Plant Sci* 9:167.

Bashir M, Ahmad Z, Ghafoor A (2005) Sources of genetic resistance in mungbean and blackgram against urdbean leaf crinkle virus (ULCV). *Pak J Bot* 37:47–51.

Bashir M, Malik BA (1988) Diseases of major pulse crops in Pakistan—a review. *Int J Pest Managt* 34(3):309–314.

Bergot M, Cloppet E, Pérarnaud V, Déqué M, Marçais B, Desprez-Loustau ML (2004) Simulation of potential range expansion of oak disease caused by *Phytophthora cinnamomi* under climate change. *Glob Chang Biol* 10(9):1539–1552.

Beyene AT, Derera J, Sibiya J, Fikre A (2016) Gene action determining grain yield and chocolate spot (*Botrytis fabae*) resistance in faba bean. *Euphytica* 207:293–304.

Bhadauria V, Ramsay L, Bett KE, Banniza S (2017) QTL mapping reveals genetic determinants of fungal disease resistance in the wild lentil species *Lens ervoides Sci Rep* 7(1):1–9.

Bhairappanavar SB, Fakrudin B, Sambasivarao KRS (2014) Inheritance studies of sterility mosaic disease (SMD) resistance cross Gullyal white X BSMR736 in pigeonpea (*Cajanus cajan* (L) Millsp). *Plant Gene Trait* 5:27–32.

Bhanu NA, Singh MN, Rathore AS, Hemantaranjan A (2018) Study of inheritance of Fusarium wilt resistance in pigeonpea (*Cajanus cajan* L. Mills). *Int J Life Sci* 7:53–56.

Bindra S, Mittal RK, Sood VK, Sharma PN (2016) Inheritance of resistance in urdbean (*Vigna mungo*) to anthracnose caused by *Colletotrichum truncatum. Indian Phytopathol* 69:311–313.

Bohra A, Dubey A, Saxena RK, Penmetsa RV, Poornima KN, Kumar N et al (2011) Analysis of BAC-end sequences (BESs) and development of BES-SSR markers for genetic mapping and hybrid purity assessment in pigeonpea (*Cajanus* spp.). *BMC Plant Biol* 11:56.

Bohra A, Jha R, Pandey G, Patil PG, Saxena RK, Singh IP et al (2017) New hypervariable SSR markers for diversity analysis, hybrid purity testing and trait mapping in pigeonpea (*Cajanus cajan* (L.) Millspaugh). *Front Plant Sci* 8:377.

Borlaug NE (1958) The use of multilineal or composite varieties to control airborne epidemic diseases of self-pollinated crop plants. *First International Wheat Genetics Symposium*, Winnipeg, pp. 12–27.

Braich S, Sudheesh S, Forster JW, Kaur S (2017) Characterisation of faba bean (*Vicia faba* L) transcriptome using RNA-seq: sequencing, de novo assembly, annotation and expression analysis. *Agronomy* 7:53.

Cai K, Yin J, Chao H, Ren Y, Jin L, Cao Y et al (2018) A C_3HC_4-type RING finger protein regulates

Carrillo E, Boucherot K, Satovic Z, Rubiales D, PAubert G, Fondevilla S (2014) Identification of quantitative trait loci and candidate genes for specific cellular resistance responses against *Didymella pinodes* in pea. *Plant Cell Rep* 33:1133–1145.

Chakraborty S (2005) Potential impact of climate change on plant-pathogen interactions. *Australas Plant Pathol* 34(4):443–448.

Chakraborty S (2013) Migrate or evolve: options for plant pathogens under climate change. *Glob Chang Biol* 19(7):1985–2000.

Chakraborty S, Datta S (2003) How will plant pathogens adapt to host plant resistance at elevated CO_2 under a changing climate? *New Phytol* 159(3)733–742.

Chankaew S, Somta P, Isemura T, Tomooka N, Kaga A, Vaughan DA et al (2013) Quantitative trait locus mapping reveals conservation of major and minor loci for powdery mildew resistance in four sources of resistance in mungbean (*Vigna radiata* (L) Wilczek). *Mol Breed* 32:121–130.

Chankaew S, Somta P, Sorajjapinun W, Srinives P (2011) Quantitative trait loci mapping of cercospora leaf spot resistance in mungbean, *Vigna radiata* (L) Wilczek. *Mol Breed* 28:255–264.

Chen HM, Ku HM, Schafleitner R, Bains TJ, Kuo CG, Liu CA et al (2013) The major quantitative trait locus for mungbean yellow mosaic Indian virus resistance is tightly linked in repulsion phase to the major bruchid resistance locus in a cross between mungbean (*Vigna radiata* (L.) Wilczek) and its wild relative *Vigna radiata* ssp. *sublobata. Euphytica* 192:215–216.

Chopra R, Saini R (2014) Transformation of blackgram (*Vigna mungo* (L.) Hepper) by barley chitinase and ribosome-inactivating protein genes towards improving resistance to Corynespora leaf spot fungal disease. *Appl Biochem Biotechnol* 174:2791–2800.

Choudhary S, Choudhary AK, Sharma OP (2011) Screening of mungbean (*Vigna radiata*) genotypes to identify source of resistant to dry root rot. *J Food Leg* 24:117–119.

Coakley SM, Scherm H, Chakraborty S (1999) Climate change and plant disease management. *Annu Rev Phytopathol* 37(1):399–426.

Collins M, Knutti R, Arblaster J, Dufresne JL, Fichefet T, Friedlingstein P et al (2013) Long-term climate change: projections, commitments and irreversibility. In: Stocker TF, Qin D, Plattner GK, eds. *Climate Change 2013-the Physical Science Basis: Contribution of Working Group I to the Fifth assessment report of the Intergovernmental Panel On Climate Change*. Cambridge University Press: New York. pp. 1029–1136.

Cook J, Lewandowsky S (2016) Rational irrationality: modeling climate change belief polarization using Bayesian networks. *Top Cogn Sci* 8(1):160–179.

Cruz ARR, Aragão FJL (2014) RNAi-based enhanced resistance to cowpea severe mosaic virus and cowpea aphid-borne mosaic virus in transgenic cowpea. *Plant Pathol* 63:831–837.

Curtin SJ, Tiffin P, Guhlin J, Trujillo DI, Burghart LT, Atkins P et al (2017) Validating genome-wide association candidates controlling quantitative variation in nodulation. *Plant Physiolo* 173: 921–931.

Dadu RHR, Ford R, Sambasivam P, Gupta D (2018) Evidence of early defence to Ascochyta lentis within the recently identified *Lens orientalis* resistance source ILWL180. *Plant Pathol* 67:1492–1501.

Dadu RHR, Ford R, Sambasivam P, Gupta DA (2017) Novel *Lens orientalis* resistance source to the recently evolved highly aggressive Australian Ascochyta lentis Isolates. *Front Plant Sci* 8:1038.

Daspute A, Fakrudin B (2015) Identification of coupling and repulsion phase DNA marker associated with an allele of a gene conferring host plant resistance to pigeonpea sterility mosaic virus (PPSMV) in pigeonpea (*Cajanus cajan* L. Millsp.). *Plant Pathol J* 31:33–40.

Daspute A, Fakrudin B, Bhairappanavar SB, Kavil SP, Narayana YD, Muniswamy Kaumar A et al (2014) Inheritance of pigeonpea sterility mosaic disease resistance in pigeonpea. *Plant Pathol J* 30:188–194.

Dey SK, Singh G (1993) Resistance to Ascochyta blight in chickpea – genetic basis. *Euphytica* 68:147–153.

Dhole V, Kandali S (2012) Genetic analysis of resistance to mungbean yellow mosaic virus in mungbean (*Vigna radiata*). *Plant Breed* 131:414–417.

Dhole VJ, Reddy KS (2013) Development of a SCAR marker linked with a MYMV resistance gene in mungbean (*Vigna radiata* L. Wilczek). *Plant Breed* 132:127–132.

Erskine W, Bayaa B (1991) Screening for resistance to Ascochyta blight. ICARDA Annual Report: 98-102.

FAO Bulletin of Statistics (2010) Food and Agricultural Organizations of the United Nations, Rome. https://faostat.fao.org/site/384/ default.aspx.

FAOSTAT (2018) https://www.faostat/fao.org/en/#data.QC.

Fondevilla S, Kuster H, Krajinski F, Cubero JI, Rubiales D (2011) Identification of genes differentially expressed in a resistant reaction to *Ascochyta pinodes* in pea using microarray technology. *BMC Genomics* 12:28.

Fondevilla S, Rubiales D (2012) Powdery mildew control in pea. A review. *Agron Sustain Dev* 32(2):401–409.

Fondevilla S, Satovic Z, Rubiales D, Moreno MT, Torres AM (2008) Mapping of quantitative trait loci for resistance to *Ascochyta pinodes* in *Pisum sativum subsp. syriacum*. *Mol Breed* 21:439–454.

Fondevilla S, Torres AM, Moreno MT, Rubiales D (2007) Identification of a new gene for resistance to powdery mildew in *Pisum fulvum*, a wild relative of pea. *Breed Sci* 57:181–184.

Gamas P, Debellé F, Berges H, Godiard L, Niebel A, Journet E (2006) Medicago truncatula cDNA and genomic libraries. In: Mathesius U, Journ *Medicago truncatula handbook* et E, Sumner LW, eds. *The Medicago Truncatula* Handbook.

Ganapathy KN, Byregowda M, Venkatesh SC, Ramachandra R, Gnanesh BN, Girish G (2009) Identification of AFLP markers linked to sterility mosaic disease in pigeonpea *Cajanus cajan* (L.) Millsp. *Int J Integr Biol* 7:145–149.

Gautam NK, Kumar K, Prasad M (2016) Leaf crinkle disease in urdbean (*Vigna mungo* L. Hepper): an overview on causal agent, vector and host. *Protoplasma* 253(3):729–746.

Gayacharan C, Rani U, Singh S, Basandrai AK, Rathee VK et al (2020) Identification of novel resistant sources for ascochyta blight (*Ascochyta rabiei*) in chickpea. *PLoS One* 15(10): e0240589.

Ghafoor A, McPhee K (2012) Marker assisted selection (MAS) for developing powdery mildew resistant pea cultivars. *Euphytica* 186:593–607.

Glazebrook J (2005) Contrasting mechanisms of defense against biotrophic and necrotrophic pathogens. *Annu Rev Phytopathol* 43:205–227.

Gnanesh BN, Bohra A, Sharma M, Byregowda M, Pandey S, Wesley V et al (2011) Genetic mapping and quantitative trait locus analysis of resistance to sterility mosaic disease in pigeonpea (*Cajanus cajan* (L.) Millsp.). *Field Crops Res* 123:53–61.

Graham PH, Vance CP (2003) Legumes: importance and constraints to greater use. *Plant Physiol* 131(3):872–877.

Gupta O (2003) Resistance to mungbean yellow mosaic virus, phenotypic characters and yield components in urdbean. *Indian Phytopathol* 56:110–111.

Gupta S, Gupta DS, Anjum KT, Pratap A, Kumar J (2013) Transferability of simple sequence repeat markers in blackgram (*Vigna mungo* L. Hepper). *Aust J Crop Sci* 7:345–353.

Gupta SK, Gopalakrishna T (2010) Development of unigene-derived SSR markers in cowpea (Vigna unguiculata) and their transferability to other Vigna species. *Genome* 53:508–523.

Gupta SK, Souframanien J, Reddy KS (2015) Validation of molecular markers linked to yellow mosaic virus disease resistance in diverse genetic background of black gram (*Vigna mungo* (L.) Hepper). *Electron J Plant Breed* 6:755–763.

Halila MH, Strange RN (1997) Screening of Kabuli chickpea germplasm for resistance to Fusarium wilt. *Euphytica* 96:273–279.

Hanounik SB, Robertson LD (1989) Resistance in *Vicia faba* germplasm to blight caused by Ascochyta fabae. *Plant Dis* 73:202–205.

Hatta R, Ito K, Hosaki Y, Tanaka T, Tanaka A, Yamamoto M et al (2002) A conditionally dispensable chromosome controls host-specific pathogenicity in the fungal plant pathogen *Alternaria alternata*. *Genet* 161:59–70.

Haware MP, Nene YL, Pundir RPS, Narayanarao J (1992) Screening of world chickpea germplasm for resistance to Fusarium wilt. *Field Crops Res* 30:147–154.

Humphry ME, Magner T, McIntyre CL, Aitken EAB, Liu CJ (2003) Identification of a major locus conferring resistance to powdery mildew (*Erysiphe polygoni* DC) in mungbean (*Vigna radiata* L. Wilczek) by QTL analysis. *Genome* 46:738–744.

Indira N, Gayatri S (2003) Management of blackgram root rot caused by *Macrophomina phaseolina* by antagonistic microorganisms. *Madras Agric J* 90(9):490–494.

Iqbal SM, Ghafoor A, Arshad M, Bashir M (2003) Screening of urdbean (Vigna mungo L) germplasm for resistance to charcoal rot disease. *Pak J Plant Pathol* 2:107–110.

Jendoubi W, Bouhadida M, Millan T, Kharrat M, Gil J, Rubio J et al (2016) Identification of target region including the Foc01 /foc01 gene and development of near isogenic lines for resistance to Fusarium wilt race 0 in chickpea. *Euphytica* 210:119–133.

Jha AB, Gali KK, Taran B, Warkentin TD (2017) Fine mapping of QTLs for ascochyta blight resistance in Pea using heterogeneous inbred families. *Front Plant Sci* 8:765.

Jha AB, Taran B, Diapari M, Sindhu A, Shunmugam A, Bett K et al (2015) Allele diversity analysis to identify SNPs associated with ascochyta blight resistance in pea. *Euphytica* 202:189–197.

Jha AB, Taran B, Stonehouse R, Warkentin TD (2016) Identification of QTLs associated with improved resistance to ascochyta blight in an interspecific pea recombinant inbred line population. *Crop Sci* 56:2926–2929.

Johansen C, Bakr MA, Sirajul Islam M, Mondal NA, Afzal A, MacLeod WJ et al (2008) Integrated crop management of chickpea in environments of Bangladesh prone to Botrytis grey mould. *Field Crops Res* 108:238–249.

Jones JDG, Dangl JL (2006) The plant immune system. *Nature* 444:323–329.

Kankanala P, Nandety RS, Mysore KS (2019) Genomics of plant disease resistance in legumes. *Front Plant Sci* 10:1345.

Kant P, Materne M, Rodda MS, Slater AT (2017) Screening lentil germplasm for stemphylium blight resistance. *Australas Plant Pathol* 46:129–136.

Karthikeyan A, Shobhana VG, Sudha M, Raveendran M, Senthil N, Pandiyan M et al (2014) Mungbean yellow mosaic virus (MYMV): a threat to green gram (*Vigna radiata*) production in Asia. *Int J Pest Manag* 60(4):314–324.

Kasettranan W, Somta P, Srinives P (2010) Mapping of quantitative trait loci controlling powdery mildew resistance in Mungbean (*Vigna radiata* (L.) Wilczek). *J Crop Sci Biotech* 3:155–161.

Kaur L, Singh P, Sirari A (2011) Biplot analysis for locating multiple disease resistant diversity in mungbean germplasm. *Disease Res* 26:55–60.

Kido E, Pandolfi V, Houllou-Kido L, et al (2010) Plant antimicrobial peptides: an overview of SuperSAGE transcriptional profile and a functional review. *Curr Protein Pept Sci* 11:220–230.

Kim HJ, Kim MJ, Pak JH, Jung HW, Choi HK, Lee YH et al (2013) Characterization of SMV resistance of soybean produced by genetic transformation of SMV-CP gene in RNAi. *Plant Biotechnol Rep* 7:425–433.

Kitsanachandee R, Somta P, Chatchawankanphanich O, Akhtar KP, Shah TM, Nair RM et al (2013) Detection of quantitative trait loci for mungbean yellow mosaic India virus (MYMIV) resistance in mungbean (*Vigna radiata* (L.) Wilczek) in India and Pakistan. *Breed Sci* 63:367–373.

Konečná E, Šafářová D, Navratil M, Hanáček P, Coyne C, Flavell A et al (2014) Geographical gradient of the *eIF4E* alleles conferring resistance to potyviruses in pea (Pisum) germplasm. *PLoS One* 9(3):e90394.

Kotresh H, Fakrudin B, Punnuri SM, Rajkumar BK, Thudi M, Paramesh H et al (2006) Identification of two RAPD markers genetically linked to a recessive allele of a Fusarium wilt resistance gene in pigeonpea (*Cajanus cajan* L. Millsp.). *Euphytica* 149:113–120.

Kulkarni NK, Reddy AS, Kumar PL, Vijaynarasimha J, Rangaswamy KT, Muniyappa V et al (2003) Broad-based resistance to pigeonpea sterility mosaic disease in accessions of *Cajanus scarabaeoides* (L.) Benth. *Indian J Plant Prot* 31:6–11.

Kumar A, Bal RS (2012) Screening of urdbean germplasm against MYMV and Cercospora leaf spot diseases. *Trends Biosci* 5:127–128.

Kumar S, Tanti B, Patil BL, Mukherjee SK, Sahoo L (2017) RNAi derived transgenic resistance to Mungbean yellow mosaic India virus in cowpea. *PLoS One* 12:e0186786.

Kumari R, Ashraf S, Bagri GK, Khatik SK, Bagri DK, Bagdi DL (2018) Impact of seed treatment from bio-agents and fungicides on growth, biomass and yield of lentil (*Lens culinaris* Medik). *J Pharmacogn Phytochem* 7(3):251–253.

Kumbar B, Balol G, Lokesh BK, Kukanur S, Nayak H (2021) Field evaluation of different fungicides against *Uromyces ciceris-arietini* causing rust in chickpea in northern Karnataka. *J Pharmacogn Phytochem* 10(1):896–900.

Kusmenoglu I (1990) Ascochyta blight of chickpea: inheritance and to Ascochyta blight in chickpea. *Euphytica* 24:209–211.

Kwon SJ, Smykal P, Hu J, Wang M, Kim SJ, McGee RJ et al (2013) Userfriendly markers linked to Fusarium wilt race 1 resistance Fw gene for marker-assisted selection in pea. *Plant Breed* 132:642–648.

Landa BB, Navas-Cortés JA, del Mar Jimenez-Gasco M, Katan J, Retig B, Jiménez-Díaz RM (2006) Temperature response of chickpea cultivars to races of *Fusarium oxysporum* f. sp. ciceris, causal agent of Fusarium wilt. *Plant Dis* 90(3):365–374.
Maalouf F, Ahmed S, Shaaban K, Bassam B, Nawar F, Singh M et al (2016) New faba bean germplasm with multiple resistances to Ascochyta blight, chocolate spot and rust diseases. *Euphytica* 211:157–167.
Maiti S, Basak J, Kundagrami S, Kundu A, Pal A (2011) Molecular marker assisted genotyping of mungbean yellow mosaic India virus resistant germplasm of mungbean and urdbean. *Mol Biotechnol* 47:95–104.
Maphosa Y, Jideani VA (2017) The role of legumes in human nutrition in. In: Hueda MC, ed. *Functional Food-Improve Health through Adequate Food.* Intech Open Publishing: Croatia, p. 13.
Marappa N (2008) Screening of mungbean genotypes and its wild relatives for resistant sources to Cercospora leaf spot disease. *Asian J BioSci* 3:324–326.
McClendon MT, Inglis DA, McPhee KE, Coyne CJ (2002) DNA markers linked to Fusarium wilt race 1 resistance in pea. *J Am Soc Horti Sci* 127:602–607.
McPhee KE, Inglis DA, Gundersen B, Coyne CJ (2012) Mapping QTL for Fusarium wilt race 2 partial resistance in pea (*Pisum sativum*). *Plant Breed* 131:300–306.
Meehl GA, Covey C, Delworth T, Latif M, McAvaney B, Mitchell JF et al (2007) The WCRP CMIP3 multimodel dataset: a new era in climate change research. *Bull Am Meteorol Soc* 88(9):1383–1394.
Mehrtens F, Kranz H, Bednarek P, Weisshaar B (2005) The Arabidopsis transcription factor MYB12 is a flavonol-specific regulator of phenylpropanoid biosynthesis. *Plant Physiol* 138:1083–1096.
Millan T, Clarke HJ, Siddique KHM, Buhariwalla HK, Gaur PM, Kumar J et al (2006) Chickpea molecular breeding: new tools and concepts. *Euphytica* 147:81–103.
Mishra RR, Sahu AR, Rath SC, Panigrahi J (2016) Genetic linkage mapping of loci conferring resistance to Blue butterfly (*Lampides boeticus* L.) and Plume moth (*Exelastis atomosa* Wals.) on chromosome 2 (CcLG02) in Pigeonpea. *Bot Lett* 163:217–230.
Mishra SP, Asthana AN, Yadav L (1988) Inheritance of cercospora leaf spot resistance in mungbean, Vigna radiata (L.) Wilczek. *Plant Breed* 100:228–229.
Miyamoto Y, Masunaka A, Tsuge T, Yamamoto M, Ohtani K, Fukumoto T et al (2010) ACTTS3 encoding a polyketide synthase is essential for the biosynthesis of ACT-toxin and pathogenicity in the tangerine pathotype of *Alternaria alternata*. *Mol Plant-Microbe Interact* 23: 406–414.
Mondol MEA, Rahman H, Rashid MH, Hossain MA, Islam MM (2013) Screening of mungbean germplasm for resistance to mungbean yellow mosaic virus. *Int J Sustain Crop Prod* 8:11–15.
Munawwar MH, Ali A, Malik SR (2014) Identification of resistance in mungbean and mashbean germplasm against mungbean yellow mosaic virus introduction. *Pak J Agric Res* 27:129.
Munawwar MH, Iqbal SM, Malik SR, Chattha MR, Ali A (2011) Identification of resistant sources in mungbean to bacterial leaf spot disease. *Mycopathologia* 9:71–72.
Nair RM, Gotz M, Winter S, Giri RR, Boddepalli VN, Sirari A et al (2017) Identification of mungbean lines with tolerance or resistance to yellow mosaic in fields in India where different begomovirus species and different Bemisia tabaci cryptic species predominate. *Eur J Plant Pathol* 149:349–365.
Nene YL, Haware MP, Reddy MV (1981) Chickpea diseases: resistance-screening techniques. *ICRISAT Information Bull* 10:1–10.
Nene YL, Kannaiyan J (1982) Screening pigeonpea for resistance to Fusarium wilt. *Plant Dis* 66:306–307.
Nisar M, Ghafoor A (2011) Linkage of a RAPD marker with powdery mildew resistance er-1 gene in Pisum sativum L. *Russ J Genet* 47:300–304.

Noble TJ, Young AJ, Douglas CA, Williams B, Mundree S (2019) Diagnosis and management of halo blight in Australian mungbeans: a review. *Crop Pasture Sci* 70(3):195–203.

Nongmaithem N, Basudha Ch, Sharma SK (2017) Incidence of rust, powdery mildew and wilt in pea and broad bean plant of Manipur, India. *Int J Curr Microbiol Appl sci* 6(8):2611–2616.

O'Sullivan DM, Angra D (2016) Advances in faba bean genetics and genomics. *Front Genet* 7:150.

Ocana-Moral S, Gutierrez N, Torres A, Madrid E (2017) Saturation mapping of regions determining resistance to Ascochyta blight and broomrape in faba bean using transcriptome-based SNP genotyping. *Theo App Genet* 130:2271–2282.

Okiror MA (2002) Genetics of wilt resistance in pigeonpea. *Indian J Genet Pl Br* 62:218–220.

Pande S, Sharma M, Gaur PM, Gowda CLL (2010) *Host Plant Resistance to Ascochyta Blight of Chickpea*. Information Bull. No. 82. Project Report International Crops Research Institute for the Semi-Arid Tropics.

Pande S, Sharma M, Kumari S, Gaur PM, Chen W, Kaur L et al (2009) Integrated foliar diseases management of legumes. In: Ali M et al, ed. *Grain Legumes: Genetic Improvement, Management and Trade*. Indian Society of Pulses Research and Development, Indian Institute of Pulses Research: Kanpur, pp. 143–161.

Pande S, Singh G, Narayanarao J, Bakr MA, Chaurasia PCP, Joshi S et al (2002a) *Integrated Management of Botrytis Grey Mold of Chickpea*. Information Bull ICRISAT, Patancheru, India.

Pande S, Singh G, Rao JN, Bakr MA, Chaurasia PCP, Joshi S et al. (2002b) Integrated management of botrytis gray mold of chickpea. Information bulletin no. 61, ICRISAT, Andhra Pradesh, India.

Patil B, Kumar PL (2015) Pigeonpea sterility mosaic virus: a legume-infecting emaravirus from South Asia. *Mol Plant Pathol* 16(8):775–786.

Patil BS, Ravikumar RL, Bhat JS, Soregaon CD (2014) Molecular mapping of QTLs for resistance to early and late Fusarium wilt in chickpea. *Czech J Genet Plant Breed* 50:171–176.

Patil PG, Dubey J, Bohra A, Mishra RK, Saabale PR, Das A et al (2017) Association mapping to discover significant marker-trait association for resistance against fusarium wilt variant 2 in pigeon pea (*Cajanus cajan* (L.) Millsp.) using SSR markers. *J Appl Genet* 58:307–319.

Pavishna M, Kannan R, Arumugam Pillai M, Rajinimala N (2019) Screening of blackgram genotypes against mung bean yellow mosaic virus disease. *J Pharmacogn Phytochem* 8(3):4313–4318.

Podder R, Banniza S, Vandenberg A (2013) Screening of wild and cultivated lentil germplasm for resistance to stemphylium blight. *Plant Genet Resour* 11(1):26–35.

Prasad AVSD, Murugan E, Vanniarajan C (2015) Inheritance of resistance of mungbean yellow mosaic virus in Urdbean (*Vigna mungo* (L.) Hepper). *Current Biotica* 8:413–417.

Raguchander T, Samiappan R, Arjunan GJIP (1993) Biocontrol of Macrophomina root rot of mungbean. *Indian Phytopathol* 46(4):379–382.

Rahul C, Singh IS, Gupta AK (1995) Inheritance of resistance to botrytis grey mould in chickpea (*Cicer arietinum* L.). *Legume Res* 18:1–4.

Rai A, Irulappan V, Kumar MS (2022) Dry root rot of chickpea: a disease favored by the drought. *Plant Dis* 106:346–356.

Raju NL, Gnanesh BN, Lekha P, Jayashree B, Pande S, Hiremath PJ et al (2010) The first set of EST resource for gene discovery and marker development in pigeonpea (*Cajanus cajan* L. Millsp). *BMC Plant Biol* 10:45.

Rakshit S, Mohapatra T, Mishra SK, Dasgupta SK, Sharma RP, Sharma B (2001) Marker assisted breeding for powdery mildew resistance in pea (*Pisum sativum* L.). *J Genet Breed* 55:343–348.

Ramakrishnan CKD, Savithramma DL (2014) Screening of mungbean germplasm for powdery mildew disease resistant. *Int J Agronomy Agric Res* 4:16–21.
Ratnaparkhe MB, Gupta VS (2007) Pigeonpea. In: Kole C, ed. *Genome Mapping and Molecular Breeding in Plants: Pulses, Sugar and Tuber Crops*. Springer: Berlin, pp. 133–142.
Ravikumara BM, Ramanagouda G, Naik MK, Telangre R, Sharma M (2022) Distribution and pathogenic diversity in *Fusarium udum* Butler isolates: the causal agent of pigeonpea Fusarium wilt. *BMC Plant Biol* 22:147.
Ray H, Georges F (2010) A genomic approach to nutritional, pharmacological and genetic issues of faba bean (*Vicia faba*): prospects for genetic modifications. *GM Crop* 1:99–106.
Reddy KS (2009) A new mutant for yellow mosaic virus resistance in mungbean (*Vigna radiata* L. Wilczek) variety SML-668 by recurrent gamma-ray irradiation. In: Shu QY ed. *Induced Plant Mutation in the Genomics Era*. Food and Agriculture Organization of the United Nations: Rome, pp. 361–362.
Rispail N, Rubiales D (2016) Genome-wide identification and comparison of legume MLO gene family. *Sci Rep* 6(1):1–12.
Rubiales D, Avila CM, Sillero JC et al (2012) Identification and multi-environment validation of resistance to Ascochyta fabae in faba bean (Vicia faba). *Field Crops Res* 126:165–170.
Sato S, Nakamura Y, Kaneko T, Asamizu E, Kato T, Nakao M (2008) Genome Structure of the Legume, Lotus japonicus. *DNA Res Int J Rapid Publ Rep Genes Genomes* 15:227–239.
Savithramma DL, Divya Ramakrishnan CK (2016) Single marker analysis in mungbean (*Vigna radiata* (L) Wilczek) for powdery mildew disease resistance and yield attributing traits. *Proteomics Bioinform* 9(12):45.
Saxena KB (2008) Genetic improvement of pigeon pea – a review. *Tropical Plant Biol* 1:159–178.
Saxena R, Thudi M, Varshney RK (2016) Genomics, trait mapping and molecular breeding in pigeonpea and chickpea. *Ind J Genet* 76:504–511.
Shah N, Hirakawa H, Kusakabe S, Sandal N, Stougaard J, Schierup MH (2016) High-resolution genetic maps of Lotus japonicus and L. burttii based on re-sequencing of recombinant inbred lines. *DNA Res* 23:487–494.
Sharma HC, Pampapathy G, Lanka SK, Ridsdill-Smith TJ (2005) Exploitation of wild *Cicer reticulatum* germplasm for resistance to *Helicoverpa armigera*. *J Econ Entomol* 98(6):2246–2253.
Sharma M, Ghosh R (2016) An update on the host plant resistance to pigeonpea diseases. *Legume Perspect* 11:21–23.
Sharma M, Ghosh R (2017) Heat and soil moisture stress differentially impact chickpea plant infection with fungal pathogens. In: Senthil-Kumar M, ed. *Plant Tolerance to Individual and Concurrent Stresses*. Springer: New Delhi, pp. 47–57.
Sharma M, Ghosh R (2018) Isolation, identification, and pathogenicity of Phytophthora Blight of pigeonpea, *Plant Health Progress*, 19:233–236.
Sharma M, Ghosh R, Rameshwar T, Pande S (2015) Multi-environment field testing to identify broad, stable resistance to sterility mosaic disease of pigeonpea. *J Gen Plant Pathol* 81:249–259.
Sharma M, Ghosh R, Tarafdar A, Rathore A, Chobe DR, Kumar AV et al. (2019) Exploring the genetic cipher of chickpea (*Cicer arietinum* L.) through identification and multi-environment validation of resistant sources against Fusarium wilt (*Fusarium oxysporum f.* sp. ciceris). *Front Sustain Food Syst* 3: 78.
Sharma M, Rathore A, Mangala UN, Ghosh R, Sharma S, Upadhyay HD et al (2012) New sources of resistance to Fusarium wilt and sterility mosaic disease in a mini-core collection of pigeonpea germplasm. *Eur J Plant Pathol* 133:707–714.

Sharma M, Sengupta A, Ghosh R, Agarwal G, Tarafdar A, Nagavardhini A et al. (2016) Genome wide transcriptome profiling of *Fusarium oxysporum f sp. ciceris* conidial germination reveals new insights into infection-related genes. *Sci Rep* 6:37353.

Sharma M, Tarafdar A, Chandran USS, Chobe DR, Ghosh R (2018) Molecular markers as tools for identification and introgression of virus resistant genes. In: Patil BL, ed. *Genes, Genetics and Transgenic for Virus Resistance in Plants.* Caister Academic Press: Poole, UK, pp. 87–102.

Sharma V, Basandrai AK, Basandrai D, Gautam NK, Sharma PN (2014) Identification of stable sources of resistance against anthracnose (Colletotrichum truncatum) in urdbean (Vigna mungo). *J Food Leg* 27:218–220.

Sharma, M, Ghosh, R, Pande S (2013) Occurrence of *Alternaria alternata* causing Alternaria blight in pigeonpea in India. *Adv Biosci Biotechnol* 4:702–705.

Sillero JC, Avila CM, Moreno MT, Rubiales D (2008) Identification of resistance to Ascochyta fabae in Vicia faba germplasm. *Plant Breed* 120:529–531.

Sillero JC, Villegas-Fernandez AM, Thomas J, Rojas-Molina MM, Emeran AA, Fernandez-Aparicio M et al (2010) Faba bean breeding for disease resistance. *Field Crop Res* 115:297–307.

Singh D, Sinha B, Rai VP, Singh MN, Singh DK, Kumar R, Singh AK (2016) Genetics of Fusarium wilt resistance in pigeonpea (*Cajanus cajan*) and efficacy of associated SSR markers. *The Plant Pathol* 32(2): 95–101.

Singh KB, Reddy MV (1993) Resistance to six races of *Ascochyta rabiei* in the world germplasm collection of chickpea. *Crop Sci* 33:186–189.

Singh M, Sharma SK, Singh B, Malhotra N, Chandora R, Sarker A et al (2018) Widening the genetic base of cultivated gene pool following introgression from wild Lens taxa. *Plant Breed* 137:470–485.

Smitha KP, Rajeswari E, Alice D, Raguchander T (2015) Assesment of vascular wilt and dry root rot of pigeonpea in Tamil Nadu. *Int J Trop Agri* 33(3):2145–2151.

Souframanien J, Gopalakrishna T (2006) ISSR and SCAR markers linked to the mungbean yellow mosaic virus (MYMV) resistance gene in blackgram (*Vigna mungo* (L.) Hepper). *Plant Breed* 125:619–622.

Srivastava AK, Kashyap PL, Chakdar H, Kumar M, Srivastava AK, Yadav J et al (2018) First de novo draft genome sequence of the pathogenic fungus *Fusarium udum* F02845, associated with pigeonpea (*Cajanus cajan* L. Millspaugh) wilt. *Microbiol Resour Announc* 7:e01001–18.

Srivastava RK, Mishra SK, Singh K, Mohapatra T (2012) Development of a coupling-phase SCAR marker linked to the powdery mildew resistance gene er1 in pea (*Pisum sativum* L.). *Euphytica* 186:855–866.

Subramanyam K, Arun M, Mariashibu TS, Theboral J, Rajesh M, Singh NK et al (2012) Overexpression of tobacco osmotin (Tbosm) in soybean conferred resistance to salinity stress and fungal infections. *Planta* 236:1909–1925.

Sudha M, Karthikeyan A, Anusuya P, Ganesh NM, Pandiyan M, Senthil N et al (2013) Inheritance of resistance to mungbean yellow mosaic virus (MYMV) in inter and intra specific crosses of mungbean (*Vigna radiata*). *Am J Plant Sci* 4:1924–1927.

Sugha SK, Sharma BK, Tyagi PD (1991) Screening of chickpea germplasm for resistance to collar rot caused by *Sclerotium rolfsiis acc. Legume Res* 14:191–192.

Sun S, Fu H, Wang Z, Duan C, Zong X, Zhu Z (2016) Discovery of a novel er1 allele conferring powdery mildew resistance in Chinese Pea (*Pisum sativum* L.) landraces. *PLoS One* 11(1):e0147624.

Sunil K, Shobharani M (2019) Integrated management of dry root rot of blackgram caused by *Rhizoctonia bataticola. Int J Curr Microbiol App Sci* 8(08):853–858.

Tadege M, Wen J, He J, Tu H, Kwak Y, Eschstruth A (2008) Large-scale insertional mutagenesis using the Tnt1 retrotransposon in the model legume *Medicago truncatula. Plant J* 54:335–347.

Talekar SC, Viswanatha KP, Lohithaswa HC (2017) Assessment of genetic variability, character association and path analysis in F2 segregating population for quantitative traits in chickpea. *Int J Curr Microbiol App Sci* 6(12):2184–2192.

Tarafdar A, Rani TS, Chandran US, Ghosh R, Chobe DR, Sharma M (2018) Exploring combined effect of abiotic (soil moisture) and biotic (*Sclerotium rolfsii* Sacc.) stress on collar rot development in chickpea. *Front Plant Sci* 17(10):1161–1167.

Temesgen TT, Keneni G, Mohammad H (2015). Genetic progresses from over three decades of faba bean (Vicia faba L.) breeding in Ethiopia. *Aust J Crop Sci* 9(1):41.

Teshome E, Tegegn A (2017) Comparative study of powdery mildew (*Erysiphe polygoni*) disease severity and its effects on yield and yield components of field pea (*Pisum sativum* L.) in the south eastern Oromia, Ethiopia. *J Plant Pathol Microbiol* 8:5.

Thompson GB, Drake BG (1994) Insects and fungi on a C3 sedge and a C4 grass exposed to elevated atmospheric CO_2 concentrations in open-top chambers in the field. *Plant Cell Environ* 17:1161–1167.

Timmerman-Vaughan GM, Frew TJ, Butler R, Murray S, Gilpin M, Falloon K, Khan T (2004) Validation of quantitative trait loci for Ascochyta blight resistance in pea (*Pisum sativum* L.), using populations from two crosses. *Theor Appl Genet* 109:1620–1631.

Timmerman-Vaughan GM, Moya L, Frew TJ, Murray SR, Crowhurst R (2016) Ascochyta blight disease of pea (Pisum sativum L.): defence-related candidate genes associated with QTL regions and identification of epistatic QTL. *Theor Appl Genet* 129:879–896.

Tiwari KR, Penner GA, Warkentin TD (1998) Identification of coupling and repulsion phase markers for powdery mildew resistance genes er1 in pea. *Genome* 41:440–444.

Tiwari S, Dhar V (2011) Prevalence of new variants of Fusarium udum in India. *Indian Phytopathol* 64:243–246.

Torres AM, Roman B, Avila CM, Satovic Z, Rubiales D, Sillero JC et al (2006) Faba bean breeding for resistance against biotic stresses: towards application of marker technology. *Euphytica* 147:67–80.

Trivedi AS, Sharma SK, Ameta OP, Sharma S (2014) Management of viral and leaf spot disease complexes in organic farming of blackgram. *Indian Phytopathol* 67:97–101.

Trujillo DI, Silverstein KAT, Young ND (2019) Nodule-specific PLAT domain proteins are expanded in the Medicago lineage and required for nodulation. *New Phytol* 222:1538–1550.

Tullu A, Banniza S, Tar'an B, Warkentin T, Vandenberg A (2010) Sources of resistance to ascochyta blight in wild species of lentil (*Lens culinaris* Medik.). *Genet Resources Crop Evil* 57:1053–1063.

Turcotte MM, Araki H, Karp DS, Poveda K, Whitehead SR (2017) The eco-evolutionary impacts of domestication and agricultural practices on wild species. *Philos Trans R Soc B: Proc Biol Sci* 372(1712):20160033.

Urbanski DF, Malolepszy A, Stougaard J, Andersen SU (2013) Highthroughput and targeted genotyping of Lotus japonicus LORE1 insertion mutants. *Methods Mol Biol* 1069:119–146.

Varshney RK, Chen W, Li Y, Bharti AK, Saxena RK, Schlueter JA et al (2012) Draft genome sequence of pigeonpea (Cajanus cajan), an orphan legume crop of resource-poor farmers. *Nat Biotechnol* 30:83–89.

Varshney RK, Mohan M, Gaur PM, Gangarao NVPR, Pandey MK, Bohra A et al (2013) Achievements and prospects of genomics-assisted breeding in three legume crops of the semiarid tropics. *Biotechnol Adv* 31:1120–1134.

Varshney RK, Mohan SM, Gaur PM, Chamarthi SK, Singh VK, Srinivasan S et al (2014) Marker-assisted backcrossing to introgress resistance to Fusarium wilt (FW) race 1 and Ascochyta blight (AB) in C 214, an elite cultivar of chickpea. *Plant Genome* 7:1.

Vavilov NI (1926) Centers of origin of cultivated plants. NI Vavilov origin and geography of cultivated plants. *Hum Evol* 29:285–301.

Verma RPS, Singh DP (1986) The allelic relationship of genes giving resistance to mungbean yellow mosaic virus in blackgram. *Theo Appl Genet* 72:737–738.

Vijayan S, Kirti PB (2012) Mungbean plants expressing BjNPR1 exhibit enhanced resistance against the seedling rot pathogen, *Rhizoctonia solani. Transgenic Res* 21:193–200.

Villegas-Fernández AM, Sillero JC, Emeran AA, Winkler J, Raffiot B, Tay J (2009) Identification and multi-environment validation of resistance to *Botrytis fabae* in *Vicia faba*. *Field Crop Res* 114:84–90.

Vishalakshi B, Umakanth B, Shanbhag AP, Ghatak A, Sathyanarayanan N, Madhav MS et al. (2017) RAPD assisted selection of black gram (*Vigna mungo* L. Hepper) towards the development of multiple disease resistant germplasm. *3 Biotech* 7:1.

Wang J, Kuang H, Zhang Z, Yang Y, Yan L, Zhang M et al (2021) Generation of seed lipoxygenase-free soybean using CRISPR-Cas9. *The Crop J* 8:432–439.

Wang L, Rubio MC, Xin X, Zhang B, Fan Q, Wang Q et al (2019a) CRISPR/Cas9 knockout of leghemoglobin genes in *Lotus japonicus* uncovers their synergistic roles in symbiotic nitrogen fixation. *New Phytol* 224:818–832.

Wang M, Fu H, Shen XX, Ruan R, Rokas A, Li H (2019b) Genomic features and evolution of the conditionally dispensable chromosome in the tangerine pathotype of *Alternaria alternata*. *Mol Plant Pathol* 20:1425–1438.

Webb A, Cottage A, Wood T, Khamassi K, Hobbs D, Gostkiewicz K et al (2016) A SNP-based consensus genetic map for synteny-based trait targeting in faba bean (*Vicia faba* L.). *Plant Biotechnol J* 14:177–185.

Young ND, Cannon SB, Sato S, Kim D, Cook DR, Town CD (2005) Sequencing the Genespaces of Medicago truncatula and Lotus japonicus. *Plant Physiol* 137:1174–1181.

Young ND, Debelle F, Oldroyd GED, Geurts R, Cannon SB, Udvardi MK (2011) The Medicago genome provides insight into the evolution of rhizobial symbioses. *Nature* 480:520–524.

Zhang H, Yasmin F, Song BH (2019). Neglected treasures in the wild—legume wild relatives in food security and human health. *Curr Opinion Plant biol* 49:17–26.

Zhang HM, Wheeler S, Xia X, Radchuk R, Weber H, Offler CE et al (2015) Differential transcriptional networks associated with key phases of ingrowth wall construction in trans-differentiating epidermal transfer cells of *Vicia faba* cotyledons. *BMC Plant Biol* 15:103.

Zhang SC, Zhang ZN, Wen ZX, Gu CH, An YQC, Bales C et al (2017) Fine mapping of the soybean aphid-resistance genes Rag6 and Rag3c from Glycine soja 85-32. *Theor Appl Genet* 130:2601–2615.

Zhao M, Running S, Heinsch FA, Nemani R (2010) MODIS-derived terrestrial primary production. In: Ramachandran B, Justice CO, Abrams, MJ, eds. *Land Remote Sensing and Global Environmental Change*. Springer: New York, pp. 635–660.

Zhou Z, He H, Ma L, Yu X, Mi Q, Pang J et al (2015) Overexpression of a GmCnx1 gene enhanced activity of nitrate reductase and aldehyde oxidase, and boosted mosaic virus resistance in soybean. *PLoS One* 10:e0124273.

Zhu H, Choi HK, Cook DR, Shoemaker RC (2005) Bridging Model and crop legumes through comparative genomics. *Plant Physiol* 137:1189–1196.

13 Integrated Management of Emerging Phytoplasma Diseases

Govind P. Rao

13.1 INTRODUCTION

Phytoplasmas are unculturable, cell-wall-less prokaryotes causing economically important diseases in many crops worldwide. These phloem-inhabiting plant pathogenic bacteria are classified in the genus *Candidatus* (*Ca.*) Phytoplasma and the class *Mollicutes* (Bertaccini et al. 2014). All phytoplasmas are transmitted by phloem-feeding insects, mostly leafhoppers, planthoppers, and psyllids belonging to suborder Auchenorrhyncha (Weintraub 2007).

Phytoplasma are minute bacteria with pleiomorphic or filamentous shape of 200–800 µm and inhabit phloem sieve elements and insect hemolymph. The genome size of phytoplasmas is very small ranges between 530 and 1,350 kb (Kube et al. 2018). Phytoplasma cause diseases in about 700 plant diseases and cause significant losses (Rao et al. 2018). They cannot be successfully cultured *in vitro*. Comparison of 16S rDNA sequences showed that phytoplasmas are phylogenetically close to the acholeplasma/anaeroplasma group of the Mollicutes and are classified into 34 phylogenetic groups (Bertaccini and Lee 2018).

Phytoplasma genomes contain clusters of transposon genes and insertion sequences, which are called potential mobile units (PMUs) (Bai et al. 2006; Hogenhout et al. 2008) or sequence-variable mosaics (SVMs) (Jomantiene and Davis 2006; Jomantiene et al. 2007). They may enable phytoplasmas to survive in diverse environments of plants and insects (Bai et al. 2006; Jomantiene and Davis 2006). Due to their unique ability to survive and multiply in plants and insect vectors, phytoplasmas can spread into new hosts in a short timespan (Suzuki et al. 2006). Phytoplasmas also modulate and regulate plant hosts genes, hormones, and secondary metabolite biosynthesis (MacLean et al. 2011; Sugio and Hogenhout 2012; Sugio et al. 2011). Phytoplasmas can also produce specific proteins (called effectors) that help them to overcome physical and biochemical plant defenses, multiply in plant hosts, disperse by insect vectors, and modulate plant host growth (Weintraub et al. 2019).

The unique nature of phytoplasmas requires different management strategy based on polyphasic integrated approach combining components of cultural, physical, biological, resistance, and chemical control (Bertaccini 2019; Kumari et al. 2019; Marcone et al. 1999; Marcone and Rao 2019) (Figure 13.1). In the field, phytoplasma diseases management has been stood mainly Ron eradication of infected plants and

DOI: 10.1201/9781003531944-13

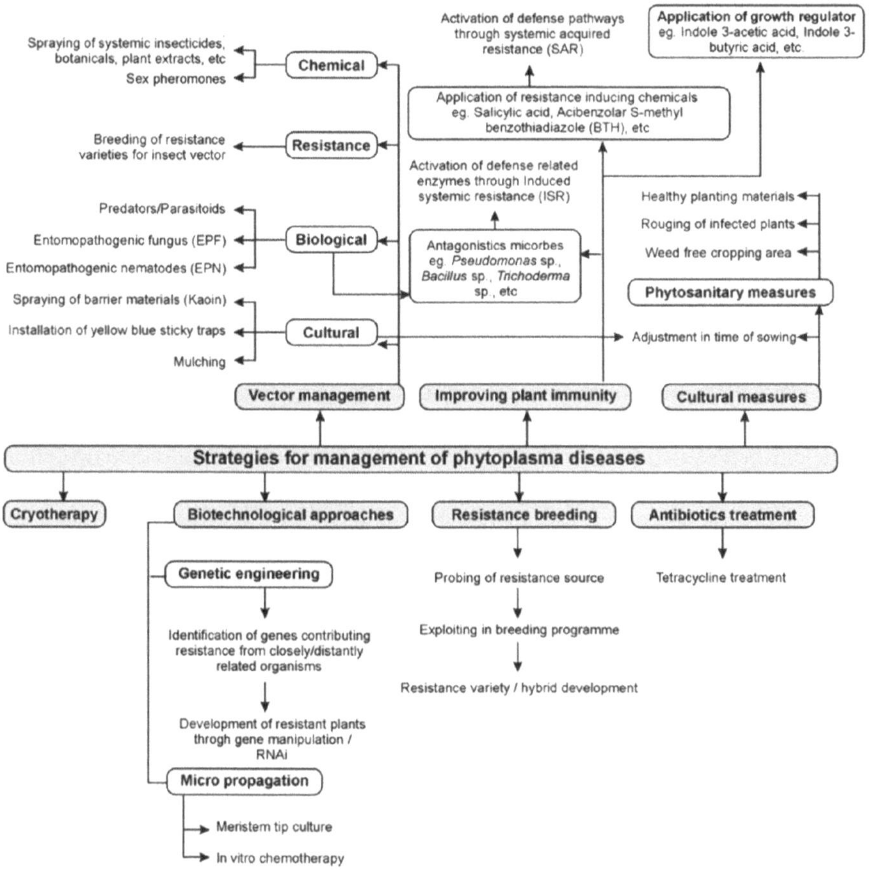

FIGURE 13.1 Strategies for the management of phytoplasma diseases.

control of insects (Weintraub and Wilson 2010; Weintraub and Beanland 2006). All other methods including quarantine measures, certification, thermotherapy or meristem culture, host resistance, and innovative control strategies preventing spread or reducing inoculum sources of phytoplasmas contribute to their sustainable management (Figure 13.1). However, less attention has been paid to the integrated management of phytoplasmas. This chapter presents an up-to-date overview on the progress of practices being applied for the management of emerging phytoplasma diseases.

13.2 HEALTHY PLANTING AND PROPAGATION MATERIAL

The use of phytoplasma-free clean planting material is one of the major tools in the management of phytoplasma diseases (Roddee et al. 2018). Establishing of new plantations with certified pathogen-free planting materials is of primary importance for a successful phytoplasma prevention and control (Seemüller et al. 1998a,b). Adequate sampling and use of highly sensitive Polymerase Chain Reaction (PCR) and recently developed isothermal assays would be required to ensure that nuclear stocks are free from phytoplasma infections.

In addition, if the materials would transfer to the new area, then heating or cooling treatment, dipping in hot water, and use of radiation would decrease the chance of new disease introduction to other countries. Heat therapy, hot water treatments, *in vitro* meristem tip cultures, tetracycline in combination with *in vitro* techniques are suggested useful in producing clean vegetative propagative materials in most of horticultural crops and recorded efficient in checking further spread of the disease in establishment of new orchards and/or in cultivation practices (Linck et al. 2019; Tiwari et al. 2011; Singh et al. 2007). On the other hand, phytoplasma-infected plants shouldn't be transported to areas where the potential or natural vectors are present. The infected *Limonium* sp. transferred to Israel where the potential vectors were distributed, and consequently, the phytoplasma symptoms were observed in seedlings of other *Limonium* plantations within the area (Weintraub et al. 2004).

Tissue culture and thermotherapy are techniques recommended for eliminating the phytoplasmas from the diseased plants. For eradication of "Ca. P. prunorum" infections in diseased *Prunus* plants, *in vitro* thermotherapy and meristem tip culture have been recommended (Laimer 2003; Laimer and Bertaccini 2008). Several tissue culture techniques coupled with thermotherapy have been suggested for production of "*Ca* P. pnoenicum" phytoplasma-free almond plantlets (Laimer and Bertaccini 2019). Stem cutting culture combined with thermotherapy was also reported to be effective technique against phytoplasma (Chalak et al. 2005). Subsequently, shoot tip and stem cutting cultures associated with heat treatment were found to be all suitable for *Candidatus* Phytoplasma phoenicium elimination from regenerated shootlets in almonds (Chalak and Choueiri 2015). Meristem tip culture is also used to produce phytoplasma-free grape and stone fruit nursery stock in Turkey. In India, sugarcane is also produced through tissue culture to obtain virus and phytoplasma-free material (Viswanathan and Rao 2011; Tiwari et al. 2011), while disease-free planting materials have also been used for the management of sandalwood spike disease (SSD) in India (Mondal et al. 2020). The selection of certified seeds or planting material is one of the better options for the management of witches' broom disease of acid lime in Oman and Iran (Al-Sadi et al. 2012; Siampour et al. 2019). Another effective method for pathogen eradication from infected tissue is cryotherapy of shoot tips (Wang and Valkonen 2008). This method has been succeeded to eliminate several phytoplasma strains and some other graft-transmissible pathogens from different crop plants. Sweet potato little leaf phytoplasma have been successfully eradicated from a sweet potato line using cryotherapy of shoot tip (Wang and Valkonen 2008, 2009). But this technique is not prophylactic, and when plants are exposed to nature, the insect vector again makes them infected.

13.3 CULTURAL CONTROL OF PHYTOPLASMA DISEASES

Cultural control is commonly practiced for the management of various emerging phytoplasma diseases. Several cultural and conventional practices have an indirect effect on vector populations and phytoplasma disease spread. Implementation of appropriate prophylactic measures also supports the disease management. Removal of diseased plants or wild hosts from natural habitat as inoculum reservoirs contributes to eradication of the disease from the plantations, especially in areas where the disease has limited distribution. In Lebanon, eradication of almond trees infected with "*Ca.* Phytoplasma phoenicium" has been implemented to reduce the foci of infection (Lova et al. 2014). Some growers practiced to remove the infected trees and replace them with other citrus species or cultivars (Al-Sadi et al. 2012). In Oman, growers usually remove severely diseased acid lime trees that are affected by witches' broom disease. The regular removal of symptomatic branches also reported to reduce phytoplasma titers and vector attraction to the symptomatic branches (Al-Subhi et al. 2021). In addition, this practice also helps citrus trees to get rid of the symptomatic and non-productive branches, since they do not produce any fruits or bear only small non-marketable fruits (Hemmati et al. 2021). Destruction of infected vines as well as uprooting of wild *Vitis* plants in contaminated vineyards is also effective prophylactic activity for the management with Flavescence dorée (FD).

Weeds also can be the potential reservoir of the vectors as well as phytoplasmas, and their control is considered as important tool in the phytoplasma disease control (Weintraub and Wilson 2010; Duduk et al. 2018; Rao 2021).

13.4 RAPID DIAGNOSIS APPROACH

Most of the symptoms caused by plant viruses and phytoplasmas are common, and therefore, there is always confusion to identify the real causal organisms associated with the diseases especially in horticultural crops. Recently developed molecular biology diagnosis techniques such as nested Polymerase Chain Reaction assays, Reverse Transcription – Polymerase Chain Reaction, isothermal assays (Loop-mediated isothermal Amplification, RPA, etc.), multilocus group specific designed primers, and Recombinase Polymerase Amplification-based diagnosis are reliable and can be applied efficiently in quick and authentic detection of phytoplasmas infection at early stage of the crop. Another problem in diagnosis of phytoplasmas is their non-cultural nature, and hence, only molecular diagnosis is the best option to detect and confirm phytoplasma's presence in any plants at early stages even without symptoms. Field-based molecular and serological diagnostic kits (lateral flow) should be developed and validated for the diagnosis of emerging phytoplasma diseases in horticultural/agricultural crops so that farmers and growers could detect phytoplasma diseases at early stage (Bertaccini et al. 2019).

13.5 ANTIBIOTICS

Tetracyclines including oxytetracycline act as an inhibitor of protein synthesis interfering with the ability of procaryotes to produce proteins that are essential for growth and multiplication (Borgi and Palma 2014). Injection or spray of tetracycline (oxytetracycline) antibiotics into the trunk or on plants has been used to control phytoplasma infection in individual host plants. After antibiotic treatments, the disease symptoms can disappear for a period up to 2 years; however, tetracycline is not able to eliminate the pathogen completely in the host plant and has only a temporary remission effect of symptoms (Raychaudhuri et al. 1970; Schmid 1983).

Tetracycline (oxytetracycline) has been registered earlier against coconut lethal yellowing, pear decline, lethal decline of pritchardia palm, peach X-disease, etc., in the USA (McCoy 1983). Oxytetracycline used for *in vitro* and in vivo chemotherapy at concentration of 60–100 mg/L was found to be successful in freeing infected tissues of ornamentals (chrysanthemums, marigolds, roses, etc.), vegetables (brinjal), oil (sesame), and several fruits species infected with phytoplasmas (Singh et al. 2007b; Rao et al. 2017; Taloh et al. 2018).

The method of tetracycline treatment is not easily applicable, feasible, and cost-efficient for commercial production; however, it can be employed for high-valuable ornamental plants at tourist sites (McCoy et al. 1976; Eziashi and Omamor 2010; Taloh et al. 2018). The use of tetracycline can also induce resistant bacterial populations. Moreover, antibiotics in agriculture are prohibited in several countries in the world (Europe, Turkey, etc.) because of health and environmental issues (Been 1995; Musetti et al. 2013b). All these factors make tetracycline antibiotics impractical for phytoplasma disease management.

Apart from tetracycline, some other molecules such as biophenicol, chloramphenicol, enteromycelin, lycercelin, paraxin, roscillin, and camphicillin also showed promising results for phytoplasma management but with limited success (Kumari et al. 2019).

13.6 HABITAT MANAGEMENT

Habitat management is an important agro-ecological approach in integrated pest management (IPM) that employs border crops, trap cropping, mulching, intercropping, multi-cropping, pheromone trapping, insectary plants, and weed strips techniques. These techniques are very effective to minimize pest pressure by improving the fitness of host or unfavoring the habitat for living. Vegetation composition that is adjunct to orchards/fields can have a vital role on the presence and dispersal of phytoplasma vectors (Weintraub and Beanland 2006). Weed species can have a profound effect on vector distribution and infection levels. However, plant density might affect vector distribution. The fact that the infection is started from the margins showed that the vector over-wintering on perennial weeds influx the fields. The primary direction of *S. titanus* movement was recorded from the wild to the cultivated vegetation in grapevine orchards. In another experiment, the traps in vineyards and those placed at 12 and 24 m beyond the vineyard border captured a large number of insects (Lessio and Alma 2004).

The application of several synthetic or organic mulches is another means of habitat manipulation. *Hyalestes obsoletus* lays eggs at or just below the soil surface. Plastic sheeting as a synthetic mulch is able to prevent the movement of vectors into the soil. Reflective mulches can repel the vector from the plants such as maize leafhopper, *Dalbulus maidis* was better controlled by plastic reflective mulch (Summers and Stapleton 2002). The use of aluminum foil mulch could effectively control the aster yellows phytoplasma vector, *Macrosteles quadrilineatus* (Forbes) in carrot fields. The type of mulching materials used in controlling is important. For example, the coconut mulched with coarse materials such as pine bark nuggets attracted fewer

Haplaxius crudus, the vector of lethal yellowing, in comparison with coconut frond and eucalyptus. It can be noted that mulch application is too expensive but can be a sustainable approach.

13.7 COVERING PLANTS

The most consistent way for controlling phytoplasma vectors can be achieved by covering crops. However, due to the logistics of large-scale agriculture in major crops such as rice, wheat, maize, sugarcane, and tress, its application is limited that its use cannot even be contemplated. However, the plants that can be grown in the greenhouse much less suffer from the phytoplasma diseases showing the efficacy of this method. Bananas and date fruits are now covered by plastic or papers for controlling other pests and pathogens that can be applied worldwide for other trees (Laimer and Bertaccini 2019).

A survey was conducted in Australia to compare the effect of covering papayas by insect-exclusion screening (IES) with spraying systemic insecticide such as imidacloprid, for management of phytoplasma diseases (Elder et al. 2002; Walsh et al. 2006) performed a survey in Australia. They revealed that covering the trees could control 100% vectors in comparison with insecticide and non-treated control as the phytoplasma symptoms were significantly reduced in covered trees.

This method was also applied in vineyards that controlled phytoplasma disease in vineyards (Mannini 2007). The clonal or mother plants that were continuously covered by IES showed no phytoplasma disease. Blua et al. (2005) studied the effect of a 5-m-high barrier screen to prevent the movement of the sharp shooter, *Homalodis cavitripennis* (Germar), a vector of the plant pathogen, *Xylella fastidiosa*, to high-value vineyards and nursery stock. The results showed that the vector behavior changed as they moved away from the barriers to surrounding plants and few flew over the barriers. These findings suggested that a barrier could add significantly to vector control in several crops.

13.8 INSECT VECTOR CONTROL STRATEGIES

Insect vector control using pesticides is one of the best methods recommended for limiting outbreaks of phytoplasma diseases (Bertaccini and Duduk 2009), but it is not always as efficient as expected. The insecticides do not act immediately, which allows phytoplasmas rapidly to be acquired (Weintraub and Wilson 2010). On the other hand, conventional insecticides are not very effective because pathogen transmission occurs faster than the action of insecticides, and there is always possibility of a constant influx of new efficient vector populations from surrounding habitats (Wally et al. 2004). At best, use of insecticides might help control vector populations and thus reduce intra-crop transmission (Weintraub 2007; Rao et al. 2014). Systemic and contact insecticides can also be applied to protect plants from carrying migratory vectors and suppress the vector population, respectively. Sarnaik et al. (1986) showed that three sprays of fenvalerate @ 0.01% gave the best control in minimizing the incidence of sesame phyllody in India. In addition, Reddy et al. (2019) tested some insecticides against *O. albicinctus* in sesame fields in India and revealed that pymetrozine spraying showed the highest percent reduction of the nymphal population (91.5%) followed by imidacloprid + ethioprole (80%), acephate (73%), and dinotefuron (70%) (Tiwari et al. 2016, 2017). Pilkington et al. have demonstrated that phytoplasma disease incidence can be effectively reduced by treating crop fields' border with systemic insecticides.

The use of insecticides independently or in conjunction with other methods needs more attention to arrive at the appropriate management tactics. Other limiting aspects of using insecticides in insect control are related to insecticide resistance (Marcone and Rao 2019), increasing the costs of production, and negative effect on environment (due to their soil persistence and water contamination) and human health. Elimination of phytoplasmas and insect eggs from grapevine is possible by hot water treatment (Mannini 2007). It would also be important not to transport phytoplasma-infected plant materials to new areas where potential vectors may be present (Weintraub et al. 2004).

The most reliable means of controlling vectors is by covering or mulching of crop with insect-proof screening or plastics (Weintraub 2007; Marcone and Rao 2019). For example, in Australia, phytoplasma diseases associated with papaya have been very well managed by using insect-proof screening. Nets protect the plants from the visit of insect vectors and the inoculum of phytoplasmas. They are more efficient than insecticides but are expensive and inapplicable for large-scale application over agriculture crops (sugar cane, corn, rice, fruit trees, and grapes etc). It can be suggested for cash crops, organically grown fruits, or when a premium price can be obtained for the precious vegetative propagative seedlings of horticultural crops (Walsh et al. 2006).

Mineral coating on plant surface is another effective means of insect vector control through physical prevention. Kaolin, a nonabrasive fine-grained aluminosilicate mineral, when applied as a particle film, inhabits the vector feedings from the plants (Weintraub 2007). Puterka et al. (2003) demonstrated that physical coating of plant surfaces with mineral film protected grape plants from feeding, oviposition, and infestation by the glassy-winged sharpshooter, *Homalodisca coagulata* (Tubajica et al. 2007).

Insect vectors of phytoplasmas utilize chemical cues for identification of their hosts. Phytoplasma infections may influence the production of plant volatiles, which indirectly influenced the behavior of vector insects. Insects-species-specific attractive and repellent compounds identified for many phytoplasma vectors can be used in traps as lures for monitoring and mass trapping purposes. In this way, semiochemicals provide an innovative approach for phytoplasma vector controls (Gross et al. 2011). Manipulation of symbiotic bacteria in the vectors is a new approach for controlling pathogen transmission. The effective symbiotic bacterial strains can be used to compromise phytoplasma transmission by reducing vector competence and a high offspring mortality rate (Weintraub 2007). For example, bacterium, identified as *Cardinium hertigii* and localized in the fat bodies and salivary glands of the leafhopper *Scafoideus titanus*, the vector of Flavescence doree phytoplasma is considered as promising in symbiotic control efforts of FD (Bigliardi et al. 2006; Marzorati et al. 2006). One of the major challenges is the delivery of the transgenic bacteria under field conditions to the target insect vectors without adversely affecting the environment or other insect populations, which needs further study.

Also, genetic manipulation of the plants that can secrete the defensive compounds against vectors is another effective way. *Sagotella furcifera*, *Nilapar vatalugens* planthoppers, and *Nephotettix virescens* leafhoppers are the main phytoplasma transmitters of rice crops. Nagadhara et al. (2004) revealed that genetically modified plants, expressing lectin especially *Galanthus Nivalis Agglutinin* (GNA), were highly toxic to plant hoppers as the survival, development, and fecundity of *S. furcifera* significantly reduced. In addition, they revealed that the vectors feeding on GNA rice caused 90% mortality in *S. furcifera* and 29% and 53% mortality in *N. virescens* and *N. lugens*, respectively (Nagadhara et al. 2004). As a consequence, reduction in planthopper population contributed to a reduction in disease spread. The insertion of this gene into other plants could have significant effects, which needs further investigations.

Systemic acquired resistance (SAR) is a response of plants that can be activated when challenged by either a pathogen or arthropods. It is also elicited by a number of chemicals. Studies on *Colladonus montanus* (Van Duzee), the vector of X-disease in fruit trees, showed that treatment with benzothiadiazole (BTH) protected plants from the feeding of this vector and consequently phytoplasma infection (74%). It may be due to plant phloem that has been morphologically modified to prevent phytoplasma from establishing or replicating as well as production of a substance inhibiting vector feeding. In addition, the survival of the leafhopper significantly decreased on plants treated with benzothiadiazole than non-treated ones (Weintraub and Wilson 2010).

Integrated environmentally safe strategies for vector control would be proper survey schedules including mating disruption, pheromones and plant-based volatiles, attractant-and-kill, and push–pull strategies (Reddy and Guererro 2010). Pheromones are chemicals produced by insects to induce a behavioral response on conspecific individuals. Some pheromones function as attractants, allowing individuals to detect and locate mates (Witzgall et al. 2010). Insect control by pheromones alone has precincts, but they can be used in integrated control in conjunction with other practices (Howse et al. 1998). In the past four decades, there have been successful stories on insect sex pheromones in managing psyllid vectors associated with pome and stone fruits infected with phytoplasma in Europe (Jarausch and Jarausch 2010). Mayer et al. (2008a,b) concluded that phytoplasma-infected apple fruit trees increased their attractiveness to *Cacopsylla picta* by release of sesquiterpene. A synthetic β-caryophellen was also found highly attractive to newly emerged adults of psyllids and can be successfully used for mass trapping of *C. picta* for minimizing the incidence of phytoplasma in the apple orchards. The management of pear decline has been achieved through the control of the insect vector *Cacopsylla pyri*. Several other insecticides have been proposed for the management of psyllids transmitting pear decline and apple proliferation phytoplasmas (Laimer and Bertaccini 2019).

Models for prediction of phytoplasma and vector spread may provide an insight in disease epidemiology and control. However, a few studies incorporating information on climate change and its relation to disease and vector spread are available. Lessio et al. (2015) developed an FD epidemiological model with integration of different parameters of the transmission process and showed the risks of establishing new vineyards in severely contaminated locations (Oliviera et al. 2018).

13.9 HOST RESISTANCE

Host resistance and tolerance are terms associated with different plant responses to the pathogens. Use of resistant plants seems to be the best approach for preventing of phytoplasma diseases. Phytoplasma-resistant or -tolerant genotypes provide an effective and environmentally safe tool for protection of crops from these pathogens. Resistance against vectors based on non- preference for a plant host was also suggested as a good phytoplasma management tool (Beech 1981; Laimer et al. 2009).

Quantitative Recombinase Polymerase Amplification (qRecombinase Polymerase Amplification) diagnostic assay provides opportunity not only to identify the pathogen but also to phenotype phytoplasma disease severity more objectively (Huang et al. 2014). This method is used to quantify phytoplasmas even in early identification of tolerant and resistant genotypes in different crops, which has also a great potential in enhancing plant breeding (Huang et al. 2014; Ikten et al. 2016; Ustun et al. 2017). Besides inoculation experiments (Sinclair et al. 2000) and aquantitaice Recombinase Polymerase Amplification, other molecular DNA methods such as the detection of RAPD markers associated with the resistance (Cardeña et al. 2003) were used for an adequate determination of genotype resistance level. Simple sequence repeats (SSRs), amplified fragment length polymorphism (AFLP), and long terminal repeats (LTRs) markers were utilized in phytoplasma resistance mapping (Lenz and Dai 2017).

For now, the molecular, anatomical, and physiological basis of phytoplasma resistance is still poorly understood (Marcone and Rao 2019). Additional studies are necessary to reveal clearly its physiological and molecular mechanisms. The resistance may be associated with morphological and physiological differences in the plants, which are designated by the genetic background of each cultivar and their responses to the environment (Oliviera et al. 2018). It may also be linked to their reaction to specific phytoplasma effectors, playing a role in the host–pathogen interaction (Eveillard et al. 2016; Oliviera et al. 2018). In some host plants such as Chinese jujube, genes associated with phytoplasma resistance through suppressive subtraction hybridization were identified (Liu et al. 2014).

Genotype resistance to phytoplasma is influenced by environmental conditions, aggressiveness of the phytoplasma strains, rootstocks, etc. Atkinson and Urwin (2012) argued that genotypes selected as disease-resistant under field conditions showed diverse response of disease resistance under various environmental conditions. This can be due to the fact that the environmental conditions, particularly the interaction between temperature and moisture in the host, act directly upon the phytoplasma disease severity (Krishnareddy 2013) and the population and distribution of vectors (Gurr et al. 2016). The genetically diverse phytoplasma populations and appearance of new phytoplasma strains in the different environments may also reduce the level of resistance or may cause the total breakdown of resistance (Quaicoe et al. 2009; Gurr et al. 2016). In addition, research on tagging disease-resistant genes should employ molecular markers, which are stable and dependable, which may lead to the long-cherished objective of identification of resistant progenies in nursery stage itself (Marcone and Rao, 2019).

The scientific efforts on phytoplasma resistance have been ended up with the identification of some resistant genotypes and progenies. The high resistance to apple proliferation was determined in apomictic *M. sieboldii* progenies and in some selections having *M. sieboldii* in their ancestors (Seemüller et al. 2008). The genotype "Xingguang" (clone of the cultivar Junzao) crossed and released in 2005 in China (Zhao et al. 2009; Liu et al. 2004) has very high resistance to jujube witches' broom (Jwb) phytoplasma, one of the most destructive diseases of Chinese jujube (*Ziziphus jujuba* Mill.) in Asia (Liu et al. 2014). Resistant genotype was also developed against brinjal little leaf phytoplasma inducing over 40% yield losses in *Solanum melongena* L. in subtropic and tropic areas using a resistance source from wild relatives of the brinjal, *Solanum gilo,* and *S. integrifolium* (Chakrabarti and Choudhury 1975). Genotypes with different level of resistance against sesame phylody responsible for the most destructive disease of *Sesamum indicum* in some Asian countries (Akhtar et al. 2009; Rao et al. 2015) were also developed in India, Pakistan, and Turkey (Singh et al. 2007a; Akhtar et al. 2013; Ustun et al. 2017), and inheritance of phyllody resistance was reported (Singh et al. 2007a; Shindle et al. 2011; Utsun et al. 2017; Yadav et al. 2022).

Currently, management of coconut lethal yellowing is based on detection, immediate destruction of infected plants along with replanting using any available palm varieties with resistance (Baudouin et al. 2009; Myrie et al. 2011). A major limitation of this strategy is the nonspecific or delay appearance of the symptoms after infection, by the time the vectors may have spread the pathogen widely. This highlights the need for fast and sensitive field diagnostics, particularly Loop-mediated isothermal Amplification and Recombinase Polymerase Amplification field kits and remote sensing, to support current management efforts, while success in plant breeding to provide longer-term control is likely to be well supported by CRISPR-based tools in phytoplasma disease management.

Many carrot (*Daucus carota*) varieties showed significant resistance against the aster yellows phytoplasma in the USA and Canada (Gabelman et al. 1994). The recovery of symptomatic plants after infection, especially in apple, stone fruits, and grapevine (Musseti et al. 2014), has been found to produce resistant plants, which was attributed to stimulating jasmonate (JA) and Ca^{2+}-signal-related defense mechanisms (Musseti et al. 2013a, 2003b). In addition, the use of resistance stimulants proved to be effective in managing grapevine phytoplasma diseases in Iran (Ghayeb Zamharir and Taheri 2019). An excellent new cultivar with very high level of resistance to jujube witches' broom (JWB), named "Xingguang," was bred and released in China (Zhao et al. 2009; Liu et al. 2012).

Many economically important phytoplasma diseases are very difficult to manage through resistant plants. For example, resistance screening to lethal yellowing disease caused by "*Ca*. Phytoplasma palmae" poses a significant threat to global coconut production and is not so easy because of difficulties in genetic improvement of coconut (Cardeña et al. 2003) and the time gap in symptom expression of phytoplasma in the genotype. Up to now, some coconut palm varieties and hybrids less susceptible to lethal yellowing are known; however, no genotype of coconut was declared as resistant (Baudouin et al. 2009). Also resistance or tolerance to phytoplasmas associated with Grapevine yellows (GY) disease has not been found in the *Vitis* species examined up to now (Laimer et al. 2009). Unfortunately, most cultivars have been reported as highly susceptible or intermediately susceptible with severe or medium severe symptoms, and only a very limited number were referred to as low-susceptible cultivars. Nonetheless, many economically relevant cultivars have not yet been studied for tolerance/susceptibility to FD (Oliviera et al. 2018). No fully rresistant cultivars are known to "*Ca* P. phoenicium" (subgroup 16SrIX-B), which is the aetiological agent of almond witches'-broom (AlmWB), a severe disease affecting almond, peach, and nectarine trees in Lebanon and Iran (Ghayeb Zamhariri et al. 2011).

Development of genotypes with durable resistance is required for successful phytoplasma management. For durability of resistance over time, breeding of disease-resistant cultivars carrying not single but multiple sources of resistance is more prefarable (Luo et al. 2020). So far, some identified resistant genotypes have been unsatisfactory and with low agronomic value. However, they can be further exploited as genitors in breeding programs aiming to develop plants with a sufficient agronomical properties and stable inherited phytoplasma resistance (Marcone and Rao 2019). For example, the productivity of apple tree on M. sieboldii-based AP-resistant apomictic rootstocks was enchanced by crossing and backcrossing *M. sieboldii* and its apomictic hybrids with M9 and trees with high resistance and good properties were developed (Marcone and Rao 2019). CRISPR-based new technology for gene editing has also a great potential and can be further used for progress and better understanding and exploatation of phytoplasma resistance (Belhaj et al. 2013).

13.10 HOST–PATHOGEN INTERACTIONS

Transcriptome analysis for several phytoplasma–host combination has been initiated and was carried out to ascertain the intricate host–pathogen interactions with respect to coconut yellow decline disease in Malyasia (Nejat et al. 2015), Cranberry false blossom disease in the USA (Pradit et al. 2019), *Paulownia* infected by paulownia witches'-broom phytoplasma in China (Mou et al. 2013), and other important phytoplasma strains (Kube et al. 2018). The differentially expressed transcripts could be sorted out into pathways such as formation of cell wall, plant–pathogen interactions, primary and secondary metabolism, biosynthesis of hormones, and signaling cellular transport. The expression patterns of a set of genes, both upregulated and downregulated, were validated by quantitative real-time Recombinase Polymerase Amplification (qReverse Transcription – Polymerase Chain Reaction), the results of which were comparable to those observed by RNA-Seq analysis. The use of new-generation sequencing techniques has witnessed large-scale generation of ESTs in several phytoplasma-infected hosts, which has provided insights into plant–pathogen interactions and facilitated elucidation of genes involved in somatic embryogenesis. The available genome sequence forms a potential resource to be utilized in genetic enhancement of the crops concerned.

13.11 INDUCED RESISTANCE

Another strategy for the control of phytoplasma disease is spraying the plant canopy with resistance inducers (Devendra et al. 2008). These molecules, called also elicitors, can be of abiotic or biotic nature and challenge the plant, leading to a reaction, often linked to the production of antimicrobial compounds and/or the elicitation of plant defense mechanisms. The most common elicitors identified in controlling phytoplasma diseases are: Phosetyl-Al, benzothiadiazole, indole-3-acetic acid, indole-3-butyric acid, chitosan, salicylic acid, mixture of glutathione, prohexadione calcium, and oligosaccharines (Romanazzi et al. 2009). One of the commercial forms of phosetyl-Al is Previcur EnergyTm (31% fosetyl-Al plus 53% propamocarb), which is a new systemic synthesized molecule with protective action. It is easy to use (drip application, drench, spray) and also shows root growth stimulant effect (Leljak-Levanic et al. 2010).

SA was successfully applied in the management of tomatoes infected with potato purple top phytoplasma (Wu et al. 2012) and lime with witches' broom phytoplasma (Ghayeb Zamharir et al. 2020). Application of humic and fulvic acids and algae extracts also provided promising results (Mazio et al. 2008). Significant reduction in incidence of grapevine phytoplasmas in cv Chardonnay was achieved by weekly applying commercial resistance inducers, based on chitosan, Phosetyl-A1, benzothiadiazole, and two glutathione and oligosaccharides formulation (Romanazzi et al. 2009).

Application of resistance-inducer compounds played an important role in plant diseases control against witches' broom diseases of lime in Iran, Oman, and United Arab Emirates (Romanazzi et al. 2009; Schmidt et al. 2015). The lime trees under the treatment of 31% Phosetyl-Al plus 53% propamocarb showed consistently high yield increase during 2 years of field trails (Ghayeb Zamharir et al. 2020). Similar results were reported by D'Amelio et al. (2007), who observed a delay in symptom development on chrysanthemums infected by "*Ca.* P. asteris" after chitosan application. Treatments with resistance inducers carried out on phytoplasma-infected *Catharanthus roseus* showed different responses of the plants sprayed with Phosetyl-Al and chitosan (Chiesa et al. 2007).

13.12 BIOLOGICAL CONTROL AND BIOTIC INDUCERS

Planthoppers and leafhoppers are fed by a range of predators. Spiders that feed nymphs and adults are very important predators especially in the grasslands ecosystem. Bagheri et al. (2017) revealed that two species of predatory spiders, namely *Plexippus iranus*Longunov and *Thyene imperialis*, belonging to Salticidae family, voraciously fed *H. phycitis* in citrus orchards of southern Iran. Mirids are known as egg predators in ecosystems. Auchenorrhyncha are attacked by specialist parasitoids such as Mymaridae (Hymenoptera), Dryinidae (Hymenoptera), and Pipunculidae (Diptera). There are also specialist parasitoids associated with Auchenorrhyncha (Waloff and Jervis 1987). Mymaridae, known as fairy flies, are a common egg parasitoids, which complete the whole life span in eggs. However, there are little reports on leafhoppers and planthoppers eggs (Bagheri et al. 2017).

The role of arbuscular mycorrhizal (AM) fungi in "stolbur" phytoplasma infection of tomato has been investigated, and agglutinations and degeneration of phytoplasma cells, coupled to reduced symptom expression, were reported in plants treated with AM fungi. Inoculation with *Glomus intraradices* in infected pear trees also reported to increase tolerance to pear decline disease (Romanazzi 2013). Endophytic bacteria also promote plant growth and protect them against pathogen infections. Endophytes–plant and endophytes–pathogens interactions are poorly explored. Recently, *Pseudomonas putida* strain S1Pf1Rif was tested, alone or in combination with the mycorrhizal fungus, *Glomus mosseae* BEG12, against chrysanthemum yellows phytoplasma. Fungal endophytes establish mutualistic relationships with plants and induce physiological modifications in their hosts making them more resistant. Fungal endophyte strains have been also identified from grapevines and apple plants grown in areas where recovery phenomenon was recurrent. But further works are required for their future practical applications.

The fungal strains of *Epicoccum nigrum* and *Aureobasidium pullulans* were chosen for further research activities because they are extensively reported as biocontrol agents or resistance inducers. Using the model plant, *Catharanthus roseus* infected with *Ca*. P. mali, it was observed that reduction in symptom severity and lower phytoplasma titer in host tissues occurred when the plants were previously inoculated with endophytic strain of *E. nigrum* (Musetti et al. 2011).

13.13 SYMBIOTIC CONTROL

Phytoplasmas are not the only microorganisms living in the vectors' body, and they are involved in complex microbiomes enclosing bacteria, viruses, and fungi, in turn, influencing such associations. Insect symbionts including bacteria, fungi (yeast and yeast-like), and viruses are famous effectors of ecology, life history, and development of their hosts. Similar to all sap feeders, phytoplasma vectors depend on bacterial symbionts supplementing their unbalanced diet (Zchori-Fein and Bourtzis 2011). Insect microbial symbionts accompanied by supplying the host with nutrients offer protection against pathogens or other stress factors while manipulating reproduction. Such multifaceted interaction could be used for developing microbe-based control approaches against phytoplasma-borne diseases (Dale and Moran 2006).

Psyllids are the only groups of Sternorrhyncha that transmit the phytoplasma. Two genus *Cacopsylla* and *Diaphorina* are known as the vector of all phytoplasma diseases in the world. The primary symbiont of the psyllids is "*Ca*. Carsonella ruddii" (Gonella et al. 2019), which has a long evolution with the host and provides the essential amino acids (Cooper et al. 2015). A study conducted by Hosseinzadeh et al. (2019) on endosymbionts of *D. citri* revealed that two obligate symbionts, namely "*Ca*. carsonella ruddii" and "*Ca*. proftella armatura," and a reproductive manipulator symbiont, namely *Wolbachia pipentis*, were present in Asian psyllids. This information can be used in symbiotic controls of the vector.

The evolutionary and ecological success of the insects might rely on yeast and yeast-like symbiont (YLS) (Douglas 2011). Yeast-like relationships with insects are recognized as obligate and common symbiotic relationships. Hemmati et al. (2017) concluded that *Hishimonus phycitis* harbored two yeast-like symbionts. These complex symbionts (bacterial and fungal symbionts) living in *H. phycitis* may help us find a novel way for controlling Witches' Broom Disease of Lime (WBDL).

13.14 LEGISLATITAVE MEASURES AND QUARANTINE

Many phytoplasma diseases have spread from one place to another with the active help of trained human personnel and establishment of modern and equipped molecular diagnostic labs. The unrestricted movement of cane from one country or state to another has been responsible for the migration of phytoplasma diseases endemic to an Asian country to other neighboring countries in the past and still continues for many phytoplasma-associated disease especially in vegetative cutting of ornamentals and fruit crops. Earlier, pathogens had spread to new areas, and now virulent strains are spreading to new pastures. Legislatitave measures are important for prevention of the entry, spread, and establishment of exotic phytoplasma strains into new areas. Using certified pathogen-free nursery materials is essential for exclusion of the phytoplasmas from the production areas. Phytoplasmas can enter into new areas by infected vegetative plant materials and insect vectors. To prevent the spread and entry into new areas, several mandatory quarantine requirements, rules, and laws are usually implemented into force by plant health authorities worldwide. Various phytoplasma strains with high phytosanitary risks and their vectors categorized as quarantine pest should be strictly taken care of. Flavescence Doree, European stone fruit yellows, lime wicthes' broom, apple proliferation, stolbur phytoplasmas, coconut lethal yellowing, and sugarcane white leaf are the most emerging and harmful quarantine phytoplasma strains for many Asian countries (https://www.eppo.int/ACTIVITIES/quarantine_activities). For these emerging phytoplasma diseases, strict phytosanitary regulations and restrictions on their seedlings/vegetative propagative materials should be followed (Gurr et al. 2016). But in Asian countries, still strict quarantine regulations are not being implemented that allow chances of introducing new quarantine phytoplasma strains into newer locations and chances for introduction of new phytoplasma strains are at high risk. In Australia, Europe, and the USA, the local legislations are very strict and being implemented systematically in the contaminated regions of the country and for international border entry ports. For example, in France, directives from the local authorities regulate the mandatory control of not only FD disease but also its vector *S. titanus* for particular areas where the disease is present. (https://www.cabi.org/isc/datasheet/26184). For example, citrus is the major plant in China; however, the leafhopper *H. phycitis* has not been reported from China. The government should apply very strict rules in transporting citrus materials and fruits from where the leafhopper is distributed. Such a scenario was also reported in Europe as *Schaphoideus titanus* was introduced to Europe as a vector of flavescence doréé in 1958 by the eggs deposited under the bark of grapevine canes (Vidano 1964).

13.15 CONCLUSIONS AND FUTURE PROSPECTS

Phytoplasma disease management is complicated and can be achieved only via multiple approaches combining different practices. Several tools for exclusion of the pathogens from production sites, eliminating or reducing the inoculum sources, and prevention of disease and insect vectors spread have to be integrated. Other preventive and phytosanitary measures such as the use of healthy plant material and removal of diseased plants are not fully satisfactory as new infections do occur.

The same applies for treatments with tetracycline antibiotics, which are not allowed in many countries. The most promising approach to control phytoplasma disease would be through the use of natural resistant or tolerant plants. Intra- and inter-specific differences in the response of plants to various phytoplasma diseases have been observed under both experimental inoculation using differently virulent strains and natural infection conditions. Detailed epidemiological data are always necessary for development of the best management strategy. The deep knowledge on pathogen strains and diversities, host resistance and wild hosts, insect species and their behavior, environmental conditions and their effect, and existing control options are essential to take appropriate control measures. Moreover, host–phytoplasma interactions are important to be clearly explained. Over the past decades, attempts to understand the phytoplasma diseases have resulted in accumulation of a lot of knowledge on these pathogens. Unfortunately, there are still gaps in many aspects regarding pathogens, host–pathogen interactions, host resistance, insect vectors, etc. For better understanding, host–pathogen interactions and host resistance mechanisms have to be studied at genomic, transcriptomic, proteomic, and metabolomic level, and the functions of membrane or secreted proteins and effectors in the phytoplasma genome and also phytoplasma defense-related proteins in the plant have to be further investigated.

The future research should concentrate on identification of more resistant/tolerant genotypes proving the most environmentally safe tool for protection of crops from these pathogens. Insect control strategies based on semiochemicals, physical prevention, or other innovative approaches have to be further improved and vector monitoring systems to be developed for more efficient control. Application of plant resistance inducers, elicitors, or microorganisms triggering the defense mechanisms of the plants may be another promising alternative for phytoplasma containments. However, the studies indicating the potential of inducers for phytoplasma disease protection are still limited; more research is needed to optimize and expand their implementation. In the future, exploitation of new gene modulating technologies is expected to pioneer better understanding of phytoplasma diseases and development of new managing methodologies. In addition, implementation of domestic and international quarantine regulations in movement of plant germplasm would result in a significant reduction in disease spread of serious phytoplasma diseases.

Although chemical control has been used as the first way of control and will be applied in the predictable future, we anticipate that vector management gradually moves to numerous genetic manipulations of plants to produce truthfully resistant plants. At least, the researchers can produce transgenic plants expressing some chemicals to allow them to be tolerant of phytoplasma infection. Less attention has been paid to applying biological control, although some specific parasitoids have been reported. The awareness of microbial relationships in phytoplasma insect vectors has considerably increased in the past few years. More work is still needed to fully understand the mechanisms and interactions between the endosymbiont (both fungi and bacteria) and insect vectors. This way can be the most reliable way to achieve the best control way in the near future.

In the emerging scenario on climate change and consequent upsurge of insects and phytoplasma disease problems, need on strengthening quarantine, systematic surveillance, and constant vigilance in transborders are suggested to counter biosecurity risks. The best and viable option in the future will be the management of phytoplasma coconut diseases mostly relying on integrated management practices centered on the exploitation of sources of disease resistance. Modern digital technologies such as computer-aided, remote-controlled drones should be applied for plant protection operations. Remote sensing using advanced technologies can help in regular surveillance for emerging phytoplasma disease incidence, intensity, and spread, which would be important for planning control strategy.

REFERENCES

Akhtar, K.P.; Sarwar, G.; Dickinson, M. et al. Sesame phyllody disease: symptomatology, etiology and transmission in Pakistan. *Turkish Journal of Agriculture and Forestry* **2009**, 33, 477–486.

Akhtar, K.P.; Sarwar, G.; Sarwar, N.; Elahi, M.T. Field evaluation of sesame germplasm against sesame phyllody disease. *Pakistan Journal of Botany* **2013**, 45, 1085–1090.

Al-Sadi, A.M.; Al-Moqbali, H.S.; Al-Yahyai, R.A.; Al-Said, F.A. AFLP data suggest a potential role for the low genetic diversity of acid lime (*Citrus aurantifolia* Swingle) in Oman in the outbreak of witches' broom disease of lime. *Euphytica* **2012**, 188, 285–297. https://doi.org/10.1007/s10681-012-0728-7

Al-Subhi, A.M.; Al-Sadi, A.M.; Al-Yahyai, R.A. et al. Witches' broom disease of lime contributes to phytoplasma epidemics and attracts insect vectors. *Plant Disease* **2021**, 105, 2637–2648. https://doi.org/10.1094/PDIS-10-20-2112-RE.

Atkinson, N.J.; Urvin, P.E. The interaction of plant biotic and abiotic stresses: from genes to the field. *Journal of Experimental Botany* **2012**, 63, 3523–3543.

Bagheri, A.; Hemmati, C.; Askari, S.M.; Modarres, N.S.S.; Nikooei, M. Preliminary study on natural enemies of Hishimonusphycitis (Distant, 1908) (Hemi., Cicadellidae), the vector of lime witches' broom phytoplasma in Hormozgan province. *2nd Iranian International Congress of Entomology*, 2–4 September, University of Tehran, Karaj, Iran, **2017.**

Bai, X.; Zhang, J.; Ewing, A. et al. Living with genome instability: the adaptation of phytoplasmas to diverse environments of their insect and plant hosts. *Journal of Bacteriology* **2006,** 188, 3682–3696.

Baudouin, L.; Philippe, R.; Quaicoe, R.; Dery, S.; Dollet, M. General overview of genetic research and experimentation on coconut varieties tolerant/resistant to Lethal Yellowing. *Oléagineux, Corps Gras, Lipides* **2009**, 16, 127–131. https://doi.org/10.1051/ocl.2009.0244

Beech, D.F. Phyllody – Its impact on yield and possible control measures. In: *Sesame: Status and Improvement*; Ashri, A., Poetiary, P., Eds.; FAO Plant Production and Protection Paper: Italy, **1981**; pp. 73–80.

Been, B.O. Integrated pest management for the control of lethal yellowing: quarantine, cultural practices and optimal use of hybrids. In: *Lethal Yellowing: Research and Practical Aspects*; Oropeza, C., Howard, F.W., Ashburner, G.R., Eds.; Kluwer Academic Publishers: Dordrecht, **1995**; pp. 101–109.

Belhaj, K.; Chaparro-Garcia, A.; Kamoun, S.; Nekrasov, V. Plant genome editing made easy: targeted mutagenesis in model and crop plants using the CRISPR/Cas system. *Plant Methods* **2013**, 9, 1–10. https://doi.org/10.1186/1746–48119-39

Bertaccini, A. Plant pathogens, minor (phytoplasmas). In: *Encyclopedia of Microbiology (Fourth Edition)*; Thomas, M.S., Ed.; Academic Press: New York, **2019**; pp. 627–638. https://doi.org/10.1016/B978-0-12-809633-8.13097-3

Bertaccini, A.; Duduk, B. Phytoplasma and phytoplasma diseases: a review of recent research. *Phytopathologia Mediterranea* **2009**, 48, 355–378.

Bertaccini, A; Duduk, B.; Paltrinieri, S.; Contaldo, N. Phytoplasmas and phytoplasma diseases: a severe threat to agriculture. *American Journal of Plant Sciences* **2014**, 5, 1763–1788. doi: 10.4236/ajps.2014.512191

Bertaccini, A.; Lee, M. Phytoplasmas: an update. In: *Phytoplasmas: Plantpathogenic Bacteria-I*; Rao, G.P., Bertaccini, A., Fiore, N., Liefting, L.W., Eds.; Springer: Singapore, **2018**; pp. 1–29.

Bigliardi, E.; Sacchi, L.; Genchi, M. et al. Ultrastructure of a novel Cardinium sp. symbiont in Scaphoideus titanus (Hemiptera: Cicadellidae). *Tissue and Cell* **2006**, 38(4), 257–261.

Blua, M.J.; Campbell, K.; Morgan, D.J.W.; Redak, R.A. Impact of a screen barrier on dispersion behavior of *Homalodiscacoagulata* (Hemiptera: Cicadellidae). *Journal of Economic Entomology* **2005**, 98(5), 1664–1668.

Borgi, A.A.; Palma, M.S.A. Tetracycline: production, waste treatment and environmental impact assessment. *Brazilian Journal of Pharmaceutical Sciences* **2014**, 50, 1. https://doi.org/10.1590/S1984-82502011000100003

Cardeña, R.; Ashburner, G.R.; Oropeza, C. Identification of RAPDs associated with resistance to lethal yellowing of the coconut (Cocos nucifera L.) palm. *Scientific Horticulture* **2013**, 98, 257–263. https://doi.org/10.1016/S0304-4238(02)00162-0

Casanova, R; Llacer, G.; Sanchez-Capuchino, J.A. Remission of symptoms of apple proliferation, after injection of concentrated tetracycline solutions. *Acta Phytopathological Academy Sciences Hungarica,* **1980**, 15, 273e277.

Chakrabarti, A.K.; Choudhury, B. Breeding brinjal resistant to little leaf disease. *Proceedings of the Indian National Science Academy (Section B)* **1975**, 41, 379–385.

Chalak, L.; Choueiri, E. Contribution to the production scheme of local certified propagating material of almond: in vitro sanitation and micropropagation. *Acta Horticulturae* **2015**, 1083, 163–168.

Chalak, L.; Elbitar, A.; Rizk, R. et al. Attempts to eliminate Candidatus phytoplasma phoenicium from infected Lebanese almond varieties by tissue culture techniques combined or not with thermotherapy. *European Journal of Plant Pathology* **2005**, 112(1), 85–89. https://springerlink.metapress.com /link.asp?id=100265. https://doi.org/10.1007/s10658-004-7953-4

Chiesa, S.; Prati, S.; Assante, G.; Maffi, D.; Bianco, P.A. Activity of synthetic and natural compounds for phytoplasma control. *Bulletin of Insectology* **2007**, 60, 313–314.

Cooper, S.A.; McLean, G.; Guthrie, B et al. Multiple physical and mental health comorbidity in adults with intellectual disabilities: population-based cross-sectional analysis. *BMC Family Practice* **2015**, 16, 110.

D'Amelio, R.; Massa, N.; Gamalero, E. Preliminary results on the evaluation of the effects of elicitors of plant resistance on chrysanthemum yellows phytoplasma infection. *Bulletin of Insectology* **2007**, 60, 317–318.

Dale, C.; Moran, N.A. Molecular interactions between bacterial symbionts and their hosts. *Cell* **2006**, 126(3), 453–465.

Devendra, K.C.; Prakash, A.; Johri, B.N. Induced systemic resistance (ISR) in plants: mechanism of action. *Indian Journal of Microbiology* **2008**, 47(4), 289–297.

Douglas, A.E. Lessons from studying insect symbioses. *Cell Host & Microbe* **2011**, 10(4), 359–367.

Duduk, B.; Mejia, J.F.; Calari, A.; Bertaccini, A. Identification of 16SrIX group phytoplasmas infecting Colombian periwin-kles and molecular characterization on several genes. 2008. In: 17th Congress of the International Organization of Mycoplasmology, 6–11 July, Tienjin, China, 2008, 112, 83.

Elder, R.J.; Milne, J.R.; Reid, D.J.; Guthrie, J.N.; Persley, D.M. Temporal incidence of three phytoplasma-associated diseases of *Caricapapaya* and their potential hemipteran vectors in central and south-east Queensland. *Australasian Plant Pathology* **2002**, 31(2), 165–176.

Eveillard, S.; Jollard, C.; Labroussaa, F. et al. Contrasting susceptibilities to flavescence dorée in vitis vinifera, rootstocks and wild vitis species. *Frontiers in Plant Science* **2016**, 7, 1762. https://doi.org/10.3389/fpls.2016.01762

Eziashi, E.; Omamor, I. Lethal yellowing disease of the coconut palms (*Cocos nucifera* L.): an overview of the crises. *African Journal of Biotechnology* **2010**, 9, 9122–9127.

Francis, I., Holsters, M., Vereerke, D. The Gram-positive side of plant-microbe intercations.

Gabelman, W.; Goldman, I.; Breitbach, D. Field evaluation and selection for resistance to aster yellows in carrot (*Daucus carota* L.). J. Am. Soc. Hortic. Sci. 1994, 119, 1293–1297. doi: 10.21273/JASHS.119.6.1293.

Galetto, L.; Bosco, D.; Balestrini, R.; Genre, A.; Fletcher, J.; Marzachì, C. The major antigenic membrane protein of 'Candidatus Phytoplasma asteris' selectively interacts with ATP synthase and actin of leafhopper vectors, 2011. PLoS One 2011, 6, e22571.

Garnier, M.; Zreik, L.; Bové, J-M. Witches' broom, a lethal mycoplasmal disease of lime trees in the sultanate of Oman and the United Arab Emirates. *Plant Disease* **1991**, 75(6), 546e551.

Ghayeb Zamharir, M.; Majeed, A.S.; Mostafa, P. "Witches' broom" disease of lime suppressed by some resistance Inducers. *Indian Phytopathology* **2020**, 73, 517–525. https://doi.org/10.1007/s42360-020-00222-1

Ghayeb Zamharir, M.; Mardi, M.; Zade, N.H.H. et al. Identification of genes differentially expressed during interaction of acid lime infected with *Candidatus* Phytoplasma aurantifolia. *BMC Microbiology* **2011**, 11, 1.

Ghayeb Zamharir, M.; Taheri, M. Effect of new resistance inducers on grapevine phytoplasma disease. *Archives of Phytopathology and Plant Protection* **2019**, 52, 1207–1214.

Gonella, E.; Musetti, R.; Crotti, E.; et al. Microbe relationships with phytoplasmas in plants and insects. In: *Phytoplasmas: Plant Pathogenic Bacteria-II*; Bertaccini, A., Weintraub, P., Rao, G., Mori, N. Eds., Springer: Singapore, **2019**; pp. 207–305.

Gross, J.; Mayer, C.J. ; Eben, A. Development of innovative meth-ods for trapping phytoplasma vectors by attractive infochemicals. *Bulletin of Insectology* **2011**, 64, S141–S142.

Gurr, G.M.; Johnson, A.C.; Ash, G.J et al. Coconut lethal yellowing diseases: a phytoplasma threat to palms of global economic and social significance. *Frontiers in Plant Science* **2016**, 7, 1521. https://doi.org/10.3389/fpls.2016.01521

Hemmati, C.; Moharramipour, S.; Askari Siahooei, M.; Bagheri, A.; Mehrabadi, M. Identification of yeast and yeast-like symbionts associated with *Hishimonusphycitis* (Hemiptera: Cicadellidae), the insect vector of lime witches' broom phytoplasma. *Journal of Crop Protection* **2017**, 6(4), 439–446.

Hemmati, C.; Nikooei, M.; Al-Sadi, A.M. Five decades of research on phytoplasma-induced witches' broom diseases. *CAB Reviews* **2021**, 16, 1–16. https://doi.org/10.1079/PAVSNNR202116002

Hogenhout, S.A.; Oshima, K.; Ammar, E.D. et al. Phytoplasmas: bacteria that manipulate plants and insects. *Molecular Plant Pathology* **2008**, 9, 403–423.

Hosseinzadeh, S.; Shams-Bakhsh, M.; Mann, M. et al. Distribution and variation of bacterial endosymbiont and "*Candidatus* Liberibacter asiaticus" titer in the huanglongbing insect vector, Diaphorina citri Kuwayama. *Microbial Ecology* **2019**, 78(1), 206–222.

Howse, P. E.; Stevens, I. D. R.; Jones, O. T. Insect pheromones and their use in pest management, **2018**, Chapman and Hall, London.

Huang, D.; Walla, J.A.; Dai, W. Quantitative phenotyping of X-disease resistance in chokecherry using real-time PCR. *Journal of Microbiological Methods* **2014**, 98, 1–7. https://doi.org/10.1016/j.mimet.2013.12.016

Ikten, C.; Ustun, R.; Catal, M.; Yol, E. Multiplex real-time qPCR assay for simultaneous and sensitive detection of phytoplasmas in sesame plants and insect vectors. *PLoS One* **2016**, 11, e0155891.

Ishii, Y.; Kakizawa, S.; Hoshi, A.; Maejima, K.; Kagiwada, S.; Yamaji, M.; Oshima, K.; Namba, S. In the non-insect-transmissible line of onion yellows phytoplasma (OY-NIM), the plasmid-en-coded transmembrane protein ORF3 lacks the major promoter region. *Microbiology* **2009**, 155, 2058–2067.

Jarausch, B.; Jarausch, W. Psyllid vectors and their control. In: *Phytoplasmas: Genomes, Plant hosts and Vectors*; Weintraub, P.G., Jones, P., Eds.; CABI Publishing: Wallingford, **2010**; pp. 250–271.

Jomantiene, R.; Davis, R.E. Clusters of diverse genes existing as multiple, sequence-variable mosaics in a phytoplasma genome. *FEMS Microbiology Letters* **2006**, 255, 59–65.

Jomantiene, R.; Zhao, Y.; Davis, R.E. Sequence-variable mosaics: composites of recurrent transposition characterizing the genomes of phylogenetically diverse phytoplasmas. *DNA and Cell Biology* **2007**, 26, 557–564.

Krishnareddy, M. Impact of climate change on insect vectors and vector-borne plant viruses and phytoplasma. In: *Climate-Resilient Horticulture: Adaptation and Mitigation Strategies*; Singh, H.C.P., Rao, N.K.S., Shivashankar, K.S., Eds.; Springer: New Delhi, **2013**; pp. 255–277.

Kube, M.; Mitrovic, J.; Duduk, B.; Rabus, R.; Seemu¨ller, E. Current view on phytoplasma genomes and encoded metabolism. *Science World Journal* **2012**, 185942.

Kumari, S.; Nagendran, K.; Rai, B.A.; Singh, B.; Rao, G.P.; Bertaccini, A. Global status of phytoplasma diseases in vegetable crops. *Frontiers in Microbiology* **2019**, 10, 1349. https://doi.org/10.3389/fmicb.2019.01349

Laimer, M. Detection and elimination of viruses and phytoplasmas from pome and stone fruit trees. *Horticultural Reviews* **2003**, 28, 187–236.

Laimer, M.; Bertaccini, A. European stone fruit yellows. In: *Characterization, Diagnosis and Management of Phytoplasmas*; Harrison N.A., Rao, G.P., Marcone C., Eds.; Studium Press: Texas, USA, **2008**; pp. 73–92.

Laimer, M.; Bertaccini, A. Phytoplasma elimination from perennial horticultural crops. In: *Phytoplasma: Plant Pathogenic Bacteria II: Transmission and Management of Phytoplasma-Associated Diseases*; Bertaccini, A., Weintraub, P.G., Rao, G.P., Mori, N., Eds.; Springer: Singapore, **2019**; pp. 185–206.

Laimer, M.; Lemaire, O.; Herrbach, E. et al. Resistance to viruses, phytoplasmas and their vectors in the grapevine in Europe: a review. *Journal of Plant Pathology* **2009,** 91(1), 7–23.

Leljak-Levanic´, D.; Ješic´, M.; Cesar, V. et al. Biochemical and epigenetic changes in phytoplasmarecovered periwinkle after indole-3-butyric acid treatment. *Journal of Applied Microbiology* **2010**, 109, 2069–2078.

Lenz, R.R.; Dai, W. Mapping X-disease phytoplasma resistance in *Prunus virginiana. Frontiers in Plant Science* **2017**, 8, 2057. https://doi.org/10.3389/fpls.2017.02057

Lessio, F.; Alma, A. Dispersal patterns and chromatic response of *Scaphoideustitanus* Ball (Homoptera Cicadellidae), vector of the phytoplasma agent of grapevine flavescence dorée. *Agricultural and Forest Entomology* **2004**, 6(2), 121–128.

Lessio, F.; Portaluri, A.; Paparella, F.; Alma, A. A mathematical model of Flavescence dorée epidemiology. *Ecological Modelling* **2015,** 312, 41–43.

Linck, H.; Lankes, C.; Krüger, E.; Reineke, A. Elimination of phytoplasmas in Rubus mother plants by tissue culture coupled with heat therapy. *Plant Disease* **2019**, 103(6), 1252–1255.

Liu, M.J.; Zhou, J.Y.; Zhao, J. Screening of Chinese jujube germplasm with high resistance to witches' broom disease. *Acta Horticulturae* **2004**, 663, 575–580.

Liu, Z.; Wang, Y.; Xiao, J.; Zhao, J.; Liu, M. Identification of genes associated with phytoplasma resistance through suppressive subtraction hybridization in Chinese jujube. *Physiological and Molecular Plant Pathology* **2014**, 86, 43–48. https://doi.org/10.1016/j.pmpp.2014.03.001

Lova, M.M.; Abou-Jawdah, Y.; Choueiri, E. et al. Almond witches' broom phytoplasma: disease monitoring and preliminary control measures in Lebanon. In: *Phytoplasmas and Phytoplasma Disease Management: How to Reduce their Economic Impact*; Bertaccini, A., Ed.; International Phytoplasmologist Working Group: Bolonga, Italy, **2014**; pp. 71–74.

Luo, F.; Evans, K.; Norelli, J.L. et al. Prospects for achieving durable disease resistance with elite fruit quality in apple breeding. *Tree Genetics & Genomes* **2020**, 16, 21. https://doi.org/10.1007/s11295-020-1414-x

MacLean, A.M.; Sugio, A.; Makarova, O.V. et al. Phytoplasma effector SAP54 induces indeterminate leaf-like flower development in Arabidopsis plants. *Plant Physiology* **2011**, 157(2), 831–841.

Mannini, F. Hot water treatment and field coverage of mother plant vineyards to prevent propagation material from phytoplasma infections. *Bulletin of Insectology* **2007**, 60(2), 311.

Maramorosch, K.; Loebenstein, G. Plant disease resistance: natural, non-host innate or inducible. *Encyclopedia of Microbiology* **2009**, 589, 596.

Marcone, C.; Neimark, H.; Ragozzino, A.; Lauer, U.; Seemüller, E. Chromosome sizes of phytoplasmas composing major phylogenetic groups and subgroups. *Phytopathology* **1999**, 89, 805–810.

Marcone, C.; Rao, G.P. Control of phytoplasma diseases through resistant plants. *Phytoplasmas: Plant Pathogenic Bacteria – II* **2019**, 8, 165–184. https://doi.org/10.1007/978-981-13-2832-9-8

Marzorati, M.; Alma, A.; Sacchi, L. et al. A novel Baceroidetes symbiont is localized in Scaphoideus titanus, the insect vector of "flavescence dorèe" in *Vitis vinifera. Applied and Environmental Microbiology Journal* **2006**, 72(2), 1467–1475.

Mayer, C.J.; Jarausch, B.; Jarausch, W.; Jelkmann, W.; Vilcinskas, A.; Gross, J. Pathogen induced release of plant allomone manipulates vector insect nbehaviur. *Journal of Chemical Ecology* **2008b**, 34, 1518–1522.

Mayer, C.J.; Vilcinskas, A.; Gross, J. Phytopathogen lures its insect vector by altering host plant odor. *Journal of Chemical Ecology* **2008a**, 34, 1045–1049.

Mazio, P.; Montermini, A.; Brignoli, P. Preliminary trials to test the effectiveness of biological promoters for the control of grapevine yellows symptoms. *Giornate Fitopatologiche Phytopathology Molly* **2008**, 3(1), 593–600.

McCoy, R.E.; Thomas, D.L.; Tsai, D.H. Lethal yellowing: a potential danger to date production. In: *Date Growers Institute Report* **1976**, 4e8, 53.

McCoy, R.E.; Howard, F.W.; Tsai, J.H. et al. "Lethal Yellowing of Palms," in Bulletin 834 (Gainesville, FL: University of Florida Institute of Food and Agricultural Sciences), 100, **1983**.

Mondal, S.; Sundararaj, R.; Yashavantha, R.H.C. A critical appraisal on the recurrence of sandalwood spike disease and its management practices. *Forest Pathology* **2020**, 50, 12648. https://doi.org/10.1111/efp.12648

Mou, H-Q.; Lu, J.; Zhu, S-F. et al. Transcriptomic aalysis of paulownia infected by paulownia witches'-broom phytoplasma. *PLoS One*, **2013**, 8(10): e77217. https://doi.org/10.1371/journal.pone.0077217

Musetti, R.; Ermacora, P.; Martini, M.; Loi, N.; Osler, R. Current knowledge about recovery from phytoplasma diseases. In: Phytoplasma and Phytoplasma Disease Management: How to Reduce Their Economic Impact; International Phytoplasmologist Working Group: Bologna, Italy, **2014**; pp. 250–258.

Musetti, R.; Ermacora, P.; Martini, M.; Loi, N.; Osler, R. What can we learn from the phenomenon of "recovery"? *Phytopathogenic Mollicutes* **2013a**, 3, 63–65.

Musetti, R.; Farhan, K.; De Marco, F. et al. Differentially-regulated defence genes in *Malus domestica* during phytoplasma infection and recovery. *European Journal of Plant Pathology* **2013b**, 136, 13–19.

Musetti, R.; Grisan, S.; Polizzotto, R.; Martini, M.; Paduano, C.; Osler, R. Interactions between '*Candidatus* Phytoplasma mali' and the apple endophyte *Epicoccum nigrum* in *Catharanthus roseus* plants. *Journal of Applied Microbiology* **2011**, 110, 746–756.

Myrie, W.; Oropeza, C.; Saenz, L. et al. Reliable improved molecular detection of coconut lethal yellowing phytoplasma and reduction of associated disease through field management strategies. *Bulletin of Insectology* **2011**, 64, S203–S204.

Nagadhara, D.; Ramesh, S.; Pasalu, I.C. et al. Transgenic rice plants expressing the snowdrop lectin gene (gna) exhibit high-level resistance to the whitebacked planthopper (*Sogatellafurcifera*). *Theoretical and Applied Genetics* **2004**, 109(7), 1399–1405.

Oliveira, D.V.; do Nascimento Júnior, J.R.A.; da Cunha, P.M.; Siqueira, J.F.; da Silva, É.C.F.; Cavaglieri, C.R. Factors associated with body image dissatisfaction of weight training practitioners. *Motriz, Rio Claro* **2018**, 24(2), e101852.

Pilkington, L.J.; Gurr, G.M.; Fletcher, M.J.; Elliott, E.; Nikandrow, A.; Nicol, H.I. Reducing the immigration of suspected leafhopper vectors and severity of Australian lucerne yellows disease. *Australian Journal of Experimental Agriculture* **2004**, 44(10), 983e992.

Pradit, N.; Rodriguez-Saona, C.; Kawash, J.; Polashock, J. Phytoplasma infection influences gene expression in American cranberry. *Frontiers in Ecology*, **2019**, 7, 178. https://doi.org/10.3389/fevo.2019.00178

Puterka, G.J.; Reinke, M.; Luvisi, D. et al. Particle film, surround WP, effects on glassy-winged sharpshooter behaviour and its utility as a barrier to sharpshooter infestations in grapes. *Plant Health Progress*, **2003**, 4, 1. https://doi.org/10.1094/PHP2003-0321-01-RS

Quaicoe, R.N.; Dery, S.K.; Philippe, R. et al. Resistance screening trials on coconut varieties to cape saint paul wilt disease in Ghana. *Oléagineux* **2009**, 16, 132–136. https://doi.org/10.1051/ocl.2009.0249

Rao, G.P.; Bertaccini, A.; Fiore, N.; Lieftling, L. (Eds) Phytoplasmas: plant pathogenic bacteria-I. In: *Characterization and Epidemiology of Phytoplasma-Associated Diseases*. Springer Nature: Singapore, **2018**; p. 345.

Rao, G.P.; Nabi, S.; Priya, M. Overview on a century progress in research on sesame phyllody disease. *Phytopathogenic Mollicutes* **2015**, 5, 74–83.

Rao, G.P.; Priya, M.; Thorat, V.; Manimekalai, R.; Kumar, A.J.; Yadav, A. A century progress of research on phytoplasma diseases in India. *Phytopathogenic Mollicutes* **2017**, 7 (1), 1–38.

Rao, G.P.; Priya, M.; Tiwari, A.K.; Kumar, S.; Baranwal, V.K. Identification of Sugarcane grassy shoot-associated phytoplasma and one of its putative vectors in India. *Phytoparasitica* **2014**, 42, 349–354.

Rao, G.P. Our understanding about phytoplasma research scenario in India. *Indian Phytopathology* **2021**, 74: 371–401

Raychaudhuri, S.P.; Varma, A.; Chenulu, V.V.; Prakash, N.; Singh, S. Association of mycoplasma- like bodies with little leaf of Solanum melongena L. In: *Proceedings of the X International Congress of Microbiolpgy Mexico HIV-6*, Mexico, **1970**.

Reddy, G.V.P.; Guerrero, A. New pheromones and insect control strategies. In: *Vitamins and Hormones*; Litwack, G. Ed.; Vol. 83, Academic Press: Burlington, **2010**; pp. 493–520.

Reddy, T.V.; Prasad, K.H.; Chalam, M.; Viswanath, K. Management of Leafhopper (Orosius albicintus) of sesamum with certain insecticides. *Andhra Pradesh Journal of Agricultural Sciences* **2019**, 5(3), 181–186.

Roddee, J.; Kobori, Y.; Hanboonsong, Y. Multiplication and distribution of sugarcane white leaf phytoplasma transmitted by the leafhopper, matsumuratettix hiroglyphicus (Matsumura) (Hemiptera: Cicadellidae), in infected sugarcane. *Sugar Technology* **2018**, 20, 445–453. https://doi.org/10.1007/s12355-017-0559-x

Romanazzi, G. Perspectives for the management of phytoplasma diseases through induced resistance: what can we expect from resistance inducers? *Phytopathogenic Mollicutes* **2013**, 3(1), 60–62.

Romanazzi, G.; D'Ascenzo, D.; Murolo, S. Field treatment with resistance inducers for the control of grapevine bois noir. *Journal of Plant Pathology* **2009**, 3, 677–682.

Romanazzi, G; Murolo, S; Feliziani, E. A new approach to manage phytoplasma diseases: field treatments with resistance inducers to contain grapevine "bois noir". Phytopathology 2013, 103, 785–791.

Salehi, M.; Bagheri, A.; Faghihi, M.M; Izadpanah, K. Study of partial biological and behavioral traits of Hishimonus phycitis, vector of lime witches' broom, for management of the disease.

Sarnaik, D.N.; Ghode, R.N.; Peshkar, L.N.; Satpute, U.S. Insecticidal control of sesamum jassids Orosius albicinctus (Distant). *PKV Research Journal* **1986**, 10(1), 41–43.

Schmid, G. Effects of tetracycline injection against apple proliferation. *Acta Horticulturae* **1983**, 130,237–241.

Schmidt, S.; Baric, S.; Massenz, M. et al. Resistance inducers and plant growth regulators show only limited and transient effects on infection rates, growth rates and symptom expression of apple trees infected with 'candidatus Phytoplasma mali'. *Journal of Plant Diseases and Protection* **2015**, 122, 207–214.

Seemüller, E.; Lorenz, K.H.; Lauer, U. Pear decline resistance in Pyrus communis rootstocks and progenies of wild and ornamental Pyrus taxa. *Acta Horticulturae* **1998a**, 472, 681–691.

Seemüller, E.; Moll, E.; Schneider, B. Apple proliferation resistance of Malus sieboldii-based rootstocks in comparison to rootstocks derived from other Malus species. *European Journal of Plant Pathology* **2008**, 121, 109–119.

Seemüller, E.; Stolz, H.; Kison, H. Persistence of the European stone fruit yellows phytoplasma in aerial parts of Prunustaxa during the dormant season. *Journal of Phytopathology* **1998b**, 146, 407–410.

Shindle, G.G.; Lokesha, R.; Naik, M.K.; Ranganath, A.G.R. Inhertance study on phyllody resistance in sesame. *Plant Archives* **2011**, 11, 777–778.

Siampour, M.; Izadpanah, K.; Salehi, M.; Afsharifar, A. Occurrence and distribution of phytoplasma diseases in Iran. In: *Sustainable Management of Phytoplasma Diseases in Crops Grown in the Tropical Belt*; Olivier, C., Dumonceaux, T., Pérez-López, E. Eds.; Springer: Singapore, **2019**; pp. 47–86.

Sinclair, W.A.; Townsend, A.M.; Griffiths, H.M.; Whitlow, T.H. Responses of six eurasian ulmus cultivars to anorth american elm yellows phytoplasma. *Plant Disease* **2000**, 84, 1266–1270.

Singh, P.K.; Akram, M.; Vajpeyi, M.; Srivastava, R.L.; Kumar, K.; Naresh, R. Screening and development of resistant sesame varieties against phytoplasma. *Bulletin of Insectology* **2007a**, 60, 303304.

Singh, S.K.; Amminudin, S.P.; Singh, B.R.; Khan, J.A. Production of phytoplasma-free plants from yellow leafdiseased Catharanthus roseus L. (G.) Don. *Journal of Plant Diseases and Protection* **2007b**, 114, 2–5.

Sugio, A.; Hogenhout, S.A. The genome biology of phytoplasma: modulators of plants and insects. *Current Opinion in Microbiology* **2012**, 15(3), 247–254.

Sugio, A.; Kingdom, H.; MacLean, A.; Grieve, V.; Hogenhout, S. Phytoplasma protein effector SAP11 enhances insect vector reproduction by manipulating plant development and defense hormone biosynthesis. *Proceedings of the National Academy of Sciences* **2011**, 108(48), E1254–E1263.

Summers, C.G.; Stapleton, J.J. Management of corn leafhopper (Homoptera: Cicadellidae) and corn stunt disease in sweet corn using reflective mulch. *Journal of Economic Entomology* **2002**, 95(2), 325–330.

Taloh, A.; Dantuluri, D.S.; Kumar, G.; Padaria, J.C.; Kumar, S.; Rao, G.P. Elimination of '*Candidatus* Phytoplasma aurantifolia' from Chrysamthemum morifolium cv. Ajay Orange by application of oxytetracycline by in vivo and in vitro treatments. *Phytopathogenic Mollicutes* **2018**, 8(1), 1–7.

Tiwari, A.K.; Kumar, S.; Mall, S.; Jadon, V.; Rao, G.P. New efficient natural leafhopper vectors of sugarcane grassy shoot phytoplasma in India. *Sugar Technology* **2017**, 19(2), 191–197.

Tiwari, A.K.; Priya, M.; Srivastava, V.K.; Pandey, K.P.; Sharma, B.L.; Rao, G.P. Detection of sugarcane grassy shoot phytoplasma (16SrXI-B subgroup) in Pyrilla perpusilla Walker in Uttar Pradesh, India. *Phytopathogenic Mollicutes* **2016**, 6, 56–59.

Tiwari, A.K.; Tripathi, S.; Lal, M.; Sharma, M.L.; Chiemsombat, P. Elimination of sugarcane grassy shoot disease through apical meristem culture. *Archives of Phytopathology and Plant Protection* **2011**, 44(20), 1942–1948.

Tubajikaa, K.M.; Civerolob, E.L.; Puterkac, G.J.; Hashimd, J.M.; Luvisid, D.A. The effects of kaolin, harpin, and imidacloprid on development of Pierce's disease in grape. *Crop Protection* **2007**, 26, 92–99.

Ustun, R.; Yol, E.; Ikten, C.; Catal, M.; Uzun, B. Screening, selection and real-time qPCR validation for phytoplasma resistance in sesame (*Sesamum indicum* L.). *Euphytica* **2017**, 213, 159.

Vidano, C. Scoperta in Italia dello Scaphoideus littoralis Ball cicalina Americana collegata alla "Flavescence dorée" della Vite. *L'Italia Agricola* **1964**, 101(10), 1031–1049.

Viswanathan, R.; Rao, G.P. Disease scenario and management of major sugarcane diseases in India. *Sugar Technology* **2011**, 13, 336–353. https://doi.org/10.1007/s12355-011-0102-4

Wally, O.; Daayf, F.; Khadhair, A.H. et al Incidence and molecular detection of yellows-type disease in carrots, associated with leafhoppers in southern Manitoba, Canada. *Canadian Journal of Plant Pathology* **2004**, 26(4), 498–505

Waloff, N.; Jervis, M.A. Communities of parasitoids associated with leafhoppers and planthoppers in Europe. *Advances in Ecological Research* **1987**, 17, 281–376.

Walsh, K.B.; Guthrie, J.N.; White, D.T. Control of phytoplasma diseases of papaya in Australia using netting. *Australasian Plant Pathology* **2006**, 25(1), 49–54.

Wang, Q.C.; Valkonen, J.P.T. Efficient elimination of sweetpotato little leaf phytoplasma from sweetpotato by cryotherapy of shoot tips. *Plant Pathology* **2008**, 57, 338–347.

Wang, Q.C.; Valkonen, J.P.T. Cryotherapy of root tips: Novel pathogen eradication method. *Trends in Plant Science* **2009**, 14, 119–122.

Weintraub, P.G. Insect vectors of phytoplasmas and their control – an update. *Bulletin of Insectology* **2007**, 60(2), 169–173.

Weintraub, P.G.; Beanland, L. Insect vectors of phytoplasmas. *Annual Review of Entomology* **2006**, 51, 91–111.

Weintraub, P.G.; Pivonia, S.; Rosner, A.; Gera, A. A new disease in *Linonium latifolium* hybrids. II. Investigating insect vectors. *HortScience* **2004**, 39, 1060–1061.

Weintraub, P.G.; Trivellone, V.; Krüger, K. The biology and ecology of leafhopper transmission of phytoplasmas. In: *Phytoplasmas: Plant Pathogenic Bacteria-II, Transmission and Management of Phytoplasma Associated Diseases*; Bertaccini, A., Weintraub, P., Rao, G.P., Mori, N. Eds.; Springer: Singapore, **2019**; pp. 27–51.

Weintraub, P.G.; Wilson, M.R. Control of phytoplasmas diseases and vectors. In: *Phytoplasmas: Genomes, Plant hosts and Vectors*; Weintraub, P.G., Jones, P., Eds.; CABI Publishing: Wallingford, **2010**; pp. 233–249.

Witzgall, P.; Kirsch, P.; Cork, A. Sex pheromones and their impact on pest management. *Journal of Chemical Ecology* **2010**, 36, 80–100. https://doi.org/10.1007/s10886-009-9737-y

Wu, W.; Ding, Y.; Wei, W. et al. Salicylic acid-mediated elicitation of tomato defence against infection by potato purple top phytoplasma. *Annals of Applied Biology* **2012**, 161, 36–45.

Yadav, R.; Kalia, S.; Rangan, P.; et al. Current research trends and prospects for yield and quality improvement in sesame, an important oilseed crop. *Frontiers in Plant Science* **2022**, 13, 863521. https://doi.org/10.3389/fpls.2022.863521

Zchori-Fein, E.; Bourtzis, K. (Eds) *Manipulative Tenants: Bacteria Associated with Arthropods*. CRC Press, Boca Raton, FL, **2011**.

Zhao, J.; Liu, M.J.; Liu, X.Y.; Zhao, Z.H. Identification of resistant cultivar for jujube witches' broom disease and development of management strategies. *Acta Horticulturae* **2009**, 840, 409–412. https://doi.org/10.17660/ActaHortic.2009.840.57

14 Novel and Strategic Approaches for Management of Plant Parasitic Nematodes in Sustainable Agriculture

Raman Kumar Walia, Matiyar Rahaman Khan, and Vishal Singh Somvanshi

14.1 NEMATODE DIVERSITY AND PLANT PARASITISM

Nematodes are classified in the phylum Nematoda that includes a diverse group of free-living and parasitic forms. The plant parasitic nematodes co-inhabit free-living forms in arable soils and stable uncultivated vegetative ecosystems. Barring most animal parasitic species, the majority of soil-dwelling forms are microscopic. The body architecture is typically organized for a life in edaphic systems. It is basically tubular without the external manifestation of limbs, etc., to facilitate movement within the soil pore spaces. All soil forms are essentially aquatic and spend most part of their life in thin films of water covering the soil particles. The obligatory plant parasitic nematodes (PPNs) are characterized by a hollow needle-like structure called a stylet that is the primary organ to cause infection of plant tissues as well as to ingest plant cell cytoplasm and is connected to the alimentary canal. The plant cell cytoplasm constitutes the food of the nematode, and they do not survive on anything else. Being multicellular invertebrates, nematodes possess well-developed digestive, nervous, excretory, and reproductive systems. The outermost component of the body wall, the cuticle, is flexible and facilitates sinusoidal movement within the soil pore spaces. Generally, males and females are separate so the mode of reproduction is sexual; however, some nematodes reproduce parthenogenetically also. Some PPNs that infect aerial plant parts survive by quiescence (anhydrobiosis) during water stress conditions, while others that have very restricted plant hosts (e.g., cyst nematodes) have protective stages to survive during off seasons.

DOI: 10.1201/9781003531944-14

14.2 GLOBAL SCENARIO OF NEMATODE PROBLEMS AND EMERGING PROBLEMS

Agriculture is extremely vulnerable to climate change. Higher temperatures eventually reduce yields of desirable crops while encouraging weed and pest proliferation. Higher soil temperatures will increase the rate of development of PPNs resulting in greater multiplication of some and hence more damage to host crops. The global scenario of nematode problems in agronomic and horticultural crops has witnessed an unprecedented change and is likely to remain a dynamic process. The sufficiency in food grain production in most parts of the world is inclining shift toward more remunerative horticultural crops. The unpredictable climatic vagaries are leading to more stable protected cultivation systems. The meager availability of water resources in most geographies has ushered in an era of microirrigation systems. Rapid development of seed replacement rates and widespread use of soilless growth media have enhanced crop yields due to disease-free planting materials and high-yielding crop cultivars. All these developments have had their negative impacts and some positive impacts as far as plant protection problems are concerned. PPNs now occupy a significant niche among the pest and disease scenario globally.

14.2.1 The Ten Most Important Genera of Plant Parasitic Nematodes

Based on scientific and economic importance, Jones et al. (2013) listed ten most important plant parasitic nematodes; here we present a brief modified account of these nematode genera in the descending order of their importance.

14.2.1.1 Root-Knot Nematodes (*Meloidogyne Spp.*)

More than 100 species constitute the genus *Meloidogyne*; however, four species, namely *Meloidogyne incognita*, *M. javanica*, *M. arenaria*, and *M. hapla*, are considered most important; the latter is restricted to temperate areas only. The global distribution, vast host range of *Meloidogyne* spp., and sedentary parasitism on vascular tissues make them economically most important among all the PPNs. In the recent past, *M. graminicola* has emerged as a major pest of rice in most parts of the world. Also, a relatively little known species, the guava root-knot nematode, *M. enterolobii* has been recognized as a major problem on guava and some other plant hosts in many countries. A few more species such as *M. chitwoodi*, *M. thamesi*, barley root-knot nematode, *M. naasi* are also considered important.

The life cycle of *Meloidogyne* spp. consists of six stages—an egg, four larval stages, and adult (females and males). The first juvenile stage (J1) is completed within the egg, and the second juvenile stage (J2) hatches out from the egg into the soil. J2 is the only infective stage and invades the plant root near the root tip. Larval hatching is generally not influenced by plant root exudates and happens freely in response to favorable temperature (usually $25°C \pm 5°C$) and adequate moisture. Soon after root invasion, the J2s move toward undifferentiated vascular tissues. Their movement within the plant tissues is facilitated by the esophageal gland secretions that dissolve the middle lamella. The J2 seeks to establish a permanent feeding site in the vicinity of vascular tissues through proteinaceous secretions via their stylet. About 8–12 cells around the nematode head are transformed into multinucleate giant cells due to karyokinesis without cytokinesis. Maintenance of giant cells is a must for the further development of the J2; failure to produce giant cells in non-hosts and resistant cultivars leads to atrophy of the nematode. Successful induction of giant cells ensures continuous availability of plant nutrients for nematode growth and development. Within a few days, the J2 molts to J3 and further to J4 *in situ*. The nematode continues to grow width-wise. The last molt results in much enlarged pear-shaped females. A few J2s develop into vermiform adults after the final molt; they leave the roots and emerge into the soil and may or may not be required for reproduction depending upon species. The female lays on an average 400 eggs held together in a gelatinous material secreted by the female itself. This eggmass may be located on the root surface in case of primary infection, but is usually embedded inside the galled roots in secondary infection. Nematode may complete multiple generations in a crop season, but this is governed by the ambient temperature. Root-knot nematodes are often involved in disease complexes as the galled roots generally attract the secondary infection in the rhizosphere to accentuate the wilt and rotting pathogens.

The above-ground symptoms due to root-knot nematode infection are not diagnostic but appear as stunted plant growth with pale foliage, undersized and fewer fruiting. The roots, however, exhibit clear symptoms of galling that vary in intensity and manifestation depending on plant species, growth stage of the crop, and nematode densities.

14.2.1.2 Cyst Nematodes (*Heterodera and Globodera*)

The globally important species of cyst nematodes include the soybean cyst nematode, *Heterodera glycines*, potato cyst nematodes (PCNs), *Globodera rostochiensis* and *G. pallida*, and cereal cyst nematodes, *Heterodera avenae*, *H. filipijevi*, etc. Other cyst nematodes of regional importance are: pigeonpea cyst nematode, *Heterodera cajani*, maize or corn cyst nematode, *Heterodera zeae,* and sugarbeet cyst nematode, *Heterodera schachtii.*

Cyst nematodes have several common biological features as root-knot nematodes. Cyst nematodes are generally host-specific, and hatching occurs in response to plant root exudates in most species. The reproduction is generally sexual as males are present. The adult females protrude out of the roots, lay eggs inside their bodies, and the females turn into a protective case (cyst) to protect the eggs during off season. Variable proportions of the eggs can survive inside the cysts for several years in the absence of a host crop. Induction of a permanent feeding site "syncytium" is obligatory for the continued development of infective J2. The syncytium is formed by the partial dissolution of walls of adjacent vascular cells around the nematode head, resulting in a big multinucleate cellular mass on which the nematode feeds.

Cyst nematodes do not form any galls on the roots. Symptoms on shoots are not diagnostic; patches of stunted plants, pale foliage, reduced tillering, etc., are commonly visible.

14.2.1.3 Lesion Nematodes (*Pratylenchus Spp.*)

Lesion nematodes are a group of about 70 species. *Pratylenchus* species are mostly polyphagous, but some species cause more damage to particular hosts. The economically important species specifically of temperate climate regions are *Pratylenchus penetrans*, *P. vulnus* that damage many a fruit and ornamental plants, and stone fruits, respectively. *P. minyus* damages tobacco, and *P. brachyurus* is important pest of potato. *P. zeae* is important on maize and sugarcane in the tropical regions, while *P. thornei* seriously damages wheat in Australia, Israel, the USA, and Mexico. *P. coffeae* causes damage to banana and citrus.

Lesion nematodes are migratory endoparasites, and feeding is mainly confined to the root cortex. J2, J3, J4, and adults can cause infection, initial infection originates from soil. Once root infection happens, the nematodes migrate in cortical tissues intercellularly while continuing their biological activities. Nematodes lay eggs scattered within the cortex, where they hatch, molt, feed, and reproduce. The life cycle can take 3–8 weeks depending on the prevailing conditions. The nematodes can come out in the soil, particularly at the time of crop senescence, only to re-infect the next crop. Nematode enzymes breakdown complex glycosides in cortical parenchyma cells to release hydrogen cyanide that kills the localized cells to form lesions. Necrotic lesions merge with one another as the infection spreads, ultimately girdling the roots. Secondary infection follows the rotten tissues. Above-ground symptoms are not diagnostic.

14.2.1.4 Burrowing Nematode (*Radopholus Similis*)

Genus *Radopholus* comprises about 30 species, but only *Radopholus similis* is economically important. Its distribution is confined to tropical regions of South America, the Caribbean, Africa, Asia, and the Pacific. It can parasitize a wide range of crops, but is most damaging to banana and banana-based cropping systems that include crops such as peppervine, coconut, arecanut, coffee, cocoa, cardamom, ginger, and turmeric.

Like lesion nematodes, *R. similis* is a migratory plant parasitic nematode and resembles the former in most respects biologically. The infective stages, tissue parasitization, development, reproduction are similar to lesion nematodes. The nematode migration and feeding in cortical tissues result in extensive cavity formation that enlarge to form burrows. The nematode lays eggs in cortex, but oviposition can happen is soil as well when they emerge from roots due to overcrowding. Thousands of nematodes may be present in a small root bit. Further infection cracks open the burrows at epidermis, and this phenomenon causes secondary infection leading to extensive root rot. Banana plants at bearing stage often topple due to poor root anchorage. The above-ground symptoms include paleness of leaves in banana, and defoliation in peppervines is very prominent.

14.2.1.5 Stem and Bulb Nematode (*Ditylenchus Dipsaci*)

The genus *Ditylenchus* is a conglomerate of about 60 species that have diverse parasitic capabilities. The most prominent species is the stem and bulb nematode, *Ditylenchus dipsaci*. Basically a temperate climate nematode, it parasitizes many crops ranging from cereals (oats, rye), tuberous ornamental (iris, daffodils, narcissus), leguminous crops (field beans, peas), vegetables crops (onion, garlic), besides others. The unique characteristic of this nematode is the occurrence more than 30 host races, some polyphagous, while others attack specific plant hosts. An array of pectinase enzymes imparts the production of varied symptoms by different races on their respective plant hosts. The nematode is widespread in Europe.

D. dipsaci is a migratory endoparasite and infects above-ground plant parts. All life cycle stages are capable of infection. The nematode enters plants through stoma or wounds, pectinases released by the nematode macerate the middle lamella that facilitates feeding on the cortical parenchyma cells. A single female lays about 250 eggs, and the nematode completes multiple generations at optimum temperature of 15°C–20°C. At the end of the crop season and under water stress conditions, the J4 stages of the nematode aggregate to form clusters of anhydrobiotic nematodes that appear as "nema wool." In this desiccated stage, the nematode can survive for several years in crop residues. The diversity of symptoms produced by races of *D. dipsaci* are known by different names, e.g., sagging disease in tulips, bloat disease in onion, and spikkles in narcissus.

The rice stem nematode, *D. angustus* causes *Ufra* disease in deep water rice in India and South-East Asia. The nematode feeds ectoparasitically on meristematic tissues of rice stems and leaves. The potato rot nematode, *D. destructor* is a pest of potato in Europe, whereas the peanut pod nematode, *D. africanus* damages groundnuts in South Africa. *D. myceliophagus* is fungivorous that destroys button mushroom (*Agaricus bisporus*) throughout the world.

14.2.1.6 The Aphelenchids (*Aphelenchoides and Bursaphelenchus*)

The aphelenchids have diverse parasitic relationships, ranging from typical plant parasitism to insect-vectored plant parasitism to exclusive fungivory.

Aphelechoides besseyi is a typical seed-borne plant parasite that attacks rice shoot. The nematode can survive in a dormant stage inside the rice husk and kernel. Upon contact with water, the nematode becomes active, ascends up the rice stem, multiplying on the way. The life cycle duration is very short for aphelenchids in general, 8–10 days in this case. Finally, the nematodes reach the rice panicle where they lodge in between the developing kernels. As the crop matures, the nematode enters dormant phase. The nematode causes white tip disease of rice, characterized by the off white dried 2–3 cm portion of the leaf tip. The foliar nematodes, *A. fragariae* and *A. ritzemabosi* are temperate species that parasitize the ornamentals belonging to family Compositae. The nematodes enter the leaf through stomatal openings and destroy the mesophyll tissues. The infection spreads from lower to upper leaves and ends in the apical bud that is partially or completely destroyed, consequently no blossom or distorted flower appears. The leaves exhibit typical angular brown spots. *A. composticola* is fungivorous, which also destroys button mushroom extensively. The short life cycle of 4–5 days results in enormous population build-up that feeds on fungal mycelium, showing barren patches in mushroom beds. The yield may be completely lost.

Bursaphelenchus spp. are generally fungivorous. *B. xylophilus* attacks pine in Japan, North America, and some European and SE Asian countries. The dauer larvae of the nematode are transmitted from one tree to another by beetle, *Monochamus alternatus* during maturation feeding. The nematode enters the tree trunk through wounds and spreads to the entire tree through resin canals. The tree dies within a few months of initial infection. The nematode appears in epidemics in Japan. Brown needles of dying/dead pine trees are symptoms of pine·wilt caused by this nematode. A similar example of nematode–insect relationship under tropical conditions is evidenced by *B. cocophilus* (=*Rhadinaphelenchus cocophilus*) that is transmitted by palm weevil, *Rhynchophorus palmarum*. The disease known as red ring disease of coconut is restricted to Caribbean counties and is of quarantine importance. The diseased trees show reddish brown ring in the cambial tissues of the stem upon dissection.

14.2.1.7 Reniform Nematode (*Rotylenchulus Reniformis*)

The genus *Rotylenchulus* has only nine species, and *R. reniformis* is the only economically important one. The nematode is widespread throughout the tropical and subtropical regions. A wide range of hosts covering vegetables, pulses, fruits, and ornamental are attacked. Cotton is another important host crop.

This nematode has some unique features. J2s emerge from eggs in the soil and develop to young (immature) female and male by three superimposed molts in quick succession, retaining the shed cuticles. Young vermiform females and males emerge in soil upon shedding the old cuticles. The immature females are the only infective stages; they invade the host roots partially, with only one-third of the anterior portion imbedded in the root and reach up to the stelar region. The nematode induces a permanent feeding site, more or less similar to cyst nematodes. A few endodermal and pericycle cells in the vicinity of the nematode head coalesce through partial dissolution of cell walls forming a prosyncyte. Many other cells beyond this region also show hypermetabolic reactions but do not merge. The hind portion of the female outside the root swells into a kidney-shape within a week. The nematode lays eggs in a gelatinous mass on the root surface, and each female deposits about 40–80 eggs.

The symptoms of infection are not diagnostic; shoots show stunted growth and chlorosis, while roots exhibit necrosis. There is no galling on the roots. In frequent combined infections with root-knot nematode, reniform nematode is often masked due to lack of symptoms on the roots.

14.2.1.8 Citrus Nematode (*Tylenchulus Semipenetrans*)

Genus *Tylenchulus* superficially resembles *Rotylenchulus*, but the two are completely dissimilar in all respects. Genus *Tylenchulus* has only one important species, i.e., *T. semipenetrans* that is prevalent throughout the world wherever citrus is grown. The nematode is specific to citrus and can infect all citrus types. It is semiendoparasite with J2 stage being infective. Upon root invasion, only the anterior-most portion is inside the root cortex. The nematode gradually gets into deeper layers of cortex, but does not reach up to the vascular tissues. About 8–12 cortical parenchyma cells around the nematode head show signs of hypermetabolism; they provide nutrition to the nematode and are collectively called as Nurse Cells. The female body portion outside the root becomes swollen, egg laying is on the root surface, each female lays about 40–60 eggs in a mass, which are held together in a gelatinous matrix secreted by the nematode. Males develop in soil only, remain vermiform, and do not infect the roots. Nematode completes one generation in 6–8 weeks.

The above-ground symptoms of nematode infection are not discernible till about 7–8 years of plant age; subsequently, the infected trees exhibit decline symptoms. The leaves turn pale, die, and fall off. Gradually, bare twigs appear more prominently year after year. Undersized and fewer fruits ultimately set the plantation uneconomical prematurely. Feeder roots show characteristic symptoms of browning that intensifies with severity of infection. In extreme cases, the root cortex tends to slough off. Disease complex involving fungal pathogens and nematode is common and is referred to as citrus decline.

14.2.1.9 The Anguinids (*Anguina Tritici*)

The wheat seedgall nematode, *Anguina tritici* is the most prominent species in this group that specializes in forming galls in the above-ground plant parts. *A. agrostis* forms galls on grasses in pasturelands and is important for cattle grazing. *Subanguina* also forms galls at the basal stem portion.

The seedgalls of wheat appear as black or brownish distorted seeds that contain 8–10 thousand J2s of the nematode in a quiescent stage. The nematodes can survive in this condition for about 30 years. The wheat seed contaminated with seedgalls is the source of nematode infection. The nematodes rehydrate once in soil, become active, and infect the germinating wheat seedlings while still in soil. The J2s are mechanically carried up the growing plant, concealed inside the leaf sheaths. At the time of panicle formation, the J2s enter the floral primordial and replace them with galls. Inside the green tissues, the nematodes gain adulthood quickly, mate, oviposit, and die. The eggs hatch, and J2s emerge inside the galls. As the crop matures, the galls become dry, brown, and the J2s enter quiescent phase.

The nematode alone causes earcockle disease but in association with bacterium, *Rathayibacter tritici* causes *tundu* or yellow ear rot of wheat in which case no grains are formed. The disease symptoms in both cases start as basal swelling of stem, crinkling, and twisting of leaves. In case of earcockle, the infected earheads appear broader, the spikelets are loose, and grains are replaced by cockles. *Tundu* disease is characterized by yellow slimy ooze that appears during prolonged low-temperature and high-humidity conditions; the ooze covers panicles and other upper plant parts.

14.2.1.10 The Dorylaims (*Xiphinema, Longidorus*, etc.)

While majority of the PPNs belong to Order Tylenchida, only a few genera belonging to Order Dorylaimida are also plant parasites. Among these, *Xiphinema* and *Longidorus* are more prominent. These nematodes are very long, possess much longer stylets, and feed ectoparasitically on the roots of a variety of plants, but are more commonly associated with woody plants such as grapevines, roses. This group of nematodes along with enoplids (*Trichodorus* and *Paratrichodorus*) is more prominently known as virus vectors. *Xiphinema index*, *X. diversicaudatum*, *X. americanum*, and *Longidorus elongatus* transmit NEPO (polyhedral) viruses while enoplids transmit on NETU (tubular) or TOBRA viruses. *Xiphinema* and *Longidorus* spp. also induce gall formation on the root tips of grapes, roses; however, these galls are a result of hyperplasia and hypertrophy of cortical parenchyma cells that is very different from root-knot nematodes.

14.3 MAJOR NEMATODE PROBLEMS OF CROPS IN INDIA

The first report of a PPN, the wheat seedgall nematode (now known as *Anguina tritici*), was made in 1743 in England. However, the economic importance of PPNs was realized during the 1940s, with the availability of some nematicidal fumigants that could demonstrate the harmful effects of nematodes on crops. In India, it was only in 1901 when root-knot nematode was first recorded on tea in south India. The interception of many more PPNs happened gradually, but the detection of burrowing nematode on banana in South India and cereal cyst nematode on wheat and barley in North India led to the realization of importance of PPNs on agricultural and horticultural crops in India. Table 14.1 depicts the major nematode problems of India, along with their diagnosis, hosts, and distribution in the country.

TABLE 14.1
Major Plant Parasitic Nematodes of India – at a Glance

Nematode	Disease Diagnosis	Major Hosts	Distribution
Root-knot nematode, *Meloidogyne* spp. (*M. incognita, M. javanica, M. arenaria, M. hapla*)	Induce gall formation on roots, stunted plant growth, yellowing of leaves, day time wilting during hot days	Tomato, brinjal, okra, chilli, cucurbits, jute, cotton, potato, pulses, groundnut, banana, pomegranate, gladiolus, tuberose, gerbera	Throughout the country
Rice root-knot nematode, *Meloidogyne graminicola*	Induce hook-shaped terminal galls, patchy yellowish growth of seedlings in nursery stage	Rice	Throughout the country
Guava root-knot nematode, *Meloidogyne enterolobii*	Severe galling on the roots that starts rotting; yellowing and defoliation of leaves, young plantations show slow decline symptoms	Guava	Most parts of the country
Cyst nematode, *Heterodera* spp.	Stunting in patches, reduced tillering, yellowing of leaves, white females of the size of pinhead can be seen attached to roots near crop maturity	Wheat and barley -*H. avenae* Pulses and sesame -*H. cajani* Maize - *H. zeae* Rice - *H. oryzicola*	Rajasthan, Haryana Throughout the country Throughout the country Kerala
Potato cyst nematode, *Globodera rostochiensis, G. pallida*	Yellowing of leaves, only a tuft of green leaves remain at top, roots poorly developed	Mainly potato	Nilgiri hills in Tamil Nadu, and bordering areas of Karnataka and Kerala; Higher altitudes in Himachal Pradesh, Uttarakhand, and Jammu region of J&K

(*Continued*)

TABLE 14.1 (*Continued*)
Major Plant Parasitic Nematodes of India – at a Glance

Nematode	Disease Diagnosis	Major Hosts	Distribution
Reniform nematode, *Rotylenchulus reniformis*	Non-specific above-ground symptoms, stunting in patches, roots show necrotic appearance	Vegetables, pulses, cotton, papaya, grapes	Throughout the country, except hilly regions
Citrus nematode, *Tylenchulus semipenetrans*	Dieback symptoms in trees above 5–7 years of age, undersized fruits, yields decline progressively over years, infected feeder roots appear dirty due to adhering soil particles	All citrus types	Throughout the country
Burrowing nematode, *Radopholus similis*	Severe yellowing and defoliation, fruiting reduced drastically, cessation of growth, and premature death of plants	Banana, plantation crops like black pepper, coconut, arecanut, ginger, turmeric, tea, coffee, cocoa etc	Mostly southern states; also recorded in western and some eastern parts of country
Lesion nematode, *Pratylenchus* spp.	Stunting, yellowing of foliage, roots show elliptical lesions, that girdle roots in heavy infection	Vast host ranges. Important species are:	Northern and central India
		Wheat, chickpea – *P. thornei*	Throughout the country
		Maize, sugarcane – *P. zeae*	Throughout the country
		Citrus, coffee, banana - *P. coffeae*	Throughout the country
		Ornamentals – *P. delatterii*	Southern parts
Rice root nematode, *Hirschmanniella* spp.	Stagnation of growth, roots show necrosis as growth picks up	Rice	Throughout the country
Rice stem nematode, *Ditylenchus angustus*	Yellow stripes on leaves, earheads distorted, may or may not come out of boot leaf, panicles may be sterile	Deep water rice	Some parts of Assam
White tip nematode, *Aphelenchoides besseyi*	Leaf tips become necrotic, off white, boot leaf twisted, panicles may be sterile, chaffiness of grains	Rice	South, east and central parts of country

(*Continued*)

TABLE 14.1 (*Continued*)
Major Plant Parasitic Nematodes of India – at a Glance

Nematode	Disease Diagnosis	Major Hosts	Distribution
Tuberose foliar nematode, *Aphelenchoides besseyi*	Brown leaf spots, distorted/twisted flower stalks, brown to rusty brown streaks on calyx and petals	Tuberose	West Bengal, Odisha
Wheat seedgall nematode, *Anguina tritici*	Basal swelling of stem, crinkling, curling and twisting of leaves, earheads smaller and broader, cockles form in place of grains, under humid and low temperature conditions, tundu disease appears due to bacteria carried by nematodes, tundu symptoms appear as yellow slimy ooze on upper plant parts	Wheat	Rajasthan, Madhya Pradesh, eastern UP, Bihar, Chhattisgarh
Mushroom nematode, *Aphelenchoides composticola, Ditylenchus myceliophagus*	Barren patches appear on compost beds, surface sinks and emits foul smell, sporophore yields decline gradually and cease prematurely	Button mushroom	Throughout the country

Source: Modified after Walia and Khan (2018).

14.4 THE IMPACT OF PLANT PARASITIC NEMATODES ON CROPS

14.4.1 Global Losses

Although more than 4,100 species of PPNs have been recorded (Decraemer and Hunt, 2006), not all of these are considered economically important. According to Nicol et al. (2011), an estimated crop loss of US $80–$118 billion per year is attributed to nematodes. *Meloidogyne* species are of major economic importance worldwide. An overview of the damage potential of some *Meloidogyne* species has been given by Wesemael et al. (2011). The readers are referred to Bernard et al. (2017) for detailed information on the impact of the PPNs on agriculture.

Soil-borne PPNs may account for annual yield losses of 10%–25% globally in rice (Trudgill and Blok, 2001). *Meloidogyne graminicola* is the major species attacking both upland (rainfed) and lowland (irrigated) rice. It is well adapted to flooded conditions, and yield losses of up to 87% have been reported. *Aphelenchoides besseyi* is an important and widespread pathogen that causes “white tip” disease on rice. It makes a significant contribution to the estimated $US 16 billion worth of damage caused by nematodes to rice crops (Jones et al., 2013). Cereal cyst nematodes (mainly *Heterodera avenae*) cause an estimated loss of $US 3.4 million in wheat yield in the USA (Smiley and Guiping, 2010); the losses can range from 30% to 100% in some fields (Bonfil et al., 2004; Nicol et al., 2004). *Pratylenchus* spp. are economically serious and often underestimated pests of many crops, such as cereals, causing yield losses of up to 30% in wheat in Australia (Vanstone et al., 2008). The needle nematode *Longidorus breviannulatus*, which causes stunting in corn, may cause yield losses up to 60% (Norton and Hoffman, 1975). Soybean cyst nematode (*Heterodera glycines*) has been calculated to be responsible for losses in excess of $US1.5 billion each year in the USA alone (Chen et al., 2001).

PCNs have been estimated to cause losses of 9% of total potato production worldwide (Turner and Rowe, 2006); in the United Kingdom alone, an estimated £ 50 million per year in profits is lost due to these nematodes (Jones et al., 2009). Yield losses reported for *Nacobbus aberrans* average 65% for potato in Andean Latin America, 55% and 36% for tomato and bean, respectively, in Mexico and 10%–20% for sugarbeet in the USA, but, for most crops, loss estimates are lacking (Manzanilla-López et al., 2002). Sweet potato is damaged by *Ditylenchus destructor*, and 100% losses have been recorded in major producing areas (Xu et al., 2015; Zhang et al., 2006); overall, PPNs reduce about 10.2% yields in sweet potato (Decraemer and Hunt, 2006).

Financial losses are significant due to pine wood nematode, with losses of $2 \times 10^6 m^3$ of timber under some environmental conditions (Nose and Shiraishi, 2007), but the nematode can also cause devastation on an ecosystem scale, as all infected trees in an area can die when conditions are suitable for the disease.

14.4.2 Crop Losses in India

Data generated through ICAR-AICRP on nematodes over the years reveal that overall economic losses in 30 crops amount to Rs. 102,039.79 million annually; the losses in 19 horticultural crops were assessed at Rs. 50,224.98 million, while for 11 field crops, it was estimated at Rs. 51,814.81 million (Kumar et al., 2020). The mean percent losses were higher in horticultural crops, i.e., 23.03% (fruits 25.5%, vegetables 19.6%, spices 29.5%) than in field crops, i.e., 18.23% (cereals 18.8%, pulses 23%, oilseeds 11.8%, fiber crops 19.75%). The economic losses in rice due to *Meloidogyne graminicola* alone were maximum (Rs. 23,272.32 million) among all the crops and nematodes considered. In wheat, avoidable losses due to *Heterodera avenae*, in Rajasthan and Haryana (the major *molya* disease affected states), were calculated at Rs. 8,967.52 million. Citrus (Rs. 9,828.22 million), banana (Rs. 9,710.46 million) among fruit crops; and tomato (Rs. 6,035.2 million), brinjal (Rs. 3,499.12 million), and okra (2,480.86 million) among the vegetable crops suffered comparatively more losses that are partly attributable to areas of production in respect of these crops. *Globodera rostochiensis* and *G. pallida* are restricted but widespread in the

Nilgiri Hills of South India. As per current information, potato crop is grown in about 1,440 ha in the Nilgiris, and the same has been considered for loss assessment that amounts to Rs. 127.04 million. *Meloidogyne* spp. alone are responsible for Rs. 77,373.87 million losses in different crops that constitute about 75.83% of the total estimated losses, thus proving to be the economically most important of all the PPNs (Kumar et al., 2020).

14.5 NEMATODE MANAGEMENT PRACTICES

14.5.1 Nematode Management Knowledge Prerequisites – The Five Pillars

14.5.1.1 Correct Disease Diagnosis

The biggest bane of the nematode disease diagnosis is the lack of clear diagnostic symptoms. Most nematodes are soil-borne and damage the roots; consequently, the roots are unable to take up available water and nutrients from soil and translocate to the shoots. The resulting symptoms exhibited on the shoots mimic the nutrient deficiencies and wilting. The partial to complete unawareness about the nematode problems in the developing and underdeveloped world also contributes largely to the non-detection of nematode diseases. Most often, the visual inspection for detection of pests and diseases is confined to above-ground plant parts; rarely, the roots and other under-ground parts are examined. In such cases, only those nematode problems are detected that produce clear symptoms on roots, e.g., galling by root-knot nematodes. A large majority of nematodes feeding on roots go undetected. Usually, the first time detection in a root-knot nematode-infested field is recorded late in the cropping season, no control measures are possible at that stage. But being a soil-borne nematode, it is going to persist in the soil; therefore, nematode management factually commences with the next crop.

14.5.1.2 Nematode Assays

The plant–nematode relationship is quantitative; it implies that some general idea can be developed about forecasting the crop losses based on the nematode population (initial, Pi) at the time of planting. The threshold levels (Pt) are also known in majority of the nematode–crop combinations. However, sampling soil from the infested fields/orchards is critical because PPNs are not distributed uniformaly in the field. Extensive research on this aspect has led to certain recommended procedure for sample collections in different situations that are depicted in Figure 14.1.

14.5.1.3 Only Pre-Planting Solutions

"The earlier, the better" principle applies for all nematode diseases. Since majority of PPNs are soil-borne, they can infect immediately after sowing or planting. Therefore, all control operations have to taken before sowing/planting or immediately after that. Usually, nematode eggs are more resistant to chemicals than the juvenile/adults. Some cyst nematodes behave diiferently; PCNs in particular are dependent upon root exudates for hatching; therefore, the emergance of infective juveniles happens a few weeks after sowing. In such cases, the application of chemical nematicides may be coincided with peak hatching periods.

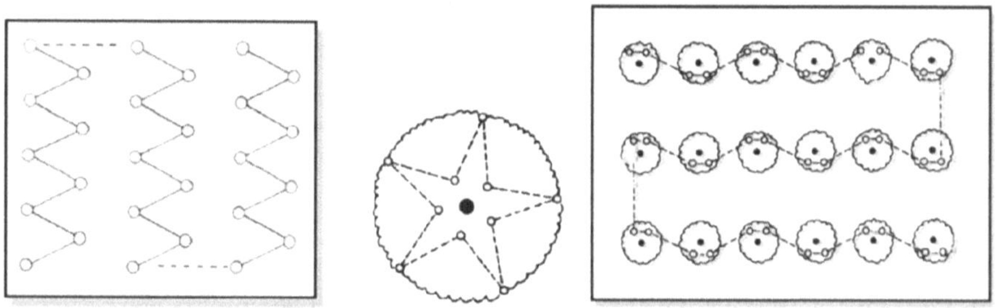

FIGURE 14.1 Recommended sampling patterns from (L to R) a field crop, single tree, and an orhard.

14.5.1.4 Eradication Impossible

Barring a few cases involving seed-borne diseases such as white tip dusease of rice and earcockle disease of wheat, eradication of nematodes is impossible. The maximum achievable level of nematode population reduction should be undertaken with the available management technologies. Usually, population reductions below threshold levels (Pt) should be targetted. But remanent populations do manage to survive on weed hosts, dormant stages, left over root bits, etc., and again shoot up as and when the susceptible host is planted.

14.5.1.5 Fitness of Nematode Management Technologies in Farming System

While devising nematode management technologies, care must be taken that the intended technology fits into the Integrated Pest Management (IPM) of cropping system of that geography. Validation of innovative nano and micro technologies (INM) technologies must be undertaken for its suitability in the overall Intergrated Crop Management and Farm Management; otherwise, it becomes redundant.

14.5.2 Exclusion

14.5.2.1 Regulatory Methods

The worldwide dissemination of pests and diseases, including PPNs, has been going on inadvertently due to human activities. Legal enforcement of quarantine regulations has been employed in case of several nematode diseases. The objective of such enforcements is to check the introduction of pests/diseases into new areas, or if already introduced, to reduce the chances of their further spread. Different countries have their own priorities and have to make legislations accordingly. Such legislations are: (i) external or international quarantine for those exotic pests yet to be introduced in a specified country, and in such cases, tolerance limit is zero; (ii) internal or domestic quarantine for those pests that are already present in some but not all parts of country.

Although quarantine measures and phytosanitary regulations have been in force for over last 100 years, these were based primarily on *ad hoc* consideration of risk by experts, and it was not until 2001 that the procedures for analyzing risk began to be formalized. Pest risk analysis (PRA) is a process to evaluate scientific and economic evidences to determine whether a nematode is a pest or not, whether it should be regulated or not, and the strength of phytosanitary measures that may be required to meet the challenge. PRA is a key element of all the three major international agreements related to plant protection, namely International Plant Protection Convention (IPPC), World Trade Organization's Agreement on Application of Sanitary and Phytosanitary Measures, and Convention on Biological Diversity. Procedures of PRA analysis are now undertaken worldwide and follow IPPC's International Standards on Phytosanitary Measures. Government of India's Plant Quarantine Order, 2003 (Regulation of Import into India) and its subsequent amendments regulate import and prohibition of import of plants and plant products into India. In schedules IV, V, and VI of this order are included plants/planting materials and countries from where import is prohibited or restricted (permissible only by authorized institutions with additional declarations and subject to special conditions) due to pests including nematodes. Several nematode species have been included in Schedule VI that enlists the plants/plant materials permitted to be imported with additional declarations and special conditions.

Countries determine their lists of regulated pests according to guidelines set by the IPPC. The quarantine listing of nematode species varies between countries. According to Kahn (1983), who surveyed regulations of 125 countries, 46 species of nematodes have been cited, out of which the 10 most frequently cited nematode species are: *Globodera rostochiensis*, *Ditylenchus dipsaci*, *Heterodera schachtii*, *Ditylenchus angustus*, *Aphelenchoides fragariae*, *Ditylenchus destructor*, *Radopholus similis*, *Meloidogyne javanica*, *Aphelenchoides ritzemabosi,* and *Aphelenchoides besseyi.* Besides these, *Globodera pallida*, *Anguina tritici*, *Heterodera glycines*, *Nacobbus abberans*, *Xiphinema americanum*, *X. index*, *Bursaphelenchus xylophilus*, and *B. cocophilus* are the most widely regulated PPNs. All lists of regulated pests are dynamic.

Domestic quarantine law in India: In India, the first law, called Destructive Insects and Pests (DIP) Act, 1914, was passed to regulate the introduction and release of all materials likely to contain pests and pathogens. This was later supplemented by a more comprehensive act in 1917. The custodian of the act is the Directorate of Plant Protection, Quarantine and storage (PPQ & S), Government of India (headquarters at Faridabad, Haryana), which is headed by Plant Protection Adviser (PPA) to the Govt. of India. According to the act, no consignment of seeds, fruits, or plants whether for consumption or for sowing/planting purposes can be imported into India without a valid import permit. The act is also applicable to nematodes and empowers the PPA to adopt appropriate measures against the introduction of potentially harmful exotic nematodes. To meet the requirements of the FAO Plant Protection Convention of 1951, PPQ & S maintains a chain of 26 quarantine stations in the country, nine at major seaports, ten at airports, and seven on land frontiers with neighboring countries. All plant and plant materials have to be imported only through these prescribed ports of entry. The domestic quarantines are enforced from time to time through the DIP Act and the Agricultural Pests, Weeds and Diseases Acts of different states in the country, i.e., regulating the inter-state movement of plants and plant materials to prevent the spread of destructive pests and pathogens established in localized area of the country. After the first authentic report on the PCN, *G. rostochiensis* from the Nilgiris in 1961, domestic quarantine was implemented against transport of seed potatoes to other parts of the country under the Madras Agricultural Pests and Disease Act, 1919. Under this Act, (i) golden nematode is declared an insect pest, (ii) the Nilgiri district is declared as a notified area, (iii) no owner of any notified field can harvest a potato crop on it, no part of produce will be allowed for seed purposes. Produce will not be sold to anyone except to the Nilgiri Marketing Co-operative Society, Ooty. Infested land must be treated with nematicides as per recommendations for three crop seasons. After three crop seasons, if potatoes are to be grown, the field must be inspected through soil sampling and should be declared free of nematodes. The DIP Act mentions the name golden nematode (*Globodera rostochiensis*) and red ring nematode (*Bursaphelenchus cocophilus*), but any nematode can be stopped if warranted under discretion of quarantine officials. The development of the new Plant Quarantine order, 2003 reflects the primary plant quarantine concerns of the Government of India.

14.5.2.2 Use of Nematode-Free Planting Materials

Nematode transmission through planting materials happens in two ways: true seed and vegetative planning materials. Seed-borne nematodes are very few, e.g., white tip nematode of rice (*Aphelenchoides besseyi*). Another nematode, *Anguina tritici* known to cause earcockle disease of wheat, spreads when seed is contaminated with cockles. It is not truly seed-borne since cockles do not germinate into plants, and a normal seed does not harbor nematodes. In both these cases, nematode eradication is possible by using clean certified seed material. Rapid seed replacement in favor of high yielding and agronomically superior wheat cultivars has nearly eradicated *A. tritici* from most parts of India, save a few pockets in Madhya Pradesh, Rajasthan, and Bihar.

Nematode dissemination through vegetatively propagated planting materials, particularly those of fruit nurseries, has international and national ramifications. It is believed that this method alone is solely responsible for most cases of intercontinental, international, and national nematode dissemination. The most recent example of guava root-knot nematode, *Meloidogyne enterolobii* is cited in this context. The nematode has been intercepted in many public and private nurseries across India and traced to destined places in 11 states so far. The nematode is highly virulent, and cases of post-planting mortality of infected guava planting materials are on record (Poornima and Walia, 2021). The root cause of this phenomenon can be attributed to unawareness about nematodes among all stakeholders, i.e., the nurserymen, the orchardists, the phytosanitary inspectors, the quarantine personnel, etc. While grafts, clones, tissue culture plants, and air layers are usually free from nematode infection, it is their subsequent transfer to media (usually nematode-infested garden soil) contained in the polybags for hardening that causes nematode infection. The following measures are suggested to prevent this phenomenon.

- Use of soil-less medium such as sterilized coir pith, coco peat, or vermiculite instead of native (garden) soil for hardening of rooted planting materials will add a little to the cost, but will provide absolute nematode as well as disease-free planting materials. However, application of nutrition will be required whenever soil-less media are used.

- The bulk soil to be used as hardening medium can be alleviated of PPNs by several methods, each having its own advantages and disadvantages.
  - Use of chemicals such as formaldehyde, hydrogen peroxide to treat the bulk soil.
  - Solarization of soil in the open. Use of thin transparent polythene sheets for about 2–3 months during peak summer is effective and may be used as a routine good horticultural practice.
  - Solarization of soil/polythene bags filled with soil by covering with polythene sheets (double solarization).
  - Use of virgin soil.
- Nursery certification: The certification of nursery planting materials must be enforced in letter and spirit by government agencies to ensure movement of nematode-free plants only. The infected materials must be discarded and destroyed.

14.5.3 Agro-Technological Methods

14.5.3.1 Crop Rotations

The principle of crop rotation for nematode management is primarily to prevent nematode multiplication on the crops grown in the cropping sequence. Thus, the initial or pre-plant population densities (Pi) of nematodes remain below the damage threshold level (Pt), which allows the subsequent crop to grow without being heavily attacked by the resident population. At the end of each crop growing season, the carryover/final population (Pf) must progressively decline during the rotational cycle. This can be achieved by growing non-host/poor host crops and/or resistant varieties to the target pathogenic nematode species. Crop rotation also conserves antagonistic and predaceous nematodes and mites at population densities to reduce the plant-feeding nematodes. Okra–cowpea–cabbage and okra–cucumber–mustard sequences were found to reduce *M. incognita* populations most effectively. Growing crops such as cabbage, mustard, and rice in a sequence demonstrated suppressive effects on *M. incognita* while okra, brinjal, cowpea, and tomato supported multiplication of this nematode (Chandra and Khan, 2011). Following crop rotation with non-host crops and intercropping with marigold at early growth stages of banana lessened root-knot and burrowing nematode populations. Crop rotation with resistant varieties or non-host crops such as mustard, sesame, maize, and wheat is highly successful to bring down the soil population of nematodes below the damage threshold level. Crop rotation with non-hosts such as *sarson, toria, raya, taramira*, gram, *berseem*, carrot, and coriander with wheat is effective to reduce the population of *cereal cyst nematode*. In the PCN-affected areas, it is suggested to grow non-host and non-solanaceous vegetable crops such as cabbage, cauliflower, beetroots, carrots, garlic, radish, and turnips. Practicing crop rotation such as potato–cabbage–carrot reduces PCN problems in the Nilgiri hills. Inter-cropping with antagonistic plants such as marigold (*Tagetes* spp.) reduces the soil population of many soil nematodes including root-knot nematodes. Incorporation of such crop in cropping system either as inter-crop or as alternative crop should be considered whenever feasible. Brassicas (*Brassica juncea, B. campestris*), marigold (*Tagetes patula*), maize, and sorghum are non-hosts of

R. reniformis, and growth of these crops reduces the soil nematode population. The success of crop rotation depends on the local situations, growers' socioeconomic conditions, resources, and assessment of the nematode problems. If crop rotation is considered a feasible option for the growers, it is one of the best nematode management options in terms of cost, soil fertility benefits, and lessening other biotic stress in the production system.

14.5.3.2 Tillage Practices

Tillage has a direct effect; it kills nematodes mechanically or by exposing nematodes to solar heat (desiccation). Several studies have reported the effects of tillage regimes on *Heterodera avenae*. It reduces the nematode population in soil directly by mechanical action (discing or drilling) and indirectly by weed control. Repeated summer plowings in May–June under Indian conditions reduces nematode populations, but it makes the soil prone to wind and water erosion. Two to three summer plowings (20 cm deep) during May–June at 15-day intervals expose nematodes, weeds, pathogen propagules, and hibernating stages of insect pests to sun rays. Deep summer plowing (2–3) at an interval of 10–15 days during hot summer months reduces the population of *H. avenae* under North India conditions. Tilling of soil breaks the compact layer of soil and makes the soil loose, which allows the plant roots to penetrate deeper. The plants grown in well-tilled soil have large root biomass that draws more nutrients from a larger volume of soil, making the plant capable of tolerating greater biotic stress including herbivore nematodes. The practice of conservation tillage (the tillage practice that builds up crop residues in soil) requires the management of crop residues, which play a key role in improving soil health conditions and reducing weed and nematode problems. Many cover crops release root exudates and organic compounds, for example, brassicas release nematode toxic compounds, when grown under the conservation tillage practices also reduce the harmful PPNs and enhance beneficial nematodes (fungivores, bacterivores, and predatory nematodes).

14.5.3.3 Fallowing

Fallowing helps to reduce the plant-feeding nematode population because they are killed due to the non-availability of required food, water, and protection from the solar heat. It becomes most effective for the nematodes, which are unable to survive in the absence of a host or live in anhydrobiotic conditions. Some nematodes could be controlled by the adoption of bare fallow. Several reports have appeared on the effects of bare fallow on nematode populations in soil. For example, *Rotylenchulus reniformis*, *Pratylenchus penetrans*, *Radopholus similis*, and some species of *Meloidogyne* have been successfully controlled following bare fallows. However, this practice is difficult to follow as it increases the risk of soil erosion and loss of crops for several months.

14.5.3.4 Flooding

Flooding is not a common practice to control nematodes because it requires a large amount of water. It does reduce the nematode population in soil because the death occurs due to starvation for food, asphyxiation, and toxic compounds generated under microaerobic to anaerobic conditions. The field should be inundated for at least 25 days to get rid of the nematodes, although the required period of flooding may vary depending on the nematode species and their adaptive capabilities in different growing conditions. Flooding cannot be recommended as a certain means of exterminating nematodes from soil. More than 7 weeks of flooding effectively controls *Ditylenchus dipsaci*, *Radopholus similis*, and some *Meloidogyne* spp. Flooding is not a generally feasible option where water is scarce, particularly in rain-fed agriculture and light soil in hilly areas. Root-knot nematode (*M. incognita*) could be substantially lessened by flooding soil for 15 days at 30°C. Reduced root galling on rice root due to *M. graminicola* has been found under a longer period of flooding.

14.5.3.5 Time of Planting

The timely planting of crops is essential for obtaining optimum yield. However, in the nematode-infested areas, some nematode is a potential threat to the cultivation of crops. They can be planted/sown early when the plant can tolerate the low temperature, but nematode activity is minimum at that temperature. Thus, it would give the plant some time to develop a root system capable of tolerating nematode infection. Further, early planting allows early harvest before the nematode becomes active and causes substantial damage at the susceptible stage of the crop. Early planting of rice in June or middle of July is recommended for *Hirschmanniella oryzae.*

14.5.3.6 Stubble Management

Many nematodes complete a part of their life cycle inside the root, and destruction of the infected roots is a viable option to keep the nematode outside the field and kill conveniently. For example, root-knot nematode, a sedentary endoparasite of roots for many crops. Removal and safe destruction of old stubbles after harvest of jute eliminate residual nematode population from the infested field. Thus, it helps to grow the crops in crop sequence where the previous crop has huge nematode infection. Some ectoparasitic nematodes such as white tip nematode (*Aphelenchoides besseyi*) and rice stem nematode (*Ditylenchus angustus*) can survive in the rice stubbles (facultative fungal feeding nematodes also multiply on the fungi grown on stubbles) and carryover the population for the next season. However, it does not ensure that nematodes that live freely in soil are not removed for the destruction of crop roots. Further, the large quantity of crop residues that adds soil organic matter back to soil in every season for improving and maintaining soil fertility is not insured.

14.5.3.7 Sanitation/Hygiene

Field hygiene includes clean cultivation and burying of rice stubbles, straw, crop debris, weeds, and new growths after harvest, which reduces nematode infection. The spread of PCNs occurs through soil particles adhering to tubers, farm implements, gunny bags, farmers' feet, and irrigation water or rainwater running down the hill slope, which carry the cysts from the infested field to uninfested fields. The white tip nematode (*Aphelenchoides besseyi*) of rice is seed-borne, and the nematode can disseminate through infected seeds and also through irrigation water across the fields. Often the old stubbles, ratoon rice, and infected grains lodged in the field could be the sources of fresh infection in the rice monocropping system. Therefore, the use of the nematode-free healthy seed, adoption of seed disinfection processes, and field hygiene have been suggested to control the nematode. Likewise, infection of tuberose foliar nematode and contamination of the field occur through infected bulbs, flower stalks, and irrigation water under the rice–tuberose cultivation system. In storage conditions, the nematode undergoes anhydrobiosis and remains in a quiescent state. Therefore, field inspection and destruction of infected leaves, flower stalk, and maintaining overall field hygiene are recommended for managing the nematode problem in tuberose.

14.5.3.8 Cultivar Selection

Plant resistance plays a key role in nematode management. However, the availability of nematode-resistant varieties of all crops is still not much; majority of them are not acceptable to the farmers for poor yielding capability and marketability compared with susceptible cultivars. For example, tuberose cultivar Prajwal and Phule Rajani are tolerant to *Aphelenchoides besseyi,* but they are not much preferred due to poor fragrance, less number of flower stalks, poor adaptability under local growing conditions and marketability, as compared with Calcutta Single. Two other tuberose cultivars Shringar and Suvasini have been reported as resistant/tolerant to *M. incognita*. Resistant wheat cultivar Raj Molya Rodhak1 (Raj MR1), and barley cultivars Rajkiran, C-164, BH-72 are promising against *Heterodera avenae*.

14.5.3.9 Organic Amendments of Soil

Amending soil with organic matter changes physical and chemical soil properties, soil structure, and soil nutrient content; all these are beneficial for optimum plant growth. Besides, the composition of soil biota including soil nematodes changes and makes the soil more resilient for growing crops. The addition of organic materials increases the number of beneficial nematodes such as bacterial or fungal feeders and decreases or suppresses the plant-feeding nematodes (Renco, 2013). Most commonly used organic amendments include animal dung and green manure processed locally as farmyard manure, compost, and vermicompost. The impact of organic amendments on PPNs has been realized over several decades, and its multi-facet mode of actions has been documented (Thoden et al., 2011) Application of organic manure, compost, and farmyard manure at 18–20 t/ha reduces nematode population by multiple actions: releasing toxic substances, enhancing crop tolerance, and encouraging soil microbial antagonists. Several oilcakes such as neem (1.9–2.5 t/ha), castor (1.8 t/ha), groundnut (1.3–1.8 t/ha) reduce root-knot nematode populations. Use of Vesicular-Arbuscular Mycorrhiza (VAM), oil cakes of neem, pongamia, and castor either alone or in combination with *Purpureocillium lilacinum* (an egg parasitic fungus) is effective to manage root-knot nematode problems in tuberose (Nagesh et al., 1998). Application of oilcakes such as neem, karanj, mustard, and Farm Yard Manure (FYM) and two to three deep summer plowings during hot months (May–June) help to reduce soil populations of PPNs. Incorporation of neem cake at 100 g/m^2 in the nursery area 15 days before the sowing lessens the build-up of *M. graminicola*. The application of neem cake @ 100 g/pit during mid-February to mid-March when the vine starts sprouting in pointed gourd (*parwal*) prevents the attack of *M. incognita* and increases the crop yield.

14.5.3.10 Physical Methods

Control of nematodes using solar heat, hot water, steam, irradiation, electricity, etc., has been suggested in the literature for certain crops and growing conditions, particularly for de-nematization of planting materials (bulbs, corms, rhizomes, tubers, seeds, etc.), disinfestation of nursery beds, greenhouse, microplots, pot soils, mushroom beds, etc. Most nematodes cannot tolerate temperatures more than 50°C. However, the duration of exposure to heat is also important to achieve the substantial killing of nematodes. The hot-water treatment of the tuberose bulbs is most effective for managing the foliar nematode. The thermal death point of *A. besseyi* is 48°C ± 2°C for 5 minutes while the tolerance of the bulbs is 50°C ± 2°C for 30 minutes for germination (Khan et al., 2005). Rice seeds can be disinfected by presoaking in water followed by hot-water treatment at 52°C–54°C for 10–15 minutes. Infected seeds can also be made nematode-free following a simple method of spreading the rice seed on a concrete floor on bright sunny days for at least 4 hours for six consecutive days. Alternatively, overnight seed-soaking in water (1:2 ratio), followed by 10 minutes soaking in boiling water can be done before drying of seeds in shade, which could effectively clear the seeds of foliar nematode infection. Paring+hot water treatment of banana suckers at 55°C for 20 minutes + carbofuran 3G @ 0.5 g a.i./plant+neem cake @1 kg/plant in the pit before planting is effective for reduction of major phytonematodes and enhancement of fruit yield of banana.

14.5.4 Botanicals

14.5.4.1 Antagonistic Plants

Antagonistic plants are non-hosts that suppress nematode infection and multiplication due to physiological and biochemical incompatibility between the plant and nematode. The most common nematode antagonistic plants are marigold, sunhemp, mustard, asparagus, etc. They are useful as cover crops, intercrop, border crops, biofumigation, and rotation crops. Some antagonistic crops produce root exudates, for example, marigold containing α-terthienyl, which is highly toxic to *Pratylenchus penetrans,* and *Meloidogyne* spp. African (*Tagetes erecta*) and French (*T. patula*) marigolds have been widely used against many PPNs either as a sole crop in rotation or as intercrops. French marigold is most effective against most of the nematodes. The cultivation of marigold controls nematodes by multiple mechanisms: as a poor or a non-host, acting as a trap crop, producing antinematode compounds, and enhancing nematode antagonistic microbial populations in soil (Maniruzzaman and Vestergård, 2020). *Sesbania rostrata* acts as a trap crop for *Hirschmanniella oryzae.* Trap crops are not effective for ectoparasites and migratory endoparasites because they cannot trap the nematode into the host tissue. Sunhemp (*Crotalaria spectabilis*) can serve as a good trap because it allows nematode infection only and does support nematode multiplication. Trap crops are most effective for the nematodes such as PCNs, which require host root exudates for hatching. If a susceptible crop is used as a trap, it must be removed from the field before the nematode multiplication.

14.5.4.2 Nematicidal Plant Products

Natural sources of plant and microbial-based products are a potential alternative to synthetic nematicides as they offer several advantages such as selective mode of action, biodegradable, safe for non-target organisms, and relatively safe for the environment. The nematicidal properties of plants and their secondary metabolites (phenolic compounds, terpenoids, saponins, benzoxazinoids, organosulfur compounds, alkaloids, and glucosinolates) have been documented. The most widely studied compounds against nematodes are azadirachtin (a limonoid secondary metabolite) obtained from neem (*Azadirachta indica* L.), α-terthienyl (polythienyl compound) from marigolds (*Tagetes* sp.). Several commercial formulations (Aza 1,000–50,000 ppm) of Azadirechtin are available commercially for the control of insects and nematodes. Besides commercial formulations of neem, fresh extracts of leaves and seed kernel and neem cake reduce nematode populations in the soil.

14.5.4.3 Biofumigation

Biofumigation is a volatile action of brassicas that produces isothiocyanates, almost similar to methyl isothiocyanate released from metham sodium. It is practiced in the field by incorporating brassicas (e.g., cabbage, cauliflower, kale, mustard) and some other plants (e.g., horse-radish) into the soil and upon decomposition produces nematoxic substances such as isothiocyanates (ITCs). In all such cases, the myrosinase enzyme is partitioned in the plant tissues, and its disruption initiates the myrosinase-glucosinolate reaction and releases the ITCs as a defense compound of the plant system. This action of biocidal compounds suppresses soil-borne pathogens including weeds and nematodes in the soil. They are usually grown as cover crops and mowed at flowering, chopped, and then incorporated into the soil. Incorporation followed by irrigation and covering fields with polythene sheet increase the efficacy of the biofumigation. Crop residues such as leaves, stems, and roots of cabbage and cauliflower (5–8 kg/m^2 of green matter) reduced the soil nematode population. The use of brassicas can also provide other benefits including improved soil structure, health, and fertility, suppression of various soil-borne diseases and pests, and increased soil microbial activity leading to induction of plant systemic resistance (Meyer et al., 2011; Valdes et al., 2012).

The biofumigation using non-brassica plants such as sudangrass, sorghum, white clover, and flax has also shown promising results against nematodes. All these crops have been used as cover/rotation crops and green manures. The soil incorporation of sorghum/sudangrass young biomass at the age of 1–3 months has been suggested because older plant tissues (more than 3 months old) contain lesser dhurrin as compared to younger plants (Dutta et al., 2019). Residues of *T. patula* and *T. erecta* are often left in the field, and incorporation of their plant biomass reduces *M. incognita* populations in soil.

14.5.5 Genomics for Nematode Management

Plant nematology is currently undergoing a revolution with the availability of several genome sequences from a range of nematode species (Table 14.2). The model nematode *Caenorhabditis elegans* was the first nematode genome to be sequenced

TABLE 14.2
A comparison of Published Plant Parasitic Nematode Genomes

Sr. No.	Nematode	Sequencing Platform	Assembly Approach/ Assembler/ Assembly-Pipeline	Assembly Size	No. of Scaffolds	N50 Value (kbp)	BUSCO/CEGMA Score (Complete/Partial %) or Complete%	Year
1.	*Meloidogyne incognita*	Sanger, ABI3730×1 DNA analyzer	De-novo, Arachne	86.1	2,995	62.5	77/80.6	2008
2.	*Meloidogyne hapla*	ABI3730, Megabase Sequence analyzer	De-novo, Arachne v2.0.1	53.0	3,452	37.6	94.8/96.8	2008
3.	*Bursaphelenchus xylophilus*	454 FLX, Illumina GAI	De-novo, Newbler v2.3, Velvet v 1.0.12, AbacasII, IMAGE, iCORN	74.6	5,527	949.8	97.6/98.4	2011
4.	*Meloidogyne floridensis*	Illumina HiSeq2000	De-novo, Velvet v1.1.04	96.7	58,696	3.7	58.1/77.4	2014
5.	*Globodera pallida*	ABI 3730 Capillary DNA Analyzer, 454 GS-20 and GS-FLX sequencer, Illumina GAIIx	De-novo, Celera, Newbler, Abyss v1.2.7, SSPACE v1	124.6	6,873	122	74.19/80.65	2014
6.	*Pratylenchus coffeae*	Roche 454	De-novo, Newbler	19.7	5,821	10	NA	2015
7.	*Globodera rostochiensis*	Illumina HiSeq2000	De-novo, SGA v0.9.7, Velvet v1.3.7, SSPACE, Gapfiller	95.9	4,377	89	93.55/95.56	2016
8.	*Meloidogyne enterolobii* L30	Illumina	De-novo, Platanus	162.4	46,090	9.2	81/NA	2017
9.	*Meloidogyne floridensis* SJF1	Illumina	De-novo, Platanus	74.9	9,134	13.2	84	2017

(Continued)

TABLE 14.2 (*Continued*)
A comparison of Published Plant Parasitic Nematode Genomes

Sr. No.	Nematode	Sequencing Platform	Assembly Approach/ Assembler/ Assembly-Pipeline	Assembly Size	No. of Scaffolds	N50 Value (kbp)	BUSCO/CEGMA Score (Complete/Partial %) or Complete%	Year
10.	*Meloidogyne incognita* W1	Illumina	De-novo, Platanus	122.1	33,735	16.4	83	2017
11.	*Globodera ellingtonae*	HISeq, MiSeq, PacBio	De-novo assembler Allpaths-LG, PBJelly	119.1	2,248	360	92/96	2017
12.	*Meloidogyne javanica*VW4	Illumina	De-novo, Platanus	142.6	34,394	14.2	90	2017
13.	*Meloidogyne arenaria* HarA	Illumina	De-novo, Platanus	163.7	46,509	10.5	91	2017
14.	*Ditylenchus destructor*	Illumina HiSeq2500, PacBio RSII	De-novo, ALLPATHS-LG, SSPACE, pb-jelly, Gapfiller	112	1,761	570	91	2016
15.	*Meloidogyne incognita*	454, Illumina HiSeq2000	De-novo, MIRA, SSPACE, GapCloser (SOAPdenovo 2)	183.5	12,091	38.6	97	2017
16.	*Meloidogyne javanica*	454, Illumina HiSeq2000	De-novo, MIRA, SSPACE, GapCloser (SOAPdenovo 2)	256.3	31,341	10.4	96	2017
17.	*Meloidogyne arenaria*	454, Illumina HiSeq2000	De-novo, MIRA, SSPACE, GapCloser (SOAPdenovo 2)	235.5	26,196	16.4	95	2017
18.	*Meloidogyne arenaria Genotype A2-O*	PacBio RS II	Canu v1.3	284.05	NA	204.5	94.76	2018

(*Continued*)

TABLE 14.2 (*Continued*)
A comparison of Published Plant Parasitic Nematode Genomes

Sr. No.	Nematode	Sequencing Platform	Assembly Approach/ Assembler/ Assembly-Pipeline	Assembly Size	No. of Scaffolds	N50 Value (kbp)	BUSCO/CEGMA Score (Complete/Partial %) or Complete%	Year
19.	*Meloidogyne graminicola*	Illumina GAIIx	De-novo, Platanus, Redundans	38.18	4,304	20.4	84.27/90.73	2018
20.	*Ditylenchus dipsaci*	PacBio RSII	Canu v1.7	227.23	1,394	287	87.9	2019
21.	*Heterodera glycines*	PacBio	FALCON	129	NA	304	91.1	2019
22.	*Radopholus Similis*	Roche 454	Newbler	65	8,133	13.4	89.11% complete (CEGMA)	2019
23.	*Meloidoygne luci*	Illumina HiSeqX,PacBio	HGAP4 pipeline, Pilon	209.16	327	1,711.905	88.1% (BUSCO)/ 95.16% complete (CEGMA)	2020
25.	*Meloidogyne graminicola*	Oxford Nanopore Technologies and Illumina	Canu	41.5	283	294	CEGMA 95.97%	2021
26.	*Meloidogyne graminicola*	Pacbio	Falcon pipeline	36.86	514	105	BUSCO 88.6%	2021
27.	*Meloidogyne enterolobii*	Pacbio, Illumina	Canu, Pilon	240	4,437	143	CEGMA C: 94.76 (3.30) P: 96.77	2020
28.	*Heterodera glycines*	PacBio, Chicago, and Hi-C	FALCON, Dovetail, Juicer etc.	158	9	17,908.190	87.9% complete, 12.1% missing (BUSCO)	2021

in 1998. It is the only nematode for which the genome sequence is complete. Since then, genomes of 157 nematodes have been sequenced and published on WormBase Parasite, the primary parasitic-nematode genome repository. In 2019, the *C. elegans* genome was re-sequenced using long-read sequencing approach, and newer genes were discovered. Several strategies are being used to exploit this wealth of genomic information. Comparative genomics is being used to explore the genes associated with parasitic lifestyles. Functional analyses of nematode genes are moving toward larger-scale studies including global transcriptome profiling. RNA interference (RNAi), a popular functional genomics tool has been shown to reduce expression of a range of PPN genes and is a powerful tool for functional analysis of nematode genes. Nematode genomes show several interesting features. Nematode genomes are compact, range from 53 to 265 Mb, and contain only marginally fewer genes than that of humans. The average gene density is higher in nematodes. The nematode genomes contain a large number of genes with no recognizable homologs in other nematodes outside the same genus, indicating high rates of gene gain and loss. In addition, the nematode genes are organized in gene clusters called operons, and no unifying genomic features were identified in parasitic nematodes that would distinguish them from free-living nematodes. Several novel genes are known to arise in Phylum Nematoda.

A list of the sequenced PPNs, the technologies used for sequencing, and their genome statistics is provided in Table 14.1. Recent advancements in PPN genomics and transcriptomics have made it possible to identify promising molecular targets and metabolic choke points that may be exploited for their management through transgenic approaches or target-based drug discovery. The molecular targets for nematode management have been broadly classified into nematode effectors, housekeeping genes, genes involved in reproduction and development, and nematode responsive host plant genes. Disruption of nematode housekeeping genes such as *M. incognita* integrase, splicing factor, *Heterodera glycines* homologs of small ribosomal proteins RPS-3a, RPS-4, synaptobrevin (SNB-1) and a spliceosomal SR protein SPK-1, a *H. glycines* β-subunit of the coatomer (COPI) complex and a pre-mRNA splicing factor, *M. incognita* spliceosome subunit Pv010, *M. incognita* RPN-7, etc., have shown to affect the reproductive fitness and parasitic success of the invading nematode species. However, not all the housekeeping genes can be utilized as target(s). Although new drugs or chemicals designed using genomic information are still awaited, the transgenic crop-based approaches, such as host delivered RNAi, have been successfully used for nematode management. It is expected that increased use of genomic information would give a major boost to find smarter options for animal- and PPN management.

14.5.6 Plant Resistance against Nematodes

Host–nematode interactions are classified as "compatible" when the plant supports reproduction of the nematode, and "incompatible" when the plant does not support growth and reproduction by the parasite. In pursuit of identifying resistance against important nematodes, wild germplasm has been screened, and many R genes have been identified and cloned. The earliest reports of selection for plant resistance to nematodes date back to the late 19th century and were based on phenotypic selection for plants that had fewer galls on roots when infected with root knot nematodes. From these selections, varieties of cowpea, sugar beet, cotton, and coffee were reported with improved resistance to root knot nematodes. Almost all resistances reported against cereal cyst nematodes in commercial cultivars have been based on introgressions of single dominant genes, since these are generally most convenient to select and to use in breeding programs. Single dominant genes are also favored for engineering resistance as they are more amenable for identification and manipulation than the multiple genes. However, single gene resistance is vulnerable to breakdown due to repeated exposure to resistant host plants.

14.5.6.1 Plant R-Genes

Based on the amino acid motif organization and membrane spanning domains, eight broad groups of plant resistance genes have been identified (Gururani et al., 2012). First category of these genes includes NBS-LRR genes—genes that encode cytoplasmic proteins with a nucleotide-binding site (NBS), a C terminal leucine rich repeat (LRR), and a putative coiled coil domain (CC) at the N-terminus. Second class contains cytoplasmic proteins with LRR and NBS motifs and an N-terminal domain with homology to the mammalian toll-interleukin-1- receptor (TIR) domain. Third family is devoid of NBS motif but consists of extra cytoplasmic leucine rich repeats (eLRR), attached to a transmembrane domain (TrD). Fourth class of resistance genes consists of an extracellular LRR domain, a transmembrane domain (TrD), and an intracellular serine-threonine kinase (KIN) domain. Fifth class of resistance genes contain the putative extracellular LRRs, along with a PEST (Pro-Glu-Ser-Thr) domain for protein degradation, and short protein motifs (ECS) that might target the protein for receptor-mediated endocytos. The sixth major class of resistance genes contains a membrane protein domain (TrD), fused to a putative coiled coil domain (CC), whereas, the seventh class of resistance genes includes TIReNBSeLRR R protein class. Last and eighth major class of resistance genes includes the enzymatic R-genes, which contain neither LRR nor NBS groups. The R-genes that have been described for providing nematode resistance are provided in Table 14.3.

Some of these nematode-resistant genes have been used by transferring those into other crops, e.g., tomato and potato cultivars with the Mi or H1 gene. However, the efficacy of these genes in various crops has not been as expected and indicates that the molecular ecosystem required for expected expression of these genes needs to be understood better. However, the advances in biotechnology have enabled rapid transfer and use of these genes in many different ways.

14.5.6.2 Transgenic Technologies

Identification of natural resistance genes through genotype screening and using them through marker-assisted breeding to improve disease and insect pest resistance in crops are still the standard way to impart biotic stress resistance to popular cultivars. However, this is a slow approach, and for many crops, adequate resistance sources are not available. Transgenic technologies have been extensively used for various crop plants and offer a fast way of transferring resistance to widely accepted cultivars. Resistance genes exhibiting various modes of action can be pyramided to

TABLE 14.3
R Genes Described for Various Nematodes in Different Crops

Crop	Gene/QTLs	Nematode
Lycopersicum esculentum	*Mi-1, Mi-9*	*Meloidogyne incognita*
Solanum peruvianum	*Mi-1, Mi-3, Mi-9*	*Meloidogyne incognita*
Piper nigrum	*Me3*	*Meloidogyne incognita*
Arachis hypogaea	*Mae, Mag, Rma*	*Meloidogyne arenaria*
Capsicum annum	*CaMi*	*Meloidogyne incognita*
Solanum tuberosum	*Hero, Gpa2*	*Globodera pallida*
Solanum tuberosum	*Gro1, H1, Hero, Gpa1-4*	*Globodera rostochiensis*
Beta vulgaris	$Hs1^{pro-1}$, *Hs2*	*Heterodera schachtii*
Triticum spp.	*Cre1, cre3, cre8*	*Heterodera avenae*
Lycopersicum esculentum	*Hero*	*Globodera rostochiensis*
Hordeum vulgare	*Ha2, Ha3, Ha4*	*Heterodera avenae*
Glycine max	*rhg1, rhg4*	*Heterodera glycines*
Triticum aestivum	QTLS for e.g., QRlnt.lrc-6D.2, QRlnt.lrc-6D.1	*Pratylenchus thornei*
Triticum aestivum	QTLs for e.g., QRlnn.lrc-4D.1, QRlnn. lrc-6D.1, R1nn1	*Pratylenchus neglectus*
Hordeum vulgare	*Pne3H-1, Pne3H-2, Pne5H, Pne6H, Pne7H*	*Pratylenchus neglectus*
Triticum aestivum	*R1nn1*	*Pratylenchus penetrans*
Triticum aestivum	*R1nnp6H*	*P. neglectus, P. penetrans*
Triticum aestivum	*Xbarc 183*	*P. thornei, P. neglectus*

Source: Modified from Williamson and Kumar (2006), Gururani et al. (2012), and Fosu-Nyarko and Jones (2015).

obtain superior insect/disease resistance. It is also possible to limit the expression of these genes to specific plant parts. Acceptance of genetically modified crops is a concern in various countries; however, the majority of the world's population lives in countries where genetically modified crops are grown. As the population grows, the acceptance of transgenic crops might increase to ensure global food security.

In recent years, several advancements in the genetic modification technology have happened, such as cisgenesis, RNAi, reverse breeding, genome editing using CRISPR-Cas systems or virus-delivered ZFN genome editing, epigenetics, agroinfection, virus-induced gene silencing and genomics-enabled technologies, e.g., ectopic delivery of dsRNA, and lastly, grafting of nontransgenic scions onto genetically modified rootstocks (e.g., for vines or fruit trees). In the traditional sense, a transgenic plant genome contains a gene integrated from another organism. However, new genome editing techniques have emerged allowing us to make precise changes to the genome. These techniques enable the modification of existing genes by causing targeted changes in the genome. In addition, the recent explosion in the availability of genomic information for various nematodes and nematode–plant interactions has provided unprecedented insights into nematode parasitism and biology. This knowledge makes it possible to exploit specific aspects of nematode–host plant interactions to develop nematode management strategies that may enable plants to prevent nematode invasion, hinder nematode migration through plant tissues, prevent successful establishment, and reduce nematode's feeding ability or nematode fecundity. These genes vital for nematode parasitism are potent targets that can be used for development of new chemical molecules for nematode management. Following are some specific examples of use of transgenic technologies for nematode management.

14.5.6.3 Plants Expressing Protein Toxins, Protease Inhibitors, and Lectins

Various bacterial exotoxins and endotoxins can kill insects and nematodes, e.g., *Bacillus thuringiensis* delta-endotoxins encoded by *Cry* genes. These toxins have been used for conferring resistance against insects and nematodes in crop plants. One limitation of using *Cry* proteins for nematode management is that these toxins are big in size and can't pass through feeding tubes of cyst, root-knot, and other nematodes. Protein engineering to reduce the size of the toxin molecule may provide an alternate way to use these toxins.

Proteases (=proteinases or proteolytic enzymes) are a group of enzymes that hydrolyze peptide bonds of proteins. There are four classes of proteases—cysteine, serine, aspartyl, and metalloproteases, which hydrolyze various peptide bonds. Nematodes use proteinases to digest protein in food to obtain amino acids for protein synthesis. Blocking the nematode protease inhibitors can result in starvation and death of the nematodes. A few proteinase inhibitors such as cowpea trypsin inhibitor (CpTI), cystatins, and serine based PIs have been used against nematodes (Hepher and Atkinson, 1996). Expression of a cowpea trypsin inhibitor (*CpTi*) in transgenic potato hindered early growth of *Globodera pallida* and influenced its sexual fate toward male juveniles. A modified rice cysteine PI, Oc-IΔD86 (cystatin) expressed as a transgene in *Arabidopsis thaliana* significantly reduced the size and fecundity of females of both *Heterodera schachtii* and *M. incognita* and inhibited egg production. A sunflower cystatin and a protein engineered variant of a rice cystatin (Oc-IΔD86) under the control of CaMV35S promoter conferred resistance against *Globodera* spp. to potato plants similar to chicken egg white cystatin (CEWC). The modified cystatin reduced the reproductive success of *H. schachtii*, *M. incognita*, and *Rotylenchulus reniformis* by ~70% in transgenic *Arabidopsis*, transgenic potatoes, rice, and banana. PIN2, a type of serine protease inhibitor from *Solanum tuberosum* into *Triticum durum*, was found effective in inducing resistance against *H. avenae* (Vishnudasan et al., 2005). Even this protease inhibitor does not impart bio-safety issues for humans as these are cystacin is part of human diet and a non-allergen. Also these studied PIs are found to be digested by gastric juices. Proteinase inhibitors have been studied in many different crops (Lilley et al., 1999; Ali et al., 2017).

Lectins are protein or glycoprotein substances of non-immunoglobulin nature, capable of specific recognition and specific binding to carbohydrates. Lectins may bind with free sugar or with sugar residues of polysaccharides, glycoproteins, or glycolipids, which can be free or bound. Expression of snowdrop (*Galanthus nivalis*) lectin (GNA) in potato and oilseed rape confers partial resistance to cyst nematodes and in *Arabidopsis* to root-knot nematodes (Ripoll et al., 2003). Lectin may bind glycoproteins localized on the surface of the nematodes, on the chemoreceptors from the amphids, or in the amphidial secretions and interfere with the nematode sensory perception and its ability to establish feeding cells.

14.5.6.4 Disruption of Feeding Site Formation or Function

The reproduction of sedentary endoparasitic nematodes (*Heterodera* spp., *Globodera* spp., *Meloidogyne* spp., *Rotylenchulus* spp., *Nacobbus* spp., and *Tylenchulus* spp.) depends on successful formation and function of giant cells, syncytia, or similarly modified host cells. If the feeding site formation or function is disrupted, the nematode reproduction and development can be hampered. In order for this strategy to become successful, plant promoters specifically expressing in feeding cells need to be linked to a gene that when expressed in feeding cells results in disruption of feeding sites. Opperman et al. (1994) reported that the truncated promoter of the water channel protein TobRB7 expressed specifically in root-knot giant cells, when linked to the cytotoxic ribonuclease barnase, resulted in cell death. Although a series of genes highly upregulated or downregulated in nematode feeding cells have since been identified, such as the heat shock promoter Hahsp17.7G4. It appears that most of these promoters show leaky expression. Matthews et al. (2012) selected 100 soybean genes with modified expression in syncytia for overexpression in a composite hairy root soybean system. Nine of these genes reduced the number of females by 50% or more when overexpressed, whereas some enhanced the number of females. The overexpression may confer a level of nematode resistance, but there is the risk that in a field situation, an abnormal phenotype or reduced yield might result.

14.5.6.5 RNAi-Based Nematode Resistance

RNA interference is a powerful tool to silence genes in plants and animals. With the transgenic technology, it is possible to develop plants that can express double-stranded RNA against nematode target genes and silence them. This is a sustainable and environment-friendly strategy for nematode management. dsRNA feeding/soaking has been used to assess the effects of downregulation of over 30 essential and parasitism genes of various plant nematode species (Reviewed by Lilley et al., 2012; Ali et al., 2017). It appears that for *H. glycines* and some *Pratylenchus* spp., silencing does not always produce stable phenotypic effects for some genes. Nevertheless, the soaking method is useful for initial screening of gene function and for discovery of candidate target genes suitable for plant-delivered RNAi for nematode management. Host-delivered RNAi provides continuous dsRNA if controlled by a constitutive promoter, which is ideal for management of plant parasitic nematodes. In-planta delivery of dsRNA of two target genes (an integrase and a pre-mRNA splicing factor) was demonstrated by Yadav et al. (2006) to reduce replication of *M. incognita* on transgenic tobacco plants, and this was quickly followed by the work of Huang et al. (2006), who expressed dsRNA to an *M. incognita* effector protein in transgenic plants and also showed reduced nematode reproduction. Since then, a series of experimental and crop plants have been engineered to generate inverted repeats of dsRNA targeting genes expressed in pharyngeal gland cells or those essential for development and reproduction in cyst and root-knot nematodes. To date, reports of significant reductions in the number of females of soybean cyst nematode (81%–93%) and eggs (68%–95%) produced by female cysts developing on hairy roots or composite transgenic soybean expressing dsRNA of genes involved in RNA and protein synthesis provide a level of confidence that RNAi can be an important tool for nematode control (reviewed by Dutta et al., 2015). A recent study employed HIGS to target *galectin*, *cathepsinL*, *vap1*, *serpin*, *flp12*, *RanBPM*, and *chitinase* gene of *H. avenae* in wheat resulting in 33.24%–72.4% reduction in the nematode multiplication (Dutta et al., 2020). The cysts formed on these transgenic plants showed developmental retardation in the form of smaller size along with translucent cuticle. This study is evident of how HIGS can be employed to manage cyst nematodes in wheat and barley in future. However, despite tantalizing results of RNAi that point to viable control strategies, this is yet to be realized in farm settings. It is our contention that a better understanding of key aspects of basic nematode biology is a prerequisite for translation into novel and effective nematode management strategies. Another non-transgenic approach to employ RNAi-based management is use of topical RNAi through non-transgenic means. The researchers are investigating ectopic delivery of dsRNA (spray formulations) as a modern approach against nematode-pest management.

14.5.7 Chemical Nematicides

Chemical nematicides can be credited with having provided the first convincing results to showcase the enormous crop losses caused by them and still provide formidable field solutions to combat nematodes quickly and provide demonstrable results. Historically, the nematicidal chemicals have witnessed several generations of products (Oka, 2020) and are being discussed below.

14.5.7.1 Bygone/Banned Molecules

Fumigants: Carbon disulfide was the first nematicide launched in Germany to control sugarbeet cyst nematode in 1871. Subsequently (1943–45), a mixture of 1,3-dichloropropene and 1,2-dichloropropane (D-D) and ethylene dibromide (EDB) were successfully used to demonstrate the damage of nematodes to crops in the USA. More fumigants such as 1,3-dichloropropene (1,3-D)-, 1,2-dibromo-3-chloropropane (DBCP), and methyl isothiocyanate (MITC)-generating compounds were developed and used extensively. Some other fumigants, such as metam sodium/potassium and dazomet, are still being used in some countries. The fumigants though exhibited remarkable nematicidal acitivity besides being general soil sterilants, their high application rates, cumbersome application methods, and use only as pre-plant application deterred their acceptance. However, their adverse effects on humans and the environment led to their ban. Among all the fumigants, methyl bromide had been widely used due to its broad spectrum activity against multiple soil pathogens, weeds, and soil arthropods. It stands phased out in most parts of the world since 2015.

The organophosphates and carbamates: Soon after the withdrawal of fumigants, another class of chemical nematicides belonging to the organophosphate and carbamate groups replaced them from 1960s onwards. The lower doses, ease of application, feasibility of pre- and post-planting applications attracted their attention and acceptability. Basically insecticides, these were promoted as nematicidal products as well. Among the organophosphates, cadusafos, ethoprophos, fenamiphos, fensulfothion, terbufos, etc., were developed as the major nematicidal products, and aldicarb, carbofuran, and oxamyl constituted the carbamates. However, the low application rates against nematodes do not cause acute nematode mortality; instead, "nematostatic" action prevails till the chemical is metabolized or degraded in the environment, and the nematodes recover and resume their biological activities. Inhibition of acetylcholinesterase activity and accumulation of acetylcholine at the neuromuscular junctions cause nematode paralysis and death. High toxicity to the environment and non-target organisms are the main reasons for their gradual phasing out. It is also realized that the "biodegradation" of nematicides as substrate by the soil microorganisms for their energy generation has led to the decreased efficacies of such nematicides over time (Cabrera et al., 2010).

14.5.7.2 New Generation of Nematicides – The 3F Nematicides

The restrictions imposed on most of the widely used nematicides (methyl bromide, fenamiphos, aldicarb, etc.), development of environmentally benign products, growing awareness about nematode damage to the crops, shift toward high-value horticultural crops, all put together, have ushered in era of new opportunities for industry to develop green molecules.

An overview on the characteristics and mode of action of new generation of nematicides in respect of the most significant new chemical nematicides to emerge in the last decade is provided in Table 14.4. The readers are referred to Oka (2020) for further detailed information about the new molecules.

TABLE 14.4
Characteristics of New Synthetic Nematicides as Compared to Older Products (Fumigants and Carbamate Nematicides)

Chemical Name	Chemical Structure	Product Type/ Chemistry	Water Solubility	Soil 1/2 Life	Mode-of-Action	Signal Words
Fumigant (1,3-D)	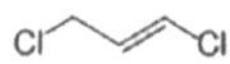	Fumigant	Gas	Short < 14 days	Unknown	Danger
Carbamate (oxamyl)	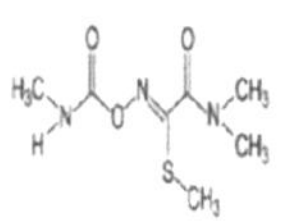	Carbamate	240,000 ppm	Short 7 days	AChE	Danger
Fluensulfone	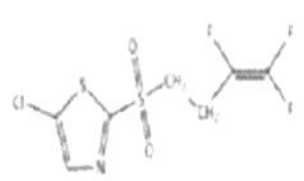	Thizaole	545 ppm	Short 7–17 days	Beta oxidation inhibitor	Caution
Fluopyram	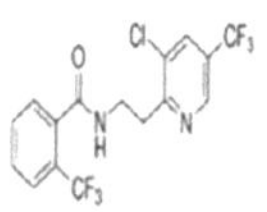	Benzamide	10 ppm	Long > 200 days	SDHI	Caution
Fluazaindolizine	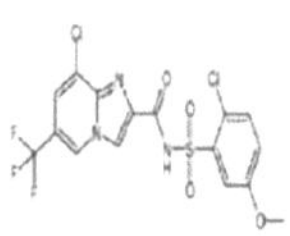	Carboxamide	2,000 ppm	Medium 30 days	Unknown	TBD

Source: Modified after Desaegar et al. (2020).
ACHE, acetyl cholinesterase inhibition; SDHI, succinate dehydrogenase inhibition; ACC, Acetyl-CoA carboxylase.

Fluensulfone, fluopyram, and fluazindolizine are new nematicides that all have a trifluoro (3-F) group in their molecular structures and are hereby referred to as **3-F nematicides**. These nematicides have a much safer toxicity profile than the older nematicides (fumigants, organophosphates, carbamates). However, they are quite different in terms of their chemical and physical properties and their modes of action.

Fluensulfone (Nimitz®, Adama) is a member of the class of 1,3-thiazoles; this compound is a specific nematicide and was the first of its kind to become registered in the USA in 2014. Fluensulfone has a soil half-life up to 36 days. Under field conditions, fluensulfone had a 50% dissipation rate between 23.7 and 24.3 days depending on soil type, similar to that of fosthiazate. The unpublished mode of action of fluensulfone is that it is a fatty acid beta oxidation inhibitor. Several *in vitro* studies have been conducted to determine the effects of fluensulfone on *Meloidogyne* species. Doses as low as 25 ppm impacted *M. incognita* J2 activity after 24 hour exposure, and egg hatch was reduced at 95 ppm.

Fluopyram (Verango®, Velum®, Indemnify®, ILeVo®, Bayer Crop Science) is a member of the class of benzamides and a Fungicide Resistance Action Committee code 7 fungicide. It is an inhibitor of the enzyme succinate dehydrogenase in fungi. Heiken (2017) confirmed this to be the likely mode of action in nematodes, demonstrating that succinate dehydrogenase knockdown mutants of *C. elegans* had a roughly 2.6-fold increase in sensitivity to fluopyram. Unlike the other 3-F nematicides, fluopyram has a very long soil half-life, up to 746 days depending on soil type. Fluopyram is a very fast-acting and potent nematicide, with *M. incognita* J2 exposed to fluopyram at 5.58 ppm for 2 hour showing reduced mobility by 80%, and the 24 hour ED_{50} for *M. incogntia* being as low as 1 ppm. Fluopyram can be called a "real nematicide," which causes irreversible immobilization and leads to nematode death after short exposure periods at relatively low concentrations. However, due to its nematicidal and fungicidal activity, this nematicide may have an advantage for fungal diseases associated with nematodes.

Fluazaindolizine (Salibro®, DuPont/Corteva) is a member of the class of carboxamides. Its mode of action is unknown. *M. incognita* J2 exposed to 5–50 ppm of fluazaindolizine were increasingly immobile and eventually dead 24–96 hour after exposure. Fluazaindolizine has irreversible effects on *M. incognita* even after 24 hour of exposure. However, based on nematode motility, a 24-hour EC_{50} for *M. incognita* J2 for this compound is over 2× that of fluensulfone and over 200× that of fluopyram, making this a slower-acting nematicide. Fluazaindolizine was also a poor ovicide, with reductions in *M. incognita* egg hatch only at concentrations over 250 ppm for 7 days of exposure.

Two more products, namely *Tioxazafen* (Nemastrike®, Monsanto/Bayer) and *Spirotetramat* (Movento®, Monsanto/Bayer), are in the process of development as nematicides. *Tioxazafen* is a new systemic nematicide belonging to the chemical class oxadiazoles. Its mode of action may be linked to disruption of ribosomal activity of nematodes. *Spirotetramat* is a spirocyclic tetramic acid derivative and was first marketed as a fully systemic insecticide. Spirotetramat has unique translocation properties in that it is translocated within the entire vascular system (upward and downward through the xylem and phloem, respectively). To be active against nematodes, the compound needs to be hydrolyzed in the plant to the active spirotetramat-enol form.

Efficacy of new chemical nematicides in greenhouse and field experiments: Lot of information is available on the greenhouse and field efficacy results of 3-F compounds, and the results have been reviewed by Desaeger et al. (2020). Rates of application for the new nematicides range from 1 to 2 kg ai/ha for fluensulfone and fluazaindolizine and less than 0.5–0.7 kg ai/ha for fluopyram and spirotetramat. The 3-F nematicides (fluensulfone, fluopyram, fluazaindolizine) have been evaluated on a variety of crops for their ability to suppress a diversity of PPNs. However, most have focused on their efficacy against *Meloidogyne* species in vegetable crops. As indicated, the bulk of efficacy data of 3-F nematicides is on *Meloidogyne* spp., and all products seem to be quite effective against these nematodes. Fluopyram overall shows more broad spectrum activity, including against non-PPNs. Water solubility is important in that it will determine how well the molecule moves through the soil profile. Higher water solubility will give better soil coverage and distribution of the active but will also increase the risk of leaching. Soil half-life determines how long the molecule stays active in the soil. Longer soil half-life will give longer soil residual activity and more extended nematode control, but also increases the risk of soil accumulation. Fluensulfone and fluazaindolizine are relatively similar in terms of solubility and soil half-life, but fluopyram is quite different, having low water solubility, and a much longer soil half-life.

14.5.7.3 Development of New Nematicides – Industry Perspectives

The future impact of the new nematicides will depend on (i) consistent field efficacy and (ii) cost of the new nematicides—with many growers facing increasingly shrinking margins, the price tag will be more important than ever. If the new nematicides are to replace fumigants, they will have to be integrated with a weed and soil disease management program. In the long term, improved soil health and more resilient soils may be one of the greatest benefits of moving away from soil fumigants and adopting more selective and safer nematicides.

The global market for nematicides ($1.3 billion in 2019) is much smaller than that for insecticides ($16.4 billion in 2019), fungicides ($13.4 billion in 2019), and herbicides ($32.6 billion in 2019). The cost of bringing a new chemical active ingredient to market is increasing every year and is now estimated to be more than US$250 million, about 10-fold what it was in the 1960s (Sparks, 2013).

14.5.8 Bioagents

Bioagents have been and will remain the most researchable materials for the sustainable management of plant parasitic nematodes. Generally designed to provide long-term, eco-friendly solutions through inoculative releases, most bioagent applications nowadays target inundative applications in the high-input horticultural cropping systems. Bioagents and organic matter augmentations of the field soils complement each other and provide durable efficacy for nematode management through multiple modes of action. Organic amendments of soils help in the establishment of bioagents to build suppressive soils. However, as a component of IPM, bioagents greatly decrease the use of chemical pesticides.

14.5.8.1 Potential Bioagents Identified

A few most potential bioagents are listed below along with their major characteristic features; however, the readers may refer to Tian et al. (2007), Tranier et al. (2014), Abd-Elgabad (2016) for more details.

Purpureocillium lilacinum (=*Paecilomyces lilacinus*)

- Soil-borne fungus
- Good root colonizer and rhizosphere competitor
- Parasitic on nematodes eggs
- Antagonistic to juvenile stages
- Suppress nematode population density and increase crop yields

Pochonia chlamydosporia (=*Verticillium chlamydosporium*)

- Soil-borne fungus
- Colonize on plant roots or confined to rhizoplane
- Parasitic on nematodes eggs
- Produce toxins that inhibit hatching or kill eggs of nematodes
- Parasite of female root-knot nematodes

Trichoderma viride and *Trichoderma harzianum*

- Active rhizosphere colonizers
- Compete for nutrients
- Produce antibiotics that affect nematodes

Pseudomonas fluorescens and *Bacillus subtilis*

- Colonize on roots and exclude nematode niche
- Produce antibiotic compounds
- Induce systemic resistance in plant

Pasteuria penetrans

- Nematode antagonists and endospore-forming bacteria
- Obligate parasites of PPNs
- Endoparasite, reduce infectivity and also fecundity

14.5.8.2 Bioagents Commercialized

Sustained research over the past few decades has led to commercialization of many potential fungal and bacterial antagonists of PPNs (Tables 14.5 and 14.6).

TABLE 14.5
Commercial Products of Fungal Biocontrol Agent in the Management of Nematodes

Fungus	Product	Formulation	Company/Institution	Country
Pochonia chlamydosporia	KlamiC®	Granulate	Rothamsted Research and Centro Nacional de Sanidad Agropecuaria UK	Cuba
P. chlamydosporia	PcMR-1 strain	Liquid	Clamitec, Myco solutions, lda	Portugal
P. chlamydosporia	Xianchong bike	Liquid	Laboratory for Conservation and Utilization of Bio-resources, Yunnan Univ.	China
Purpureocillium lilacinum	BIOACT®WG	Water-dispersible granulate	Bayer Crop Science	USA
P. lilacinum	BIOACT®WP	Water-dispersible powder	Bayer Crop Science	USA
P. lilacinum	PL Gold	Wettable powder	BASF Worldwide, Becker Underwood	South Africa
P. lilacinum	PL 251	Water-dispersible granulate	Biological Control Products	South Africa
P. lilacinum	BIOCON	Wettable powder	Asiatic Technologies Incorporation	Philippines
P. lilacinum	Shakti Paecil	Wettable powder	Shakti Biotech	India
P. lilacinum	Yorker	Wettable powder	Agriland Biotech	India
P. lilacinum	Miexianning	Unknown	Agricultural Institute, Yunan Academy of Tobacco Science	China
P. lilacinum	Pl Plus® (P. lilacinus strain 251)	Wettable powder	Biological Control Products	South Africa
P. lilacinum	Melocon®WG	Water-dispersible granulate	Prophyta GmbHCertis Germany	USA
P. lilacinum	BioNematon		T. Stanes	India
Trichoderma harzianum (isolate DB 104)	Romulus	Wettable powder	Dagutat Biolab	South Africa
Myrothecium verrucaria	DiTera®	Dry flowable	Valent Biosciences Corporation	USA
T. harzianum	IIHR-Th-2Ecosom-TH	Wettable powder, liquid, lyophilized	Agri Life SOM Phytopharma (India) Limited	India
T. harzianum	Commander Fungicide	Unknown	HTC Impex Private Limited	India

(Continued)

TABLE 14.5 (*Continued*)
Commercial Products of Fungal Biocontrol Agent in the Management of Nematodes

Fungus	Product	Formulation	Company/Institution	Country
T. harzianum	Trichobiol	Wettable powder	Control Biológico Integrado; Mora Jaramillo Arturo Orlando – Biocontrol	Columbia
T. viride (strain 2684)	Trifesol	Wettable powder	BioCultivos S.A., Bogotá, Colômbia	Columbia
T. lignorum	Mycobac	Unknown	Laboratórios Laverlam	Columbia

Source: https://www.researchgate.net/publication/281933557_Nematophagous_Fungi_Commercialization.

TABLE 14.6
Commercially Formulated Bacterial Biocontrol Agents Against Plant Parasitic Nematodes

Bacteria	Name/Producer	Target	Observations/Findings
Pasteuria pentrans	Econem—Syngenta	*Belonolaimus longicaudatus*	Ineffective
P. penetrans	Econem—Pasteuria Bioscience, USA	*Meloidogyne incognita, M. arenaria*	Effect depends on plant culture and growing conditions
P. penetrans	Econem—Nematech, Japan	*M. incognita*	Biological efficiency 89%. Increase in marketable yield of sweet potato. The results are similar to Dichloropren
P. nishizawae Pn1	Clariva PN—Syngenta, Brazil	*Heterodera glycines*	Ineffective
Pasteuria sp. *Ph3*	Naviva ST—Syngenta	*Rotylenchulus reniformis*	Inhibition of nematodes in cotton, soybean, vegetables
P. usage Bl1 + *Pasteuria* sp. *Ph3*	NewPro—Syngenta	*B. longicaudatus*	Inhibition of lance and sting nematodes in turf
B. amyloliquefaciens FZB42	RhizoVital—AbiTep GmbH, Berlin, Germany	*M. incognita*	Reduction in J2, enhanced of root weight
Burkholderia cepacia	Deny—Stine Microbial Products, Madison, WI	*M. incognita*	Suppression of root-knot nematode population on bell pepper. Biological efficiency 60%
Bacillus subtilis	Stanes Sting—Imported from T. Stanes and Co. Ltd, India, by Gaara Co., Egypt	*M. arenaria*	Reduction of J2 both in soil and roots as well as root galling, egg masses, biological efficiency 50%. Enhanced potato yield.

(*Continued*)

TABLE 14.6 (*Continued*)
Commercially Formulated Bacterial Biocontrol Agents Against Plant Parasitic Nematodes

Bacteria	Name/Producer	Target	Observations/Findings
B. firmus I-1582	Poncho—Votivo Crop Science, Raleigh, NC	*Meloidogyne* spp., *Heterodera glycines*	Biological efficiency against *Meloidogyne luci* over 50%. Triggering ISR. Degradation of *Meloidogyne* eggs, colonization of plant roots.
B. firmus I-1582	Poncho—Votivo Crop Science, Raleigh, NC	*R. reniformis*	Nematode control and yield similar to Avicta (Abamectin)
B. firmus GB-126	VOTiVO—Bayer, Germany	*R. reniformis*	Reduction in number of females, eggs, and juvenile life stages. Cotton yield similar to aldicarb.
B. firmus	BioNem-WP—BioSafe—AgroGreen, Israel	*M. incognita*	Field efficiency 75%–84%. Enhanced shoot height (29%–31%) and tomato weight (20%–24%)
B. firmus	BioNemaGon—Agri-Life, India	*Meloidogyne* spp., *Heterodera* spp., *Helicotylenchus* spp.	Reduction in nematode population and root infestation by nematodes in vegetables and herbs
B. megaterium	Bio-Arc	*Meloidogyne* spp.	Reduction in J2, egg masses, eggs and reproduction factor. Enhanced shoot weight.
B. megaterium	Bio-Arc	*Tylenchulus semipenetrans*	Biological efficiency 88%,89% on baladi orange and lime
B. methylotrophicus	Onix—Laboratorio de Bio Controle Farroupilha S.A., Brazil	*M. javanica*	Ineffective on tomato plants
Pseudomonas fluorescens	Sheathgua (Sudozone)—Agriland Biotech, India	*Meloidogyne* spp., Cyst nematodes	–
Serratia marcescens	Nemaless—Agricultural Research Centre, Egypt	*M. incognita*	Reduction in J2, egg masses, egg numbers and reproduction factor. Improvement of tomato growth criteria: fresh weight of shoots and roots, length of both systems.
B. amyloliquefaciens IN937a, *B. subtilis* GB03	BioYield—Gustafson LLC, USA	*M. incognita*	Significant reduction in nematode eggs, juveniles and galls on tomato. Enhanced root weight.
B. licheniformis, *B. subtilis*	Nemix C	*Meloidogyne* spp.	–

(*Continued*)

TABLE 14.6 (*Continued*)
Commercially Formulated Bacterial Biocontrol Agents Against Plant Parasitic Nematodes

Bacteria	Name/Producer	Target	Observations/Findings
B. licheniformis FMCH001, *B. subtilis* FMCH002	Presense—FMC Química do Brasil Ltd.a., Brazil	Plant parasitic nematodes	–
B. subtilis, *B. licheniformis*, *B. megaterium*, *B. coagulans*, *P. fluorescens*, *Streptomyces* spp.	Pathway Consortia—Pathway Holdings, USA	Plant parasitic nematodes	–
B. chitinosporus, *B. laterosporus*, *B. licheniformis*	BioStart—Bio-Cat, USA	*Meloidogyne* spp.	Inconsistent effect
Serratia spp., *Pseudomonas* spp., *Azotobacter* spp., *B. circulans*, *B. thuringiensis*	Micronema—Agricultural Research Centre, Egypt	*M. incognita*	Significant reduction in J2, galls and egg masses (97%, 80% and 88%). Enhanced growth parameters: length of shoots, fresh and dry weight of shoots and roots.
47 strains of bacilli	Equity—Naturize Biosciences LLC, Jacksonville, FL, USA	*M. incognita*	Significant reduction in nematode eggs, juveniles and galls on tomato. Enhanced root weight
Rhizobacteria and microbial metabolites produced during anaerobic fermentation of a microbial community	Ag-Blend—Advanced Microbial Solutions LLC, Pilot Point, TX, USA	*M. incognita*	Reduction of gall numbers, enhanced root weight

Source: Modified after Migunova and Sasanelli (2021).

14.5.9 Nanoproducts

Nanotechnology is an emerging area that could provide potential solutions to problems encountered in agriculture. Nanoformulations of currently used chemical and biological products are in vogue to enhance their efficacies. Silver ion (Ag^+) and its compounds are highly toxic to microorganisms, exhibiting strong biocidal effects on many species of bacteria and fungi. Recently, Ag nanoparticles (AgNP) have shown evidence of being a potentially effective nematicide. Studies have explicitly revealed promising results of AgNP against *Meloidogyne incognita* on eggplant, tomato, and okra, and *M. graminicola* on rice (Baronia et al., 2020). Nanoproducts of chemical and biological pesticides have proved to be more efficacious in combating pests and diseases leading to reduced dosages and are, therefore, more economical. AgNP may provide alternative to currently used chemical nematicides and resolve nematode problems on other crops.

14.5.10 Integrated Nematode Management

INM is not a set of fixed practices, but rather a philosophy, and will have many different shapes and forms depending on the cropping system and region (Sikora and Roberts, 2018). The philosophy and concepts of INM not withstanding, these often become redundant when applied in the context of small, marginal, and medium landholders. The various components of INM vary from region to region and cropping systems. PPNs have to be seen as one of the factors in the overall scenario of pest and disease spectrum of the area, crop, cropping system. Depending upon which pest/disease/nematode dominates, the IPM has to be designed accordingly. For example, if a nematode problem is the major cause of crop loss in a particular cropping system, the IPM has to be nema-centric. Blending of management solutions should be offered as a package that addresses all pests and diseases of that ecosystem. INM should fit into IPM, and IPM should be molded within Integrated Crop Management (ICM); finally, the ICM should be a component of Integrated Farm management (IFM). The acceptability of INM derives its relevance within IFM. These various components of INM to IFM should address the issues of environment, ecosystem, sustainability, long-term solutions of pests and diseases, leading ultimately to economic profitability to the growers.

A couple of graphics depicting the INM protocols developed for a open-field vegetable cropping system and a protected cultivation system, respectively, under north Indian conditions, are reproduced in Figure 14.2a and b.

14.6 FUTURE PROSPECTS

Nematode problems are bound to accentuate in the context of changes in cropping patterns with inclinations toward high-value horticultural crops, intensification of agriculture, and climate change. The global warming phenomena will impact PPNs leading to their expeditious development, shortened life cycles, more number of generations, and higher nematode population built-ups. The phenomenon is already evident in the protected cultivated systems vis-a-vis open field systems. The growing application of microirrigation systems is necessity but at the same system amplifies nematode intensities by ensuring adequate moisture levels in soil for their biological activities.

Clean cultivation, farm hygiene, nematode-free planting materials, strict monitoring of horticultural nurseries, and availability of tissue culture plants (wherever possible) have to be adopted on priority basis. The traditional agronomic practices of nematode management will continue to find acceptability. However, the disease diagnosis should find the use of remote sensing techniques for identifying nematode hot

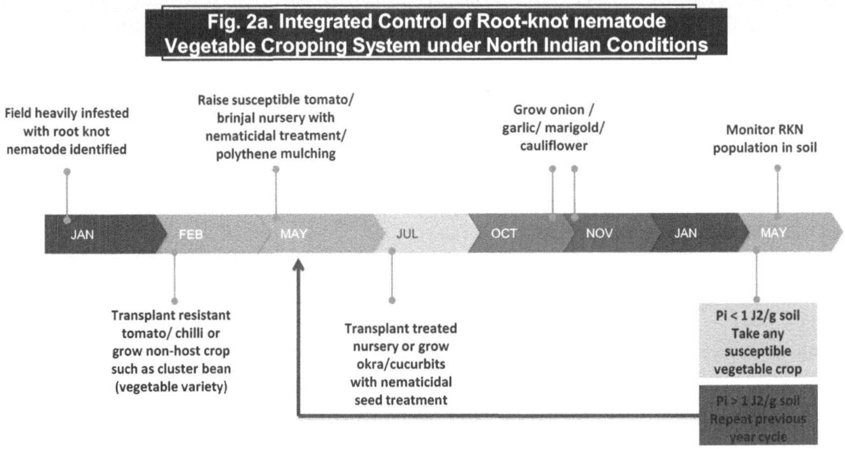

FIGURE 14.2a Integrated control of root-knot nematode vegetable cropping system under North Indian conditions.

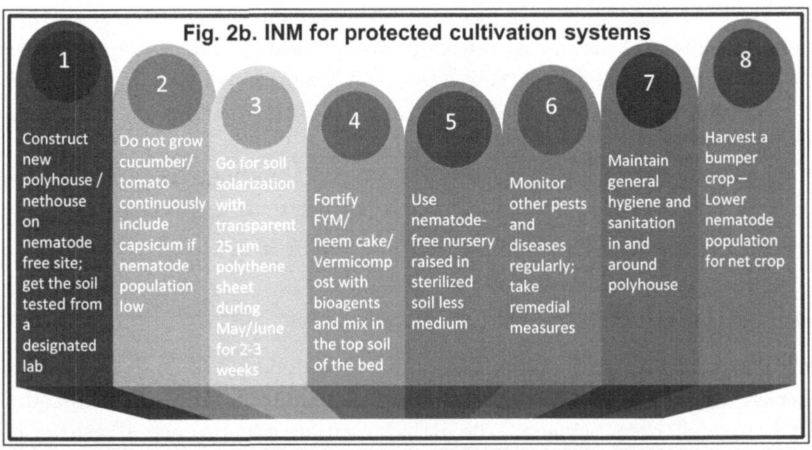

FIGURE 14.2b INM for protected cultivation systems.

spots and application of precision management technologies. Drones may find routine use in the near future for fertilizer and pesticide/herbicide sprays. Nematicidal formulations have to be developed in tune with the drone system. Development of environmentally safe liquid formulations with basipetal mode of action could ideally suit management of soil-borne nematodes and prevent soil contaminations. Application of DNA-based diagnostics could become more relevant in quarantine systems. Recent advances in genomic and proteomic studies may help nematicide development process from screening chemical libraries in vitro to computerized molecule screening and design for some target nematode species. Such a change might reduce the time and cost of development and increase the nematode control efficacy and may even lead to the ideal nematicide. INM systems with more dependence on biological and cultivars with resistant genes are desired to be more effective.

REFERENCES

Abd-Elgabad, M. M. M. Biological control agents of plant parasitic nematodes. *Egyptian Journal of Biological Pest Control* **2016**, 26(2), 423–429.

Ali, M. A.; Azeem, F.; Abbas, A.; Joyia, F. A.; Li, H.; Dababat, A. A. Transgenic strategies for enhancement of nematode resistance in plants. *Frontiers in Plant Science* **2017**, 8, 750.

Baronia, R.; Gupta, P.; Singh, S. P.; Walia, R. K. Silver nanoparticles as a potential nematicide against *Meloidogyne graminicola. Journal of Nematology* **2020**, 52, 1–9. https://doi.org/10.21307/jofnem-2020–002

Bello, A.; Manuel, A.; López-Pérez, A. et al. Biofumigation, fallow, and nematode management in vineyard replant. *Nematropica* **2004**, *34*, 53–64.

Bernard, G. C.; Egnin, M.; Bonsi, C. The impact of plant-parasitic nematodes on agriculture and methods of control. In: Shah, M. M.; Mahamood, M. (Eds), *Nematology: Concepts, Diagnosis and Control*, **2017**, p. 192. https://doi.org/10.5772/intechopen.68958

Bonfil, D.; Dolgin, B.; Mufradi, I.; Asido, S. Bioassay to forecast cereal cyst nematode damage to wheat in fields. *Precision Agriculture* **2004**, 5, 329–344. https://doi.org/10.1023/B:PRAG.0000040804.97462.02

Cabrera, J. A.; Kurz, A.; Sikora, R.A.; Schouten, A. Isolation and characterization of fenamiphos degrading bacteria. *Biodegradation* **2010**, 21, 1017–1027.

Chandra, B.; Khan, M.R. **2011**. Dynamics of soil nematodes in vegetable based crop sequences in West Bengal, India. *Journal of Plant Protection Research* **2011**, 51(1), 7–13.

Chen, S.; MacDonald, D. H.; Kurle, J. E.; Reynolds, D.A. The soybean cyst nematode. University of Minnesota Extension Service, **2001**, https://www.soybeans.umn.edu/pdfs/DC3935.pdf.

Decraemer, W.; Hunt, D. Structure and classification. In: Perry, R.; Moens, M. (Eds), *Plant Nematology*, Oxfordshire: CAB International, **2006**, pp. 3–32. https://doi.org/10.1079/9781845930561.0000

Desaeger, J.; Wram, C.; Zasada, I. New reduced-risk agricultural nematicides—Rationale and review. *Journal of Nematology* **2020**, 52, 1–16. https://doi.org/10.21307/jofnem-2020–091

Dutta, T. K.; Banakar, P.; Rao, U. The status of RNAi-based transgenics in plant nematology. *Frontiers in Microbiology* **2015**, 5, 760. https://doi.org/10.3389/fmicb.2014.00760

Dutta, T. K.; Khan, M. R.; Phani, V. Plant-parasitic nematode management via biofumigation using brassica and non-brassica plants: current status and future prospects. *Current Plant Biology* **2019**, 17, 17–32.

Dutta, T. K.; Papolu, P. K.; Singh, D.; Sreevathsa, R.; Rao, U. Expression interference of a number of *Heterodera avenae* conserved genes perturbs nematode parasitic success in *Triticum aestivum. Plant Science* **2020**, 301, 110670.

Fosu-Nyarko, J.; Jones, M. G. Application of biotechnology for nematode control in crop plants. *Advances in Botanical Research* **2015**, 73, 339–376.

Gururani, M. A.; Venkatesh, J.; Upadhyaya, C. P.; Nookaraju, A.; Pandey, S. K.; Park, S. W. Plant disease resistance genes: current status and future directions. *Physiological and Molecular Plant Pathology* **2012**, 78, 51–65.

Heiken, J. A. **2017**. *The Effects of Fluopyram on Nematodes*. Raleigh: North Carolina State University, https://repository.lib.ncsu.edu/bitstream/handle/1840.20/33746/etd.pdf?sequence=1&isAllowed=y.

Hepher, A.; Atkinson, H. J. *Nematode Control with Proteinase Inhibitors*. United States patent, Patent Number: 5,494,813, Feb. **1996**.

Huang, G.; Allen, R.; Davis, E. L.; Baum, J. T.; Hussey, S. R. Engineering broad root-knot resistance in transgenic plants by RNAi silencing of a conserved and essential root-knot nematode parasitism gene. *PNAS USA* **2006**, 103, 14302–14306.

Jones, J.; Haegeman, A.; Danchin, E. et al. Top 10 plant-parasitic nematodes in molecular plant pathology. *Molecular Plant Pathology* **2013**, 14, 946–961. https://doi.org/10.1111/mpp.12057

Jones, J. T.; Kumar, A.; Pylypenko, L. A. et al. Identification and functional characterization of effectors in expressed sequence tags from various life cycle stages of the potato cyst nematode *Globodera pallida*. *Molecular Plant Pathology* **2009**, 10, 815–828.

Kahn, R. P. Safe guarding the international exchange of plant germplasm. In *10th International Congress of Plant Protection vol 2. Proceedings of a conference held at Brighton*, England, 20–25 November **1983**; 866–872.

Khan, M.R., Shit, S., Pal, A.K. and Biswas, B. Integrated approach for management of foliar nematode, *Aphelenchoides bessyi* infecting tuberose in West Bengal, India. *International Journal of Nematology* **2005**, 15(1), 30–34.

Kumar, V.; Khan, M.R.; Walia, R.K. Crop loss estimations due to plant-parasitic nematodes in major crops in India. *National Academy Science Letters* **2020**, 43, 409–412.

Lilley, C. J.; Davies, L. J.; Urwin, P. E. RNA interference in plant parasitic nematodes: a summary of the current status. *Parasitology* **2012**, 139, 630–640.

Lilley, C. J.; Devlin, F.; Urwin, P. E.; Atkinson, H. J. Parasitic nematodes, proteinases and transgenic plants. *Parasitology Today* **1999**, 15, 414–417. https://doi.org/10.1016/S0169-4758(99)01513-6

Maniruzzaman, S. M.; Vestergård, M. Impacts of root metabolites on soil nematodes. *Frontiers in Plant Science* **2020**, 10, 1792. https://doi.org/10.3389/fpls.2019.01792

Manzanilla-López, R. H.; Costilla, M. A.; Doucet, M. et al. The genus *Nacobbus* Thorne and Allen, 1944 (Nematoda: Pratylenchidae): systematics, distribution, biology and management. *Nematropica* **2002**, 32, 149–227.

Matthews, B. F.; Beard, H.; MacDonald, M. H. et al. Engineered resistance and hypersusceptibility through functional metabolic studies of 100 genes in soybean to its major pathogen, the soybean cyst nematode. *Planta* **2012**, 237, 1337–1357.

Meyer, S. L. F.; Zasada, I. A.; Orisajo, S. B.; Morra, M. J. Mustard seed meal mixtures: management of Meloidogyne incognita on pepper and potential phytotoxicity. *Journal of Nematology* **2011**, 43, 7–15.

Migunova, V. D.; Sasanelli, N. Bacteria as biocontrol tool against phytoparasitic nematodes. *Plants (Basel)* **2021**, 10(2), 389. https://doi.org/10.3390/plants10020389

Nagesh, M.; Reddy, P. P.; Kumar, M. V. V.; Nagaraju, B. M.; Kumar, N. K. K.; Verghese, A. **1998**. Management of *Meloidogyne incognita* on *Polianthes tuberosa* using parasitic fungus, *Paecilomyces lilacinus* in combination with oil cakes. Advances in IPM for horticultural crops. *Proceedings of the First National Symposium on Pest Management in Horticultural Crops: environmental implications and thrusts*, Bangalore, India, **15–17 October 1997, (publ. 1998)**, pp. 339–344.

Nicol, J.; Elekçioğlu, H.; Bolat, N.; Rivoal, R. The global importance of the cereal cyst nematode (*Heterodera* spp.) on wheat and international approaches to its control. *Communica tions in Agricultural and Applied Biological Sciences* **2004**, 72, 677–686.

Nicol, J.; Turner, D.; Coyne, L.; den Nijs, L.; Hockland, S.; Maafi, Z. Current nematode threats to world agriculture. In: Jones, J., Gheysen, G., Fenoll, C. (Eds), *Genomics and Molecular Genetics of Plant-Nematode Interactions*, Berlin: Springer Science Business Media, **2011**, pp. 21–43. https://doi.org/10.1007/978-94-007-0434-3_2

Norton, D.; Hoffmann, J. *Longidorus breviannulatus* n. sp. (Nematoda: Longidoridae) associated with stunted corn in Iowa. *Journal of Nematology* **1975**, 7, 168–171.

Nose, M.; Shiraishi, S. Breeding for resistance to pine wilt disease In: Futai, K.; Sutherland, J. (Eds), *Pine Wilt Disease*, Tokyo: Springer, **2007**, pp. 334–350.

Oka, Y. From old-generation to next-generation nematicides. *Agronomy* **2020**, 10, 1387. https://doi.org/10.3390/agronomy10091387

Opperman, C. H.; Taylor, C. G.; Conkling, M. A. Root-knot nematode-directed expression of a plant root-specific gene. *Science* **1994**, 263, 221–223.

Poornima, K.; Walia, R. K. Horticultural nurseries and nematode dissemination. *Current Science* **2021**, 120(2), 278–284.

Renco, M. Organic amendments of soil as useful tools of plant parasitic nematodes control. *Helminthologia* **2013**, 50(1), 3–14.

Ripoll, C.; Favery, B.; Lecomte, P. et al. Evaluation of the ability of lectin from snowdrop (*Galanthus nivalis*) to protect plants against root-knot nematodes. *Plant Science* **2003**, 164(4), 517–523. https://doi.org/10.1016/S0168-9452(02)00448-X

Sikora, R. A.; Roberts, P. **2018**. Management practices: an overview of integrated nematode management technologies. In: Sikora, R. A.; Coyne, D.; Hallmann, J.; Timper, P. (Eds), *Plant Parasitic Nematodes in Subtropical and Tropical Agriculture*, Wallington, UK: CAB International, **2018**, pp. 795–838.

Smiley, R.; Guiping, Y. *Cereal Cyst Nematodes. Biology and Management in Pacific Northwest Wheat, Barley, and Oat Crops*. Washington, DC: A Pacific Northwest Extension Publication Oregon State University, **2010**. https://ir.library.oregonstate.edu/xmlui/bitstream/handle/1957/18917/pnw620.pdf

Sparks, T. C. Insecticide discovery: an evaluation and analysis. *Pesticide Biochemistry and Physiology* **2013**, 107, 8–17.

Thoden, T. C.; Korthals, G. W.; Termorshuizen, A. J. Organic amendments and their influences on plant-parasitic and free-living nematodes: a promising method for nematode management? *Nematology* **2011**, 13(2), 133–153.

Tian, B.; Yang, J.; Zhang, K. Bacteria used in the biological control of plant-parasitic nematodes: populations, mechanisms of action, and future prospects. *FEMS Microbiology Ecology* **2007**, 61, 197–213. https://doi.org/10.1111/j.1574–6941.2007.00349.x

Tranier, M-S.; Pognant-Gros, L.; Quiroz, R. C.; González, C.N.A.; Mateille, T.; Roussos, S. Commercial biological control agents targeted against plant-parasitic root-knot nematodes. *Brazilian Archives of Biology and Technology* **2014**, 57(6), 831–841. https://doi.org/10.1590/S1516-8913201402540

Trudgill, D.; Blok, V. Apomictic, polyphagous root-knot nematodes: exceptionally successful and damaging biotrophic root pathogens. *Annual Review of Phytopathology* **2001**, 39, 53–77. https://doi.org/10.1146/annurev.phyto.39.1.53

Turner, S. J.; Rowe, J. A. Cyst nematodes. In: Perry, R. N.; Moens, M. (Eds), *Plant Nematology*, Wallingford, Oxfordshire: CAB International, **2006**, pp. 90–122.

Valdes, Y.; Viaene, N.; Moens, M. Effects of yellow mustard amendments on the soil nematode community in a potato field with focus on *Globodera rostochiensis*. *Applied Soil Ecology* **2012**, 59, 39–47.

Vanstone, V. A.; Hollaway, G. J.; Stirling, G. R. Managing nematode pests in the southern and western regions of the Australian cereal industry: continuing progress in a challenging environment. *Australasian Plant Pathology* **2008**, 37, 220–234.

Vishnudasan, D.; Tripathi, M. N.; Rao, U.; Khurana, P. Assessment of nematode resistance in wheat transgenic plants expressing potato proteinase inhibitor (pin2) gene. *Transgenic Research* **2005**, 14(5), 665–675. https://doi.org/10.1007/s11248-005-5696-4

Walia, R. K.; Khan, M. R. *A Compendium of Nematode Diseases in Crop Plants*. Project Coordinating Cell, All India Coordinated Research Project on Nematodes in Cropping Systems, New Delhi, **2018**, 106 pp.

Wesemael, W.; Viaene, N.; Moens, M. Root-knot nematodes (*Meloidogyne* spp.) in Europe. *Nematology* **2011**, 13(1), 3–16. https://doi.org/10.1163/138855410X526831

Williamson, V. M.; Kumar, A. Nematode resistance in plants: the battle underground. *Trends in Genetics* **2006**, 22(7), 396–403.

Xu, Z.; Zhao, Y.; Yang, D.; Sun, H.; Zhang, C.; Xie, Y. Attractant and repellent effects of sweet potato root exudates on the potato rot nematode, *Ditylenchus destructor*. *Nematology* **2015**, 17, 117–124. https://doi.org/10.1163/15685411-00002856

Yadav, B. C.; Veluthambi, K.; Subramaniam, K. Host-generated double stranded RNA induces RNAi in plant-parasitic nematodes and protects the host from infection. *Molecular and Biochemical Parasitology* **2006**, 148, 219–222.

Zhang, S.; Zhang, S.; Wang, H.; Chen, Y. Characteristics of sweet potato stem nematode in China. *Acta Phytopathology Sinica* **2006**, 36, 22–27.

15 Plant Biostimulants
Reinvigorating Agriculture

Pranjib K. Chakrabarty, Yasmin, and Selvakumaran S.

15.1 INTRODUCTION

Deeply rooted in Indian society, agriculture plays a crucial role in both the country's economy and food security. Modern Indian agriculture grapples with a multitude of challenges—a growing population, environmental degradation, and changing climate. The economic damage due to abiotic stress is severe and often underestimated. Climate-change-induced stresses can result in a drastic reduction of up to 70% crop yield, while edaphic stresses such as nutrient deficiencies and salinity further compound challenges in crop production. Besides, crop pests cause significant losses to production and productivity of crops. As per FAO estimate, the global losses due to biotic stresses range between 20% and 40% annually (FAO, 2025). While solutions that address abiotic stress are often overlooked, the management of pests is done predominantly using agrochemicals. At a time when every now and then one or the other concerns keep arising in agricultural front from using agrochemicals, it is time to prevent the abiotic and biotic losses through the effective use of biologicals. Moreover, under the clarion call of the Government of India, whereby natural farming and organic farming are being actively advocated, the biostimulant is an ecofriendly technology and is expected to be compliant in stress alleviation and plant growth promotion. While farming under challenging conditions, biostimulant is a compelling case for science-based development, which may lead to identification of novel molecules and phenomenon, pathways and processes that would not have been discovered, if the category of biostimulant did not exist and legitimized by Govt of India. It's a matter of pride for the country to be among the first in the world to develop regulatory guidelines for registration of biostimulants. However, the mainstreaming of biostimulants for their seamless and profitable use in agriculture can be done by bridging the gaps in the regulation of the biostimulant notification (No 818 dated 22.02.2019). Based upon a series of scientific deliberations, discussions, suggestions among scientists from the public and private industries, international experts on biostimulants and public opinion, certain logical considerations have been done by the regulators in third amendment (S.O. 1963 (E). dated 08.05.2024). The requirement of long-term chronic studies as in case of pesticides was omitted from the initial guidelines for biostimulants registration. For instance, in case of protein hydrolysates, seaweed (*Ascophyllum nodosum, Saragassum, Kappaphycus, Gracilaria,* and *Ulva*), amino acids, vitamins, humic and fulvic acid, no toxicology data are required to be submitted.

DOI: 10.1201/9781003531944-15

The manufacturer/importer should submit an affidavit to the effect that their product is nontoxic and shall be safely used as biostimulants. Such amendments and other considerations from time to time would make this promising technology more farmer- and industry-friendly and the regulatory conditions more enabling to promote the sector in increasing agricultural production and enhancing farmer's income. This chapter on biostimulants discusses about their use in agriculture in India, their registration, mode of action, basic difference between biostimulants, biopesticides, biofertilizers, and PGPR, challenges, and opportunities associated with them.

15.2 WHAT ARE BIOSTIMULANTS?

Biostimulants are legally defined as

> A substance or microorganism or a combination of both whose primary function when applied to plants, seeds or rhizosphere is to stimulate physiological processes in plants and to enhance its nutrient uptake, growth, yield, nutrition efficiency, crop quality and tolerance to stress, regardless of its nutrient content, but does not include pesticides or plant growth regulators which are regulated under the Insecticide Act, 1968.

These are naturally occurring, microbially derived, and other substances of biological origin, which are known to stimulate the metabolic pathways for increasing growth of plants, the quality of produce and enhance resilience against abiotic, edaphic, and other stresses. The exact mode of action of biostimulants varies depending on the specific type and its composition. These mechanisms are depicted in Figure 15.1, and some of them are indicated below:

- **Enhanced nutrient uptake:** Biostimulants can improve root development and function, facilitate nutrient mobilization in the soil, and influence the expression of genes related to nutrient transporters.

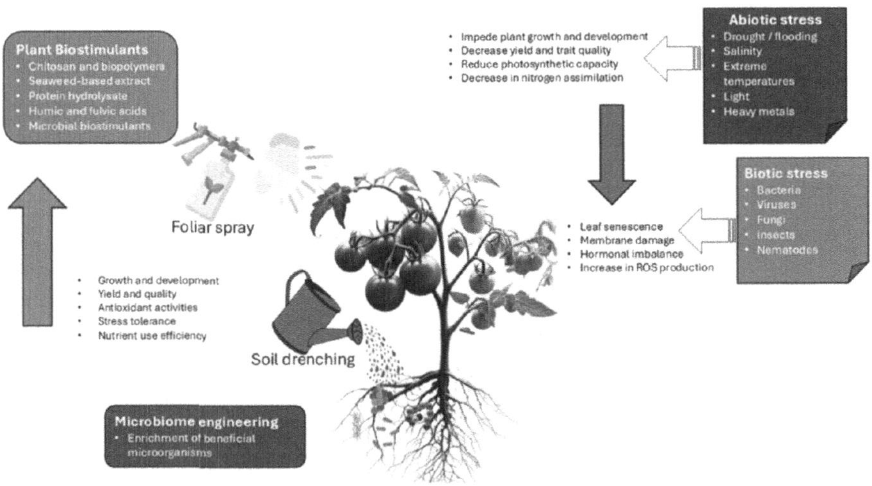

FIGURE 15.1 Biostimulants, mode of action and potential to mitigate plant stresses.

- **Improved stress tolerance:** By influencing hormone production and antioxidant activity, biostimulants can enhance plant tolerance to abiotic stresses such as drought, salinity, and heat stress.
- **Boosted plant defense mechanisms:** Biostimulants can trigger the innate defense mechanism of plants, enhancing tolerance against biotic stresses (pest and diseases).
- **Modulating plant hormone production and signaling pathways:** They can stimulate innate metabolic functions including increase in chlorophyll production, enhancing CO_2 assimilation, and improving overall photosynthetic efficiency, thereby enhancing quality and productivity.

15.3 CLASSIFICATION OF PLANT BIOSTIMULANTS

India's draft guidelines, outlined in the FCO Amendment Order 2021, categorized biostimulants based on their mode of action specified in Schedule VII (Biostimulants Draft Guidelines, 2021). These are broadly classified as below (Table 15.1):

a. Botanical extracts, seaweed extracts
b. Biochemicals
c. Protein hydrolysates and amino acids
d. Vitamins
e. Cell-free microbial products
f. Antioxidants
g. Anti-transpirants
h. Humic and fulvic acid and their derivatives
i. Live microbes

TABLE 15.1
Categories, Subgroups and Active Ingredients of Plant Biostimulants[a]

S. No	Biostimulants Category	Subgroups	Description of the Category with Active Ingredients
1	Botanical extracts, including seaweed extracts	Botanical extracts	• Extracts from plants (belonging to the kingdom plantae with five subgroups: (i) Thallophyta, (ii) Bryophyta, (iii) Pteridophyta, (iv) Gymnosperms and (v) Angiosperms) • Active ingredients: Several different molecules consisting of mixtures of primary and secondary metabolites and plant hormones.
		Seaweed extracts	• Seaweed extracts are a sub-category of botanical extracts. Seaweeds are plants belonging to subgroup of plantae called Thallophyta.

(Continued)

TABLE 15.1 (*Continued*)
Categories, Subgroups and Active Ingredients of Plant Biostimulants[a]

S. No	Biostimulants Category	Subgroups	Description of the Category with Active Ingredients
			• It includes Red (Rhodophyceae), brown (Phaeophycea) and green (Chlorophycea) algae. Composition of biostimulant depends upon the species from which it was prepared and can also vary to some extent with the culture conditions used for cultivation of seaweeds. • Active ingredients: Several different molecules consisting of mixtures of primary and secondary metabolites, polymers and plant hormones.
2	Biochemicals	Specific biomolecule and their combinations	• Biochemicals are the molecules that are synthesized/produced in the metabolism of the living organisms. • These naturally occurring biochemicals can also be synthesized by chemical synthesis in laboratories. • Active ingredients: One or more biochemicals and their combinations.
3	Protein hydrolysates and amino acids	Protein hydrolysates	• Protein hydrolysates are mixtures of amino acids and peptides obtained from chemical or biological break-down of proteins from plant and animal sources. • Active ingredients: free amino acids and polypeptides
		Amino acids	• An amino acid is a carboxylic acid-containing an aliphatic primary amino group in the α position to the carboxyl group, and an organic R group (or side chain) that is unique to each amino acid. • The following 20 amino acids that are building blocks of proteins in plants and animals: 1. Alanine; 2. Arginine; 3. Asparagine;4. Aspartic Acid; 5. Cysteine; 6. Glutamic acid; 7. Glutamine; 8. Glycine; 9. Histidine; 10. Isoleucine; 11. Leucine; 12. Lysine; 13. Methionine; 14. Phenylalanine; 15. Proline; 16. Serine; 17. Threonine; 18. Tryptophan; 19. Tyrosine; 20. Valine • Active ingredients: One or more Amino acids including NPAAs.
4	Vitamins (The specific vitamin listed are the active ingredient in each subgroup)	Coenzyme Q10 (CoQ10)	• CoQ10is 2,3-dimethoxy, 5-methyl, 6-decaprenyl benzoquinone, anisoprenylated benzoquinone and is also known as ubiquinone or ubiquinone-10. • It is abundant in microbes, plants and animals. Involved in electron transport in mitochondrial and chloroplast electron transport chain. It is a lipid soluble antioxidant.
		Vitamin A	• Fat-soluble. • This group includes retinol, retinal (also known as retinaldehyde), retinoic acid, and several provitamin A, carotenoids (βcarotene); Plants are rich source of β-carotene, the pro-vitamin A.

(*Continued*)

TABLE 15.1 (*Continued*)
Categories, Subgroups and Active Ingredients of Plant Biostimulants[a]

S. No	Biostimulants Category	Subgroups	Description of the Category with Active Ingredients
		Vitamin B1 (thiamine)	• Water-soluble, Function as coenzyme in essential factor in central carbon metabolism. It is produced by all plants and most microbes. • This group consists of Thiamine, TMP (Thiamine monophosphate), TDP (Thiamine diphosphate), TTP (Thiamine triphosphate), ATDP (Adenosine thiamine diphosphate) and ATTP (Adenosine thiamine triphosphate).
		Vitamin B2 (riboflavin)	• Water-soluble, Function as coenzyme. • It is a precursor for the coenzymes, flavin adenine dinucleotide (FAD) and flavin mononucleotide (FMN), which are mostly involved in redox reactions. • It is produced by all plants and most microbes.
		Vitamin B3 (niacin)	• Water-soluble, Function as coenzyme. • It consists of niacin and niacinamide, also known as nicotinic acid (NA) and nicotinamide (Nam), respectively. • Vitamin B3 are biosynthetic precursors to nicotinamide adenine dinucleotide (NAD+) and nicotinamide adenine dinucleotide phosphate (NADP+). • It plays a key role in intermediary metabolism, mitochondrial respiration (Electron acceptor in electron transport chain and the Krebs' cycle), reactive oxygen species generation and inhibition, and post-translational protein modifications, protein regulation and second messengers' generation.
		Vitamin B5 (pantothenic acid)	• Water-soluble, Function as coenzyme. • Pantothenic acid is the precursor of the 4'-phosphopantetheine moiety of coenzyme A and acyl-carrier protein. • It is synthesized in plants and microbes. It is a dietary requirement for animals.
		Vitamin B6 (pyridoxine, pyridoxal, pyridoxamine)	• Water-soluble, Function as coenzyme. • It consists of six vitamers: pyridoxal (PL), pyridoxine (PN), pyridoxamine (PM), and their phosphorylated derivatives. Pyridoxal-5′-phosphate (PLP) is a cofactor for over 140 biochemical reactions involved in amino acid, sugar, and fatty acid metabolism. • Plants and microbes synthesize vitamin B6, but it is a dietary requirement for animals.
		Vitamin B8 (H-biotin)	• Water-soluble, Function as coenzyme. • It is covalently bound to a Lys residue of a group of enzymes that catalyze carboxylation, decarboxylation or trans-carboxylation reactions.

(*Continued*)

TABLE 15.1 (*Continued*)
Categories, Subgroups and Active Ingredients of Plant Biostimulants[a]

S. No	Biostimulants Category	Subgroups	Description of the Category with Active Ingredients
			• Biotin containing carboxylases of plants are acetyl-CoA carboxylases, 3-methylcrotonyl-CoA carboxylase and geranoyl-CoA carboxylase. • Plants also contain a noncatalytic biotin protein that accumulates in seeds.
		Vitamin B9 (folate)	• Water-soluble, Function as coenzyme. • Naturally occurring folates are dihydrofolate (DHF) and tetrahydrofolate (THF), and the latter is ready to use form. • Folates are synthesized in microbes and plants. But humans and other vertebrates are fully depending on their diet for folate supply.
		Vitamin B12 (cobala-min)	• Water-soluble, function as coenzyme. • It is synthesized by certain bacteria and archaea. Plants and animals do not synthesize B12.The red algae Porphyra sp. and green alga Chlorella sp. produce biologically active B12. • Vitamin B12 (cyanocobalamin) is a corrinoid that contain a corrin ring. Hydroxocobalamin, methylcobalamin, and 5′-deoxyadenosylcobalamin also act as coenzymes.
		Vitamin C (Ascorbic acid)	• Water-soluble, Plants synthesizes Vitamin C, but humans cannot. • It is an antioxidant, and a cofactor of peptidylglycine alpha-amidatingmonooxygenase involved in the biosynthesis of many signalling peptides, such as oxytocin, vasopressin, cholecystokinin, and calcitonin. • It is also a cofactor for many dioxygenases (reducing the iron in the active site of these enzymes to Fe^{2+}), methylcytosinedioxygenases.
		Vitamin D	• Fat-soluble, the two forms of vitamin D are cholecalciferol (vitamin D3) and ergo-calciferol (vitamin D2). • Vitamin D2 is produced in fungi and yeast from ergosterol upon UVB- exposure, while vitamin D3 from 7-dehydrocholesterol in the skin upon UVB-exposure (290–315 nm). 7-DHC is also synthesized by some plants such as tomato, which can be converted to Vitamin D3 upon UV exposure.
		Vitamin E (α-Tocopherol)	• Fat-soluble, it consists of tocopherols and tocotrienols. Vitamin E is synthesized solely by photosynthetic organisms. α-Tocopherol is the predominant form of vitamin E green parts γ-tocopherol is the major in non-photosynthetic tissues of plants.

(*Continued*)

TABLE 15.1 (*Continued*)
Categories, Subgroups and Active Ingredients of Plant Biostimulants[a]

S. No	Biostimulants Category	Subgroups	Description of the Category with Active Ingredients
		Vitamin K	• Fat-soluble, it consists of Vitamin K1 (Phylloquinone) and K2 (menaquinone). • Phylloquinone is a prenylated naphthoquinone synthesized by plants, green algae, and some cyanobacteria; It act as electron carrier in photosystem I and electron acceptor for the formation of protein disulfide bonds. • Menaquinone is essential in electron transport and ATP generation in all Gram-positive, and anaerobic Gram-negative bacteria.
5	Cell free Microbials products	Cell free Microbials products	• Extracts from microbes which are non-pathogenic and microbes that are not having bio-control activity. Consists of large number of different molecules. CFU should be NIL. • Active ingredients: Several different molecules consisting of mixtures of primary and secondary metabolites and hormones
6	Antioxidants	Enzymatic antioxidants	• Antioxidants are molecules that scavenge reactive oxygen species (ROS) such as superoxide and hydroxyl radicals. It can be classified in to enzymatic and non-enzymatic antioxidants. • Total antioxidant capacity of such biostimulants can be determined by Trolox equivalence antioxidant capacity (TEAC) assay, ferric ion reducing antioxidant power (FRAP) assay, oxygen radical absorbance capacity (ORAC) assay, inhibiting the oxidation of low-density lipoprotein (LDL) assay, cellular antioxidant activity assay, etc. • Active ingredients: Enzymatic antioxidants: SOD, Catalase, Peroxidase.
		Nonenzymatic antioxidats	• Active ingredients: One or more of the following molecules - Glutathione, polyphenols (phenolic acids, flavonoids, anthocyanins, lignans and stilbenes), carotenoids (xanthophylls and carotenes), vitamins [vitamin E (α-Tocopherol) and C (Ascorbic acid)], etc.
7	Anti-transpirants	Molecules that close stomata	• Anti-transpirants are molecule that reduce transpiration through stomatal closure, increasing resistance to water vapor loss and/or reducing the heat load on leaf. • Active ingredients: - Molecules that close stomata: Abscisic acid (ABA); Phenyl mercuric acetate (PMA)
		Film forming molecules that increase resistance to water vapor loss and increase reflectance	• Active ingredients: One or more of the following molecules Chitosan; Waxes; Alumino-silicate (kaolin); Calcium carbonate ($CaCO_3$); Calcium oxide (CaO); di-1-p-menthene (pinolene); poly-1-p-menthene; Acrylic polymers; Potassium sulfate (K_2SO_4)

(Continued)

TABLE 15.1 (*Continued*)
Categories, Subgroups and Active Ingredients of Plant Biostimulants[a]

S. No	Biostimulants Category	Subgroups	Description of the Category with Active Ingredients
8	Humic & fulvic acid and their derivations	Humic acids (HA) Fulvic acids (FA) Humin	• Humic substances (HS) are components of the natural organic matter in soil, water and geological organic deposits formed by the process of humification. • HS are grouped in to three groups: humic acids (HA or HAs), fulvic acids (FA or FAs) and humin. HA are insoluble below pH 2, fulvic acids are soluble at any pH, and humin are insoluble in water. • Active ingredients: humic acids (HA or HAs), fulvic acids (FA or FAs) and humin.
9	Live Microbes		

[a] Microorganisms such as mycorrhizal and non-mycorrhizal fungi, bacterial endosymbionts, and plant growth promoting rhizobacteria).

Microbial biostimulants often function through modulating growth hormone and increasing efficiency of signal transduction and other signaling mechanisms (Figure 15.2). However, there is a thin line of difference between Plant Growth Regulators (PGRs), Plant Growth Promoting Rhizobacteria (PGPR), Plant Growth Promoting Phyllosphere (PGPP), endophytes, etc., all of which play a role in enhancing plant productivity, product quality, nutrient use efficiency and mitigate abiotic stresses, etc., but they work in distinct ways (Table 15.2).

Many microbes can function under multiple categories. For example, strains of *Pseudomonas fluorescens* can act as PGPR, PGR, PGPP, producing plant growth promoting substances. In such cases, different strains of a microbe are often categorized

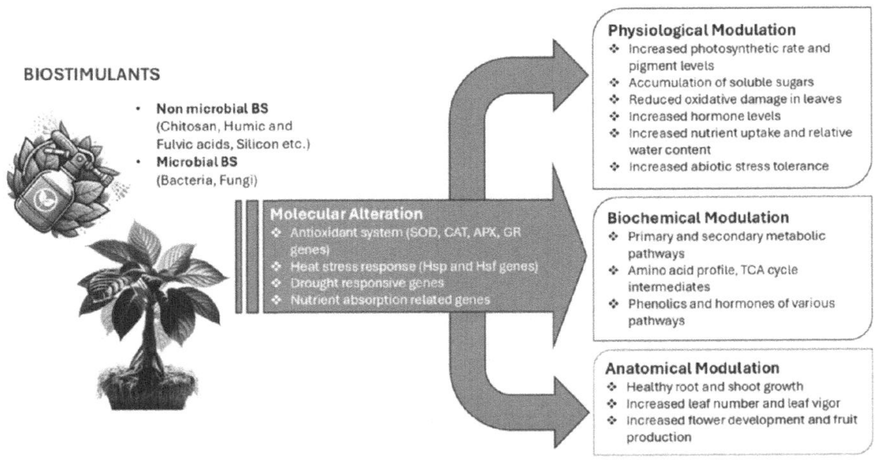

FIGURE 15.2 Interplay of metabolic functions in expression of various phenotypes by the biostimulants.

TABLE 15.2
Basic Differences between Biostimulants, PGRs, and PGPRs

Feature	Biostimulants	PGR	PGPR
Function	Enhance plant growth, stress tolerance, and nutrient uptake indirectly by stimulating natural physiological processes. They don't directly provide nutrients.	Directly influence specific plant growth processes such as cell division, stem elongation influences higher branching, early and increased flowering, fruit formation etc.	Enhance plant growth indirectly through beneficial interactions with the plant's root system. They can fix nitrogen, solubilize phosphorus, or promote root development.
Composition	Diverse range of organic materials such as humic substances, seaweed extracts, or beneficial microbes (not directly supplying nutrients).	Synthetic or naturally occurring hormones or hormone-like substances.	Specific strains of live bacteria that colonize the plant's rhizosphere (root zone).
Impact	Improved overall plant health and resilience, potentially leading to higher yields and better-quality produce.	Precise control over specific aspects of plant growth and development. Used for tasks such as delaying fruit ripening, promoting branching, or inducing flowering.	Increased nutrient availability and potentially improved plant growth and stress tolerance.
Mode of Action	Varied depending on the specific biostimulant. Examples include promoting root development, improving nutrient uptake efficiency, enhancing stress tolerance mechanisms.	Mimic or manipulate natural plant hormones to achieve desired effects.	Form symbiotic relationships with plants, providing them with nutrients or stimulating growth-promoting mechanisms.
Examples	Seaweed extracts, *Trichoderma viride*, *Bacillus subtilis* etc.	*Gibberella fujikuroi*, *Agrobacterium tumefaciens* etc.	*Pseudomonas fluorescens*, *Rhizobium leguminosarum*, *Azospirillum brasilense* etc.

based on their function. However, a controversy may arise where a particular strain possesses dual mode of action, for example, pesticidal and plant growth promotion. This may leave a chance where a registrant may choose to register the strain as a "biostimulant," ignoring its dual function of "biopesticide." The former category in India is registered under FCO, 1985 schedule VI while the biopesticides are registered under the Insecticides Act, 1968, which needs stringent toxicological tests.

Also, there could be confusion between the microbes that qualify themselves to be classified in different categories, namely biostimulants, biofertilizers, and biopesticides. While the microbial biostimulants do not function as a source of nutrition for the plants as biofertilizers on the one hand, they also do not share the pesticidal

TABLE 15.3
The Major Differences of Biofertilizers and Biopesticides

Feature	Biofertilizer	Biopesticide
Primary function	Enhance soil fertility and nutrient availability for plants by introducing live microorganisms (bacteria, fungi) that fix nitrogen, solubilize phosphorus, or promote nutrient cycling	Control pests and diseases in a natural way
Composition	Live microorganisms (bacteria, fungi) that colonize the rhizosphere (root zone) and improve nutrient availability for the plant	Living organisms such as bacteria, fungi, nematodes, or plant extracts that directly kill, repel, or disrupt the development of pests
Impact	Increased nutrient uptake and potentially reduced reliance on chemical fertilizers	Reduced pest damage and a more sustainable approach to pest management compared to chemical pesticides

Source: https://www.sciencedirect.com/science/.

properties of biopesticides, on the other hand. The major differences of biostimulants with biofertilizers and biopesticides are indicated below (Table 15.3).

15.4 CURRENT SCENARIO OF BIOSTIMULANTS – WORLD VS INDIA

India, with its vast agricultural landscape, presents a significant opportunity for biostimulants.

The Indian biostimulants market is projected to grow at a CAGR of 15%–16%, driven by factors such as:

- Rising concerns about soil health and water scarcity.
- Promoting greener agriculture through judicious use of agrochemicals and adopting GAPs.
- Government initiatives promoting sustainable agriculture.

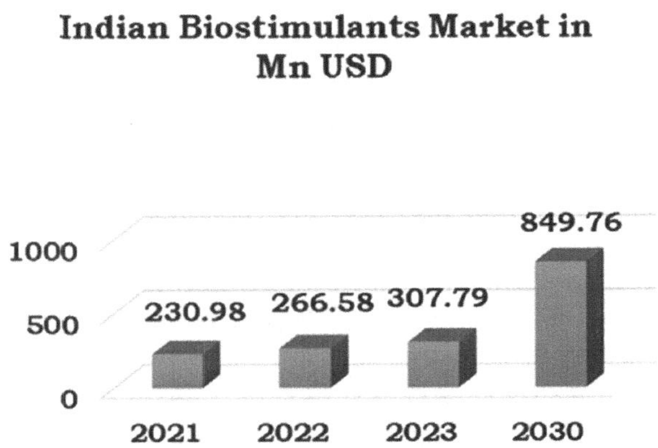

The global biostimulants market size was valued at USD 3.91 billion in 2023. It is projected to reach around USD 11.71 billion by 2033, growing at a CAGR of 11.6% in a time frame of 10 years from 2024 to 2033. While the European Union's biostimulants market size is biggest in the world, that is estimated at USD 1.49 billion in 2023 and is predicted to be worth around USD 4.63 billion by 2033. It is expected to grow at a CAGR of 12% during the same period. The region's growth is due to increased investment in organic and natural farming and a growing demand for organic food.

The US biostimulants market was valued at $0.49 billion in 2022 and is expected to grow to $0.53 billion in 2024 and $1.01 billion by 2029. In the Asia-pacific region, China's biostimulants market size is estimated at 0.21 billion USD in 2024, which is expected to reach 0.38 billion USD by 2029, growing at a CAGR of 12.10% during the forecast period. In India until 2020, biostimulants were not regulated under the statutory laws. Thereafter, the trade of biostimulants gained momentum influenced by biosafety consciousness The Indian biostimulants market is comparatively growing at a higher rate than China, size is projected to grow from $0.30 billion in 2023 to $0.84 billion by 2030, at a CAGR of 15.61%. Downloaded from: https://www.precedenceresearch.com/biostimulants-market as on 01/10/2024.

15.5 POLICIES AND REGULATIONS ON BIOSTIMULANTS IN INDIA

Biostimulants are a relatively new category of agricultural input in India, but the government has taken keen interest to regulate and promote their use. The Fertilizer (Inorganic, Organic, or Mixed) Control Amendment Order 2021, also known as the FCO Amendment Order 2021, established the regulatory framework for biostimulants in India. This order requires biostimulants to be registered before they can be sold in the country. Manufacturers must submit data on the product's composition, effectiveness on crops (bio-efficacy), and potential risks to the environment and human health (toxicity). The process of adding new category of biostimulants is dynamic and will continue to be included in the Schedule VI of the FCO, 1985. The CBC is presently engaged in examining the formulations of biostimulants that are manufactured and submitted for registration by the industries for their notification. Through fourth and fifth amendments of the FCO orders, dated 12 September 2024 and 26 May 2025 respectively, the government approved 45 different formulations under each of the 9 existing categories. Further examination of the remaining proposals submitted by the industries under provisional registration are being done for inclusion in the order and grant of permanent registration under schedule VI of the FCO, 1985.

15.6 KEY FEATURES OF BIOSTIMULANTS POLICY

- **Separate regulations:** India is one of the few countries to have specific regulations for biostimulants, highlighting its focus on promoting sustainable agriculture.

TABLE 15.4
Policies and Regulations on Biostimulants in World

Country	Regulations	Products Registered
USA	• Biostimulants are not currently regulated at the federal level in the USA. However, some states have their own regulations for biostimulants. • For example, California requires biostimulants to be registered with the California Department of Pesticide Regulation (CDPR) before they can be sold in the state. The CDPR has a list of approved biostimulant products, which is updated regularly.	According to the CDPR, there are over 300 biostimulant products registered in California
EU	• Biostimulants are regulated in the EU under the European Fertilisers Regulation (EC) No 2003/2003. • To be marketed in the EU, biostimulants must be included in the EU Fertilisers List. • The EU Fertilisers List is a list of all fertilizers that are authorized for use in the EU. It is updated regularly.	According to the Fertiliser Products Regulation, there are over 200 biostimulant products included in the EU Fertilisers List
Australia	• Biostimulants are regulated in Australia under the Agricultural and Veterinary Chemicals Act 1994. • To be marketed in Australia, biostimulants must be registered with the Australian Pesticides and Veterinary Medicines Authority (APVMA).	According to the APVMA, there are over 100 biostimulant products registered in Australia

- **Registration and data requirements:** The registration process ensures a level of quality and safety for the products used by farmers.
- **Focus on safety:** The recent amendments in 2024 established strict testing procedures to assess potential environmental and human health risks.

The government notification with amendments for registration of biostimulants is issued through Fertiliser (Inorganic, Organic or Mixed) Control (Third) Amendment Order, 2024 dated 8th May 2024.

Large number of countries in the world do not have regulatory guidelines to govern use of biostimulants. Advance countries that regulate biostimulants differ in their classification under different statutory acts. Some countries have included biostimulants under fertilizers while some placed them under the pesticides act. Some of the countries that have regulations for registration of biostimulants are indicated in the Table 15.4.

15.7 IS THERE ANY DATA PROTECTION POLICY FOR BIOSTIMULANTS?

Most countries rely on existing data protection regulations for agricultural products when dealing with biostimulants. These regulations typically focus on ensuring data accuracy and integrity of information submitted for registration or marketing purposes, protecting confidential business information (trade secrets) related to biostimulants composition or manufacturing processes, and safeguarding personal information collected during biostimulants research or field trials (e.g., farmer data).

- **EU (European Union):** The General Data Protection Regulation (GDPR) is a robust framework that applies to data protection across various sectors, including agriculture. Companies involved in biostimulants research or marketing within the EU must comply with GDPR.
- **US (United States):** The Federal Insecticide, Fungicide, and Rodenticide Act (FIFRA) governs data protection for pesticides, which can include some biostimulants. Additionally, the Toxic Substances Control Act (TSCA) might apply depending on the biostimulants' composition.
- **Australia:** There isn't currently a specific data protection regulation solely for biostimulants in Australia. However, data related to biostimulants likely fall under the purview of a couple of existing frameworks:
  - **The Gene Technology Act 2000 (and its regulations):** If a biostimulant contains genetically modified components, data related to its development and safety would be subject to this Act.
  - **The Australian Pesticides and Veterinary Medicines Authority (APVMA) and its data requirements:** If a biostimulant claims pest control abilities, data on its efficacy and safety would likely need to be submitted to the APVMA for registration, following their data requirements.
  - **General data protection principles.**
- **India:** Unlike many other countries, India doesn't have a specific data protection law for biostimulants. However, relevant regulations include:
  - **The Patents Act, 1970:** Protects intellectual property related to the development and production of novel biostimulants.
  - **The Seed Act, 1966 (and subsequent amendments):** Might be applicable if biostimulants contain microbial components and require specific testing protocols.
  - **Draft Personal Data Protection Bill, 2021:** This proposed legislation, if enacted, could establish a comprehensive data protection framework in India, potentially impacting how biostimulants data are handled.

15.8 CHALLENGES AND OPPORTUNITIES IN THE BIOSTIMULANTS INDUSTRY

- **Products of dual use:**

  Insecticide schedules contain biomolecules and chemicals such as brassinolides, blue-green algae extracts, vitamins, chitosan oligosaccharides, and growth regulators such as gibberellins, IAA, 2–4D, cytokinin, and others, which also share biostimulant properties. These substances should be excluded from pesticide schedules and allowed to be used in biostimulants to their natural limits in plants when used as biostimulants.

Their use beyond this limit for controlling pests if any, may, however, be regulated under the Insecticides Act, 1968. Until the upper limits of such innate dual-use products within the live plants are determined, biostimulants may have provisions to contain dual-use substances.

- **Fortification with micronutrients:**
  Several countries allow the fortification of biostimulants with micronutrients. This practice is becoming increasingly popular as it offers a sustainable and effective way to improve plant growth under stressed conditions, yield, quality, nutrient use efficiency, etc. Micronutrients are not always used as a direct source of plant nutrition. Most of the micronutrients form the integral components of the enzyme systems of plants influencing their metabolism. Micronutrients play important roles in redox reactions, photosynthesis, and various other biochemical pathways by modulating the function of enzymes. Some of the other functions of the micronutrients are mentioned below:
  - Chlorine (Cl) is primarily involved in osmotic and cation neutralization.
  - Copper (Cu), Iron (Fe), and Molybdenum (Mo) are active in the oxidation-reduction reactions.
  - Nitrogen (N) undergoes many transformations in plants and the soil. Zinc (Zn) and Manganese (Mn) are essential to facilitate such N transformations in plants.
  - Boron (B) is involved in cell division and seed formation.
  - Cobalt (Co) is essential for the growth of symbiotic microorganisms, such as rhizobia, free-living N-fixing bacteria.
  - Magnesium (Mg) is a cofactor for functioning of chlorophyll to facilitate its role in photosynthesis, etc.

  In EU, there is increasing recognition of the potential benefits of fortified biostimulants. In the USA, the biostimulants are regulated by the environmental protection agency that regulates both pesticides and fertilizers including biostimulants. There is flexibility for innovative products such as fortified biostimulants. India has recently implemented a stringent regulatory framework for biostimulants in its third amendment. Whereas the guidelines regarding fortification with micronutrients is vague apparently indicating that in general, such exogenous application is not a rule. Many other countries in Asia and South America have emerging regulatory frameworks for biostimulants. These frameworks often allow for the addition of micronutrients to enhance the product's efficacy. Since, the micronutrients and plant growth regulators are themselves the constituents in the raw material of the biostimulants, their fortification to the extent of their inherent levels within the plants (beyond which their effect may become detrimental to the normal growth and development of the plant) should be acceptable.

- **Pesticide contamination limits hindering biostimulants registration:**
  The Maximum Residue Limit or MRL limit (0.01 ppm) for biostimulant formulations is often unattainable and overly strict, potentially hindering the development of new technologies. Limiting a technology with stringency may kill a technology before its birth. Risk assessments have led organizations such as CODEX ALIMENTARIUS (FAO-WHO) and the Food Safety

and Standards Authority of India (FSSAI) in India to set higher MRLs for certain pesticides such as carbaryl in dried chilli and flubendiamide in tea, with limits of up to 50 ppm. It is important to note that such low levels of pesticides do not serve to protect the plants from pests and diseases neither any purpose nor any interest of the manufacturers. With enough scientific deliberations, the Central Biostimulants Committee (CBC) considered and approved to increase the limit of pesticide residues from 0.01 ppm to 1 ppm through its Third Amendment dated 8th May 2024.

- **Biostimulants registration using pooled data:**
  As most biostimulants manufacturers fall under the category of SMEs or MSMEs, they may not be able to generate all the necessary toxicity and bioefficacy data on their own due to cost constraints. To facilitate biostimulants registration, the government should consider pooled data on toxicity and bioefficacy from Good Laboratory Practice (GLP) and National Accreditation Board for Testing and Calibration Laboratories (NABL)-compliant facilities without compromising the quality of the manufactured products. A data pool would encourage the development and registration of FCO 1985-notified biostimulants, and developers could share such data with small- and medium-sized prospective manufacturers to avoid duplication, cruelty to animals, and waste of time, natural resources, and national wealth.
- **Difficulties in biostimulants label claim registration:**
  It is impractical and expensive to develop crop-specific efficacy data for each biostimulant. Since all crops may share similar physiological and metabolic pathways, biostimulants may work similarly on related crops. CIBRC has classified crops into 26 groups and subgroups based on behavior, residue accumulation, etc., for label expansion. A similar approach may be used for crop group-based biostimulants label expansion. Advanced countries do not require a bioefficacy dossier for label expansion registration. The basic premise is that industry would not unnecessarily roll out ineffective products that farmers would reject if they don't work.

15.9 CONCLUSION

The biostimulants market, although still in its infancy in many countries of the world, in India, has significant growth potential due to the increasing demand for sustainable farming practices and to promote judicious use of chemical fertilizers. Biostimulants have the potential to enhance crop yields and quality, improve soil health, and increase farmer income sustainably, making them an attractive option for the Indian agricultural sector. Plant biostimulants act by increasing the efficiency of cross talks between enzymes and metabolites, which in turn increase the innate immunity and signal transduction to withstand abiotic and/or biotic stress (indirectly). The untamed challenges in agriculture such as erratic weather patterns, soil salinity, increasing area under rainfed conditions, and declining soil health due to the climate crisis are continuing to impede the growth of agriculture. As the world's population continues to grow, sustainable agricultural practices such as use of nondestructive novel technologies such as biostimulants should be mainstreamed and made inclusive in cultivation to mitigate biotic and abiotic stresses.

REFERENCES

Biostimulants Draft Guidelines, 2021, https://agriwelfare.gov.in/Documents/Draft%20Guidelines%20for%20bio-stimulands.pdf

FAO, 2025. https://www.fao.org/plant-production-protection/about/en

Fertiliser (Inorganic, Organic or Mixed) Control (Third) Amendment Order, 2024 dated 8th May 2024 (https://uncomplycate.com/wp-content/uploads/2024/05/Fertiliser-Control-Order.pdf)

https://www.sciencedirect.com/science/

https://www.precedenceresearch.com/biostimulants-market

S.O. 3922(E) 12 September 2024. https://egazette.gov.in/WriteReadData/2024/257181.pdf

S.O. 2346(E) 26th May 2025. https://egazette.gov.in/(S(dkm5lympwnhqiyzhrskx2ue1))/ViewPDF.aspx

Index

Note: **Bold** page numbers refer to tables and *italic* page numbers refer to figures.